# Lecture Notes in Computer Science 5735

Commenced Publication in 1973
Founding and Former Series Editors:
Gerhard Goos, Juris Hartmanis, and Jan van Leeuwen

T0180973

Pierangela Samarati   Moti Yung
Fabio Martinelli   Claudio A. Ardagna (Eds.)

# Information Security

12th International Conference, ISC 2009
Pisa, Italy, September 7-9, 2009
Proceedings

 Springer

Volume Editors

Pierangela Samarati
Claudio A. Ardagna
Università degli Studi di Milano
Dipartimento di Tecnologie dell' Informazione
Via Bramante 65, 26013 Crema (CR), Italy
E-mail: {pierangela.samarati, claudio.ardagna}@unimi.it

Moti Yung
Google Inc. & Columbia University
Computer Science Department
Room 465, S.W. Mudd Building, New York, NY 10027, USA
E-mail: my123@columbia.edu

Fabio Martinelli
National Research Council (CNR)
Institute of Informatics and Telematics (IIT)
Information Security Group
Pisa Research Area, Via G. Moruzzi 1, 56125 Pisa, Italy
E-mail: fabio.martinelli@iit.cnr.it

Library of Congress Control Number: 2009934111

CR Subject Classification (1998): E.3, E.4, D.4.6, K.6.5, C.2

LNCS Sublibrary: SL 4 – Security and Cryptology

ISSN 0302-9743
ISBN 978-3-642-04473-1 Springer Berlin Heidelberg New York

springer.com

© Springer-Verlag Berlin Heidelberg 2009

Typesetting: Camera-ready by author, data conversion by Scientific Publishing Services, Chennai, India
Printed on acid-free paper     SPIN: 12761302     06/3180     5 4 3 2 1 0

# Message from the Program Chairs

These proceedings contain the papers selected for presentation at the 12th Information Security Conference (ISC 2009), held September 7–9, 2009, in Pisa, Italy.

In response to the call for papers, 105 papers were submitted to the conference. These papers were evaluated on the basis of their significance, novelty, technical quality, and practical impact. As in previous years, reviewing was "double-blind": the identities of reviewers were not revealed to the authors of the papers and author identities were not revealed to the reviewers. The Program Committee meeting was held electronically, yielding intensive discussions over a period of two weeks. Of the papers submitted, 29 full papers and 9 short papers were selected for presentation at the conference. Besides the technical program composed of the papers collated in these proceedings, the conference included two keynotes.

An event like ISC does not just happen; it depends on the volunteer efforts of a host of individuals. There is a long list of people who volunteered their time and energy to put together the conference and who deserve special thanks. Thanks to all the members of the Program Committee and the external reviewers for all the hard work they put in evaluating the papers. We are also very grateful to all the people whose work ensured a smooth organization process: the ISC Steering Committee, and Javier Lopez in particular, for their advice; Fabio Martinelli; for his support for the overall organization as General Chair; Claudio Ardagna for collating this volume, and Eros Pedrini, for taking care of publicity and for maintaining the website. A special thanks to the two keynote speakers, Jan Camenisch and Sushil Jajodia, for accepting our invitation to deliver keynote talks at the conference.

Last but certainly not least, our thanks go to all the authors who submitted papers and all the attendees. We hope you find the proceedings stimulating and a source of inspiration for your future research and development programs.

September 2009

Pierangela Samarati
Moti Yung

# Organization

The 12th International Security Conference (ISC 2009)
Pisa, Italy, September 7-9, 2009

## General Chair

Fabio Martinelli IIT-CNR, Italy

## Steering Committee

Ed Dawson Queensland University of Technology, Australia
Der-Tsai Lee Academia Sinica, Taiwan
Javier López University of Málaga, Spain
Masahiro Mambo University of Tsukuba, Japan
Fabio Martinelli IIT-CNR, Italy
Eiji Okamoto University of Tsukuba, Japan
Rebecca Wright Rutgers University, USA
Yuliang Zheng University of North Carolina at Charlotte, USA

## Program Chairs

Pierangela Samarati Università degli Studi di Milano, Italy
Moti Yung Google Inc. and Columbia University, USA

## Publication Chair

Claudio A. Ardagna Università degli Studi di Milano, Italy

## Publicity Chair

Eros Pedrini Università degli Studi di Milano, Italy

## Local Organizing Committee

Adriana Lazzaroni IIT-CNR, Italy
Raffaella Casarosa IIT-CNR, Italy

# Program Committee

| | |
|---|---|
| Vijay Atluri | Rutgers University, USA |
| Tuomas Aura | Microsoft, UK |
| Paulo S. L. M. Barreto | University of San Paolo, Brazil |
| Marina Blanton | University of Notre Dame, USA |
| Carlo Blundo | Università di Salerno, Italy |
| David Chadwick | University of Kent, UK |
| Jan Camenisch | IBM, Switzerland |
| Anupam Datta | Carnegie Mellon University, USA |
| S. De Capitani di Vimercati | Università degli Studi di Milano, Italy |
| Yves Deswarte | CNRS, France |
| Claudia Diaz | Katholieke Universiteit Leuven, Belgium |
| William Enck | The Pennsylvania State University, USA |
| Sara Foresti | Università degli Studi di Milano, Italy |
| Michael Franz | University of California Irvine, USA |
| David Galindo | University of Luxembourg, Luxembourg |
| Dimitris Gritzalis | University of Patras, Greece |
| Maribel Gonzalez | Universidad Rey Juan Carlos, Spain |
| Yong Guan | Iowa State University, USA |
| Goichiro Hanaoka | AIST, Japan |
| Javier Herranz | Universitat Politècnica de Catalunya, Spain |
| Sotiris Ioanidis | FORTH, Greece |
| Cynthia Irvine | Naval Postgraduate School, USA |
| Sushil Jajodia | George Mason University, USA |
| Aleksandar Jurisic | University of Ljubljana, Slovenia |
| Angelos Keromytis | Columbia University, USA |
| Aggelos Kiayias | University of Connecticut, USA |
| Kwangjo Kim | Information and Communication University, Korea |
| Lea Kissner | Google, USA |
| Adam J. Lee | University of Pittsburgh, USA |
| David Lee | Ohio State University, USA |
| Benoit Libert | Université Catholique de Louvain, Belgium |
| Donggang Liu | University of Texas at Arlington, USA |
| Peng Liu | Pennsylvania State University, USA |
| Javier López | University of Málaga, Spain |
| Carlos Maziero | Pontifical Catholic University, Brazil |
| Catherine Meadows | NRL, USA |
| Nasir Memon | Polytechnic University, USA |
| John C. Mitchell | Stanford University, USA |
| David Naccache | University of Paris, France |
| Mats Naslund | Ericsson Research, Sweden |
| Gregory Neven | IBM, Switzerland |
| Antonio Nicolosi | Stevens Institute of Technology, USA |

| | |
|---|---|
| Eiji Okamoto | University of Tsukuba, Japan |
| Stefano Paraboschi | Università degli Studi di Bergamo, Italy |
| Thomas Peyrin | Ingenico, France |
| Duong Hieu Phan | University of Paris VIII, Paris France |
| Indrakshi Ray | Colorado State University, USA |
| Kui Ren | Illinois Institute of Technology, USA |
| Rei Safavi-Naini | University of Calgary, Canada |
| Martijn Stam | EPFL, Switzerland |
| Angelos Stavrou | George Mason University, USA |
| Michael Steiner | IBM, USA |
| Rainer Steinwandt | Florida Atlantic University, USA |
| Joe-Kai Tsay | Ruhr University Bochum, Germany |
| Jaideep Vaidya | Rutgers University, USA |
| Ting Yu | North Carolina State University, USA |
| Sencun Zhu | Pennsylvania State University, USA |
| Huaxiong Wang | Nanyang Technological University, Singapore |
| Lingyu Wang | Concordia University, Canada |
| Bogdan Warinsschi | University of Bristol, UK |

## External Reviewers

Isaac Agudo
Elena Andreeva
Frederik Armknecht
Nuttapong Attrapadung
Kun Bai
Jens Matthias Bohli
Joppe Bos
Elie Burzstein
Kevin Butler
Christian Cachin
Sébastien Canard
Claude Carlet
Sudip Chakraborty
Haowen Chan
Yi Cheng
Sherman Chow
Baudoin Collard
Baris Coskun
Paolo D'Arco
Vasil Dimitrov
Qi Dong
Stelios Dritsas
Patrik Ekdahl
Jonathan Etrog
Lei Fan

Oriol Farras
Sebastian Faust
Nelly Fazio
Martin Gagne
Flavio Garcia
Deepak Garg
Danilo Gligoroski
Jovan Golic
Jian Guo
Matt Henricksen
Dennis Hofheinz
Hung-Yuan Hsu
Yating Hsu
Sebastiaan Indesteege
Vincenzo Iovino
Mohammad Jafari
Gregor Jerse
Yoon-Chan Jhi
Shaoquan Jiang
Charanjit Jutla
Eike Kiltz
Chiu Yuen Koo
Eric Lam
Anja Lehmann
Jin Li

Liyun Li
Stijn Lievens
Michael Liljenstam
Wenjie Lin
Wenming Liu
Hans Loehr
Mark Manulis
John Mattsson
Krystian Matusiewicz
Jonathan McCune
Stephen McLaughlin
Breno de Medeiros
Kazuhiro Minami
Djedjiga Mouheb
Pablo Najera
Duc Dang Nguyen
Jesper Buus Nielsen
Peter Nose
Machigar Ongtang
Onur Özen
Christos Papachristou
Serdar Pehlivanoglu
Ludovic Perret
Angel L. Perez del Pozo
Vassilis Prevelakis
Rodrigo Roman
Juraj Şarinay
Joshua Schiffman

Basit Shafiq
Jun Shao
Guoqiang Shu
Koen Simoens
Hong-Yeop Song
Yannis Soupionis
Ron Steinfeld
Hung-Min Sun
Yagiz Sutcu
Marianthi Theoharidou
Manachai Toahchoodee
Umut Topkara
Matjaz Urlep
Damien Vergnaud
Janos Vidali
Jorge Villar
Ivan Visconti
Jose L. Vivas
Tuan Vu
Christian Wachsmann
Shabsi Walfish
Xinran Wang
Zhi Xu
Yanjiang Yang
Fang Yu
Dazhi Zhang
Shengzhi Zhang
Hong-Sheng Zhou

# Table of Contents

## Analysis Techniques

## Hash Functions

## Database Security and Biometrics

## Algebraic Attacks and Proxy Re-Encryption

## Distributed System Security

## Identity Management and Authentication

## Applied Cryptography

## Access Control

## MAC and Nonces

## P2P and Web Services

# A New Approach to $\chi^2$ Cryptanalysis of Block Ciphers

Jorge Nakahara Jr.[1], Gautham Sekar[4,5,*], Daniel Santana de Freitas[2], Chang Chiann[3], Ramon Hugo de Souza[2], and Bart Preneel[4,5]

[1] EPFL, Lausanne, Switzerland
jorge.nakahara@epfl.ch
[2] Federal University of Santa Catarina, Brazil
{santana,ramonh}@inf.ufsc.br
[3] University of São Paulo, Brazil
chang@ime.usp.br
[4] Interdisciplinary Institute for BroadBand Technology (IBBT), Belgium
[5] Katholieke Universiteit Leuven, Belgium
{gautham.sekar,bart.preneel}@esat.kuleuven.be

**Abstract.** The main contribution of this paper[1] is a new approach to $\chi^2$ analyses of block ciphers in which plaintexts are chosen in a manner similar to that in a square/saturation attack. The consequence is a faster detection of $\chi^2$ correlation when compared to conventional $\chi^2$ cryptanalysis. Using this technique we *(i)* improve the previously best-known $\chi^2$ attacks on 2- and 4-round RC6, and *(ii)* mount the first attacks on the MRC6 and ERC6 block ciphers. The analyses of these fast primitives were also motivated by their low diffusion power and, in the case of MRC6 and ERC6, their large block sizes, that favour their use in the construction of compression functions. Our analyses indicate that up to 98 rounds of MRC6 and 44 rounds of ERC6 could be attacked.

**Keywords:** Block ciphers, $\chi^2$, square and linear cryptanalysis.

## 1 Introduction

In this paper we present a new, generic approach to $\chi^2$ cryptanalysis which combines conventional $\chi^2$ and integral techniques. In this approach, the plaintexts are chosen like in a square/saturation attack, that is, part of the input is fixed and the remaining part is varied exhaustively. Further, the attack is adaptive in the sense that we keep on generating plaintexts until $\chi^2$ correlation is detected. The advantage of this approach is that it allows faster detection of $\chi^2$ correlations in block ciphers compared to previous approaches. One drawback is that it is not straightforward to turn the chosen-plaintext (CP) setting into a known-plaintext (KP) one.

* This author is supported by an FWO project.
[1] The work described in this paper has been supported, in part, by the European Commission through the ICT programme under contract ICT-2007-216676 ECRYPT II.

We apply this new approach to the block ciphers RC6, ERC and MRC6. RC6 [1] was designed by Rivest *et al.* for the AES Development Process [2]. RC6 was one of the five finalists in the AES competition and was also submitted to NESSIE and CRYPTREC projects. ERC6 [3] is a wide-block variant of RC6, designed by Ragab *et al.* in 2001. MRC6, proposed by El-Fishawy *et al.* in 2004 [4], is another wide-block variant of RC6.

In our attacks, the choice of the plaintext bits to be chosen and the ciphertext bits to be analysed is based on prior linear analysis, which provided the bit positions with highest expected non-uniform bias. Our attacks follow a similar methodology as the mod-$n$ attacks against the block ciphers RC5P and M6 [5].

Our considerations and conclusions of the analyses in this paper are based on empirical data collected through several attack simulations. We have used $\chi^2$ threshold values corresponding to 25% significance level (or 75% specificity). See Table 12 in the appendix. This choice was based on the following reasons:

1. Our aim is to show the effectiveness of our attacks on RC6, ERC6 and MRC6 when compared to conventional $\chi^2$ cryptanalysis with randomly generated plaintexts. Hence, as long as the same significance level is used for the two types of tests, the value of the significance level is irrelevant.
2. Our attack simulations show that the number of chosen plaintexts required with a better (we considered 10%) significance level could be determined by the number corresponding to 25% level.
3. In the literature 25% seems to be an acceptable value [6].

This paper is organized as follows. Section 2 briefly describes $\chi^2$ cryptanalysis and introduces our technique; Sect. 3 gives the specifications of the RC6, ERC6 and MRC6 ciphers; Sect. 4 provides the experimental results of our $\chi^2$ attacks on the three ciphers. Also, in Sect. 4 comparisons are drawn between our attacks and previously applied methods. Section 5 concludes the paper.

## 2   The $\chi^2$ Test and Our Generic Approach

The $\chi^2$ statistical test has already been applied to a number of ciphers, such as the DES in [7], on SEAL [8], on M6, MX and RC5P [5], on RC5, RC6 and many simplified variants [9,10,11,12,13,14,1,15].

Consider an experiment $E$ with $k$ simple, mutually independent outcomes. Let $o_1, \ldots, o_k$ and $x_1, \ldots, x_k$ denote the observed and expected frequencies, respectively, of the $k$ outcomes when $E$ is performed $N$ times. Therefore, $N = \sum_{i=1}^{k} o_i = \sum_{i=1}^{k} x_i$. For each outcome, there can be a difference between the observed and the expected frequencies. The idea behind a $\chi^2$ test is to combine all these differences into one overall measure of the distance between the data and the expectations of the model. The $\chi^2$ statistic with $k-1$ degrees of freedom is defined [16] as,

$$Q = \sum_{i=1}^{k} \frac{(o_i - x_i)^2}{x_i}, \tag{1}$$

where the sum is over $x_i \neq 0$. When the observed frequency is far from the expected one, the corresponding term $o_i - x_i$ in the sum is large; when they are close, $o_i - x_i$ is small. The quantity $Q$ gives a measure of the distance between the observed and expected frequencies; large values of $Q$ indicate that the observed frequencies are far from the expected ones. In a $\chi^2$ goodness-of-fit test, one defines two hypotheses - the null hypothesis (denoted $H_0$) and the alternative hypothesis ($H_1$). The null hypothesis is the one that exists solely to be *falsified* by the sample. If the null hypothesis is rejected, the result is *positive*. When the test result tallies with the actual reality, the result is *true*. The false-negative rate of the test, that is, the fraction of *positive* instances that were falsely reported as *negative*, is denoted by $\beta$. The sensitivity (or power) of the test is the true-positive rate $(1 - \beta)$. The significance of the test is the false-positive rate ($\alpha$) and the specificity of the test is the true-negative rate $(1 - \alpha)$. Let $\chi^2_{1-\alpha,k-1}$ denote the $(1 - \alpha)$-th lower quantile of a $\chi^2$ distribution with $k - 1$ degrees of freedom. In a $\chi^2$ test, $H_0$ is rejected (in other words, $H_1$ is accepted), if $Q > \chi^2_{1-\alpha,k-1}$ with $100\alpha$ % error. We denote $\chi^2_{1-\alpha,k-1}$ simply as $\chi^2_{1-\alpha}$ when $k - 1$ is clear from the context.

In our approach, $N$ is the number of plaintexts - the parameter to be determined. Let $E'$ denote the experiment $E$ repeated $N$ times. To minimise error, we consider $q$ randomly generated keys and $E'$ is performed $q$ times. We could estimate the mean and variance of the $\chi^2$ values for the entire key space using the Student's t-distribution. But this requires that the population be normally distributed. This is nearly achieved when the number of degrees of freedom $(k - 1)$ is large since when $k \to \infty$, the $\chi^2$ variate becomes a normal variate. Finally, using $q$, the $q$-sample mean and sample variance, a confidence interval (CI) is computed, using the t-curve, for the mean of the population. We use 90% confidence interval in our tests. In other words, the chance that the population mean falls below (or above) the interval is 5%. The lower end point of the interval $(minCI)$ is taken for the population mean. This means that there is 95% chance that the actual population mean is above this value. In our experiments, we accept $H_1$ if $minCI$ is greater than $\chi^2_{1-\alpha,k-1}$. This automatically implies that the actual population mean is greater than $\chi^2_{1-\alpha,k-1}$ with 95% probability and thus, the error is small.

In this paper, we use the $\chi^2$ test under the following settings (where XRC6 denotes RC6, MRC6 or ERC6 and $r > 0$):

$H_0$: a subset of bits output by $r$-round XRC6 is uniformly distributed,
$H_1$: a subset of bits output by $r$-round XRC6 is non-uniformly distributed.

Thus, (1) becomes

$$Q = \sum_{i=1}^{k} \frac{(n_i - N/k)^2}{N/k}. \tag{2}$$

A requirement in $\chi^2$ tests is that $N \geq 5 \cdot k$, so that the computed $\chi^2$ value is valid. In conventional $\chi^2$ cryptanalysis, most of the plaintext bits are generated at random. However, plaintexts can be chosen in the following way to yield more efficient attacks. Initially, a linear analysis (LC) is performed to determine which $z$ least significant bits (lsb) of $d$ words, in an $n$-bit block are linearly correlated to the same set of bits after a certain number $r$ of rounds. This approach of using LC results prior to the $\chi^2$ analysis has already been adopted in [13]. For RC6, $d = 2$, $z \le 5$, $n = 128$ and $r$ is multiple of 2, as indicated by (3) in Sect. 4.1. This set of $d \cdot z$ plaintext bits will be fixed (to an arbitrary value), while the remaining $n - d \cdot z$ plaintext bits are free to vary. These two sets of bits are disjoint. These plaintexts are encrypted across $r$ rounds, and the $\chi^2$ value is computed for the $d \cdot z$ ciphertext bit positions given by the linear relation. If the resulting $\chi^2$ value supports acceptance of $H_0$, then we stop, record the number $N$ of plaintexts encrypted so far, and proceed the same analysis $y$ rounds farther (in this paper, $y = 2$). Otherwise, we consider the remaining $n - d \cdot z$ plaintext bits as a counter, increment it, and encrypt the corresponding plaintext for $r$ rounds. The number of degrees of freedom is $k - 1 = 2^{d \cdot z} - 1$. We look for the minimum $N$ for which $H_1$ is accepted. Each test is repeated $q$ times; we use $q = 20$. The following pseudocode describes the overall procedure.

TEST $(H_0, H_1, N, r, q, \alpha)$
(1.)    for $(i = 1; i \le q; i + +)$ {
(2.)      for $(j = 1; j \le N; j + +)$ {
(3.)          fix the given set of $d \cdot z$ bits of plaintext $P_j$
(4.)          vary the remaining bits of $P_j$ incrementally
(5.)          encrypt $P_j$ through $r$ rounds and obtain $C_j$
(6.)          let $X$ be the concatenation of given $d \cdot z$ bits of $C_j$
(7.)          increment counter $T[X]$ by 1
(8.)      }
(9.)      let $Q_i$ be the $\chi^2$ value of $T[X]$'s
(10.)   }
(11.)   let $m$ be the average over all $Q_i$, $1 \le i \le q$
(12.)   let $\sigma$ be the standard deviation over all $Q_i$, $1 \le i \le q$
(13.)   let $minCI = m - 1.729 \cdot \sigma / \sqrt{q}$ (lower limit of a 90% CI)
(14.)   let $\chi^2_{1-\alpha,k-1}$ = value at $100(1 - \alpha)\%$ in the $\chi^2$ cumulative distribution
          with $k - 1$ degrees of freedom
(15.)   if $(minCI > \chi^2_{1-\alpha,k-1})$
(16.)     choose $H_1$ and note the $j$ corresponding to $N$
(17.)     else choose $H_0$

For our target ciphers, a further consequence of the new approach is a smaller demand for chosen plaintexts, due to weak diffusion. As already pointed out in [13], too small or too large rotation amounts lead to weak diffusion across multiple rounds of RC6. The same phenomenon can be observed in ERC6 and MRC6. This is an essential weakness exploited in our attacks since the linear relations (3), (4) and (5), which indicate the $d \cdot z$ bits in lines (3.) and (6.) of TEST(), rely on these assumptions. A smaller number of plaintexts implies a smaller encryption time, and thus, faster attacks. It shall be observed that the more bits are under analysis, the better the attack outcome. Nonetheless, the data (and time) complexities increased quickly beyond our computational resources. Consequently, we used different value of $z$ for the plaintext and ciphertext, unlike in TEST() where $z$ is identical for plaintext and ciphertext (here, we have followed the approach of [12,14]).

Our attacks on RC6 and the approach used in [13] are different. We fix a number of bits to zeros and vary the remaining bits incrementally; whereas in the latter, the remaining bits are random. The result is that, with Knudsen and Meier's method, one can turn the CP setting into a KP one at the cost of a factor of $2^{d \cdot z}$ in the data and time complexities. Secondly, we used 90% confidence interval (CI) to minimise error, whereas [13] did not use CI.

## 3   The RC6, ERC6 and MRC6 Families of Block Ciphers

Initially, we provide some relevant notations: '$\oplus$' denotes bitwise exclusive-OR; '$\boxplus$' denotes addition modulo $2^w$; '$*$' denotes multiplication modulo $2^w$; $x \lll y$, where $x$ and $y$ are $w$-bit words, means that $x$ is cyclically shifted to the left by the amount given by least significant $\log_2 w$ bits of $y$. The function $F : \mathbb{Z}_2^w \to \mathbb{Z}_2^w$ is given by $F(X) = (2 * X^2 \boxplus X) \lll \log_2 w$. Notice that $F$ has only one operand, and is a bijective mapping. Thus, it behaves as a $w \times w$-bit nonlinear S-box.

### 3.1   RC6

The RC6 cipher follows a generalized Feistel Network structure, and stands for a family of ciphers formally denoted RC6-$w/r/b$, where $w$ is the word size in bits, $r$ is the number of rounds, and $b$ is the key size in bytes. For the AES competition, $w = 32$, $r = 20$, and $b \in \{16, 24, 32\}$, and RC6 is a shorthand for these parameter choices. All internal cipher operations are over $w$-bit words, where $w \in \{8, 16, 32, 64\}$. Fig. 1 depicts the RC6 encryption algorithm. Each text block contains four $w$-bit words. For instance, $A_i$, $B_i$, $C_i$, $D_i$, denote the input words to the $i$-th round. The $w$-bit round keys are indexed $S[0], \ldots, S[2r + 3]$. The key schedule algorithm generates the round keys from the $b$-byte user key. We do not exploit the key schedule algorithm in our analysis; therefore, we omit its description and refer the interested reader to [1]. Former security analyses of RC6 include differential and linear analyses [1], multiple linear relations [17], and $\chi^2$ analyses [9,13,1,15].

### 3.2   MRC6

The MRC6 cipher follows a generalized Feistel Network structure and was proposed in [4], with main focus on (software) performance. No security analysis was presented. MRC6 is a parameterized family of ciphers formally denoted MRC6-$w/r/b$, with the same meaning as for the parameters of RC6. But, nominal values of these parameters were omitted in [4]; one can find the values $w = 32$, $b = 16$ and $r = 16$ when the software performance of MRC6 is compared with that of the AES and RC6 (on Pentium-III, with the se parameters, MRC6 encrypts at about

19.5 MB/sec making it nearly twice as fast as RC6). Otherwise, these parameters are unspecified. The fact that these parameters are unrelated helps adapt MRC6 as a compression function in hash modes [18] such as Miyaguchi-Preneel and Matyas-Meyer-Oseas, where the key and text inputs have different sizes. An MRC6 text block contains sixteen $w$-bit words, denoted $A_i, B_i, \ldots, P_i$ as inputs to the $i$-th round. Moreover, the $w$-bit round keys are indexed $S[0], \ldots, S[8r+7]$. Like in RC6, there are pre-whitening and post-whitening layers. Here again, we omit the description of the key schedule algorithm and refer the reader to [4]. Fig. 3 depicts the MRC6 encryption algorithm. In our experiments, we use MRC6 with $w = 32$ and $b = 16$.

### 3.3   ERC6

The ERC6 cipher follows a generalized Feistel Network structure, and was proposed in [3], as a parameterized family of ciphers formally denoted ERC6-$w/r/b$, with $w \in \{16, 32, 64\}$, $r \in \{0, 1, 2, \ldots, 255\}$, $b \in \{0, 1, 2, \ldots, 255\}$. These parameters appear to be loosely coupled. No attacks have been reported on any version of ERC6. On Pentium-III, with parameters $w = 32$, $b = 16$ and $r = 16$, ERC6 encrypts at about 17.3 MB/sec making it about 1.7 times faster than RC6. Each text block of ERC6 contains eight $w$-bit words, denoted $A_i, B_i, C_i, D_i,$ $E_i, F_i, G_i, H_i$, as inputs to the $i$-th round. The $w$-bit round keys are indexed $S[0], \ldots, S[4r + 7]$. Here again, there are pre-whitening and post-whitening layers. Fig. 2 depicts the ERC6 encryption algorithm. In our experiments, we use ERC6 with $w = 32$ and $b = 16$.

## 4   Experimental Observations

Our $\chi^2$ attacks operate in an adaptive chosen-plaintext (CP) setting.

### 4.1   Reduced-Round RC6

For RC6, the $\chi^2$ test is motivated by an ensemble of linear relations involving up to the five least significant bits of words $A_i$ and $C_i$ for every two rounds [19]. These linear relations can be represented by

$$A_i \cdot e_{t_1} \oplus C_i \cdot e_{t_2} = A_{i+2} \cdot e_{t_3} \oplus C_{i+2} \cdot e_{t_4}, \tag{3}$$

where $A_i$ and $C_i$ denote the first and third input words to the $i$-th round. Each bitmask, $e_j = 2^j$, $0 \le j < 5$, contains only a single bit equal to one, in the $j$-th least significant bit ($j = 0$ denotes the lsb). This is the lowest possible Hamming weight. Table 1 shows the result of the experiment on reduced-round RC6 using our method in the case of ten bits: $\text{lsb}_5(A_{2i})$ and $\text{lsb}_5(C_{2i})$.

We use $\chi^2_{95} = 1098$ (95% specificity) to facilitate comparison, since [13] also uses the same threshold. Moreover, for the same comparison purpose, we did not use confidence intervals this time. In Table 1, note that with $2^2$ texts we already

**Table 1.** $\chi^2$ attack simulations on RC6, $2^{10} - 1$ degrees of freedom and 20 tests

| #Rounds | $\log_2 N$ | average $\chi^2$ | hypothesis |
|---------|-----------|-----------------|------------|
| 2 | 2 | 1071.2 | $H_0$ |
| 2 | 3 | 1169.6 | $H_1$ |
| 2 | 4 | 1398.4 | $H_1$ |
| 2 | 5 | 1561.6 | $H_1$ |
| 2 | 6 | 2009.6 | $H_1$ |
| 4 | 16 | 1039.5 | $H_0$ |
| 4 | 17 | 1066.3 | $H_0$ |
| 4 | 18 | 1094.2 | $H_0$ |
| 4 | 19 | 1151.6 | $H_1$ |
| 4 | 20 | 1267.6 | $H_1$ |
| 6 | 32 | 1030.6 | $H_0$ |
| 6 | 33 | 1036.0 | $H_0$ |
| 6 | 34 | 1020.6 | $H_0$ |
| 6 | 35 | 1018.1 | $H_0$ |
| 6 | 36 | 1028.4 | $H_0$ |
| 6 | 37 | 1009.6 | $H_0$ |

start to reach the same results of [13], whereas they needed $2^{13}$ texts to arrive at a $\chi^2$ value of 1098. For four rounds, we noticed very close approximations for the same $\chi^2$ values with $2^{18}$ texts, while [13] required $2^{29}$ texts to arrive at data with the same specificity.

The experimental results for 2-round RC6 show that our approach requires only $2^3$ texts to reach the same $\chi^2$ value that is obtained with $2^{14}$ texts using the approach in [13]. For 4-round RC6, these figures are $2^{19}$ texts using our technique against $2^{30}$ for [13]. For 6-round RC6, our method required more than $2^{37}$ texts to detect correlation. In this case (and for more rounds), it could not be concluded whether our approach was better than [13].

## 4.2   MRC6

For MRC6, our $\chi^2$ attacks were motivated by the following 2-round iterative linear relation (using Type-I approximations [19])

$$A_i \cdot e_{t_1} \oplus C_i \cdot e_{t_2} \oplus E_i \cdot e_{t_3} \oplus G_i \cdot e_{t_4} \oplus I_i \cdot e_{t_5} \oplus K_i \cdot e_{t_6} \oplus M_i \cdot e_{t_7} \oplus O_i \cdot e_{t_8} =$$
$$A_{i+2} \cdot e_{t_9} \oplus C_{i+2} \cdot e_{t_{10}} \oplus E_{i+2} \cdot e_{t_{11}} \oplus G_{i+2} \cdot e_{t_{12}} \oplus$$
$$I_{i+2} \cdot e_{t_{13}} \oplus K_{i+2} \cdot e_{t_{14}} \oplus M_{i+2} \cdot e_{t_{15}} \oplus O_{i+2} \cdot e_{t_{16}} \qquad (4)$$

where $A_i$, $C_i$, $E_i$, $G_i$, $I_i$, $K_i$, $M_i$ and $O_i$ are input words to the $i$-th round. In particular, the masks $e_j$ with highest bias are such that $0 \le j < 5$, that is, the bits in the masks are restricted to the five least significant bit (lsb) positions. Our experiments distinguish $r$ rounds of MRC6 from a random permutation, where $r$ is even. We fix up the $8 \cdot \log_2 w$ least significant bits of words $A_0$, $C_0$, $E_0$, $G_0$, $I_0$, $K_0$, $M_0$ and $O_0$ (that is, including the pre-whitening), and analyse

**Table 2.** $\chi^2$ attack simulations on MRC6, $2^8 - 1$ degrees of freedom and 20 tests

| #Rounds | $\log_2 N$ | average $\chi^2$ | minCI | #values> $\chi^2_{75}$ | hypothesis |
|---------|-----------|------------------|-------|------------------------|------------|
| 2 | 1 | 305.2 | 264.6 | 4 | $H_0$ |
| 2 | 2 | 444.0 | 376.8 | 18 | $H_1$ |
| 2 | 3 | 667.2 | 622.9 | 20 | $H_1$ |
| 4 | 3 | 299.2 | 262.8 | 8 | $H_0$ |
| 4 | 4 | 321.6 | 303.0 | 19 | $H_1$ |
| 4 | 5 | 368.8 | 353.7 | 20 | $H_1$ |
| 6 | 8 | 265.8 | 256.7 | 8 | $H_0$ |
| 6 | 9 | 282.1 | 270.6 | 11 | $H_1$ |
| 6 | 10 | 288.9 | 276.7 | 15 | $H_1$ |
| 6 | 11 | 316.5 | 298.1 | 16 | $H_1$ |
| 6 | 12 | 370.8 | 343.3 | 20 | $H_1$ |
| 8 | 18 | 273.0 | 265.0 | 11 | $H_0$ |
| 8 | 19 | 288.1 | 279.8 | 16 | $H_1$ |
| 8 | 20 | 327.5 | 311.1 | 20 | $H_1$ |
| 10 | 27 | 273.5 | 265.4 | 11 | $H_0$ |
| 10 | 28 | 288.2 | 275.7 | 12 | $H_1$ |
| 10 | 29 | 290.2 | 278.2 | 15 | $H_1$ |
| 10 | 30 | 336.2 | 317.2 | 19 | $H_1$ |
| 10 | 31 | 402.7 | 368.0 | 19 | $H_1$ |
| 10 | 32 | 407.8 | 357.0 | 19 | $H_1$ |
| 10 | 33 | 434.1 | 374.1 | 20 | $H_1$ |

the combined $8 \cdot y$ least significant bits ($y \in \{1, 2\}$) of $A_{2i}$, $C_{2i}$, $E_{2i}$, $G_{2i}$, $I_{2i}$, $K_{2i}$, $M_{2i}$ and $O_{2i}$, for $i > 0$, that is, after an even number of rounds.

Table 2 shows the result of the experiment in the case of the eight bits: $\mathrm{lsb}_1(A_{2i})$, $\mathrm{lsb}_1(C_{2i})$, $\mathrm{lsb}_1(E_{2i})$, $\mathrm{lsb}_1(G_{2i})$, $\mathrm{lsb}_1(I_{2i})$, $\mathrm{lsb}_1(K_{2i})$, $\mathrm{lsb}_1(M_{2i})$, $\mathrm{lsb}_1(O_{2i})$. We use $\chi^2_{75} = 269.85$. Starting from six rounds, the number of texts for which $H_0$ is rejected starts to increase by a factor of about $2^{10}$ every two rounds. Thus, for $r$ rounds ($r$ even and $r \geq 6$), the following is expected for $N$ (numner of chosen plaintexts) in terms of $r$: $N = 2^9 \cdot 2^{10 \cdot (r-6)/2} = 2^{5r-21}$. In TEST(), we choose plaintexts such that the $\mathrm{lsb}_5(A_0)$, $\mathrm{lsb}_5(C_0)$, $\mathrm{lsb}_5(E_0)$, $\mathrm{lsb}_5(G_0)$, $\mathrm{lsb}_5(I_0)$, $\mathrm{lsb}_5(K_0)$, $\mathrm{lsb}_5(M_0)$, $\mathrm{lsb}_5(O_0)$ are set to zero, while the remaining bits are changed incrementally. This implies at most $2^{512-40} = 2^{472}$ plaintext blocks are available. Thus, we require $2^{5r-21} \leq 2^{472}$, or $5r \leq 493$, or $r \leq 98$. The data complexity is at most $2^{472}$ plaintext blocks. It means that MRC6 would require at least $r = 99$ rounds to counter this $\chi^2$ attack.

Table 3 shows the result of the experiment in the case of 16 bits: $\mathrm{lsb}_2(A_{2i})$, $\mathrm{lsb}_2(C_{2i})$, $\mathrm{lsb}_2(E_{2i})$, $\mathrm{lsb}_2(G_{2i})$, $\mathrm{lsb}_2(I_{2i})$, $\mathrm{lsb}_2(K_{2i})$, $\mathrm{lsb}_2(M_{2i})$, and $\mathrm{lsb}_2(O_{2i})$ after an even number of rounds of MRC6. We use $\chi^2_{75} = 65779$. Starting from six rounds, the number of texts for which $H_0$ is rejected starts to increase by a factor of about $2^{10}$ every two rounds. Thus, for $r$ rounds ($r$ even and $r \geq 6$), the following is expected for the number of chosen plaintexts in terms of $r$: $N = 2^{12} \cdot 2^{10 \cdot (r-6)/2} = 2^{5r-18}$. The analysis is similar to the 8-bit case in the previous paragraph. Following the same rationale, at most $2^{512-40} = 2^{472}$ plaintext blocks

**Table 3.** $\chi^2$ attack simulations on MRC6, $2^{16} - 1$ degrees of freedom and 20 tests

| #Rounds | $\log_2 N$ | average $\chi^2$ | minCI | #values> $\chi^2_{75}$ | hypothesis |
|---------|-----------|-------------------|----------|--------------------------|------------|
| 2 | 1 | 68810.8 | 63145.2 | 1 | $H_0$ |
| 2 | 2 | 80277.6 | 72615.5 | 8 | $H_1$ |
| 2 | 3 | 90923.2 | 84597.2 | 18 | $H_1$ |
| 2 | 4 | 125731.2 | 116640.0 | 20 | $H_1$ |
| 4 | 4 | 65520.0 | 65520.0 | 0 | $H_0$ |
| 4 | 5 | 66732.8 | 65828.2 | 5 | $H_1$ |
| 4 | 6 | 67622.4 | 66920.0 | 14 | $H_1$ |
| 4 | 7 | 69913.6 | 68843.6 | 19 | $H_1$ |
| 4 | 8 | 74035.2 | 72727.0 | 20 | $H_1$ |
| 6 | 11 | 65804.8 | 65643.3 | 12 | $H_0$ |
| 6 | 12 | 66108.8 | 65941.0 | 16 | $H_1$ |
| 6 | 13 | 66850.4 | 66685.1 | 20 | $H_1$ |
| 8 | 20 | 65804.6 | 65648.0 | 11 | $H_0$ |
| 8 | 21 | 66090.0 | 65912.8 | 15 | $H_1$ |
| 8 | 22 | 66862.9 | 66608.5 | 19 | $H_1$ |
| 8 | 23 | 68275.1 | 67872.4 | 20 | $H_1$ |
| 10 | 30 | 65637.4 | 65450.0 | 9 | $H_0$ |
| 10 | 31 | 65916.7 | 65760.9 | 14 | $H_1$ |
| 10 | 32 | 65961.5 | 65778.7 | 11 | $H_1$ |
| 10 | 33 | 66128.2 | 65961.9 | 16 | $H_1$ |
| 10 | 34 | 66521.0 | 66262.4 | 18 | $H_1$ |
| 10 | 35 | 67043.6 | 66671.7 | 19 | $H_1$ |

will be available. Thus, this analysis holds as long as $2^{5r-18} \leq 2^{472}$, or $5r \leq 490$, or $r \leq 98$. Again, the data complexity is at most $2^{472}$ plaintext blocks, and MRC6 requires at least 99 rounds to counter this $\chi^2$ attack.

In order to compare the approach in Table 2 with an alternative approach used in [13], we provide Table 4.

Experimentally, we have observed that less chosen plaintexts are needed in the new approach than in the conventional approach of [13], at least for two, four and six rounds.

We point out that in Tables 2 and 3, the minimum value of $N$ for which $H_1$ is accepted may be less than $5 \cdot k$ when the number of rounds is small. For example, the values of $N$ for 2, 4 and 6 rounds in Table 2. This phenomenon is particular for a small number of rounds, and is due to the large block size and the slow diffusion in MRC6 (unlike the AES, in which diffusion is guaranteed by an MDS matrix, in MRC6 the diffusion depends on appropriate rotation amounts). Therefore, we also use these former values of $N$ to estimate the minimum $N$ for which $H_1$ is accepted, for higher numbers of rounds. For 8 or more rounds, the (minimum) values for $N$ are greater than $5 \cdot k$.

## 4.3   ERC6

For ERC6, our $\chi^2$ attacks were guided by the following 2-round iterative linear relation (using Type-I approximations [19])

**Table 4.** $\chi^2$ attack simulations on MRC6 using the approach in [13] with $2^8 - 1$ degrees of freedom

| #Rounds | $\log_2 N$ | average $\chi^2$ | minCI | #values> $\chi^2_{75}$ | hypothesis |
|---------|-----------|------------------|-------|------------------------|------------|
| 2 | 21 | 268.0 | 262.1 | 12 | $H_0$ |
| 2 | 22 | 276.8 | 271.4 | 11 | $H_1$ |
| 2 | 23 | 293.3 | 288.9 | 18 | $H_1$ |
| 2 | 24 | 326.4 | 320.1 | 20 | $H_1$ |
| 4 | 29 | 256.9 | 250.6 | 8 | $H_0$ |
| 4 | 30 | 251.1 | 245.8 | 4 | $H_0$ |
| 4 | 31 | 257.5 | 252.1 | 8 | $H_0$ |
| 4 | 32 | 258.7 | 254.0 | 5 | $H_0$ |
| 4 | 33 | 255.6 | 251.1 | 4 | $H_0$ |
| 6 | 25 | 265.5 | 258.4 | 9 | $H_0$ |
| 6 | 26 | 258.7 | 253.2 | 8 | $H_0$ |
| 6 | 27 | 253.0 | 247.8 | 6 | $H_0$ |
| 6 | 28 | 251.3 | 246.2 | 4 | $H_0$ |
| 6 | 29 | 254.9 | 248.6 | 6 | $H_0$ |

**Table 5.** $\chi^2$ attack simulations on ERC6, $2^4 - 1$ degrees of freedom and 20 tests

| #Rounds | $\log_2 N$ | average $\chi^2$ | minCI | #values> $\chi^2_{75}$ | hypothesis |
|---------|-----------|------------------|-------|------------------------|------------|
| 2 | 1 | 18.0 | 15.25 | 5 | $H_0$ |
| 2 | 2 | 26.4 | 22.44 | 8 | $H_1$ |
| 2 | 3 | 37.8 | 34.20 | 20 | $H_1$ |
| 4 | 12 | 26.9 | 22.08 | 10 | $H_0$ |
| 4 | 13 | 38.7 | 28.95 | 14 | $H_1$ |
| 4 | 14 | 65.8 | 43.67 | 17 | $H_1$ |
| 4 | 15 | 120.8 | 78.61 | 18 | $H_1$ |
| 4 | 16 | 222.5 | 138.04 | 19 | $H_1$ |
| 4 | 17 | 446.8 | 276.82 | 20 | $H_1$ |
| 6 | 23 | 26.0 | 20.23 | 10 | $H_0$ |
| 6 | 24 | 37.9 | 27.31 | 13 | $H_1$ |
| 6 | 25 | 59.4 | 42.71 | 16 | $H_1$ |
| 6 | 26 | 99.0 | 68.65 | 17 | $H_1$ |
| 6 | 27 | 196.2 | 133.90 | 18 | $H_1$ |
| 6 | 28 | 375.3 | 258.18 | 20 | $H_1$ |

$$A_i \cdot e_{t_1} \oplus C_i \cdot e_{t_2} \oplus E_i \cdot e_{t_3} \oplus G_i \cdot e_{t_4} = A_{i+2} \cdot e_{t_5} \oplus C_{i+2} \cdot e_{t_6} \oplus E_{i+2} \cdot e_{t_7} \oplus G_{i+2} \cdot e_{t_8}, \quad (5)$$

where $A_i$, $C_i$, $E_i$ and $G_i$, are input words to the $i$-th round. In particular, the masks $e_j$ with highest bias are such that $0 \leq j < \log_2 w$.

Table 5 shows the result of attack simulation in the case of 4 bits: $\mathrm{lsb}_1(A_{2i})$, $\mathrm{lsb}_1(C_{2i})$, $\mathrm{lsb}_1(E_{2i})$ and $\mathrm{lsb}_1(G_{2i})$ after an even number of rounds of ERC6. We use $\chi^2_{75} = 22.31$. Starting from four rounds, the number of texts for which $H_0$ is rejected starts to increase by a factor of about $2^{11}$ every two rounds. Thus, for $r$ rounds ($r$ even and $r \geq 4$), the following is expected for the number

**Table 6.** $\chi^2$ attack simulations on ERC6 using the approach in [13] with $2^4 - 1$ degrees of freedom and 20 tests

| #Rounds | $\log_2 N$ | average $\chi^2$ | minCI | #values> $\chi^2_{75}$ | hypothesis |
|---|---|---|---|---|---|
| 2 | 15 | 17.1 | 14.1 | 8 | $H_0$ |
| 2 | 16 | 26.8 | 22.9 | 16 | $H_1$ |
| 2 | 17 | 35.2 | 31.3 | 20 | $H_1$ |
| 4 | 32 | 14.833 | 12.03 | 6 | $H_0$ |
| 4 | 33 | 16.633 | 13.87 | 6 | $H_0$ |
| 4 | 34 | 14.549 | 12.56 | 5 | $H_0$ |
| 4 | 35 | 15.257 | 13.20 | 4 | $H_0$ |
| 4 | 36 | 13.143 | 11.25 | 2 | $H_0$ |

**Table 7.** $\chi^2$ attack simulations on ERC6, $2^8 - 1$ degrees of freedom and 20 tests

| #Rounds | $\log_2 N$ | average $\chi^2$ | minCI | #values> $\chi^2_{75}$ | hypothesis |
|---|---|---|---|---|---|
| 2 | 2 | 309.6 | 272.0 | 7 | $H_0$ |
| 2 | 3 | 356.8 | 316.6 | 15 | $H_1$ |
| 2 | 4 | 520.0 | 444.0 | 20 | $H_1$ |
| 4 | 11 | 284.5 | 270.3 | 9 | $H_0$ |
| 4 | 12 | 310.0 | 291.0 | 14 | $H_1$ |
| 4 | 13 | 366.8 | 334.6 | 17 | $H_1$ |
| 4 | 14 | 468.0 | 405.2 | 18 | $H_1$ |
| 4 | 15 | 711.6 | 579.7 | 20 | $H_1$ |
| 6 | 23 | 290.5 | 283.5 | 14 | $H_0$ |
| 6 | 24 | 325.6 | 311.9 | 18 | $H_1$ |
| 6 | 25 | 404.5 | 379.9 | 19 | $H_1$ |
| 6 | 26 | 549.3 | 491.7 | 20 | $H_1$ |

of chosen plaintexts: $N = 2^{13} \cdot 2^{11 \cdot (r-4)/2} = 2^{5.5r-9}$. The algorithm TEST(.) chooses plaintexts such that the $\text{lsb}_5(A_0)$, $\text{lsb}_5(C_0)$, $\text{lsb}_5(E_0)$, $\text{lsb}_5(G_0)$ are set to zero. This implies at most $2^{256-20} = 2^{236}$ plaintext blocks are available. Thus, this analysis holds as long as $2^{5.5r-9} \le 2^{236}$, or $5.5r \le 245$, or $r \le 44$. Since the attack effort is at most $2^{236}$ encryptions equivalent number of text blocks, it means that ERC6 would require at least 45 rounds to counter this $\chi^2$ attack.

In order to compare the approach in Table 5 with the approach used in [13], we provide Table 6. Empirically, we have observed that significantly less chosen plaintexts are needed in the new approach than in the conventional approach of [13], at least for two and four rounds.

Table 7 shows the result of analysing the 8-bit value from the concatenation of $\text{lsb}_2(A_{2i})$, $\text{lsb}_2(C_{2i})$, $\text{lsb}_2(E_{2i})$ and $\text{lsb}_2(G_{2i})$ after an even number of rounds of ERC6. We use $\chi^2_{75} = 284.34$. Starting from four rounds, the number of texts for which $H_0$ is rejected starts to increase by a factor of about $2^{12}$ every two rounds. Thus, for $r$ rounds ($r$ even and $r \ge 4$), the following behaviour is expected for the number of chosen plaintexts: $N = 2^{12} \cdot 2^{12 \cdot (r-4)/2} = 2^{6r-12}$. Following a similar reasoning as in the previous paragraph, this analysis holds as long as $2^{6r-12} \leq 2^{236}$, or $6r \leq 248$, or $r \leq 41$. Since the attack effort is at most $2^{236}$ encryption, ERC6 requires at least 42 rounds to counter this $\chi^2$ attack.

## 5   Conclusions and Further Work

This paper presented a new approach to the $\chi^2$ statistical test applied to RC6, ERC6 and MRC6 block ciphers. These attacks were preceeded by a linear crypt-analysis of these same ciphers, which provided promising bit positions to be analysed by the $\chi^2$ tests. For 2-round and 4-round RC6, our method improves the data complexity of the previously best-known $\chi^2$ attacks [13] by a factor of about $2^{11}$. Tables 8, 9, 10 and 11 summarize our attacks on ERC6 and MRC6.

Overall, our attacks reduced the number of chosen plaintexts to detect $\chi^2$ correlation when compared to conventional $\chi^2$ attacks. Consequently, we could apply and check in practice our predictions on attacks up to 10-round MRC6

**Table 8.** Summary of $\chi^2$ attacks analysing 8 bits output by MRC6

| #Rounds | Time | Data | Memory | Comment |
|---------|------|------|--------|---------|
| 2 | $2^2$ | $2^2$ CP | $2^2$ | Table 2 |
| 4 | $2^4$ | $2^4$ CP | $2^4$ | Table 2 |
| $r$ | $2^{5r-21}$ | $2^{5r-21}$ CP | $2^{5r-21}$ | $6 \leq r < 99$, $r$ even |

**Table 9.** Summary of $\chi^2$ attacks analysing 16 bits output by MRC6

| #Rounds | Time | Data | Memory | Comment |
|---------|------|------|--------|---------|
| 2 | $2^2$ | $2^2$ CP | $2^2$ | Table 3 |
| 4 | $2^5$ | $2^5$ CP | $2^5$ | Table 3 |
| $r$ | $2^{5r-18}$ | $2^{5r-18}$ CP | $2^{5r-18}$ | $6 \leq r < 99$, $r$ even |

**Table 10.** Summary of $\chi^2$ attacks analysing 4 bits output by ERC6

| #Rounds | Time | Data | Memory | Comment |
|---------|------|------|--------|---------|
| 2 | $2^2$ | $2^2$ CP | $2^2$ | Table 5 |
| 4 | $2^{13}$ | $2^{13}$ CP | $2^{13}$ | Table 5 |
| $r$ | $2^{5.5r-9}$ | $2^{5.5r-9}$ CP | $2^{5.5r-9}$ | $4 \leq r < 45$, $r$ even |

**Table 11.** Summary of $\chi^2$ attacks analysing 8 bits output by ERC6

| #Rounds | Time | Data | Memory | Comment |
|---------|------|------|--------|---------|
| 2 | $2^3$ | $2^3$ CP | $2^3$ | Table 7 |
| 4 | $2^{11}$ | $2^{11}$ CP | $2^{11}$ | Table 7 |
| $r$ | $2^{6r-12}$ | $2^{6r-12}$ CP | $2^{6r-12}$ | $4 \leq r < 42$, $r$ even |

and 6-round ERC6. The reduction in the data complexity of our attacks was influenced by the weak diffusion in the target ciphers.

In the analyses of M6, MX and RC5P in [5], the $\chi^2$ tests were supported by evidence collected from mod-$n$ analyses of these ciphers. The nonuniform distribution of residues modulo $n$ of internal cipher components, for $n$ a prime number, was corroborated by experimental $\chi^2$ tests. Likewise, in this paper, our results were supported by linear relations.

The analyses presented in this paper considered sets of randomly chosen keys, that is, no particular (weak) keys were purposefully used. This implies that even better results could have been achieved with keys that caused null rotation in some rounds under analysis (as observed in [13]). This issue of weak keys for $\chi^2$ attacks is left as a problem for future work. Analogously, we have focused on distinguishing attacks only. Key-recovery attacks are also left as further work.

# References

1. Rivest, R.L., Robshaw, M.J.B., Sidney, R., Yin, Y.L.: The RC6 block cipher (1998), http://csrc.nist.gov/encryption/aes/
2. AES: The advanced encryption standard development process (1997), http://csrc.nist.gov/encryption/aes/
3. Ragab, A., Ismail, N., Allah, O.F.: Enhancements and implementation of RC6 block cipher for data security. In: IEEE TENCON (2001)
4. El-Fishawy, N., Danaf, T., Zaid, O.: A modification of RC6 block cipher algorithm for data security (MRC6). In: International Conference on Electrical, Electronic and Computer Engineering, pp. 222–226 (2004)
5. Kelsey, J., Schneier, B., Wagner, D.: Mod n cryptanalysis, with applications against RC5P and M6. In: Knudsen, L.R. (ed.) FSE 1999. LNCS, vol. 1636, pp. 139–155. Springer, Heidelberg (1999)
6. Knuth, D.: The Art of Computer Programming, Seminumerical Algorithms. Addison-Wesley, Reading (1997)
7. Vaudenay, S.: An experiment on DES statistical cryptanalysis. Technical report, Ecole Normale Supérieure, LIENS–95–29 (1995)
8. Handschuh, H., Gilbert, H.: $\chi^2$ cryptanalysis of the SEAL encryption algorithm. In: Biham, E. (ed.) FSE 1997. LNCS, vol. 1267, pp. 1–12. Springer, Heidelberg (1997)
9. Baudron, O., Gilbert, H., Granboulan, L., Handschuh, H., Joux, A., Nguyen, P., Noilhan, F., Pointcheval, D., Pornin, T., Poupard, G., Stern, J., Vaudenay, S.: Report on the AES candidates (1999), http://csrc.nist.gov/encrypt/aes/round1/conf2/papers/baudron1.pdf
10. Gilbert, H., Handschuh, H., Joux, A., Vaudenay, S.: A statistical attack on RC6. In: Schneier, B. (ed.) FSE 2000. LNCS, vol. 1978, pp. 64–74. Springer, Heidelberg (2001)
11. Hinoue, T., Miyaji, A., Wada, T.: The security of RC6 against asymmetric chi-square test attack. IPSJ Journal 48(9), 1–10 (2007)
12. Isogai, N., Matsunaka, T., Miyaji, A.: Optimized $\chi^2$-attack against RC6. In: Zhou, J., Yung, M., Han, Y. (eds.) ACNS 2003. LNCS, vol. 2846, pp. 16–32. Springer, Heidelberg (2003)

13. Knudsen, L., Meier, W.: Correlations in RC6 with a reduced number of rounds. In: Schneier, B. (ed.) FSE 2000. LNCS, vol. 1978, pp. 94–108. Springer, Heidelberg (2001)
14. Nonaka, M., Miyaji, A.: A note on the security of RC6 against correlation attack. In: The 2002 Symposium on Cryptography and Information Security, pp. 681–686 (2002)
15. Shimoyama, T., Takeuchi, K., Hayakawa, J.: Correlation attack to the block cipher RC5 and the simplified variants of RC6 (2001), http://csrc.nist.gov/encryption/aes/
16. Bain, L., Engelhardt, M.: Introduction to Probability and Mathematical Statistics. Duxbury Press (1987)
17. Shimoyama, T., Takenaka, M., Koshida, T.: Multiple linear cryptanalysis of a reduced-round RC6. In: The 2002 Symposium on Cryptography and Information Security, pp. 931–936 (2002)
18. Menezes, A., van Oorschot, P., Vanstone, S.: Handbook of Applied Cryptography. CRC Press, Boca Raton (1997)
19. Contini, S., Rivest, R., Robshaw, M., Yin, Y.: The security of the RC6 block cipher, v. 1.0 (1998)

# A    Tables and Figures

**Table 12.** $\chi^2$ threshold values, specificities and degrees of freedom

| $\chi^2$ | | Degrees of Freedom $(k-1)$ | | | | | |
| --- | --- | --- | --- | --- | --- | --- | --- |
| | | $2^4-1$ | $2^8-1$ | $2^{12}-1$ | $2^{16}-1$ | $2^{20}-1$ | $2^{24}-1$ |
| | 0.60 | 15.73 | 260.09 | 4117.30 | 65626.10 | 1048941.26 | 16778682 |
| | 0.70 | 17.32 | 266.34 | 4141.97 | 65724.37 | 1049333.93 | 16780252 |
| | 0.75 | 18.24 | 269.85 | 4155.67 | 65778.82 | 1049551.40 | 16781122 |
| Specificity | 0.80 | 19.31 | 273.79 | 4170.96 | 65839.50 | 1049793.60 | 16782090 |
| $(1-\alpha)$ | 0.85 | 20.60 | 278.43 | 4188.84 | 65910.27 | 1050075.96 | 16783219 |
| | 0.90 | 22.31 | 284.34 | 4211.40 | 65999.39 | 1050431.31 | 16784639 |
| | 0.95 | 24.99 | 293.25 | 4244.99 | 66131.63 | 1050958.14 | 16786744 |
| | 0.99 | 30.58 | 310.46 | 4308.47 | 66380.16 | 1051946.85 | 16790690 |

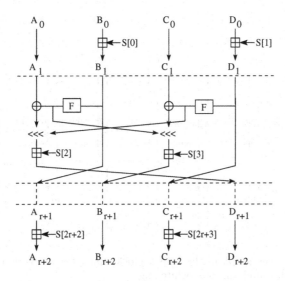

**Fig. 1.** Computational graph of the RC6 block cipher for encryption, showing pre-whitening, one full round and post-whitening

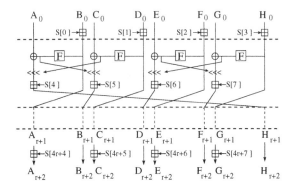

**Fig. 2.** Computational graph of the ERC6 block cipher for encryption, showing prewhitening, one full round and post-whitening

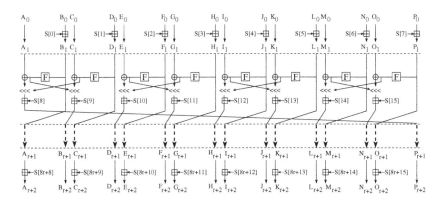

**Fig. 3.** Computational graph of the MRC6 block cipher for encryption, showing prewhitening, one full round and post-whitening

# Analysis and Optimization of Cryptographically Generated Addresses

Joppe W. Bos[1], Onur Özen[1], and Jean-Pierre Hubaux[2]

[1] EPFL IC IIF LACAL, Station 14, CH-1015 Lausanne, Switzerland
{joppe.bos,onur.ozen}@epfl.ch
[2] EPFL IC ISC LCA1, Station 14, CH-1015 Lausanne, Switzerland
jean-pierre.hubaux@epfl.ch

**Abstract.** The need for nodes to be able to generate their own address and verify those from others, without relying on a global trusted authority, is a well-known problem in networking. One popular technique for solving this problem is to use self-certifying addresses that are widely used and standardized; a prime example is cryptographically generated addresses (CGA). We re-investigate the attack models that can occur in practice and analyze the security of CGA-like schemes. As a result, an alternative protocol to CGA, called CGA++, is presented. This protocol eliminates several attacks applicable to CGA and increases the overall security. In many ways, CGA++ offers a nice alternative to CGA and can be used notably for future developments of the Internet Protocol version 6.

## 1   Introduction

Cryptographically generated addresses (CGA) is a technique that creates a fixed size address by hashing the address owner's public key with the help of a cryptographic hash function. This technique enables the address owner to assert address ownership by creating a relation between the address and the address owner's public/private key pair.

An example where this technique can be used is in Internet Protocol version 6 (IPv6) addresses. The 64-bits of these addresses known to be the *interface identifier* are then generated with the help of CGA as proposed by Aura [1,2].

The main advantage of CGAs is that they are self-certified; a trusted third party or a public-key infrastructure (PKI) is not needed to generate the IPv6 address or to verify other addresses. Self-certified addresses are extensively used in protocols such as Secure Neighbor Discovery [3], Shim6 [4] and the IPv6 mobility support [5]; they offer features such as duplicate address detection and proof of address ownership.

The idea of cryptographically generated addresses first appeared in the child-proof authentication for mobile-IPv6 (CAM) protocol by O'Shea and Roe [6]; this was later improved by Nikander [7]. An alternative method was proposed by Montenegro and Castelluccia [8] under the name "statistically unique and

P. Samarati et al. (Eds.): ISC 2009, LNCS 5735, pp. 17–32, 2009.
© Springer-Verlag Berlin Heidelberg 2009

cryptographically verifiable addresses" (SUCV). Finally, the actual model was presented by Aura [1] and appears as an RFC [9].

The drawback of the early proposals of CGA, namely the small number of bits of the address to accommodate the result of the cryptographic hash function, is solved by Aura in [1] by using hash extensions. With this technique, the resistance of the scheme against impersonation is increased at the cost of increasing the time needed for address generation while keeping the necessary operations required for the verification constant. This is realized by scaling, with the help of a relation defined by the security parameter, the effort needed to break the system in the future due to the progress of technology. This implies an increase in the cost of address generation; the exponential growth over time of computing power compensates this increase.

To the best of our knowledge, no work has been published, besides the RFC documents and the unpublished work in [2], on the analysis of CGA since the original proposal [1]. As also observed in the original proposal, CGA is susceptible to a global time-memory trade-off attack that eliminates the effect of hash extensions in the long run at the cost of storage. Such an attack is assumed to be impractical in [1]. Moreover, due to lack of authentication, CGA is vulnerable to a replay attack where an attacking node is capable of sniffing and storing signed messages from a target node and replay them later.

In this work, we present a detailed security/efficiency analysis of CGA and a proposal to solve security problems related to self-certifying address generation and verification in CGA. This protocol, mainly based on the ideas of CGA, is called CGA++. In the analysis part, a novel security framework is provided, which enables us to evaluate the security of CGA-like schemes, including CGA++. In this design, we mitigate the global time-memory trade-off attack by reducing its effect to a specific network. Furthermore, we introduce digital signatures in order to overcome the lack of authenticity in CGA and to increase the security when no hash extensions are used. CGA++ offers an alternative to CGA and can be used in practice for future development of IPv6.

The organization of the paper is as follows. Section 2 introduces the preliminaries for IPv6 addresses including the system model and the necessary notation throughout the paper. In Section 3 and 4, we provide the specification and the analysis of CGA, respectively. We introduce the design of CGA++ in Section 5 followed by its analysis in Section 6. The compatibility and the applications of CGA++ are discussed in Section 7. We conclude the paper in Section 8.

## 2   Preliminaries

The objective for using CGA is to generate self-certified IPv6 addresses. For the sake of clarity, we adhere to the conventions and the notations from [10,11].

IPv6 addresses are 128-bit data blocks constructed by the concatenation of two 64-bit words: *subnet prefix* and *interface identifier* [10]. The former is located on the most significant 64-bit, which is used to determine the nodes' location in the Internet topology. The latter, being comprised of the least significant 64-bits,

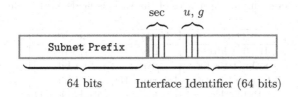

**Fig. 1.** Format of an IPv6 address. The parameters *sec*, *u* and *g* are placed at the most significant three, seventh and eighth bit of the *interface identifier*, respectively

acts as an identity of the node whose generation process is the main target of this work. See Fig. 1 for the general overview of IPv6 addresses.

In the address format, there are several parameters in the *interface identifier*, which have special semantics. The first parameter set is comprised of the $u$ and $g$ bits, located in the seventh and the eighth bit of the *interface identifier*. The combination $u = g = 1$ is unused for other purposes and suggested by Aura to indicate the use of CGA [1]. The other value in the *interface identifier* is the security parameter *sec*, a three-bit user defined parameter used to determine the length of the hash extension in the protocol. In CGA, this parameter is used to scale the relation in the hash extension. We provide the necessary notation used in the rest of the paper in Table 1.

In the original RFC [9], the proposed hash function is SHA-1 [12]. The recently discovered weaknesses of SHA-1 led to a modification of the CGA specification to enable the support of alternative hash functions [13]. Therefore, in this paper, we denote the hash function used in the protocol as $\mathcal{H}$ and assume this hash function to be ideal: it is optimally collision, preimage and second-preimage resistant.

We assume, throughout the paper, that (mobile) nodes are capable of dealing with Internet protocols and are also equipped with and capable of calculating basic cryptographic algorithms. Moreover, we assume an increase of performance of mobile nodes following Moore's law. Yet, of course, there will always be a market for low-end devices. But it is unlikely for them to be isolated and mobile, and even if they are, they will not be security sensitive.

For the attacker, we ignore the time required to generate valid public/private key pairs as these can easily be computed and stored, if necessary, in an initialization phase.

## 3   CGA Specification

The actual protocol for self-certified address generation and verification for IPv6 using CGA appears in RFC 3972 [9], which is based on the ideas from [1]. In this section, we recall the specification of CGA. CGA uses a technique called hash extension, which is realized by the security parameter *sec*; this parameter linearly scales the number of bits used in the hash extension by imposing $16 \times sec$ many bits to zero in the hash value denoted by Hash2.

**Table 1.** Notation and data sizes of the various fields used in this article

| Data | Notation | Length |
|---|---|---|
| IPv6 Address | IPv6 | 128-bit |
| Subnet Prefix | SP | 64-bit |
| Interface Identifier | Interface ID | 64-bit |
| Public-Key | $K_{pub}$ | Variable length |
| Private-Key | $K_{priv}$ | Variable length |
| Digital Signature | Sign | Variable length |
| Modifier | $m$ | 128-bit |
| Collision Count | $CC$ | 8-bit |
| Security Parameter | $sec$ | 3-bit |
| u,g flags | $u, g$ | 1-bit each |

The main idea behind CGA is to trade efficiency for security. When generating a new address, a node has to satisfy certain constraints, i.e. the hash extension, which decreases the efficiency of the address generation. Because an attacker needs to do this extra work as well, the level of security increases compared to the setting where no hash extensions are used. The verification, on the other hand, requires a constant amount of time and does not suffer from an efficiency decrease. This ensures that an attacking node needs to perform all the computational work, thereby preventing the denial of service of verifiers.

**Address Generation.** The procedure for generating an IPv6 address using CGA is illustrated in Fig. 2 which can be described as follows.

1. Set the *modifier* to a random 128-bit value. Select the security parameter sec and set the *collision count* to zero.
2. Concatenate the *modifier*, $64 + 8$ zero bits, and the encoded public-key. Execute the $\mathcal{H}$ algorithm on the concatenation. The leftmost 112 bits of the result are Hash2.
3. Compare the $16 \times sec$ leftmost bits of Hash2 with zero. If they are all zero (or if $sec = 0$), continue with Step (4). Otherwise, increment the *modifier* and go back to Step (2).
4. Concatenate the *modifier*, *subnet prefix*, *collision count* and encoded public-key. Execute the $\mathcal{H}$ algorithm on the concatenation. The leftmost 64 bits of the result are Hash1.
5. Form an *interface identifier* by setting the two reserved bits $u$ and $g$ in Hash1 both to 1 and the three leftmost bits to *sec*.
6. Concatenate the *subnet prefix* and *interface identifier* to form an 128-bit IPv6 address.
7. If an address collision with another node within the same subnet is detected, increment the *collision count* and go back to step (4). However, after three collisions, stop and report the error.

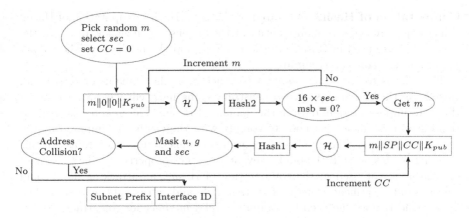

**Fig. 2.** Detailed data flow of the address generation in CGA

**Verification of Address Ownership.** The verification of address ownership is realized by the execution of the following steps. Given the IPv6 address, *collision count* and the *modifier*,

1. Check that the *collision count* is 0, 1 or 2 and that the *subnet prefix* is equal to the *subnet prefix* of the address. The CGA verification fails if either check fails.
2. Concatenate the *modifier*, *subnet prefix*, *collision count* and the public-key. Execute the $\mathcal{H}$ algorithm on the concatenation. The 64 leftmost bits of the result are Hash1.
3. Compare Hash1 with the *interface identifier* of the address. The differences in the two reserved bits $u$ and $g$ and in the three leftmost bits are ignored. If the 64-bit values differ (other than in the five ignored bits), the CGA verification fails.
4. Concatenate the *modifier*, $64 + 8$ zero bits and the public-key. Execute the $\mathcal{H}$ algorithm on the concatenation. The leftmost 112 bits of the result are Hash2.
5. Read the security parameter *sec* from the three leftmost bits of the interface identifier of the address. Compare the $16 \times sec$ leftmost bits of Hash2 with zero. If any one of these bits is nonzero, CGA verification fails. Otherwise, the verification succeeds. If $sec = 0$, verification never fails at this step.

## 4    Analysis of CGA

### 4.1    Design Rationale

In order to have a better comprehension, we explain the design rationale of CGA by going through the components inside CGA, especially the computation of Hash1 and Hash2.

**Computation of Hash2.** As stated by Aura in [1], the computation of Hash2 is introduced in order to gain security at the expense of efficiency. In CGA, the values that are used in Hash2 are the *modifier* and the public-key. The *subnet prefix* and the *collision count* are set to zero.

The idea is to use some common parameters in the domains of Hash1 and Hash2. As Hash2 requires special constraints, the most efficient way to satisfy this is to impose random data in the domain of Hash2. The *modifier* is used for this purpose, it allows the node to comply with the conditions imposed by the security parameter. The public-key is needed in this computation to assign this *modifier* to this node. If the public-key, or any other property specific to address generator, is not used in the computation of Hash2, an attacking node could simply send a verification request to this node and retrieve its *modifier* value. This would remove the need to compute a valid *modifier* for the attacker.

It was a design decision of Aura [1] not to include the *subnet prefix* in the computation of Hash2 for the sake of efficiency. A mobile node travels frequently from network to network and thus needs to regenerate its address over and over again. Assuming a mobile node does not have much computation power, it would be infeasible to search for a *modifier* every time it changes network. When not including the *subnet prefix* in the Hash2, a mobile node has to regenerate its Hash1, where the *subnet prefix* occurs, at the cost of only one hash function evaluation. Following the same reasoning, the *collision count* parameter is set to zero as well. This avoids computing Hash2 again when a collision of the *interface identifier* is detected after the creation of the Hash1.

**Computation of Hash1.** In the computation of the Hash1, all parameters are used. In the domain of Hash1, the *subnet prefix* is used in order to avoid a time-memory trade-off attack as it would be possible to store valid addresses from each network. *Collision count* is added to the domain to overcome the scenario where a duplicate address is generated. Finally, the *modifier* is used as a "proof" that the node generated a valid Hash2 and the public-key for the same reasons as in the Hash2.

## 4.2    Security Framework

Assume a "CGA-like" protocol design is to be assessed for security with a focus on the address generation and verification part. More precisely, a protocol is considered which makes use of two different hash function evaluations where the output of one is not used as the input for the other. A formal model can be useful for this task, especially for assessing the security of the considered protocol. Such a model is proposed in this section and is used in the remainder of this paper to state facts about the security and efficiency of such "CGA-like" protocols.

Let us denote the time needed for a node to generate an address as $T_G$, to verify an address as $T_V$ and to impersonate an address as $T_A$. These times are stated as a function of $T_1$ and $T_2$, which denote the time to compute Hash1 and Hash2, respectively and are expressed in hash function evaluations. The number of available bits in the address, which is the number of (truncated) output of

Hash1, is denoted by $l$. We denote the number of bits on which we put a special condition by $s$, the (truncated) output of Hash2; here $l, s \in \mathbb{N}$.

**Address Generation.** Address generation for a legitimate node is not expected to exceed $2^s \cdot T_2$ in order to meet the conditions of Hash2, plus $T_1$ to generate the address. The cost of address generation $T_G$ is therefore:

$$T_G = 2^s \cdot T_2 + T_1. \tag{1}$$

**Address Verification.** To verify an address, the conditions on Hash2 need to be validated, representing a duration $T_2$. If this validation is successful, the address needs to be checked in time $T_1$. The time needed for verification, $T_V$, is

$$T_V = T_1 + T_2. \tag{2}$$

**Impersonation.** In order to impersonate a node, an attacker can proceed in two ways: by first satisfying the constraints on Hash1 and then on Hash2 or vice versa. Beginning with Hash1, the attacker must first perform a second-preimage attack on Hash1, which is expected to take no more than $2^l \cdot T_1$ hash function evaluations to find another valid parameter set to match the *interface identifier*. Once fulfilled, the conditions on Hash2 for the generated *modifier* should be satisfied, which happens with probability $2^{-s}$ for an ideal hash function. The total cost $C_{A,H_1}$ for impersonation, when beginning with Hash1, becomes $C_{A,H_1} = (2^l \cdot T_1 + T_2) \cdot 2^s$.

Starting from Hash2, the conditions on Hash2 are met at a cost of up to $2^s \cdot T_2$ hash function evaluations. Next, Hash1 is created and verified if it collides with the target address. The probability of hitting this specific Hash1 is $2^{-l}$. Therefore, the total cost $C_{A,H_2}$, when beginning with Hash2, becomes $C_{A,H_2} = (2^s \cdot T_2 + T_1) \cdot 2^l$. An attacker can choose either of these values in order to minimize his attack cost. Hence, the time for impersonation, $T_A$, in a generic model becomes

$$T_A = \min(C_{A,H_1}, C_{A,H_2}) = \min((2^l \cdot T_1 + T_2) \cdot 2^s, (2^s \cdot T_2 + T_1) \cdot 2^l). \tag{3}$$

### 4.3    Security of CGA

In order to assess the security of CGA, we begin with discussing the basic principles of cryptographic hash functions: collision, preimage and second-preimage resistance. This allows us to evaluate certain bounds in order to attack the scheme, assuming that the underlying cryptographic hash function has no known weaknesses. With the help of the birthday- and the brute-force attack, finding a collision and a (second) preimage require $\mathcal{O}(\sqrt{2^n})$ and $\mathcal{O}(2^n)$ hash function evaluations, as the digest size $n$ tends to infinity, respectively.

A collision attack is not very powerful in this setting as it means being able to generate two nodes with the same address without having any control over the generated address. The preimage attack is equivalent to the second preimage attack since all the domain parameters of the hash function are known. The second preimage-attack is the most powerful attack model for this setting. It is equivalent to being able to generate another valid CGA parameter from a given address: i.e. impersonation. This attack model and its cost are treated in the following where the proof follows directly from the cost needed to perform a second-preimage attack when not using hash extensions (i.e. $sec = 0$) and our security framework for the case $sec > 0$ (cf. Section 4.2).

**Lemma 1.** *Given a network, assume the addresses are generated and verified by CGA with security parameter sec. Then, the number of operations required for the impersonation of a specific node is*

$$T_A = \begin{cases} 2^{59} & \text{if } sec = 0, \\ 2^{59+16 \times sec} + 2^{16 \times sec} & \text{if } 1 \leq sec \leq 3, \\ 2^{59+16 \times sec} + 2^{59} & \text{otherwise.} \end{cases}$$

*hash function evaluations.*

This shows that the resistance of CGA against impersonation is mainly dominated by the increasing values of the security parameter sec.

**A Time-Memory Trade-Off Attack on CGA.** As observed in [1], a time-memory trade-off attack can be mounted on CGA in order to impersonate a node. Details of the time and memory complexities of this attack are not stated in [1] and they are assumed to be impractical. The following lemma explains this more specifically.

**Lemma 2.** *Given a number of $k > 0$ networks each of size approximately $2^{n_i}$ nodes, for $0 < i \leq k$, assume the nodes use CGA for address generation and verification. Using a time-memory trade-off attack, an attacker needs at most $T$ calls to the hash function and comparisons of the hash-values in order to impersonate one random node in the network of size $2^{n_i}$. When the number of attacks $A \to \infty$, the number of calls $T$ per attack is asymptotically bounded by*

$$T \leq 2^{59 - min(n_i)}. \tag{4}$$

*In other words, $T$ is independent of the security parameter sec. The storage requirement is $2^{33 - min(n_i)}$ gigabytes, where $min(n_i)$ denotes the smallest value $n_i$.*

*Proof.* Assume a database is given with valid *modifier* values $m_j$, $j \in \mathbb{N}$, such that the condition on Hash2 is satisfied. When targeting a network of size $2^{n_i}$, a second-preimage for Hash1 of a random node is expected after $2^{59-n_i}$ hash function evaluations. The cost $C$, the number of hash function evaluations to create the database of *modifier* values, is expected to take no more than $C = 2^{59+16 \times sec - n_i}$. The database is independent of the used *subnet prefix* and can be computed once and used for all subsequent attacks in the future. Then, the average cost $T$ per attack becomes $T = 2^{59-n_i} + \dfrac{2^{59+16 \times sec - n_i}}{A}$. By selecting the smallest network size among $n_i$, which maximizes the attack cost, the number of hash evaluations becomes asymptotically, when the number of attacks go to infinity, $T \leq 2^{59-\min(n_i)}$. The storage cost is $128 \cdot 2^{59-\min(n_i)}$ bits which corresponds to $2^{33-\min(n_i)}$ gigabytes. □

Note that this attack cannot be used to impersonate a specific node. Instead, it can be used to impersonate a random node in the network. In order to illustrate the required storage for such an attack, assume an attacker wants to impersonate an address of a random node in a network of size $2^{16}$, this requires $2^{33-16} = 2^{17}$ gigabytes = 128 terabyte of storage. This is significant but not infeasible.

**Authentication.** One of the known limitations of CGA, as mentioned in [1], is the possibility for an attacking node to sniff and store signed messages from a target node. Once this is done, the attacker obtains the public-key and the *modifier*; with these values it can create a valid address using a different *subnet prefix*. Sending forged, correctly signed messages is computationally infeasible but by replaying the sniffed messages, an attacker can mislead legitimate nodes by convincing them that he owns an address.

This type of attack can also be used to generate an address that already exists in the network. That is, for a specific security parameter *sec*, the attacker can collect many valid *modifier* and public-keys together with a certain amount of signed messages from these nodes. Next, a *subnet prefix* is selected, and with the help of the stored values, a search is started for a hit in one of the addresses. This helps to reduce the complexity of the impersonation attack.

Another instance of such an attack is to search for nodes with a non-zero *collision count*. In the address generation of CGA, the nodes generate an address and look for a collision in the network. If there is a node with the generated address, the new node has to generate a new address by increasing the *collision count*. Hence, the attacker can look for a non-zero *collision count* in the network and use the valid *modifier* and public-key of this node to generate an existing address in the network with *collision count* zero. This helps the attacker to generate a duplicate addresses; he could even replay signed messages. Nevertheless, the probability of having a collision in the addresses is low and this attack fully depends on the non-zero *collision count*. Still, as the mobility property leads to the need to generate new addresses for the nodes while travelling from one network to another, the probability of address collision increases.

## 4.4   Efficiency of CGA

In CGA, the address generation time $T_G$ is equal to $T_G = 2^{16 \times sec} + 1$, which is dominated by the security parameter, assuming $sec > 0$, whereas the address verification time $T_V$ is constant, namely equal to two. To illustrate the actual computational demand, we make the optimistic assumption that a node has computing power comparable to a modern CPU used in a workstation. Our simple implementation of CGA, which uses the open source library OpenSSL [14], computes approximately $2^{18.5}$ digests/sec/CPU on a modern workstation (AMD64). The estimated time needed to comply with Hash2 requirements are provided in the following table

| $sec$ value | 1 | 2 | 3 | 4 | 5 | 6 |
|---|---|---|---|---|---|---|
| Required Time | 0.2 secs | 3.2 hrs | 24 yrs | $1.6 \cdot 10^6$ yrs | $1.0 \cdot 10^{11}$ yrs | $6.8 \cdot 10^{15}$ yrs |

These results correspond with the performance results from [1]. This table shows that the required time for generating a valid address with a high security parameter $sec$ is currently impractical. However, it will be feasible in the future due to the exponential growth in the computational capabilities of the nodes. A possible solution to the efficiency problem for the current use of larger $sec$ values is to generate these values off-line or search them in parallel, just as presented in the time-memory trade-off attack.

## 5    CGA++ Specification

**Design Rationale.** The main design rationale behind CGA++ follows from the fact that even if CGA offers a good protocol for self-certified address generation and verification, it has some limitations. Therefore, our main goal is to fix these weaknesses without losing too much efficiency. Considering the adoption and the extensive future use of CGA, one of our main goals is to adhere as closely as possible to CGA, thus offering an easy transition from CGA to CGA++.

As mentioned, a global time-memory trade-off attack is feasible at the cost of memory. In order to prevent this global attack the first obvious modification is to include the *subnet prefix* in the computation of Hash2. The verifier should make sure to check full IPv6 addresses and not the so-called link-local addresses as specified in IPv6 [10]. This has some efficiency loss; nevertheless, we believe this to be tolerable. Furthermore, an extra authentication mechanism is introduced by using digital signatures inside the verification process, preventing nearly all the mentioned attacks against CGA. This has the additional advantage that when no hash extensions are used ($sec = 0$) the security of the protocol is increased, compared to CGA. As a result, we propose a more secure, easy to adopt and compact alternative to CGA.

**Address Generation.** The general procedure of generating IPv6 address using CGA++ is depicted in Fig. 3 (note the similarities with Fig. 2). It can be described as follows.

1. Choose security parameter sec $\in \{0, \ldots, 7\}$. Set the *modifier* to a random 128-bit value and set the *collision count* to zero.
2. Concatenate the *modifier*, *subnet prefix* and the encoded public-key. Execute the hash algorithm on the concatenation. Check the most significant $16 \times sec$ bits of the result. Continue until $16 \times sec$ bits are zero by incrementing the *modifier*.
3. Sign the *modifier*, *collision count* and *subnet prefix* with the private-key corresponding to the public-key used.
4. Concatenate the encoded public-key and the signature. Execute the hash algorithm on the concatenation. The most significant 64 bits of the result are Hash1.

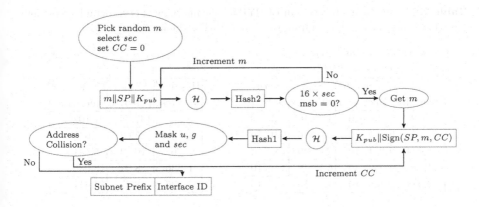

**Fig. 3.** Detailed data flow of the address generation in CGA++

5. Form an *interface identifier* by setting the two reserved bits in Hash1 both to 1 and three bits to sec.
6. Concatenate the *subnet prefix* and *interface identifier* to form an 128-bit IPv6 address.
7. If an address collision is detected, increment the *collision count* and go back to step (3). However, after three collisions, stop and report the error.

The address generation of a node begins with satisfying the constraints in the hash extension as in CGA. The *collision count* is omitted, instead of being set to zero, which makes the input to the hash function smaller. Once this is satisfied, the address owner signs the *subnet prefix*, *modifier* and the *collision count* with his private-key. The public-key is concatenated to the signature and the corresponding interface identifier is obtained by hashing this concatenation.

**Verification of Address Ownership.** After the address generation has been performed, the verification of the address ownership is realized by the execution of the following steps. Given the IPv6 address, the signature and the public-key of the node,

1. Verify the signature and obtain the *modifier*, *collision count* and *subnet prefix*.
2. Check that the *collision count* is 0, 1, or 2 and that the *subnet prefix* is equal to the *subnet prefix* of the address (not the link-local address but the full IPv6 address). The CGA++ verification fails if either check fails.
3. Read the security parameter sec from the three leftmost bits of the *interface identifier* of the address (sec is an unsigned 3-bit integer).
4. Concatenate the *modifier*, *subnet prefix* and the encoded public-key. Execute the hash algorithm on the concatenation. Check if the most significant $16 \times sec$ bits of the result are zero. The CGA++ verification fails if the check fails.

**Table 2.** Measurements taken from ECRYPT benchmarking of cryptographic systems [15]. These median results are from runs on a AMD Athlon 64 X2 (2.0 GHz).

(a) *Benchmark results, in cycles, of the RSA public-key signature system.*

| $x$-bit signature | Generate a key pair | Sign 59 bytes | Verify 59 bytes |
|---|---|---|---|
| 512 | $3.9 \cdot 10^7$ | $1.1 \cdot 10^6$ | $5.3 \cdot 10^5$ |
| 768 | $8.0 \cdot 10^7$ | $2.0 \cdot 10^6$ | $6.0 \cdot 10^5$ |
| 1,024 | $1.4 \cdot 10^8$ | $2.9 \cdot 10^6$ | $7.0 \cdot 10^5$ |
| 1,536 | $3.2 \cdot 10^8$ | $6.7 \cdot 10^6$ | $1.0 \cdot 10^5$ |
| 2,048 | $6.8 \cdot 10^8$ | $1.2 \cdot 10^7$ | $1.3 \cdot 10^5$ |

(b) *Benchmark results of the SHA-1 hash function.*

| Hashing $x$ bytes | Cycles per byte | Cycles per message |
|---|---|---|
| 8 | 137.75 | 1,102 |
| 64 | 25.62 | 1,640 |
| 576 | 10.14 | 5,841 |
| 1,536 | 8.91 | 13,686 |
| 4,096 | 8.45 | 34,611 |

5. Concatenate the encoded public-key and the signature. Execute the hash algorithm on the concatenation and compare the output with the *interface identifier*. The differences in the two reserved bits and three bits for *sec* are ignored. If the 64-bit values differ (other than in the five ignored bits), the CGA++ verification fails.

The address verification starts with the usual checks, similar to CGA, in the IPv6 address of the node to be verified. The signature is verified, then the *modifier, collision count* and *subnet prefix* are extracted. Note that, compared to CGA, CGA++ does extra authenticity checks using the signature of the address generator; in order to verify an address only the signature, public-key and the address are needed.

# 6    Analysis of CGA++

## 6.1    Security of CGA++

We analyze CGA++ in a similar fashion as we did for CGA in Section 4. With the help of digital signatures, we eliminate the lack of authentication in the verification process. Including the *subnet prefix* in both domains of Hash1 and Hash2 reduces the scope of a time-memory trade-off attack to a specific network. The following lemma introduces the computational demand for impersonation, again the proof follows from our security framework (cf. Section 4.2).

**Lemma 3.** *Given a network, assume the addresses are generated and verified by CGA++ with security parameter sec. Let $S$ denote the time required to compute a signature expressed in hash function evaluations and assume $S < 2^{16}$. Then, the number of required hash function evaluations needed for impersonation of a specific node is*

$$T_A = \begin{cases} 2^{59} \cdot (1 + S) & \text{if } sec = 0 \\ 2^{59 + 16 \times sec} + 2^{59}(1 + S) & \text{if } sec > 0. \end{cases}$$

**Table 3.** Signature and verification time expressed in SHA-1 hash function evaluations for different RSA key sizes

| RSA $x$-bit key | Signature time | $\text{Log}_2$ of the signature time | Verification time | $\text{Log}_2$ of the verification time |
|---|---|---|---|---|
| 512 | 707 | 9.5 | 35 | 5.1 |
| 768 | 1338 | 10.4 | 40 | 5.3 |
| 1024 | 1910 | 10.9 | 47 | 5.5 |
| 1536 | 4432 | 12.1 | 69 | 6.1 |
| 2048 | 7812 | 12.9 | 89 | 6.4 |

## 6.2   Attack Costs and the Efficiency of CGA++

In order to make a comparison with CGA, the timing results to measure the computational cost of signing messages in terms of hash function evaluations from ECRYPT Benchmarking of Cryptographic Systems (eBACS) [15] are used. From now on, we assume the use of the RSA [16] public-key signature scheme because this is the default in the RFC [9]. Note that CGA++ is independent from the signature scheme used. The benchmark data regarding measurements of the signature scheme are stated in Table 2(a) for different key sizes and the benchmarks on the same architecture using the SHA-1 hash digest are stated in Table 2(b).

**Address Generation.** Due to the use of digital signatures, moving from CGA to CGA++, $T_1$ increases from 1 to $(1 + S)$ and $T_2$ remains equal to 1. This increase in time is only significant when $sec = 0$ as $T_G = 2^{16 \times sec} + S + 1$. Assuming the node uses a 1024-bit RSA key, the time increases from one hash function evaluations in CGA to $T_1 \approx 2^{10.9}$ hash function evaluations in CGA++ (see Table 3). For $sec > 0$, the time increase is negligible as this is dominated by the time required to compute the hash extensions.

**Address Renewal.** In CGA++, the address renewal time is equivalent to the time needed for address generation to resist the global time-memory trade-off attack. This is a drawback compared to the constant amount of time needed by CGA. Assume a mobile node does not have much computation power, say five times less compared to a more powerful machine. The address renewal time is less than a second when using $sec = 1$. When $sec > 1$, the values of $sec$ which are currently impractical (cf. Section 4.4), we anticipate the fact that the performance of mobile nodes (capable of performing cryptographic operations) will increase accordingly in the future following Moore's law, which would reduce the efficiency problem significantly (cf. Section 2).

**Address Verification.** In both CGA and CGA++, address verification takes a constant amount of hash function evaluations. In CGA, $T_V = 1$, whereas in CGA++ this amount is increased with a signature verification: $T_V = 1 + S$. Fortunately, the signature verification time is shorter compared to the time needed to sign a message, but it still consumes the same number of CPU cycles as 47 hash function evaluations when using 1024-bit RSA keys, see Table 3. This constant increase is tolerable considering the efficiency of evaluating hash functions in practice.

**Impersonation.** The time needed for impersonation in CGA++ is roughly equivalent to the time needed in CGA when $sec > 1$, not taking the possibility of mounting the time-memory trade-off attack into account for CGA, see Lemma 1 and 3. The increase of time needed for impersonation, due to the use of digital signatures, becomes negligible with respect to the hash extension time. However, when no hash extensions are used, the digital signature time is significant. Assuming the use of 1024-bit RSA keys, an attacker would need $2^{59}$ hash function evaluations using CGA, whereas this value increases to $2^{69.9}$ in CGA++.

### 6.3  Comparison of CGA++ with CGA

Table 4 summarizes the comparison between CGA and CGA++. The overall efficiency decreases when moving from CGA to CGA++. The time needed to generate a new and verify a current address is increased by a constant amount of time, whereas the time needed to renew an address increases exponentially when hash extensions are used. The security of CGA++ is improved compared to CGA. The global time-memory trade-off attack is no longer possible, increasing the security level against impersonation attacks. Moreover, an additional authentication mechanism is introduced by using digital signatures inside the protocol. The constant amount of loss in efficiency and gain of security, when no hash extensions are used, are due to the additional computation needed for signing and verifying digital signatures.

## 7   Compatibility and Applications

To facilitate its adoption, it is desirable to design a protocol that is compatible with the current schemes. Hence, when designing CGA++, one design criterion was to adhere to CGA as closely as possible. CGA offers features for protocols that require self-certified address generation and verification, where the nodes are assumed to be capable of signing messages as they are equipped with public/private-key pairs. Therefore, our main contribution to the current design, using digital signatures, does not harm the compatibility of CGA++ because the rest of the protocol is nearly the same as CGA.

The Secure Neighbor Discovery [3], Shim6 [4] and IPv6 mobility support protocol [5] are the main protocols using CGA. The common feature of these protocols is to use CGA to prove address ownership and continue to sign additional data with the corresponding private key of the CGA. This is supported by CGA++ as well.

**Table 4.** Comparison between CGA and CGA++ for IPv6 using a 1024-bit RSA key. All timings are expressed in hash function evaluations. The parameter $sec = s$ is the security parameter used for hash extensions.

|  | CGA | CGA++ |
|---|---|---|
| Time to generate a new address when $s = 0$ | 1 | $1 + 2^{10.9}$ |
| Time to generate a new address when $s > 0$ | $2^{16 \times s} + 1$ | $2^{16 \times s} + 1 + 2^{10.9}$ |
| Time to verify an address when $s = 0$ | 1 | $1 + 2^{5.5}$ |
| Time to verify an address when $s > 0$ | 2 | $2 + 2^{5.5}$ |
| Impersonation time when $s = 0$ | $2^{59}$ | $2^{69.9}$ |
| Impersonation time when $s > 0$ | $2^{59}$ | $2^{59+16 \times s} + 2^{69.9}$ |
| Time to renew the address when moving to a different network when $s = 0$ | 1 | $1 + 2^{10.9}$ |
| Time to renew the address when moving to a different network when $s > 0$ | 1 | $2^{16 \times s} + 1 + 2^{10.9}$ |
| Resistance against the global time-memory trade-off attack | No | Yes |
| Authentication mechanism inside the verification protocol | No | Yes |

# 8   Conclusion

In this work, we have presented a detailed security/efficiency analysis of CGA together with a proposal to solve some security problems and limitations related to self-certifying address generation and verification in CGA. This new protocol, which is very similar to and based on the ideas of CGA, is called CGA++. The global time-memory trade-off attack, which eliminates the effect of hash extensions in the long run for CGA, is no longer possible. CGA++ has an efficiency drawback in that the address renewal costs as much as address generation. However, we believe that this is tolerable; the computational capabilities of mobile nodes (able to perform cryptographic operations) increase with the progress of technology and the current used hash extension values are still practical. As another improvement, we have introduced the use of digital signatures in the address generation and verification process, which provides authentication in the protocol and eliminates the effect of replay attacks. Although this leads to an increase in time required for address generation and verification, it increases the security of the system, especially when no hash extensions are used. We believe that, in many ways, CGA++ is a nice practical alternative to CGA, e.g. in IPv6.

**Acknowledgments.** We would like to thank Tuomas Aura for sharing detailed information related to CGA. We would also like to thank the anonymous reviewers of ISC 2009 for their insightful and helpful comments.

# References

1. Aura, T.: Cryptographically Generated Addresses (CGA). In: Boyd, C., Mao, W. (eds.) ISC 2003. LNCS, vol. 2851, pp. 29–43. Springer, Heidelberg (2003)
2. Aura, T., Roe, M.: Strengthening Short Hash Values, http://research.microsoft.com/en-us/um/people/tuomaura/misc/aura-roe-submission.pdf
3. Arkko, J., Kempf, J., Zill, B., Nikander, P.: SEcure Neighbor Discovery (SEND). RFC 3971, IETF (March 2005), http://www.ietf.org/rfc/rfc3971.txt
4. Nordmark, E., Bagnulo, M.: Multihoming L3 Shim Approach (July 2005), http://tools.ietf.org/html/draft-nordmark-multi6dt-shim-00.txt
5. Johnson, D., Perkins, C., Arkko, J.: Mobility Support in IPv6. RFC 3775, IETF (June 2004), http://www.ietf.org/rfc/rfc3775.txt
6. O'Shea, G., Roe, M.: Child-proof Authentication for MIPv6 (CAM). Computer Communication Review 31(2), 4–8 (2001)
7. Nikander, P.: A Scalable Architecture for IPv6 Address Ownership, Internet Draft (2001)
8. Montenegro, G., Castelluccia, C.: Statistically Unique and Cryptographically Verifiable (SUCV) Identifiers and Addresses. In: NDSS, The Internet Society (2002)
9. Aura, T.: Cryptographically Generated Addresses (CGA). RFC 3972, IETF (March 2005), http://www.ietf.org/rfc/rfc3972.txt
10. Hinden, R., Deering, S.: Internet Protocol Version 6 Addressing Architecture. RFC 4291, IETF (February 2006), http://www.ietf.org/rfc/rfc4291.txt
11. Hinden, R., Deering, S., Nordmark, E.: IPv6 Global Unicast Address Format. RFC 3587, IETF (August 2003), http://www.ietf.org/rfc/rfc3587.txt
12. National Institute of Standards and Technology: Secure hash standard. FIPS 180-1, NIST (April 1995)
13. Bagnulo, M., Arkko, J.: Support for Multiple Hash Algorithms in Cryptographically Generated Addresses (CGAs). RFC 4982, IETF (July 2007), http://www.ietf.org/rfc/rfc4982.txt
14. OpenSSL: The Open Source Toolkit for SSL/TLS (2008), http://www.openssl.org/
15. Bernstein, D.J., Lange, T. (eds.): eBACS: ECRYPT Benchmarking of Cryptographic Systems, http://bench.cr.yp.to (accessed January 7, 2009)
16. Rivest, R., Shamir, A., Adleman, L.: A method for obtaining digital signatures and public key cryptosystems. Communications of the ACM, 42–111 (February 1978)

# Security Analysis
# of the PACE Key-Agreement Protocol

Jens Bender[1], Marc Fischlin[2], and Dennis Kügler[1]

[1] Bundesamt für Sicherheit in der Informationstechnik (BSI), Germany
[2] Darmstadt University of Technology, Germany

**Abstract.** We analyze the Password Authenticated Connection Establishment (PACE) protocol for authenticated key agreement, recently proposed by the German Federal Office for Information Security (BSI) for the deployment in machine readable travel documents. We show that the PACE protocol is secure in the real-or-random sense of Abdalla, Fouque and Pointcheval, under a number-theoretic assumption related to the Diffie-Hellman problem and assuming random oracles and ideal ciphers.

## 1   Introduction

Authenticated key exchange is a fundamental cryptographic protocol in which two parties, usually called the client and the server, establish a secure key. In a password-based key-agreement protocol both parties only share a low-entropy secret, usually drawn at random from a set of size $N$. Since the security only relies on this short password an adversary can guess the right password with probability at least $1/N$ and then impersonate another party in an execution (in a so-called online dictionary attack). Ideally, this should also be an upper bound on the adversary's success probability, even if the adversary eavesdrops or actively participates in other protocol executions. In particular, the adversary should not be able to deduce the password of any party in an offline dictionary attack by successfully matching password candidates to executions afterwards.

A widely accepted model to capture the above security requirements is the real-or-random security notion of Abdalla, Fouque and Pointcheval [1], a refinement of the model of Bellare, Pointcheval and Rogaway [2]. The original and the refined model have been accepted as a profound approach to capture security of key agreement protocols and several password based protocols for authenticated key exchange (AKE) have been shown secure via this approach [2,1,3]. The real-or-random security model says that an adversary, mounting an active attack on several concurrently running instances of the key agreement protocol, cannot distinguish genuine keys from random strings.

*The PACE Protocol.* Here we investigate the security of the Password Authenticated Connection Establishment (PACE) protocol. This protocol has been specified by the German Federal Office for Information Security (BSI) to secure the communication between a chip contained in a machine readable travel document

P. Samarati et al. (Eds.): ISC 2009, LNCS 5735, pp. 33–48, 2009.

and a reader (terminal) [4]. The purpose of PACE is to establish a secure channel based on weak passwords like the personal data of the passport holder. The protocol is currently under standardization of ISO/IEC JTC1/SC17/WG3.

The PACE protocol can be roughly divided into four phases (see also Figure 1 on Page 38): In the first phase the chip sends a random nonce $s$ encrypted with the password to the terminal. In the second phase both parties execute an interactive protocol Map2Point, mapping the nonce to a random generator $\widehat{G}$ of a group, e.g., an elliptic curve (the group parameters are provided by the chip and authenticated by a governmental authority). In the third phase the two parties run a Diffie-Hellman (DH) key agreement on the agreed-upon generator $\widehat{G}$ and use the DH key to derive the actual keys for subsequent use. Finally, both parties conclude the execution by sending some authentication data.

PACE is rather a framework allowing different instantiations than a single protocol. Here we focus on the most prominent version based on elliptic curves. Still, we look at different options to implement the Map2Point protocol in which the nonce is thrown to a random generator. A candidate is the DH-based protocol advocated in [4] where both parties generate a DH key $H$ and define $\widehat{G} = sG + H$ for the generator $G$ of the elliptic curve and the nonce $s$. Another option is to use a coin-flipping protocol instead to generate $H$ jointly and then again letting $\widehat{G} = sG + H$. A third possibility is to hash into the elliptic curve directly. We discuss these options in more detail later.

*Security Result for PACE.* In this paper we provide a security analysis of the PACE framework. We remark that the purpose of this work here is not to investigate the design choices of the protocol (which are based on implementation aspects and patent issues) but to analyze PACE as a given protocol with respect to security. Some aspects of the protocol are, of course, security-related and in this case we explore them in more detail.

We analyze PACE in the random oracle model and the ideal cipher model (which have recently been shown to be equivalent [5]). These models entail idealized assumptions about the hash function and cipher deployed in the protocol. Namely, it is assumed that the hash function behaves like a random function, and that the cipher acts like a random permutation. We note that neither model may be instantiable in practice [6,5]. Yet, security shows that, in order to break the scheme, some weaknesses of these primitives must be exploited.

We also introduce a new Diffie-Hellman-like problem, called *PA*ssword-based *C*hosen-*E*lement Diffie-Hellman (PACE-DH) problem. This problem basically says that it is infeasible for an adversary to derive the final DH key of PACE, even if the adversary impersonates one of the two parties and biases the outcome of the Map2Point subprotocol. It follows that the PACE-DH problem is connected to the specific choice of the Map2Point step in PACE.

Our PACE-DH problem resembles the password-based chosen-base (PCDH) problem of Abdalla et al. [7]. Yet, while the PCDH problem is known to be equivalent to the basic DH problem [7,3], hardness of the PACE-DH is not known to imply the DH assumption. We nonetheless show that the PACE-DH problem is hard in Shoup's generic group model [8] for the choices of Map2Point discussed above.

Assuming the hardness of the PACE-DH problem we show that PACE is real-or-random secure in the sense of [1], in the random oracle model and ideal cipher model. We also discuss that the protocol provides forward security.

## 2   Security Model

We analyze the PACE protocol in the real-or-random security model of Abdalla et al. [1] which extends the model of Bellare et al. [2]. Here we provide an overview over the model, for more information and discussion about the choices see [2] and [1].

*Attack Model.* The model considers a set of honest participants, also called users. Each participant may run several instances of the key agreement protocol, and the $j$-th instance of a user $U$ is denoted by $U_j$ or $(U, j)$. Each pair of participants shares a secret password $\pi$ which may be used multiple times to generate session keys. The password $\pi$ is chosen randomly from a (public) dictionary with $N$ elements.

To obtain a session key the protocol $P$ is executed between two instances of the corresponding users. An instance is called an initiator or client (resp. respondent or server) if it sends the first (resp. second) message in the protocol. For sake of distinctiveness we often denote the client by $A$ and the server by $B$.

We consider security against active attacks where the adversary's goal is to distinguish between genuine keys, derived in executions between honest parties, and random keys. This corresponds to the so-called real-or-random setting [1], a stronger model than the original find-then-guess model of [2], where the adversary can see several test keys (instead of a single one only).

Each user instance is given as an oracle to which an adversary has access, basically providing the interface of the protocol instance. By assumption, the adversary is in full control of the network, i.e., decides upon message delivery. The adversary can make the following queries to the oracles:

*Execute*$(A, i, B, j)$. Causes the users $A$ and $B$ to run the protocol for (fresh) instances $i$ and $j$. The final output is the transcript of a protocol execution. This query simulates a passive attack where the adversary merely eavesdrops the network.

*Send*$(U, i, m)$. Causes the instance $i$ of user $U$ to proceed with the protocol when having received message $m$. The output is the message generated by $U$ for $m$ and depends on the state of the instance. This query simulates an active attack of the adversary where the adversary pretends to be the partner instance.

*Reveal*$(U, i)$. Returns the session key of the input instance. The query is answered only if the session key was generated and the instance has terminated in accepting state. This query models the case when the session key has been leaked. We assume without loss of generality that the adversary never queries about the same instance twice.

*Corrupt*($U$). The adversary obtains the party's long-term key $\pi$. This is the so-called *weak-corruption* model. In the *strong-corruption* model the adversary also obtains the state information of all instances of user $U$. The corrupt queries model a total break of the user and allow to model forward secrecy.

*Test*($U, i$). The oracle test is initialized with a random bit $b$. Assume the adversary makes a test query about $(U, i)$ during the attack and that the instance has terminated in accepting state, holding a secret key $sk$. Then the oracle returns $sk$ if $b = 0$ or a random key $sk'$ from the domain of keys if $b = 1$. If the instance has not terminated yet or has not accepted, then the oracle returns $\perp$. This query should determine the adversary's success to tell apart a genuine session key from an independent random key. We assume again without loss of generality that the adversary never queries about the same instance twice.

In addition, since we work in the random oracle and ideal ciper model where oracles providing a random hash function oracle and an encryption/decryption oracle are available, the attacker may also query these oracles.

*Partners, Correctness and Freshness.* Upon successful termination we assume that an instance $U_i$ outputs a key $sk$, the session ID sid, and a user ID pid identifying the intended partner (assumed to be empty in PACE for anonymity reasons). We note that the session ID usually contains the entire transcript of the communication but, for efficiency reasons, in PACE it only contains a fraction thereof. We discuss the implications in more detail in Section 3. We say that instances $A_i$ and $B_j$ are *partnered* if both instances have terminated in accepting state with the same output. In this case the instance $A_i$ is called a partner to $B_j$ and vice versa. Any untampered execution between honest users should be partnered and, in particular, the users should end up with the same key (this correctness requirement ensures the minimal functional requirement of a key agreement protocol).

Neglecting forward security for a moment, an instance $(U, i)$ is called *fresh* at the end of the execution if there has been no Reveal($U, i$) query at any point, neither has there been a Reveal($B, j$) query where $B_j$ is a partner to $U_i$, nor has somebody been corrupted. Else the instance is called *unfresh*. In other words, fresh executions require that the session key has not been leaked (by neither partner) and that no Corrupt-query took place.

To capture forward security we refine the notion of freshness and further demand from a fresh instance $(U, i)$ as before that the session key has not been leaked through a Reveal-query, and that for each Corrupt($U$)-query there has been no subsequent Test($U, i$)-query involving $U$, or, if so, then there has been no Send($U, i, m$)-query for this instance at any point. In this case we call the instance *fs-fresh*, else *fs-unfresh*. This notion means that it should not help if the adversary corrupts some party after the test query, and that even if corruptions take place before test queries, then executions between honest users are still protected (before or after a Test-query).

*AKE Security.* The adversary eventually outputs a bit $b'$, trying to predict the bit $b$ of the Test oracle. We say that the adversary wins if $b = b'$ and instances $(U, i)$ in the test queries are fresh (resp. fs-fresh). Ideally, this probability should be close to $1/2$, implying that the adversary cannot significantly distinguish random keys from session keys.

To measure the resources of the adversary we denote by $t$ the number of steps of the adversary, i.e., its running time, (counting also all the steps required by honest parties); $q_e$ the maximal number of initiated executions (bounded by the number of Send- and Execute-queries); $q_h$ the number of queries to the hash oracle, and $q_c$ the number of queries to the cipher oracle. We often write $Q = (q_e, q_h, q_c)$ and say that $\mathcal{A}$ is $(t, Q)$-bounded.

Define now the AKE advantage of an adversary $\mathcal{A}$ for a key agreement protocol $P$ by

$$\mathbf{Adv}_P^{\mathrm{ake}}(\mathcal{A}) := 2 \cdot \mathrm{Prob}[\mathcal{A} \text{ wins}] - 1$$

$$\mathbf{Adv}_P^{\mathrm{ake}}(t, Q) := \max \left\{ \mathbf{Adv}_P^{\mathrm{ake}}(\mathcal{A}) \,\middle|\, \mathcal{A} \text{ is } (t, Q)\text{-bounded} \right\}$$

The forward secure version is defined analogously and denoted by $\mathbf{Adv}_P^{\mathrm{ake\text{-}fs}}(t, Q)$.

# 3 The PACE Protocol

In this section we describe the PACE framework and options for its subprotocol Map2Point.

## 3.1 The Main Protocol of PACE

We describe the elliptic curve instantiation of the PACE protocol [4]. Roughly, the chip in the PACE protocol first transmits the authenticated group data $\mathcal{G}$ and a nonce $s$, encrypted with (the hash value of) the password. The receiver can recover this value with the matching password. Then both parties engage in an interactive protocol Map2Point($s$) to map $s$ to a random group element $\widehat{G}$. This generator is subsequently used to run a Diffie-Hellman key agreement to derive a common key $K$. Once this key is agreed upon, the parties derive the encryption and authentication keys by hashing $K$ appropriately.

We let $\mathcal{H}$ be a hash function, $\mathcal{C}$ be a block cipher, and $\mathcal{M}$ be a MAC. We use $\mathcal{C}(K; s)$ and $\mathcal{C}^{-1}(K; z)$ to denote the encryption and decryption of $s$ and $z$, respectively, for a secret key $K$. Let $\mathcal{G} = (a, b, p, q, G, k)$ be the description of an elliptic curve $y^2 = x^3 + ax + b \bmod p$ where $\langle G \rangle$ is a group of prime order $q$. The chip (A) and terminal (B) share a secret password $\pi$ from a dictionary with $N$ elements, chosen at random, and use some mapping to generate the secret key $K_\pi$ for the block cipher from $\pi$. Below we let $K_\pi = \mathcal{H}(\pi \| 0)$. Note that we implicitly assume that the parties know the right password when engaging in an interaction, e.g., the user may enter the PIN at the reader or the terminal optically scans the machine readable zone of the passport. The ellptic curve version of the PACE protocol is given in Figure 1.

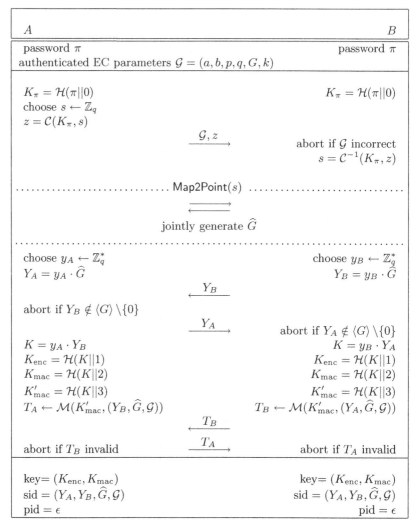

**Fig. 1.** PACE based on DH over elliptic curves (with generic Map2Point protocol)

*Remarks.* Some remarks about the changes compared to the original protocol in [4] and about underlying assumptions are in order.

*Session IDs.* In the definition of the protocol only the final values (and the group parameters) enter the session ID. This is in order to spare the parties from saving or processing the transcript data in the execution. It follows that the partner definition is "more loose" than the common definition including the whole transcript in sid. With this loose partnering approach here an adversary may now be able to run a man-in-the-middle attack making the honest parties assume they communicate with someone else, even though they hold the same key. Still, the confidentiality of the key is not affected by this.

*The final authentication step.* The original scheme uses the output key $K_{\mathrm{mac}}$ for the MAC computations in the key-agreement protocol, too. This version, however, may not be provable secure in the [2] and [1] model. The reason is that with the Test query the adversary obtains a random or the genuine secret key, including $K_{\mathrm{mac}}$. Then the adversary can possibly test whether this key part $K_{\mathrm{mac}}$ together with $Y_A$ or $Y_B$ and matches the transmitted value $T_A$ or $T_B$. Since $\widehat{G}$ also enters the MAC computation the adversary also needs to be able to compute all possible values for $\widehat{G}$ (over the password choices) for this attack. The adversary's success thus heavily depends on the specific Map2Point choice[1] and for the general analysis we therefore suggest to derive an ephemeral MAC key $K'_{\mathrm{mac}}$ as $K'_{\mathrm{mac}} = \mathcal{H}(K\|3)$ and use this key for authentication. A similar strategy is recommended in [2].

## 3.2   The Map2Point Protocol

In this section we describe possible instantiations for the Map2Point sub routine. We always implicitly assume that both parties check for the right format of received values, e.g., that $H$ is a group element. We take a closer look at the security requirements for Map2Point in Section 4.2.

*The Diffie-Hellman Mapping* DH2Point. The DH2Point mapping is based on the Diffie-Hellman key agreement. Both parties generate a DH key $H$ (relative to the generator $G$ in $\mathcal{G}$) by exchanging $X_A = x_A G$ and $X_B = x_B G$ and letting $H = x_A x_B G$. The nonce $s$ in then "added" to this DH key via $\widehat{G} = sG + H$. Note that the parties should also check that $H \neq 0$, otherwise the final output would deterministically depend on the nonce $s$ only.

*The Coin-flipping Mapping* Coin2Point. Also creates $\widehat{G}$ as $\widehat{G} = sG + H$, but both parties use a coin-flipping protocol to generate the random element $H$. Namely, party $A$ first generates $X_A = x_A G$ and sends a hash value $\mathcal{H}(X_A)$ of $X_A$, then party $B$ transmits $X_B = x_B G$ and $A$ finally reveals $X_A$ to $B$ (who checks that this value matches the initial hash). Both parties set $H = X_A + X_B$.

*The Hash-into-the-Curve Mapping* Hash2Point. Assume that we have an efficient function $s \mapsto \mathrm{hash2curve}(s)$ allowing to throw the string $s$ to the curve directly. Possible instantiations are given for example in [9,10]. Then this function can be combined with the two previous methods to generate $\widehat{G}$ as $\widehat{G} = \mathrm{hash2curve}(s) + H$. A faster approach is to have party $B$ contribute by re-encryption of the nonce $s$, i.e., $B$ sends a random key $K'$ and both parties compute $s' = \mathcal{C}(K', s)$ and set $\widehat{G} = \mathrm{hash2curve}(s')$.

*The Power-to-Group Mapping* Power2Point. This method only works for groups over $\mathbb{Z}_p^*$. Here the parties use a function $s \mapsto s^w \bmod p$ for $p = wq + 1$ for $w$ with large prime factors only to map $s$ to a sub group element of $\mathbb{Z}_p^*$. The fact that $w$ does not have small factors ensures that the mapping is statistically

---

[1] Using randomized Map2Point strategies seem to impede such attacks significantly; see also Section 4.2.

close to uniform (given that the value $s$ is uniform). One can again combine this mapping with an interactive generation of $H$, or with the re-encryption technique discussed in the previous case.

# 4   Security Assumptions

As remarked above we carry out our security analysis assuming an ideal hash function (random oracle model) and an ideal encryption scheme (ideal cipher model). Basically, the first assumption says that $\mathcal{H}$ acts like a random function to which all parties have access. The second property says that for each key $K$ the mapping $\mathcal{C}(K, \cdot)$ is an independent random permutation and one can evaluate both $\mathcal{C}(K, s)$ and $\mathcal{C}^{-1}(K, z)$ for arbitrary values $(K, s)$ and $(K, z)$.

We also require that the message authentication code $\mathcal{M}$ is unforgeable under adaptively chosen-message attacks. We denote by $\mathbf{Adv}_{\mathcal{M}}^{\mathrm{forge}}(t, q)$ a (bound on the) value $\epsilon$ for which no attacker in time $t$ can output a new message and a valid tag (after having seen at most $q$ MACs for adaptively chosen messages) with probability more than $\epsilon$. For the analysis we even assume that $\mathcal{M}$ acts like a pseudorandom function and denote by $\mathbf{Adv}_{\mathcal{M}}^{\mathrm{prf}}(t, q)$ the (maximal) advantage of an attacker running in time $t$ making at most $q$ queries to a function oracle for distinguishing $\mathcal{M}$ from a random function. Note that $\mathbf{Adv}_{\mathcal{M}}^{\mathrm{forge}}(t, q) \leq \mathbf{Adv}_{\mathcal{M}}^{\mathrm{prf}}(t + O(\ell), q + 1) + 2^{-\ell}$ where $\ell$ denotes the output size of the MAC.

## 4.1   Number-Theoretic Assumptions

*The PACE-DH Problem.* For passive adversaries, merely eavesdropping the network, security follows from the classical Diffie-Hellman (DH) assumption. Active adversaries, injecting messages, can usually contribute to the input to the DH problem and we thus require a stronger assumption based on the *PA*ssword-based *C*hosen-*E*lement (PACE) DH problem. Assume that we are given $N$ values $s_i$ from $\mathbb{Z}_q$ (each value corresponding to $\mathcal{C}^{-1}(K_\pi, z)$ for a possible password $\pi$) of which one corresponds to the actual password $s_k$. Potentially, these values $s_i$ are biased by the adversary through its choice of $z$ so we precautiously let the adversary fully determine them (with the only restriction that they are distinct).

Suppose further that we are given a random group element $H$ (generated via Map2Point and possibly known to the adversary), as well as $y_B(s_k G + H)$ for a random $y_B$ (for the value $Y_B$ sent by an honest party). Then the adversary's task is to find a group element $Y_A$ (i.e., the $Y_A$ sent in the protocol) and a key $K$ such that $K = y_B Y_A$. Since the adversary may try different possibilities for $K$, below we let the adversary output a set of $q_\ell$ possible key values $K_1, \ldots, K_{q_\ell}$.

We first remark that the setting above corresponds to the case that the adversary impersonates the chip. This means that the adversary can adaptively decide upon his choices after seeing the group elements (representing the chip's choices). The case that the adversary plays the terminal is a special case where the adversary first ignores parts of the data. We also remark that, while we consider concurrent executions of the key agreement protocol, for the analysis it suffices to consider our interactive number-theoretic problem in a somewhat isolated setting.

Note that we cannot exclude trivial guessing strategies for our problem. That is, if the adversary manages to guess $k$ it can simply set $Y_A = s_k G + H$ and later choose $K = Y_B$. Similarly, it can choose any linear transformation $Y_A = a(s_k G + H)$ and $K = aY_B$ for $a \in \mathbb{Z}_q$ (also covering the case that $a = 0$ in which case the other party aborts). Hence, there is always an adversarial strategy with success probability at least $1/N$. Yet, this should be close to optimal:

**Definition 1 (Password-Based Chosen-Element DH Problem).** *The password-based chosen-element DH problem is $(t, N, q_\ell, \epsilon)$-hard if for any adversary $\mathcal{A} = (\mathcal{A}_0, \mathcal{A}_1)$ running in total time $t$ the probability that the following experiments returns 1 is most $\frac{1}{N} + \epsilon$:*

> *pick $\mathcal{G}$ (including a generator $G$)*
> *let $(st, s_1, \ldots, s_N) \leftarrow \mathcal{A}_0(\mathcal{G}, G, N)$*
> > *where $s_1, \ldots, s_N$ are pairwise distinct and $st$ is some local state*
> *pick $H \leftarrow \langle G \rangle$, $y_B \leftarrow \mathbb{Z}_q$ and $k \leftarrow \{1, 2, \ldots, N\}$*
> *let $(Y_A, K_1, \ldots, K_{q_\ell}) \leftarrow \mathcal{A}_1(st, y_B(s_k G + H), H)$*
> *output 1 iff $Y_A \neq 0$ and $K_i = y_B Y_A$ for some $i \in \{1, 2, \ldots, q_\ell\}$*

We let $\mathbf{Adv}^{\mathrm{PACE\text{-}DH}}(t, N, q_\ell)$ denote a (bound on the) value $\epsilon$ for which the PACE-DH problem is $(t, N, q_\ell, \epsilon)$-hard.

*On the Hardness of the PACE-DH Assumption.* The hardness of the PACE-DH problem implies hardness of the discrete logarithm problem (see Section 4.2). We note that the PACE-DH problem resembles the password-based chosen-basis problem of Abdalla et al. [7,3]. Yet, while that problem has been proven to be equivalent to the DH problem [7,3] (albeit with a loose security reduction),[2] we are not aware if the PACE-DH problem here is also infeasible assuming the hardness of the DH problem. However, in the generic model of Shoup [8] the problem is also as hard as the DH problem, indicating that only "clever" attacks exploiting the group representation can make a difference in comparison to the regular DH problem. We discuss this in the full version.

*The gPACE-DH Problem.* In the PACE-DH problem above the group element $H$ is assumed to be random. In the actual protocol execution, however, it depends on the execution of protocol Map2Point in which the adversary may control one of the parties. Hence, in the general PACE-DH problem we mimic the generation of $H$ via Map2Point and thus lend the adversary more power in generating $H$:

**Definition 2 (General Password-Based Chosen-Element DH Problem).** *The general password-based chosen-element DH problem is $(t, N, q_\ell, \epsilon)$-hard (with respect to Map2Point) if for any adversary $\mathcal{A} = (\mathcal{A}_0, \mathcal{A}_1, \mathcal{A}_2)$ running in total time $t$ the probability that the following experiments returns 1 is most $\frac{1}{N} + \epsilon$:*

---

[2] Note that the similar chosen-basis *decisional* Diffie-Hellman problems of Abdalla and Pointcheval [11] have been shown to be insecure by Szydlo [12]; Szydlo's attacks do not transfer to the computational counterparts, though.

> pick $\mathcal{G}$ *(including a generator $G$)*
> let $(\mathsf{st}_0, s_1, \ldots, s_N) \leftarrow \mathcal{A}_0(\mathcal{G}, G, N)$
>     *where $s_1, \ldots, s_N$ are pairwise distinct and $\mathsf{st}_0$ is some local state*
> pick $y_B \leftarrow \mathbb{Z}_q$ *and* $k \leftarrow \{1, 2, \ldots, N\}$
> let $\widehat{G}$ *be the  output of the honest party in an execution of* Map2Point$(s_k)$,
>     *where $\mathcal{A}_1(\mathsf{st}_0)$ controls the other party (and generates the local state $\mathsf{st}_1$).*
> let $(Y_A, K_1, \ldots, K_{q_\ell}) \leftarrow \mathcal{A}_2(\mathsf{st}_1, y_B\widehat{G})$
> output 1 *iff* $Y_A \neq 0$ *and* $K_i = y_B Y_A$ *for some* $i \in \{1, 2, \ldots, q_\ell\}$

We let $\mathbf{Adv}_{\mathsf{Map2Point}}^{\mathsf{gPACE\text{-}DH}}(t, N, q_\ell)$ denote a (bound on the) value $\epsilon$ for which the gPACE-DH problem is $(t, N, q_\ell, \epsilon)$-hard (with respect to Map2Point).

*PACE-DH vs. gPACE-DH.* Using the coin flipping Coin2Point for Map2Point the output $H$ is (statistically close to) uniformly distributed and thus security holds under the basic PACE-DH problem. Next consider the DH2Point protocol which generates $H$ as the DH key from $X_A$ and $X_B$. Then the hardness of the PACE-DH problem clearly implies hardness of the gPACE-DH problem for DH2Point. That is, given adversary $\mathcal{A}_{\mathsf{PACE}}$ breaking the case of a random $H$ we can easily build an adversary $\mathcal{A}_{\mathsf{DH2Point}}$ against the gPACE problem for DH2Point by simply following the DH key agreement honestly, such that $H$ is a random element. Then any solution to the random case returned by $\mathcal{A}_{\mathsf{PACE}}$ also gives a solution to the DH2Point case. The converse is not known to hold, essentially because the DH key agreement may not yield a uniformly distributed element (if the honest party goes first). Still, we note again that in the generic group model both problems are hard and we elaborate on the relationship in the full version. Some potential advantages of the DH2Point approach over Coin2Point are discussed in the next section.

## 4.2   Requirements for the Map2Point Protocol

A necessary functional requirement for the Map2Point protocol is that for any $s \in \mathbb{Z}_q$ the output $\widehat{G}$ of an execution of Map2Point$(s)$ *between honest parties* must satisify $\widehat{G} \neq 0$. Note that for DH2Point and Coin2Point, for example, there is a small probability of $1/q$ that the output of Map2Point is 0, namely, if $H = -sG$. We ignore this small term to simplify the presentation and merely note that such cases can be easily thwarted by testing for trivial values and setting $\widehat{G} = G$ in this case.

As for security, assume that Map2Point consists of two phases, an interactive step RndPoint() where both parties jointly generate some randomness, say, a random group element $H$. This step should be independent of the nonce $s$ and only depend on the public data (including transmitted values). Only in the final local step the parties compute $\widehat{G}$ from this value $H$ and the nonce $s$ via a non-interactive algorithm NncPoint$(s, H)$. We call such Map2Point protocols *canonical*.

To be suitable for the gPACE-DH problem any canonical Map2Point protocol must guarantee that the adversary cannot bias the outcome of the Map2Point protocol such that it knows the logarithm $a = \log_{\mathsf{NncPoint}(H,s_i)} \mathsf{NncPoint}(H, s_j)$ for distinct admissible passwords $s_i, s_j$. Else, as we discuss in the full version, if the adversary succeeds with probability $\epsilon$, one can break the gPACE-DH problem with advantage $\epsilon/N$. In particular, for the DH2Point protocol it must be hard to compute $\log_G H$.

The requirement also indicates that deterministic protocols Map2Point must be treated with special care because then the outcome of the protocol may be under full control of the adversary (if it acts as the chip and chooses $z$). It must then be ensured that the adversary cannot find $s_1, \ldots, s_N$ such that it knows the discrete logarithm of $\mathsf{Map2Point}(s_j)$ with respect to $\mathsf{Map2Point}(s_i)$ for some $i \neq j$.

## 5   AKE-Security of PACE

We analyze the PACE protocol with respect to general Map2Point protocols:

**Theorem 1.** *Let* Map2Point *be canonical and assume that the password is chosen from a dictionary of size $N$. In the random oracle model and the ideal cipher model we have*

$$\boldsymbol{Adv}^{ake}_{PACE}(t, Q)$$

$$\leq \frac{q_e}{N} + q_e \cdot \boldsymbol{Adv}^{gPACE\text{-}DH}_{\mathsf{Map2Point}}(t^*, N, q_h)$$

$$+ 2q_e \cdot \boldsymbol{Adv}^{prf}_{\mathcal{M}}(t^* + O(\ell), 2q_e + 1) + \frac{2q_e N^2 + 8q_e^2 N + q_c q_e}{\min\{q, |Range(\mathcal{H})|, 2^\ell\}}$$

*where $t^* = t + O(kq_e^2 + kq_h^2 + kq_c^2 + k^2)$ and $Q = (q_e, q_c, q_h)$ and $\ell$ denotes the output length of $\mathcal{M}$.*

We remark that the time $t^*$ covers the additional time to maintain lists and perform look-ups.

*Proof.* Correctness of the protocol follows from the correctness of the MAC algorithm $\mathcal{M}$ and the fact that Map2Point does not return a trivial group element $G = 0$.

We show security via the common game based approach, gradually changing the original attack $\mathsf{Game}_0$ (with random test bit $b$) via a sequence of experiments $\mathsf{Game}_1, \mathsf{Game}_2, \ldots$ to a game where the adversary's success probability to predict $b$ is bounded by the guessing probability of $\frac{1}{2}$. Each transition from $\mathsf{Game}_i$ to $\mathsf{Game}_{i+1}$ will only change the adversary's probability only slightly (depending on cryptographic assumptions), thus showing that the success probability in the original attack cannot be significantly larger than $\frac{1}{2}$. (Formally we can condition on all "bad" events ruled out in the previous games to not happen.)

Technically, we would like to conclude that the adversary never makes a hash query about a Diffie-Hellman key from which an honest party has derived the output keys. If such a query does not occur then, because we deploy a random oracle, the final keys still look random. We show that this is essentially true under the hardness of the gPACE-DH problem (and in the course take advantage of the random oracle and ideal cipher model and the pseudorandomness of the MAC).

But we also need to take into account attacks where the adversary manages to find *unpartnered* instances but which derive the same keys. In this case the adversary could easily distinguish the answer of a Test-query by posting a Reveal-query for the unpartnered instance (if the instances are partnered then such a Reveal-query is not admissible for a success). We prove that this is guaranteed by the unforgeability of the MAC.

We also remark that we assume that no Corrupt-query takes place in this setting (or else the adversary cannot win). We cover forward security and Corrupt-queries in Section 6. We next define the games.

*Description of* Game$_0$. Corresponds to the original attack on the protocol.

*Description of* Game$_1$. As Game$_0$ but abort in case of $K_\pi$ collisions.

We abort the experiment (declaring the adversary to lose) whenever there are distinct passwords $\pi \neq \pi^*$ yielding the same hash value $K_\pi = \mathcal{H}(\pi\|0) = \mathcal{H}(\pi^*\|0)$. Since there are at most $\frac{1}{2}N^2$ admissible password pairs in total and $\mathcal{H}$ is a random oracle, the adversary's success probability decreases by at most $\frac{1}{2}N^2/|\mathrm{Range}(\mathcal{H})|$ by the birthday bound.

*Description of* Game$_2$. As Game$_1$ but abort in case of collisions among decrypted values.

We abort (again declaring the adversary to lose) if there appears some value $z$ in an execution such that for some admissible passwords $\pi \neq \pi^*$ we have $\mathcal{C}^{-1}(K_\pi, z) = \mathcal{C}^{-1}(K_\pi^*, z)$. Since $\pi \neq \pi^*$ implies $K_\pi \neq K_\pi^*$ by the first game and the cipher is ideal, the probability that for any of the at most $q_e$ values $z$ in the executions we have a collision is at most $\frac{1}{2}q_e N^2/q$.

*Description of* Game$_3$. As Game$_2$ but abort in case two keys $K \neq K^*$ of two accepting user instances yield an identical key $K_{\mathrm{enc}}$, $K_{\mathrm{mac}}$ or $K'_{\mathrm{mac}}$.

Since there are at most $\frac{1}{2}(2q_e)^2$ of such user instances and the probability that two fixed ones yield a hash collision for one of the output keys is at most $3/|\mathrm{Range}(\mathcal{H})|$, the adversary's success probability only drops by the term $6q_e^2/|\mathrm{Range}(\mathcal{H})|$.

*Description of* Game$_4$. As Game$_3$ but replace the MAC values by random values.

Instead of performing MAC computations we now let honest parties simply transmit a random value and if we receive a putative MAC from the adversary we reject this MAC as invalid. This basically cannot decrease the adversary's success probability significantly by the pseudorandomness of the MAC. This holds as long as there are no inconsistencies in our answers due to (a) sending of a new random value in an execution although the input data to the MAC and the key are identical to a previous execution (in which case the party would send the same value again for a deterministic MAC in Game$_3$); and (b) rejecting an adversarial MAC which the honest user would accept in Game$_3$.

For the analysis let THQ be the event that the adversary makes a so-called *target hash query* about $K\|n$ for some $n \in \{1, 2, 3\}$, where $K$ is the key some honest user instance has derived (before computing the MACs). We will analyze the probability of generating inconsistencies by our MAC simulation under the condition that the adversary never makes a target hash query, and later bound the probability for THQ by the hardness of the gPACE-DH problem. If the adversary does not make a target hash query then the keys $K_{enc}, K_{mac}$ and, especially, $K'_{mac}$ are unknown random values to the adversary.

In the full version of the paper we show that, conditioning on event $\neg$THQ and the fact that the adversary does not forge MACs, the difference between the games is now bounded by the pseudorandomness of the MAC (times $q_e$ for the number of executions) where we make at most $2q_e$ queries. Details are omitted from this version.

*Description of* Game$_5$. As Game$_4$ but simulate the ideal cipher.

We replace the actual ideal cipher $\mathcal{C}$ by a lazy-sampling like technique. Namely, for honest users we maintain an intially empty list of tuples $(A, B, s, z)$. For each honest party (involved in a protocol instance between $A$ and $B$) calling $\mathcal{C}$ about $(K_\pi, s)$ we check the list for an entry $(A, B, s, z)$ and, if there exists one, we return $z$. Else we pick a random element $z$, return it and store $(A, B, s, z)$ in the list. For each call of an honest party (involved in instance $(A, B)$) to $\mathcal{C}^{-1}$ about $(K_\pi, z)$ we also search for an entry $(A, B, s, z)$ and return $s$ if we find such an entry; else we pick a random $s$, store $(A, B, s, z)$ and return $s$.

For the adversary we keep a separate list. For any call of the adversary to $\mathcal{C}$ about $(K_\pi, s)$ we check if there is already an entry $(K_\pi, s, z)$ and return $z$ if so; else we return a random value $z$ and store $(K_\pi, s, z)$. For each call of the adversary to $\mathcal{C}^{-1}(K_\pi, z)$ we search for an entry $(K_\pi, s, z)$ in the list and return $s$ if such an entry exist, else we pick a random $s$ and return $s$ and store $(K_\pi, s, z)$.

Note that the two lists may cause inconsistencies between the answers to honest users and to the adversary. However, conditioning on the adversary never making a hash query about a DH key derived by an accepting user instance (event $\neg$THQ) the execution of Game$_5$ does not reveal any information about the $s$-values chosen by honest parties. In this case, the probability of making an accidental query to $\mathcal{C}$ about an $s$-value chosen by an honest party is at most $q_c q_e / q$. Analogously, if no target hash queries occur, then answering calls of the adversary to $\mathcal{C}^{-1}$ as described above, does not lead to any difference in the success probability.

We next bound the probability for event THQ by describing another game in which we abort if this happens (and then show that under the gPACE-DH assumption this cannot happen too often). To be precise we actually consider the event THQ in Game$_3$ where it occured for the first time. But since we are only interested in the first target hash query and up to the point where this target hash query is made the modifications from Game$_3$ to Game$_5$ cannot affect the adversaries success probability significantly (as shown above), it suffices to consider event THQ in Game$_5$. An important observation here is that up to the first target hash query the data in Game$_5$ is independently distributed from the actual passwords of users (because neither the simulated cipher nor the MAC computations reveal anything about the password, the interactive runs of protocol RndPoint are also password-independent, and the group elements in the final DH exchange are distributed independently of $\widehat{G}$).

*Description of* Game$_6$. As Game$_5$ but stop if the adversary makes a hash query about a DH-key of an accepting user instance $(U, i)$.

We even declare the adversary victorious if it ever submits a query $K \| n$ for $n \in \{1, 2, 3\}$ to the hash oracle $\mathcal{H}$ such that a user instance $(U, i)$ has computed this key and sent out the final MAC (i.e., we even consider instances in which the user may not accept eventually). We claim that this cannot occur with probability more than $q_e/N$, plus the advantage of breaking the gPACE-DH problem (times $q_e$). Consider a user instance $(U, i)$ in accepting state and the corresponding execution in which the DH-key $K$ is derived.

We now break the gPACE-DH problem as follows. We are given $(\mathcal{G}, G, N)$ as input. We initially make a guess for the execution number between 1 and $q_e$ for which the adversary makes the first test query and, at the same time, a target hash query. Then we simulate Game$_6$. We wait to receive $z$ in this execution and then output the (possibly then chosen) values $s_1, \ldots, s_N$ for all passwords $\pi$ and all (unique) derived keys $K_\pi$ and for each call by the adversary to $\mathcal{C}^{-1}$.

In the predicted execution we run the Map2Point algorithm with the adversary to obtain $\widehat{G}$ (relaying the communication in the execution and the external Map2Point instance in the gPACE-DH problem). We then receive $y_B \widehat{G}$ as additional input and feed these data into the execution. We finally pick random keys $K_{enc}, K_{mac}, K'_{mac}$ (instead of querying $\mathcal{H}$) and complete the protocol with the help of these data. When the adversary eventually stops we output $Y_A$, transmitted in the predicted execution by the adversary or the honest party, and the list $K_1, \ldots, K_{q_h}$ of values appearing in the at most $q_h$ hash queries of the form $K \| n$ for $n \in \{1, 2, 3\}$ . (If the adversary impersonates the chip then we output the value $Y_B$, of course.)

It remains to analyze the probability that we obtain an admissible solution to the gPACE-DH problem. Recall that we fail to win if we output $Y_A = 0$. But this case leads the honest party to abort immediately. Hence, we can assume $Y_A \neq 0$. Also note that all possible nonces $s_1, \ldots, s_N$ for the different passwords are distinct by Game$_2$ and thus comply with the requirement for the gPACE-DH game. Hence, up to the target hash query the distribution of the data is independent of the password of the user instance, and we make the right execution prediction with probability $1/q_e$, in which case we obtain a valid solution to the gPACE-DH problem whenever the adversary makes a target hash query.

Overall, the success probability cannot decrease by more than

$$\text{Prob}[\, \mathsf{THQ} \,] \leq \frac{q_e}{N} + q_e \cdot \left( \mathbf{Adv}^{\text{gPACE-DH}}_{\mathsf{Map2Point}}(t^*, N, q_h) \right)$$

From now on we can condition on the adversary not making a hash query about the DH-key of an accepting user instance.

*Description of* Game$_7$. As Game$_6$ but replace keys $K_{enc}, K_{mac}$ in Test-queries by random keys.

Note that, since we assume that the adversary never makes a hash query about a DH-key of an accepting user instance, this simulation is perfect unless there is an accepting instance $(U, i)$ having the same DH-key as another instance $(U^*, j)$ but such that the two instances are not partnered. In this case the adversary could make a Reveal-query to party $(U^*, j)$ and could notice the difference to the Test$(U, i)$ query. Note that Reveal-queries to partnered instances do not lead to a win for the adversary.

In the full version we show that, except with negligible probability, there cannot exist some user $U^*$ in execution $j$ such that this user also accepted with output $(K_{enc}, K_{mac})$, sid $= (Y_A^*, Y_B^*, \widehat{G}^*, \mathcal{G}^*)$ and pid $= \epsilon$, but such that the two instances are not partnered. Security then follows.                    □

# 6  Discussion

*On Forward Secrecy.* The above theorem remains true in the forward-secrecy setting (assuming weak corruptions). To show forward security we need a slight variant of the gPACE-DH problem in which the adversary first outputs $Y_A$, then learns $k$ and finally outputs $q_\ell$ potential keys $K_1, \ldots, K_{q_\ell}$. This (adaptive) version of gPACE-DH for parameters $(t, N, q_\ell)$ can be shown to be as hard as the (non-adaptive) gPACE-DH problem for parameters $(Nt, N, Nq_\ell)$. For this simply let the non-adaptive adversary simulate the adaptive adversary up to the point where it outputs $Y_A$. Instead of outputting $Y_A$ the non-adaptive algorithm internally completes $N$ runs of the adaptive adversary for all $N$ possible choices of $k$, yielding at most $N$ times $q_\ell$ possible keys. The non-adaptive adversary finally outputs $Y_A$ and this list of keys, and wins with the same probability as the adaptive adversary.

Forward secrecy of the protocol follows under the adaptive gPACE-DH problem. If a party gets corrupted after a Test-query (in which case an honestly or maliciously chosen $Y_A$ has already been determined, before the adversary learns the password and thus $k$) computing the DH key would require to solve the adaptive gPACE-DH problem and thus the gPACE-DH problem. If a party gets corrupted before a Test-query then this execution does not involve Send-commands and the data are thus chosen honestly. In particular, one can think of $Y_A$ as being chosen at random before the adversary learns the password. Security then also follows from the adaptive gPace-DH problem.

# Acknowledgments

We thank the anonymous reviewers of ISC 2009 for valuable comments. We also thank the participants of the WG 16 sub group for the stimulating discussions.

# References

1. Abdalla, M., Fouque, P.A., Pointcheval, D.: Password-based authenticated key exchange in the three-party setting. In: Vaudenay, S. (ed.) PKC 2005. LNCS, vol. 3386, pp. 65–84. Springer, Heidelberg (2005)
2. Bellare, M., Pointcheval, D., Rogaway, P.: Authenticated key exchange secure against dictionary attacks. In: Preneel, B. (ed.) EUROCRYPT 2000. LNCS, vol. 1807, pp. 139–155. Springer, Heidelberg (2000)
3. Abdalla, M., Bresson, E., Möller, O.C.B., Pointcheval, D.: Provably secure password-based authentication in tls. In: ASIACCS 2006, pp. 35–45. ACM Press, New York (2006)
4. Federal Office for Information Security (BSI): Advanced security mechanism for machine readable travel documents – extended access control (eac), password authenticated connection establishment (pace), and restricted identification (ri) (2008)
5. Coron, J.S., Patarin, J., Seurin, Y.: The random oracle model and the ideal cipher model are equivalent. In: Wagner, D. (ed.) CRYPTO 2008. LNCS, vol. 5157, pp. 1–20. Springer, Heidelberg (2008)
6. Canetti, R., Goldreich, O., Halevi, S.: The random oracle methodology, revisited. In: STOC 1998, pp. 209–218. ACM Press, New York (1998)
7. Abdalla, M., Pointcheval, D.: Simple password-based authenticated key protocols. In: Menezes, A. (ed.) CT-RSA 2005. LNCS, vol. 3376, pp. 191–208. Springer, Heidelberg (2005)
8. Shoup, V.: Lower bounds for discrete logarithms and related problems. In: Fumy, W. (ed.) EUROCRYPT 1997. LNCS, vol. 1233, pp. 256–266. Springer, Heidelberg (1997)
9. Shallue, A., van de Woestijne, C.: Construction of rational points on elliptic curves over finite fields. In: Hess, F., Pauli, S., Pohst, M. (eds.) ANTS 2006. LNCS, vol. 4076, pp. 510–524. Springer, Heidelberg (2006)
10. Icart, T.: How to hash into elliptic curves. In: Crypto 2009. LNCS. Springer, Heidelberg (2009)
11. Abdalla, M., Pointcheval, D.: Interactive diffie-hellman assumptions with applications to password-based authentication. In: S. Patrick, A., Yung, M. (eds.) FC 2005. LNCS, vol. 3570, pp. 341–356. Springer, Heidelberg (2005)
12. Szydlo, M.: A note on chosen-basis decisional diffie-hellman assumptions. In: Di Crescenzo, G., Rubin, A. (eds.) FC 2006. LNCS, vol. 4107, pp. 166–170. Springer, Heidelberg (2006)

# Towards Security Notions for White-Box Cryptography

Amitabh Saxena[1], Brecht Wyseur[2], and Bart Preneel[2]

[1] International University in Germany
Bruchsal 76646, Germany
amitabh123@gmail.com
[2] Katholieke Universiteit Leuven – ESAT / COSIC-IBBT
Kasteelpark Arenberg 10, 3001 Heverlee, Belgium
bwyseur,preneel@esat.kuleuven.be

**Abstract.** While code obfuscation attempts to hide certain characteristics of a program independently of an application, white-box cryptography (WBC) specifically focuses on software implementations of cryptographic primitives in an application. The aim of WBC is to resist attacks from an adversary having access to some 'executable' code with an embedded secret key. WBC, if possible, would have several applications. However, unlike obfuscation, it lacks a theoretical foundation. We present a first step towards a theoretical model of WBC via white-box security notions. We also present some positive and negative results on WBC and obfuscation. In particular, we show that for most interesting programs (such as an encryption algorithm), there are security notions that cannot be satisfied when the adversary has white-box access, while they are satisfied when it has black-box access. On the positive side, we show that there exists an obfuscator for a symmetric encryption scheme in the context of a useful security-notion (such as IND-CPA).

## 1 Introduction

White-box cryptography (WBC) aims to protect cryptographic keys embedded in a program that is in the control of an attacker. The attacker can conduct non-black-box attacks (such as code inspection, execution environment modification, code modification, etc). Practical white-box implementations of DES and AES encryption algorithms were proposed in [1,2]. However, no formal definitions of white-box cryptography were given, neither were there any proofs of security. With their subsequent cryptanalysis [3,4,5], it remains an open question whether or not such white-box implementations exist.

**Our Contribution:** The contributions of this work are two-fold: (1) we formalize white-box cryptography using a *white-box property (WBP)* that captures the security of an obfuscation with respect to an application, and (2) we present some (im)possibility results about WBP and obfuscation. We show that for most programs, there do not exist obfuscators satisfying WBP for all applications in which $P$ may be used. On the positive side, we show that there exist obfuscators satisfying WBP for a meaningful program and application.

P. Samarati et al. (Eds.): ISC 2009, LNCS 5735, pp. 49–58, 2009.

## 1.1   Notation and Preliminaries

Denote by $\mathbb{P}$ the set of all polynomials with coefficients in $[0..\infty]$ and by $\mathbb{TM}$ the set of all Turing Machines (TMs). For $X \in \mathbb{TM}$, $|X|$ is the length of the string description of $X$. A mapping $f : \mathbb{N} \ni x \mapsto f(x) \in \mathbb{R}$ is negligible in $x$ (written $f(x) \leq negl(x)$) if $\forall p \in \mathbb{P}, \exists x' \in \mathbb{N}, \forall x > x' : f(x) < 1/p(x)$. For $X, Y \in \mathbb{TM}$ we say $X = Y$ iff $\forall a : X(a) = Y(a)$. A *PPT Algorithm* is a TM with an unknown source of randomness with running time polynomial in the length of known inputs. A *Turing Machine Family* (TMF) is a TM with two read tapes: a *key* and an *input*. For any TMF $Q$, we denote $Q$'s key-space (valid strings of the key tape) of length $k$ by $\mathcal{K}_Q^k$, and the resulting TM when $Q$'s key tape contains $q$ by $Q[q] \in \mathbb{TM}$. For any $q \in \mathcal{K}_Q^k$, the input-space of $Q[q]$ is fully defined by $k$, and we denote this space by $\mathcal{I}_Q^k$. A TMF $Q$ is a *Polynomial TMF (PTMF)* if:

1. $\exists \mathbf{P}_Q \in \mathbb{P}, \forall k, \forall q \in \mathcal{K}_Q^k, \forall x \in \mathcal{I}_Q^k : |x| = \mathbf{P}_Q(k)$.
2. $\exists p \in \mathbb{P}, \forall k, \forall q \in \mathcal{K}_Q^k, \forall a \in \mathcal{I}_Q^k : Q[q](a)$ halts in at most $p(k)$ steps.

Denote by $\mathbb{PPT}$ and $\mathbb{PTF}$ the set of all PPT algorithms and PTMFs respectively.

**Definition 1.** $Q \in \mathbb{PTF}$ *is a* learnable family (LF) *if* $\exists (L, p) \in \mathbb{PPT} \times \mathbb{P}, \forall k :$

$$\Pr \left[ q \xleftarrow{R} \mathcal{K}_Q^k; X \leftarrow L^{Q[q]}(1^k) : X = Q[q] \wedge |X| \leq p(k) \right] \geq 1/p(k);$$

*and* $\forall a :$ *if* $Q[q](a)$ *halts after $t$ steps then $X(a)$ halts in $\leq p(t)$ steps.*

**Definition 2.** $Q \in \mathbb{PTF}$ *is an* approximate LF (ALF) *if* $\exists (L, p) \in \mathbb{PPT} \times \mathbb{P}, \forall k :$

$$\Pr \left[ (q, a) \xleftarrow{R} \mathcal{K}_Q^k \times \mathcal{I}_Q^k; X \leftarrow L^{Q[q]}(1^k) : X(a) = Q[q](a) \wedge |X| \leq p(k) \right] \geq 1/p(k);$$

*and if* $Q[q](a)$ *halts after $t$ steps then $X(a)$ halts in $\leq p(t)$ steps.*

Denote by $\mathbb{LF}$ and $\mathbb{ALF}$ the set of all LFs and ALFs respectively.

## 2   Obfuscation

Informally, an obfuscator $\mathcal{O}$ transforms a program $P$ into $\mathcal{O}(P)$, which is functionally equivalent to $P$ but hides certain characteristics of $P$. The following definitions are adapted from the literature [6,7,8,9,10,11,12,13,14].

Let $Q \in \mathbb{PTF}$. We consider the obfuscation of $Q[q]$ (an instantiation of $Q$ with key $q$). Let $\mathcal{O} : \mathbb{PTF} \times \{0,1\}^* \mapsto \mathbb{TM}$ be a PPT algorithm.

**Definition 3 (Correctness).** $\mathcal{O}$ *is an obfuscator for $Q$ if:*

1. *(Functionality)* $\forall k, \forall (q, a) \in \mathcal{K}_Q^k \times \mathcal{I}_Q^k : \Pr [\mathcal{O}(Q, q)(a) \neq Q[q](a)] \leq negl(k)$
2. *(Polynomial slowdown and expansion)* $\exists p \in \mathbb{P}, \forall k, \forall q \in \mathcal{K}_Q^k :$
   (a) $|\mathcal{O}(Q, q)| \leq p(k)$
   (b) $\forall a : Q[q](a)$ *halts in $t$ steps* $\Rightarrow \mathcal{O}(Q, q)(a)$ *halts in $\leq p(t)$ steps.*

Soundness is defined using a *Virtual Black-Box Property (VBBP)* [7,12,11]. let $Q[q]$ be a random instantiation of a PTMF $Q$ using key $q$. The VBBP requires that whatever information about $q$ a PPT adversary extracts from $\mathcal{O}(Q, q)$, a PPT simulator should also be able to extract with black-box access to $Q[q]$. Existing notions of VBBP fall into one of two broad categories as defined below.

**Definition 4 (Soundness).** $\mathcal{O}$ *is* sound *if at least one of following holds:*

1. *Predicate VBBP:* $\forall A \in \mathbb{PPT}, \exists S \in \mathbb{PPT} : \mathsf{Adv}^{pvbbp}_{A,S,\mathcal{O},Q}(k) \leq negl(k)$, *where*
$$\mathsf{Adv}^{pvbbp}_{A,S,\mathcal{O},Q}(k) = \left| \Pr_{q \xleftarrow{R} \mathcal{K}^k_Q} \left[ A^{Q[q]}(1^k, \mathcal{O}(Q, q)) = 1 \wedge S^{Q[q]}(1^k) \neq 1 \right] \right|.$$

2. *Indistinguishability:* $\forall A \in \mathbb{PPT}, \exists S \in \mathbb{PPT} : \mathsf{Adv}^{ind}_{A,S,\mathcal{O},Q}(k) \leq negl(k)$, *where*
$$\mathsf{Adv}^{ind}_{A,S,\mathcal{O},Q}(k) = \left| \Pr_{q \xleftarrow{R} \mathcal{K}^k_Q} \left[ A^{Q[q]}(1^k, \mathcal{O}(Q, q)) = 1 \wedge A^{Q[q]}(1^k, S^{Q[q]}(1^k)) \neq 1 \right] \right|.$$

Note that indistinguishability is too strong to yield interesting results [11,10]. On the other hand, predicate VBBP is too weak to be meaningful in practice [7,11]. Nevertheless, it is conceivable that a meaningful definition of soundness falling somewhere between the two extremes can be formulated. We show this is not the case. Specifically, we show that, under *every* definition of soundness, for every $Q \notin \mathbb{ALF}$, there exist (contrived) security notions for which white-box security fails but the corresponding black-box construction is secure.

## 3   White-Box Cryptography

We formalize white-box cryptography using a *white-box property (WBP)*, which is defined using a game-based approach [15,16,17,18]. Loosely speaking, the WBP is defined using two objects: a PTMF (such as an encryption algorithm family) and a **security notion** (such as IND-CPA). A security notion (SN) is a formal description of the security desired from a cryptographic scheme.

**Definition 5.** *A* **Security Notion** *(SN) is a 5-tuple* $(n, p_{in}, \mathbf{Q}, \mathsf{Extr}, \mathsf{Win}) \in \mathbb{N} \times \mathbb{P} \times \mathbb{PTF}^n \times \mathbb{TM} \times \mathbb{TM}$ *where* $\mathbf{Q} = (Q_1, Q_2, \ldots, Q_n) \in \mathbb{PTF}^n$ *is an n-tuple of PTMFs, and* $\mathsf{Extr}$ *and* $\mathsf{Win}$ *are TMs of the type* $\{0, 1\}^{p_{in}(k)} \mapsto \times^n_{i=1} \mathcal{K}_{Q_i}$ *and* $\{0, 1\}^* \mapsto \{0, 1\}$ *respectively. Denote by* $\mathbb{SN}$ *the set of all security notions. For any* $sn = (n, p_{in}, \mathbf{Q}, \mathsf{Extr}, \mathsf{Win}) \in \mathbb{SN}$ *and any* $Q \in \mathbb{PTF}$, *we say* $Q \in sn$ *if* $Q \in \mathbf{Q}$.

**Definition 6.** *A* **Black-box Game** *given in Algorithm 1 (*GameBB$_A$*) is a TM interacting with the adversary* $A \in \mathbb{PPT}$. *It takes as input* $(1^k, sn, r)$, *where* $sn = (n, p_{in}, \mathbf{Q}, \mathsf{Extr}, \mathsf{Win}) \in \mathbb{SN}$ *is a security notion, and* $r \in \{0, 1\}^{p_{in}(k)}$ *is a string (representing randomness). It outputs 0 or 1.*

At any instant $A$ can query at most one oracle, and each query by the adversary takes one unit time irrespective of the amount of computation involved. Queries *is a set of ordered 4-tuples of type*

$$(\mathbf{t}_j, \mathbf{i}_j, \mathbf{in}_j, \mathbf{out}_j) \in \mathbb{N} \times \{1, 2, \ldots, n\} \times \{0, 1\}^* \times \{0, 1\}^*,$$

*indicating respectively, the time, oracle, input, and the output of each query. Define* $\mathsf{AdvBB}^{sn}_A(k) = \Pr \left[ r \xleftarrow{R} \{0, 1\}^{p_{in}(k)} : \mathsf{GameBB}_A(1^k, sn, r) = 1 \right].$

---

**input** : $1^k, sn, r$
Parse $sn$ as $(n, p_{in}, \mathbf{Q}, \mathsf{Extr}, \mathsf{Win})$
Parse $\mathbf{Q}$ as $(Q_1, Q_2, \ldots, Q_n)$
$(q_1, q_2, \ldots q_n) \leftarrow \mathsf{Extr}(r)$
$s \leftarrow A^{Q_1[q_1], Q_2[q_2], \ldots, Q_n[q_n]}(1^k, sn)$
**output**: $\mathsf{Win}(r, \mathsf{Queries}, s)$

---

**Algorithm 1.** $\mathsf{GameBB}_A(1^k, sn, r)$

**Discussion.** Consider the IND-CCA2 security notion for symmetric encryption, which is defined as a game with three stages: (1) the adversary queries the encryption/decryption oracles; (2) the adversary obtains a challenge ciphertext; and (3) the adversary queries the oracles as in (1) except that decryption queries on the challenge ciphertext are disallowed. The adversary wins if it guesses some property of the challenge ciphertext. An example is given in Appendix A.

Let $\mathcal{E} = (G, E, D)$ be any IND-CCA2 secure symmetric encryption scheme with the encryption/decryption key instantiated to $K$. Observe that the adversary cannot be given an obfuscation of $D[K]$, since this will render $\mathcal{E}$ insecure under IND-CCA2 - once the adversary gets this obfuscation, we cannot prevent it from querying $D[K]$ (via the obfuscation) on the challenge ciphertext in phase (3). On the other hand, $E[K]$ is a candidate for obfuscation because the winning condition does not depend on queries to $E[K]$. We generalize this intuition in Definition 7 to describe when a family is a candidate for obfuscation.

**Definition 7.** *For any $sn \in \mathbb{SN}$ and any PTMF $Q_i \in sn$, define $\mathsf{Queries}(i)$ to be the following set: $\{(\mathbf{t}_j, \mathbf{i}_j, \mathbf{in}_j, \mathbf{out}_j) | (\mathbf{t}_j, \mathbf{i}_j, \mathbf{in}_j, \mathbf{out}_j) \in \mathsf{Queries} \wedge \mathbf{i}_j \neq i\}$.*
*$Q_i$ is **obfuscatable** in $sn$ (written $Q_i \in_{obf} sn$) if*

$$\forall r, \mathsf{Queries}, s : \mathsf{Win}(r, \mathsf{Queries}, s) = \mathsf{Win}(r, \mathsf{Queries}(i), s).$$

In other words, $Q_i \in_{obf} sn$ if: (1) $Q_i \in sn$, and (2) in the black-box game, the output of $\mathsf{Win}$ is invariant w.r.t the entries of $\mathsf{Queries}$ for $Q_i[q_i]$.

Observe that a meaningful notion of white-box security cannot exist for a family under a security notion in which it is not obfuscatable. For instance, white-boxing the decryption oracle of a symmetric encryption scheme, or the 'signing' oracle of a MAC scheme under standard security notions is not meaningful.

**Definition 8.** *For $(A, \mathcal{O}, sn) \in \mathbb{PPT} \times \mathbb{PPT} \times \mathbb{SN}$, with $sn = (n, p_{in}, \mathbf{Q}, \mathsf{Extr}, \mathsf{Win})$ s.t. $\mathbf{Q} = (Q_1, Q_2, \ldots, Q_n)$, the **white-box game** ($\mathsf{GameWB}_{A, \mathcal{O}, Q_i}$) for $i \in [1..n]$ is given in Algorithm 2. Define*

$$\mathsf{AdvWB}^{sn}_{A, \mathcal{O}, Q_i}(k) = \Pr\left[r \xleftarrow{R} \{0, 1\}^{p_{in}(k)} : \mathsf{GameWB}_{A, \mathcal{O}, Q_i}(1^k, sn, r) = 1\right].$$

**Definition 9.** *For any $(\mathcal{O}, Q, sn) \in \mathbb{PPT} \times \mathbb{PTF} \times \mathbb{SN}$ such that $Q \in sn$, define the **White-box Advantage (WBA)** $\mathsf{AdvWB}^{sn}_{\mathcal{O}, Q}(k)$ of $\mathcal{O}$ for $(Q, sn)$ as:*

$$\mathsf{AdvWB}^{sn}_{\mathcal{O}, Q}(k) = \left|\max(\mathsf{AdvWB}^{sn}_{A, \mathcal{O}, Q}(k)) - \max(\mathsf{AdvBB}^{sn}_A(k))\right|,$$

```
input  : 1^k, sn, r
Parse sn as (n, p_in, Q, Extr, Win)
Parse Q as (Q_1, Q_2, ..., Q_n)
(q_1, q_2, ... q_n) ← Extr(r)
s ← A^{Q_1[q_1], Q_2[q_2], ..., Q_n[q_n]}(1^k, sn, i, O(Q_i, q_i))
output: Win(r, Queries, s)
```

**Algorithm 2.** GameWB$_{A,O,Q_i}(1^k, sn, r)$

*where for any function $f_A(k)$, $\max(f_A(k))$ is defined as follows: Let $A' \in \mathbb{PPT}$ be such that $\forall A \in \mathbb{PPT} : \lim_{k \to \infty} f_A(k) \leq f_{A'}(k)$. Then $\max(f_A(k)) = f_{A'}(k)$.*

The following two definitions capture white-box security.

**Definition 10.** *For all $(O, Q, sn) \in \mathbb{PPT} \times \mathbb{PTF} \times \mathbb{SN}$ s.t. $Q \in sn$, $O$ satisfies* **White-box Property (WBP)** *for $(Q, sn)$ if* AdvWB$_{O,Q}^{sn}(k) \leq negl(|k|)$.

**Definition 11.** *For all $(O, Q) \in \mathbb{PPT} \times \mathbb{PTF}$, $O$ satisfies* **Universal WBP** *(UWBP) for $Q$ if $\forall sn \in \mathbb{SN} : Q \notin_{obf} sn \vee (O$ satisfies WBP for $(Q, sn))$.*

## 4 Negative Results

Barak *et al.* [7] give several impossibility results on obfuscation; their main result implies that there do not exist obfuscators satisfying UWBP for every $Q \in$ PTF. However, they do not rule out obfuscators satisfying UWBP for a useful family $Q$. Our main negative result is stronger - there do not exist obfuscators satisfying UWBP for 'interesting' families. That is, we show that for any non-approximately-learnable family, there exists a security notion that cannot be satisfied when an adversary has white-box access to the program (Theorem 1).

**Theorem 1.** *For every $(Q, O) \in \mathbb{PTF}\backslash\mathbb{ALF} \times \mathbb{PPT}$, there exists $sn \in \mathbb{SN}$ such that $Q \in_{obf} sn$ but $O$ fails to satisfy WBP for $(Q, sn)$.*

*Proof.* Let $Q \in \mathbb{PTF}\backslash\mathbb{ALF}$. Let *guess-x* $= (2, p_{in}, Q, \mathsf{Extr}, \mathsf{Win}) \in \mathbb{SN}$ be such that $\mathbf{Q} = (Q, Q_1)$; $p_{in}(k) = 2k + \mathbf{P}_Q(k)$; and other details in Algorithm 3.

Note that $Q \in_{obf}$ *guess-x*. Since $Q \notin$ ALF, therefore due to Definitions 1 and 3, the following two inequalities are guaranteed to hold:

$$\forall A \in \mathbb{PPT} : 0 \leq \mathsf{AdvBB}_A^{guess\text{-}x}(k) < \alpha(k),$$

$$\exists A \in \mathbb{PPT} : 1 \geq \mathsf{AdvWB}_{A,O,Q}^{guess\text{-}x}(k) \geq 1 - \beta(k),$$

where $\alpha, \beta$ are negligible functions. Hence, we have that $\mathsf{AdvWB}_{O,Q}^{guess\text{-}x}(k)$ is larger than $1 - \alpha(k) - \beta(k)$, which is non-negligible in $k$. This proves the theorem.

---

**Function** $Q_1[q_1](Y_1)$ :
   Parse $q_1$ as $(q, x, a)$;
   **if** $(Y_1(a) = Q[q](a))$ **then** output $x$ **else** output 0;
**Function** Extr$(r)$ :
   Parse $r$ as $(q, x, a)$;       // $x \in \{0,1\}^k, q \in \mathcal{K}_Q^k, a \in \mathcal{I}_Q^k \subseteq \{0,1\}^{\mathbf{P}_Q(k)}$,
   set $q_1 \leftarrow (q, x, a)$;
   output $q, q_1$;
**Function** Win$(r, \text{Queries}, s)$ :
   Parse $r$ as $(q, x, a)$;
   **if** $(s \neq x) \vee (more\ than\ one\ query\ to\ Q_1[q_1])$ **then** output 0 **else** output 1

---

**Algorithm 3.** $Q_1$, Extr and Win for *guess-x*

**Simultaneous Obfuscation may be Insecure.** When two families in the same SN are white-boxed, a useful question is whether the resulting implementation remains secure assuming that it was secure when each family was separately white-boxed. Theorem 2 states that simultaneous white-boxing of two families can be insecure even if white-boxing of each family separately is secure.

**Theorem 2.** *For every* $(Q_i, Q_j, \mathcal{O}) \in (\mathbb{PTF} \backslash \mathbb{ALF})^2 \times \mathbb{PPT}$, *there exists* $sn \in \mathbb{SN}$ *with* $Q_i, Q_j \in_{obf} sn$ *such that even if* $\mathcal{O}$ *satisfies WBP for* $(Q_i, sn)$ *and* $(Q_j, sn)$, *it fails to satisfy WBP for* $((Q_i, Q_j), sn)$.    (See the full version [19] for proof.)

# 5   Positive Results

Although Theorem 1 rules out obfuscators satisfying UWBP for most non-trivial families, it does not imply that meaningful security in WBC cannot exist. In fact, any asymmetric encryption scheme can be considered as a white-boxed version of some symmetric scheme. We use this observation as a starting point of our first positive result. A similar observation was used in the positive results of [12].

**Theorem 3 (WBP for "Useful" Families).** *There exists a tuple* $(\mathcal{O}, Q, sn) \in \mathbb{PPT} \times \mathbb{PTF} \backslash \mathbb{ALF} \times \mathbb{SN}$ *such that* $Q \in_{obf} sn$ *and* $\mathcal{O}$ *is an obfuscator satisfying WBP for* $(Q, sn)$ *under reasonable computational assumptions.*

We use the following definition in the proof of Theorem 3.

**Definition 12.** *(Bilinear pairing) Let* $G_1, G_2$ *be cyclic multiplicative groups of prime order* $w$ *such that the discrete logarithm problem in both is hard. A bilinear pairing is a map* $\hat{e} : G_1 \times G_1 \mapsto G_2$ *that satisfies the following properties [20,21,22]:*

1. Bilinearity: $\hat{e}(a^x, b^y) = \hat{e}(a, b)^{xy}$ $\forall a, b \in G_1$ *and* $x, y \in \mathbb{Z}_w$.
2. Non-degeneracy: *If* $g$ *is a generator of* $G_1$ *then* $\hat{e}(g, g)$ *is a generator of* $G_2$.
3. Computability: *The map* $\hat{e}$ *is efficiently computable.*

*Proof.* (of Theorem 3) We prove this by construction using an encryption scheme described in [20]. Define a *symmetric* encryption scheme $\mathcal{E} = (G, E, D)$ as follows.

1. *Key Generation:* Let $\hat{e} : G_1 \times G_1 \mapsto G_2$ be a bilinear pairing over cyclic multiplicative groups as defined above (such maps are known to exist). Let $|G_1| = |G_2| = w$ (prime) such that $\lfloor \log_2(w) \rfloor = l$. Pick random $g \xleftarrow{R} G_1 \backslash \{1\}$ and define $\mathcal{H} : G_2 \mapsto \{0,1\}^l$ to be a hash function. Pick $x \xleftarrow{R} G_1$ and define $K = (\hat{e}, G_1, G_2, w, g, \mathcal{H}, x)$. The encryption/decryption key is $K$.
2. *Encryption:* Parse $K$ as $(\hat{e}, G_1, G_2, w, g, \mathcal{H}, x)$. Let $m \in \{0,1\}^l$ be a message and $\alpha \in \mathbb{Z}_w$ be a random value. The encryption family $E$ is:

$$E[K] : \{0,1\}^l \times \mathbb{Z}_w \ni (m, \alpha) \mapsto (\mathcal{H}(\hat{e}(x^\alpha, g)) \oplus m, g^\alpha) \in \{0,1\}^l \times G_1 .$$

3. *Decryption:* Parse $K$ as $(\hat{e}, G_1, G_2, w, g, \mathcal{H}, x)$. The decryption family $D$ is:

$$D[K] : \{0,1\}^l \times G_1 \ni (c_1, c_2) \mapsto \mathcal{H}(\hat{e}(c_2, x)) \oplus c_1 \in \{0,1\}^l .$$

It can be verified that $D[K](E[K](m, \alpha)) = m$ for valid values of $(m, \alpha)$. The scheme can be proven to be CPA secure if $\mathcal{H}$ is a random oracle and $w$ is sufficiently large. We construct an obfuscation of the $E[K]$ oracle that converts $\mathcal{E}$ into a CPA secure *asymmetric* encryption scheme under a computational assumption.

**The obfuscator $\mathcal{O}$:** The input is $(E, K)$.

1. Parse $K$ as $(\hat{e}, G_1, G_2, w, g, \mathcal{H}, x)$. Set $y \leftarrow \hat{e}(x, g)$.
2. Set $K' \leftarrow (\hat{e}, G_1, G_2, w, g, \mathcal{H}, y)$ and define family $F$ with key $K'$ as:

$$F[K'] : \{0,1\}^l \times \mathbb{Z}_w \ni (m, \alpha) \mapsto (\mathcal{H}(y^\alpha) \oplus m, g^\alpha) \in \{0,1\}^l \times G_1 .$$

3. Output $F[K']$.

*Claim.* $\mathcal{O}$ is an obfuscator satisfying WBP for $(E, sn)$, where $sn \stackrel{\text{def}}{=}$ "IND-CPA security of $\mathcal{E}$", assuming that the *bilinear Diffie-Hellman assumption* [20] holds in $(G_1, G_2)$ and $\mathcal{H}$ is a random oracle.

*Proof.* Appendix A describes the IND-CCA2 security notion. IND-CPA is a restricted version where the family **D** is absent. First note that the obfuscator satisfies correctness for $E$ because $F[K'] = E[K]$. The proof of the above claim follows from the security of the BasicPUB scheme of [20].

*Claim.* If $\mathcal{H}$ is one-way then $E \in \mathbb{PTF} \backslash \mathbb{ALF}$.     (See full version [19] for proof.)

This completes the proof of Theorem 3.

*Remark 1.* To justify the particular scheme (instead of RSA/ElGamal) in the above proof, observe that RSA does not enjoy the security notion of IND-CPA, while encryption in ElGamal is learnable (to see this, consider access to the ElGamal encryption oracle and obtain encryption of 1 using randomness 1).

**Trivial Families:** Let $Q \in \mathbb{LF}$. Then it is easy to construct an obfuscator satisfying UWBP for $Q$ with a non-negligible probability (same as that of learning $Q$). We call such families *trivial*.

Although Theorem 1 rules out the possibility of obfuscators satisfying UWBP for a $Q \in \mathbb{PTF}\backslash\mathbb{ALF}$ (which includes most non-trivial families), it does not rule out the possibility of obfuscators satisfying UWBP for some non-trivial $Q \in \mathbb{ALF}$ (i.e., $Q \in \mathbb{ALF}\backslash\mathbb{LF}$), which is our next positive result.

**Theorem 4 (UWBP for a Non-Trivial Family).** *Under reasonable assumptions, there exists* $(\mathcal{O}, Q) \in \mathbb{PPT} \times \mathbb{ALF}\backslash\mathbb{LF}$ *s.t.* $\mathcal{O}$ *satisfies UWBP for* $Q$.

We refer to the full version [19] for the proof.

# 6   Conclusion

White-Box Cryptography (WBC) is of practical importance. Unfortunately, it lacks a theoretical foundation. This paper provides an initial step towards a formalization of WBC. To achieve this, we introduce the White-Box Property (WBP), which defines how much 'useful' information is leaked via the obfuscation in the context of an application. We present (im)possibility results about reductions between WBC and obfuscation. Specifically, we show that any obfuscator fails to satisfy the *Universal White-Box Property* (UWBP) for non-learnable families by presenting a (contrived) security notion that is satisfied in 'black-box' setting, but fails when an adversary has white-box access to the obfuscation. However, we show that UWBP can be achieved for non-learnable, but approximate learnable families. Further, we show that there exists a non-learnable family and an obfuscator satisfying WBP for that family under a meaningful security notion. In particular, we described an obfuscator that turns a IND-CPA secure symmetric scheme into an IND-CPA secure asymmetric encryption scheme.

# Acknowledgments

This work has been partially funded by the IAP Programme P6/26 BCRYPT of the Belgian State (Belgian Science Policy) and the European Commission through the IST Programme under contract IST-021186 RE-TRUST.

We would like to thank Amir Herzberg for motivating this work. We would also like to thank the anonymous reviewers for their valuable feedback.

# References

1. Chow, S., Eisen, P.A., Johnson, H., van Oorschot, P.C.: White-Box Cryptography and an AES Implementation. In: Nyberg, K., Heys, H.M. (eds.) SAC 2002. LNCS, vol. 2595, pp. 250–270. Springer, Heidelberg (2003)

2. Chow, S., Eisen, P.A., Johnson, H., van Oorschot, P.C.: A white-box DES implementation for DRM applications. In: Feigenbaum, J. (ed.) DRM 2002. LNCS, vol. 2696, pp. 1–15. Springer, Heidelberg (2003)
3. Billet, O., Gilbert, H., Ech-Chatbi, C.: Cryptanalysis of a White Box AES Implementation. In: Handschuh, H., Hasan, M.A. (eds.) SAC 2004. LNCS, vol. 3357, pp. 227–240. Springer, Heidelberg (2004)
4. Goubin, L., Masereel, J.M., Quisquater, M.: Cryptanalysis of White Box DES Implementations. In: Adams, C., Miri, A., Wiener, M. (eds.) SAC 2007. LNCS, vol. 4876, pp. 278–295. Springer, Heidelberg (2007)
5. Wyseur, B., Michiels, W., Gorissen, P., Preneel, B.: Cryptanalysis of White-Box DES Implementations with Arbitrary External Encodings. In: Adams, C., Miri, A., Wiener, M. (eds.) SAC 2007. LNCS, vol. 4876, pp. 264–277. Springer, Heidelberg (2007)
6. Hada, S.: Zero-Knowledge and Code Obfuscation. In: Okamoto, T. (ed.) ASIACRYPT 2000. LNCS, vol. 1976, pp. 443–457. Springer, Heidelberg (2000)
7. Barak, B., Goldreich, O., Impagliazzo, R., Rudich, S., Sahai, A., Vadhan, S., Yang, K.: On the (Im)possibility of Obfuscating Programs. In: Kilian, J. (ed.) CRYPTO 2001. LNCS, vol. 2139, pp. 1–18. Springer, Heidelberg (2001)
8. Goldwasser, S., Kalai, Y.T.: On the Impossibility of Obfuscation with Auxiliary Input. In: Proceedings of the 46th Symposium on Foundations of Computer Science (FOCS 2005), Washington, DC, USA, pp. 553–562. IEEE Computer Society, Los Alamitos (2005)
9. Lynn, B., Prabhakaran, M., Sahai, A.: Positive Results and Techniques for Obfuscation. In: Cachin, C., Camenisch, J.L. (eds.) EUROCRYPT 2004. LNCS, vol. 3027, pp. 20–39. Springer, Heidelberg (2004)
10. Wee, H.: On Obfuscating Point Functions. In: Proceedings of the 37th ACM Symposium on Theory of Computing (STOC 2005), pp. 523–532. ACM Press, New York (2005)
11. Hohenberger, S., Rothblum, G., Shelat, A., Vaikuntanathan, V.: Securely Obfuscating Re-Encryption. In: Vadhan, S.P. (ed.) TCC 2007. LNCS, vol. 4392, pp. 233–252. Springer, Heidelberg (2007)
12. Hofheinz, D., Malone-Lee, J., Stam, M.: Obfuscation for Cryptographic Purposes. In: Vadhan, S.P. (ed.) TCC 2007. LNCS, vol. 4392, pp. 214–232. Springer, Heidelberg (2007)
13. Goldwasser, S., Rothblum, G.N.: On Best-Possible Obfuscation. In: Vadhan, S.P. (ed.) TCC 2007. LNCS, vol. 4392, pp. 194–213. Springer, Heidelberg (2007)
14. Canetti, R., Dakdouk, R.R.: Obfuscating Point Functions with Multibit Output. In: Smart, N.P. (ed.) EUROCRYPT 2008. LNCS, vol. 4965, pp. 489–508. Springer, Heidelberg (2008)
15. Bellare, M., Desai, A., Pointcheval, D., Rogaway, P.: Relations Among Notions of Security for Public-Key Encryption Schemes. In: Krawczyk, H. (ed.) CRYPTO 1998. LNCS, vol. 1462, pp. 26–45. Springer, Heidelberg (1998)
16. Bellare, M., Rogaway, P.: The Security of Triple Encryption and a Framework for Code-Based Game-Playing Proofs. In: Vaudenay, S. (ed.) EUROCRYPT 2006. LNCS, vol. 4004, pp. 409–426. Springer, Heidelberg (2006)
17. Goldwasser, S., Micali, S.: Probabilistic Encryption and How to Play Mental Poker Keeping Secret All Partial Information. In: Proceedings of the 14th ACM Symposium on Theory of Computing (STOC 1982), pp. 365–377. ACM Press, New York (1982)
18. Goldwasser, S., Micali, S.: Probabilistic Encryption. Journal of Computer and System Sciences 28(2), 270–299 (1984)

19. Saxena, A., Wyseur, B., Preneel, B.: White-box cryptography: Formal notions and (im)possibility results. Cryptology ePrint Archive, Report 2008/273 (2008), http://eprint.iacr.org/
20. Boneh, D., Franklin, M.K.: Identity-Based Encryption from the Weil Pairing. In: Kilian, J. (ed.) CRYPTO 2001. LNCS, vol. 2139, pp. 213–229. Springer, Heidelberg (2001)
21. Boneh, D., Gentry, C., Lynn, B., Shacham, H.: Aggregate and Verifiably Encrypted Signatures from Bilinear Maps. In: Biham, E. (ed.) EUROCRYPT 2003. LNCS, vol. 2656, pp. 416–432. Springer, Heidelberg (2003)
22. Boneh, D., Lynn, B., Shacham, H.: Short Signatures from the Weil Pairing. In: Boyd, C. (ed.) ASIACRYPT 2001. LNCS, vol. 2248, pp. 514–532. Springer, Heidelberg (2001)

# A  The IND-CCA2 Security Notion

Let $\mathcal{E} = (G, E, D)$ be a symmetric encryption scheme. The key generation algorithm, $G$ takes in as input the security parameter $(1^k)$ and a $k$ bit random string. It outputs a $k$ bit symmetric key $K$. As an example, we describe *Indistinguishability under Adaptive Chosen Ciphertext Attack* (IND-CCA2) of $\mathcal{E}$ using security notion $ind\text{-}cca2\text{-}\mathcal{E} = (3, p_{in}, \mathbf{Q}, \mathsf{Extr}, \mathsf{Win})$, with $p_{in}(k) = 2k + 1$ and $\mathbf{Q} = (\mathbf{E}, \mathbf{D}, \mathbf{C})$ (see Algorithm 4).

---

**Function $\mathbf{E}[K](\alpha, m)$ :**
    output $E(K, \alpha, m)$;        // $(K, \alpha, m)$ = (key, randomness, plaintext)
**Function $\mathbf{D}[K](c)$ :**
    output $D(K, c)$;                // $(K, c)$ = (key, ciphertext)
**Function $\mathbf{C}[(b, K, \beta)](m_0, m_1)$ :**
            // The challenge oracle. $(b, K, \beta)$ = (bit, key, randomness)
    if $(|m_0| = |m_1|)$ then output $E(K, \beta, m_b)$ else output 0.
**Function $\mathsf{Extr}(r)$ :**
    parse $r$ as $(\gamma, \beta, b)$;    // $r \in \{0,1\}^{2k+1}$; $(\gamma, \beta, b) \in \{0,1\}^k \times \{0,1\}^k \times \{0,1\}$
    $K \leftarrow G(1^{|\gamma|}, \gamma)$;
    output $K, K, (b, K, \beta)$;
**Function $\mathsf{Win}(r, \mathsf{Queries}, s)$ :**
    Parse $r$ as $(\gamma, \beta, b)$;
    if $(\leq 1$ *query to* $\mathbf{C}) \wedge ($*no query to* $\mathbf{D}$ *on output of* $\mathbf{C}) \wedge (s = b)$ then
    output 1 **else** output 0.

---

**Algorithm 4.** Security Notion $ind\text{-}cca2\text{-}\mathcal{E}$

# A Calculus to Detect Guessing Attacks*

Bogdan Groza[1] and Marius Minea[2]

[1] Politehnica University of Timişoara
bogdan.groza@aut.upt.ro
[2] Institute e-Austria Timişoara
marius@cs.upt.ro

**Abstract.** We present a calculus for detecting guessing attacks, based on oracles that instantiate cryptographic functions. Adversaries can *observe* oracles, or *control* them either on-line or off-line. These relations can be established by protocol analysis in the presence of a Dolev-Yao intruder, and the derived guessing rules can be used together with standard intruder deductions. Our rules also handle partial verifiers that fit more than one secret. We show how to derive a known weakness in the Anderson-Lomas protocol, and new vulnerabilities for a known faulty ATM system.

## 1 Introduction and Related Work

Analyzing vulnerability to guessing attacks is of high practical relevance. A value is deemed guessable if it has small entropy (is chosen from a small cardinality set), and the guess can be verified. An adversary can perform guessing by off-line computation, or on-line, exploiting the interaction with honest participants.

Conceptually, guessing involves two steps. Any protocol must have a *generation oracle* which computes some value (the *verifier*), given the secret as input. Next, a boolean *verification oracle* compares a verifier for the guess with one computed for the actual secret. We use the term oracle for an abstract object that produces a value, regardless of how the computation is done. In particular, the adversary might use other participants as on-line oracles for this purpose.

Separating the verifier generation from the verification itself, and modeling them as oracles is key to our analysis of guessing attacks in both off-line and on-line settings. It is often argued that on-line guessing can be blocked after a threshold of incorrect guesses. However, if the adversary's guesses are cached as valid protocol interactions, relying on blocking is not a justified defense.

Our analysis identifies various guessing situations with partial or complete view over inputs and outputs of oracles and with off-line, on-line or blockable on-line oracle access. We provide inference rules which can detect guessing attacks in these situations. Once such a vulnerability is detected, it is up to further review to decide if it can be removed by limiting protocol runs. We will also

---

* This work is supported in part by FP7-ICT-2007-1 project 216471, AVANTSSAR: Automated Validation of Trust and Security of Service-oriented Architectures.

P. Samarati et al. (Eds.): ISC 2009, LNCS 5735, pp. 59–67, 2009.

distinguish pre-computed dictionary guessing as a particularly dangerous case: the adversary can build an off-line dictionary which is reusable and constitutes a time-memory trade-off.

### Related Work

A classification into off-line, detectable, and undetectable on-line guessing attacks is given in [1], recognizing that principals can be used to perform computation, without this being detectable. However, attack detection is not formalized.

Lowe's rules [2] construct new terms of intruder knowledge from the guessed value. Tracking that a term is obtained in two different ways confirms the guess. Special cases avoid false deductions. In [3], substituting the guess with a fresh term provides the second derivation; [4] has a dual set of Dolev-Yao deduction rules with an explicit comparison rule and gives complexity results. In [5], an intruder checks that two maps of terms constructed by exhaustive candidate enumeration correspond on exactly one entry. This approach can model simultaneous guesses of several message components but is limited to off-line attacks.

Equational theories and static equivalence are used in [6] for applied pi-calculus, and in [7] showing computational indistinguishability; [8] uses a constraint solving algorithm for an equational theory given as a convergent term rewriting system. Blanchet's tool ProVerif [9] detects off-line guessing attacks.

Our approach is based on Dolev-Yao-style deductions, with an adversary observing or controlling oracles for functions that are injective in the secret. Correct guessing is confirmed by checking relations on inputs and outputs of the oracle, or on the output of its inverse (e.g., in the case of encryption). We extend this to functions that match several secrets, but allow verification based on more than one observation, giving guessing a probabilistic meaning. Our guessing rules contain bounds that are sufficient to achieve a correct guess in the average case.

## 2   Adversary Relations with Oracles

We write $A \rightsquigarrow x$ if $A$ can guess $x$, and $A \rightsquigarrow_D x$ for guessing using pre-computed dictionaries. These are a space-time tradeoff: if $A \rightsquigarrow_D x$, then $A$ can also guess without pre-computed dictionaries by just repeating the dictionary construction.

A value is guessed only if it is also verified. Thus, if $A$ can guess $x$, then $A$ knows $x$, i.e., a guessed value can be used in further reasoning, together with the usual rules that describe how a Dolev-Yao intruder can acquire its knowledge.

We denote by $O^f(\cdot)$ an oracle which computes the value of function $f$ for any provided input. We define two relations between adversary and oracles: *observes* and *controls*, with different variants: off-line, on-line and on-line blockable, when the adversary presence is detected and the protocol stopped.

The adversary *observes* the output of oracle $O^f(\cdot)$, written $Adv \blacktriangleleft O^f(\cdot)$, if a protocol state is reachable where $Adv$ knows $f(T)$ for some term $T$. Thus, *observes* is a protocol property. The placeholders $\blacktriangleright$ and $\triangleright$ specify whether oracle inputs are known, completely or in part, e.g., $Adv \blacktriangleleft O^f(\triangleright)$ means that $Adv$

observes the output of an oracle but knows only part of its input. This allows us to relate the adversary's observation of oracle outputs with that of its inputs.

$Adv$ may observe $O^f(\cdot)$ using *on-line* access to the protocol, initial knowledge, standard Dolev-Yao deductions, and off-line computation. In a stronger case, $Adv$ might know the function $f$, and apply it *off-line* to any known term.

$Adv$ *controls* the oracle $O^f(\cdot)$, i.e., can compute $f(x)$ for an input $x$ of its choice if for any $x \in \operatorname{dom} f$ chosen by $Adv$ in the initial protocol state, a state where $Adv$ knows $f(x)$ is reachable, regardless of the actions chosen by the other participants. As with observations, this can occur off-line or on-line.

In the strongest case, $Adv$ knows all oracle parameters, and can compute $f$ off-line: $Adv\ \mathrm{ctl}^{off} O^f(\cdot)$. Second, $Adv$ can have *on-line* control of an oracle if a protocol participant provides this service. However, it is important to distinguish whether the protocol ends normally or adversary intervention may be detected.

Let (1) $A \to B : m$ and (2) $B \to A : E_k(m)$. $Adv$ can send $B$ any value $m$; the protocol ends normally, and $Adv$ knows $E_k(m)$. Thus, $Adv$ has unrestricted, *non-blockable* online control of the oracle, $Adv\ \mathrm{ctl}^{nb} O^f(\cdot)$. Now let (1) $A \to B : m_A$, (2) $B \to A : m_B, E_k(m_A)$, (3) $A \to B : E_k(m_B)$. Again, $Adv$ obtains $E_k(m)$, but cannot compute $E_k(m_B)$, so the protocol is not completed. $Adv$ has *blockable control* of the oracle, denoted $Adv\ \mathrm{ctl}^{bl} O^f(\cdot)$, since incorrect termination may be detected and subsequent protocol runs be blocked by the honest participant(s).

Controlling an oracle implies seeing both its inputs and outputs, therefore:

$$\frac{Adv\ \mathrm{ctl}^{off} O^f(\cdot)}{Adv \blacktriangleleft O^f(\blacktriangleright)} \qquad \frac{Adv\ \mathrm{ctl}^{nb} O^f(\cdot)}{Adv \blacktriangleleft^{nb} O^f(\blacktriangleright)} \qquad \frac{Adv\ \mathrm{ctl}^{bl} O^f(\cdot)}{Adv \blacktriangleleft^{bl} O^f(\blacktriangleright)} \tag{1}$$

Observation, off-line or non-blocking on-line control produce no different protocol behavior, thus guesses can go undetected. Yet, guessing using *blockable access* is also feasible, if a single oracle access suffices for verification.

The number of oracle accesses affects the feasibility of a guess. We specify lower bounds for *observes* and *controls*: $Adv \blacktriangleleft_b O^f(\cdot)$ and $Adv\ \mathrm{ctl}_b O^f(\cdot)$ mean that $Adv$ observes at least $b$ *distinct* independent outputs of $O^f(\cdot)$ (respectively, observes outputs of $O^f(\cdot)$ for $b$ chosen inputs) over different protocol runs. Bounds on these relations are deduced from the protocol description. If $A \to B : H(N_A)$, then since $N_A$ is randomly chosen, $Adv \blacktriangleleft_b O^g(\cdot)$ for any $b \leq |N_A|$ (set cardinality). However, if $A \to B : id_A, H(id_A.k_{AB})$, the oracle input is constant (and only partially visible, since $k_{AB}$ is unknown), and we can only state $Adv \blacktriangleleft_1 O^g(\triangleright)$. Our goal is to conservatively warn for guessing attacks, thus we do not use upper bounds to rule out attacks, although the approach could be extended.

## 3 Rules for Guessing

### 3.1 Outline of the Approach

We formalize guessing attacks done under two distinct circumstances:

1) $Adv$ knows the output of a function $f$ computed on the secret (and optionally, known additional inputs). Using an oracle $O^f(\cdot)$ for $f$, $Adv$ computes $f$ on all

possible secret values (with the known extra inputs), and verifies the guess comparing with the known output for the secret. Examples are: $Adv$ knows $H(s)$, i.e., the output of the function $f(x) = H(x)$ on $s$, or $Adv$ knows $m, MAC_s(m)$, i.e., the output of the $f(x, y) = MAC_x(y)$ on $s$ and a second known input $m$. To verify the guess, $f$ must be injective, otherwise more secrets can verify one output. In this case, we will formalize guessing using several outputs for the same secret (with different additional inputs).

2) $Adv$ knows one or more outputs of an invertible function $f$ computed on the secret and some additional, possibly unknown inputs. The adversary uses an oracle for the inverse of $f$ and computes the inputs to the known output(s) for each possible value of the secret. A guess is verified using a known property that identifies the correct input(s) to $f$. This may be: (1) a relation to a known value (a known part of the input to $f$), (2) a relation between different parts of the same input, or, (3) if there are several inputs, a relation between them. For (1), knowing $E_s(id_A.m)$, $Adv$ can guess $s$ by checking for the known value $id_A$ in the input (obtained by inverting the output, i.e. by decryption with all $s$). For (2), if $Adv$ knows $E_s(m, m)$, he can check for which value of $s$ the result of decryption (inversion) has identical halves (a relation between parts of the original input). For (3), knowing $E_s(H(m))$ and $E_s(m)$, $Adv$ inverts both outputs and checks if the two inputs are related by means of $H$.

This way of verifying the guess is valid only if the inverse of $f$ behaves as a pseudo-random function for a wrong value of $s$. Therefore, we allow this guessing rule only for encryption and decryption functions with keys dependent on $s$.

Our proposed approach works as follows: first, potential guessing opportunities are detected by matching them with one of the aforementioned situations, which are formalized in the guessing lemma of the next section. Next, oracle definition rules based on Dolev-Yao intruder capabilities are used to infer whether $Adv$ has access to the required oracles. If so, we warn that guessing is feasible.

## 3.2   The Guessing Lemma

**Definition 1.** Given $\sigma \in \{0, 1\}^k$, we call a function $f(\sigma, x)$ *distinguishing* in its first argument if there exists an algorithm $\mathcal{D}_f$, polynomial-time in $k$, that outputs a set $S = \{x_1, x_2, ..., x_{p(k)}\}$ such that the probability that there exists $s_1 \neq s_2$ such that $\forall x_i \in S . f(s_1, x_i) = f(s_2, x_i)$ is negligible, i.e., $Pr[f(s_1, x_i) = f(s_2, x_i), i = 1 .. p(k), s_1 \neq s_2] \leq v(k)$.

Here $p(k)$ is a polynomial in $k$ and $v(k)$ is a negligible function, i.e., $\forall c \geq 0$ there exists $k_c$ such that $\forall k \geq k_c . v(k) < k^{-c}$. For our calculus we use a notion that is more precise quantitatively, *strongly distinguishing* function *in q queries*:

**Definition 2.** Given $\sigma \in \{0, 1\}^k$, we call a function $f(\sigma, x)$ *strongly distinguishing* in the first argument after $q$ queries, if given any $q$ distinct queries $\{x_1, x_2, ..., x_q\}$, $\forall s_1 \neq s_2$ the probability that $f(s_1, x_i) = f(s_2, x_i)$ for all $i = 1 .. q$ is at most $2^{-k}$, i.e., $\forall s_1 \neq s_2 . Pr[f(s_1, x_i) = f(s_2, x_i), i = 1 .. q] \leq 2^{-k}$.

Note that any injective function with one argument is strongly distinguishing in one query, if we consider the second argument to be null.

The second definition allows us to give a quantitative bound on the number of attempts needed for guessing. Assuming we know $q$ outputs of $f(s_1, x)$ for the unknown secret $s_1$ and $x \in \{x_1, x_2, ..., x_q\}$, then after $q$ queries to $f(s, x)$ for each candidate value $s \in \{0, 1\}^k$, in average only the correct secret $s_1$ will match all known outputs of $f$. With our definition of *distinguishing* and *strongly distinguishing* functions we do not aim to guarantee uniqueness of the secret, but to give guessing a probabilistic meaning which addresses the average case.

**Example 1.** Let $\sharp s \sharp$ denote a term obtained by concatenating something to $s$ (to the left, right of $s$ or both). Then $E_{\sharp s \sharp}(\cdot), D_{\sharp s \sharp}(\cdot)$ are *distinguishing* and *strongly distinguishing* in one query assuming that encryption and decryption with different keys performed on the same value cannot yield the same result. Also, $H(\sharp s \sharp)$ is *distinguishing* and *strongly distinguishing* in one query if $H$ is collision-free on the argument range of $\sharp s \sharp$.

**Example 2.** Let $g(\sigma, x) = H(\sigma, x) \bmod 2^l$, $s \in \{0, 1\}^k$. If $H$ outputs more than $l$ bits, one query is not sufficient to distinguish the secret. After $q$ queries with $x \in \{x_1, x_2, ..., x_q\}$ we have $Pr[H(s_1, x) = H(s_2, x)] = 2^{-ql}$ for any $s_1 \neq s_2$. If the input space is $\{0, 1\}^k$, the average number of values for which collisions occur after $q$ queries is $2^{k-ql}$, therefore $g$ is *strongly distinguishing* in $k/l$ queries.

To express our guessing rules, we first formalize the ability of the adversary to find relations between oracle observations or subterms thereof.

**Definition 3.** Given a function $h$ and a list of terms $\alpha$, we say there is a relation under $h$ with arguments from $\alpha$, denoted $R(h, \alpha)$, if the adversary can establish an equality $h(\beta) = \gamma$ such that: i) $\beta, \gamma$ are terms constructed from the adversary knowledge and two disjoint subsets of terms from $\alpha$, with at least one subset non-empty; ii) $h(\beta)$ is injective in at least one input that comes from $\alpha$, with all other inputs kept constant.

This relation is used in the guessing lemma. Condition i) forces $Adv$ to validate a guess by using at least one term deduced after the guess, while condition ii) avoids trivial identities with terms that can result from a wrong guess.

**Lemma 1 (Guessing Lemma).** Let $s \in \{0, 1\}^k$ be a low-entropy value (i.e., $2^k$ computation steps are feasible), and $f$ an *strongly distinguishing* function in $q$ queries. The following guessing rules hold:

i) If $Adv \blacktriangleleft_{b_1} O^f(s, \blacktriangleright)$ and $Adv \operatorname{ctl}_{b_2} O^f(\cdot, \cdot)$, then $Adv$ can guess $s$ with $q$ observations of $O^f(s, \cdot)$ and $q \cdot 2^k$ queries to $O^f(\cdot, \cdot)$, i.e.

$$\frac{Adv \blacktriangleleft_{b_1} O^f(s, \blacktriangleright) \wedge Adv \operatorname{ctl}_{b_2} O^f(\cdot, \cdot)}{Adv \rightsquigarrow s} \qquad \begin{matrix} b_1 \geq q \\ b_2 \geq q \cdot 2^k \end{matrix} \qquad (2)$$

ii) If $Adv \blacktriangleleft_{b_1} O^{E_{f(s, \cdot)}}(\alpha)$, $Adv \operatorname{ctl}_{b_2} \{O^{D_{f(\cdot, \cdot)}}(\cdot), O^h(\cdot)\}$, and $R(h, \alpha)$, then $Adv$ can guess $s$ with $q$ observations of $O^{E_{f(s, \cdot)}}(\cdot)$ and $q 2^k$ queries to $O^{D_{f(\cdot, \cdot)}}(\cdot), O^h(\cdot)$.

$$\frac{Adv \blacktriangleleft_{b_1} O^{E_{f(s, \blacktriangleright)}}(\alpha) \wedge Adv \operatorname{ctl}_{b_2} \{O^{D_{f(\cdot, \cdot)}}(\cdot), O^h(\cdot)\} \wedge R(h, \alpha)}{Adv \rightsquigarrow s} \qquad \begin{matrix} b_1 \geq q \\ b_2 \geq q \cdot 2^k \end{matrix} \qquad (3)$$

iii) If $Adv \blacktriangleleft_{b_1} O^{E_{f(s, \blacktriangleright)}}(\alpha_i)$, with distinct $\alpha_i$, $Adv$ $ctl_{b_2}\{O^{D_{f(\cdot, \cdot)}}(\cdot), O^h(\cdot)\}$, and $R(h, \overline{\alpha})$, with $\overline{\alpha} = (\alpha_1, \dots, \alpha_n)$, then $Adv$ can guess $s$ with $q$ observations of $O^{E_{f(s, \cdot)}}(\cdot)$ and $q \cdot 2^k$ queries to $O^{D_{f(\cdot, \cdot)}}(\cdot)$, $O^h(\cdot)$.

$$\frac{Adv \blacktriangleleft_{b_1} O^{E_{f(s, \blacktriangleright)}}(\alpha_i) \wedge Adv\ ctl_{b_2}\{O^{D_{f(\cdot, \cdot)}}(\cdot), O^h(\cdot)\} \wedge R(h, \overline{\alpha})}{Adv \rightsquigarrow s} \quad \begin{array}{l} b_1 \geq q \\ b_2 \geq q \cdot 2^k \end{array} \quad (4)$$

**Proof sketch.** In case i) by Def. 2, for $q$ observations of $O^f(s, \cdot)$, in average only one $s$ verifies the input-output relation, so $b_1 \geq q$ suffices. Thus, $Adv$ can find $s$ by making queries to $O^f(\cdot, \cdot)$ with all $2^k$ values of $s$ and the $q$ observed inputs, then verify them against the $q$ observed outputs for $O^f(s, \cdot)$.

A sufficient bound on queries is $q \cdot 2^k$, however fewer queries are needed since each query reduces the number of candidates for $s$ in average by a factor of $2^{k/q}$.

Cases ii) and iii) are similar, but require the additional queries to $O^h(\cdot)$. If $R(h, \alpha)$ holds for the encryption input, this confirms the secret.

Case (i) is a direct match of the oracle output. Case (ii) matches the input and part of the decryption output, e.g., when $Adv$ knows $\{m, E_s(\sharp H(m)\sharp)\}$ or a relation between parts of the decrypted output, e.g., if $Adv$ knows $E_s(\sharp m \sharp H(m)\sharp)$. $Adv$ controls the decryption oracle and thus can check for $H(m)$ in the decryption result for all values of the secret. Case (iii) matches different inputs to the same oracle, e.g., when $Adv$ knows $\{E_s(\sharp m \sharp), E_s(\sharp H(m)\sharp\}$.

**Corollary 1.** In case i) of the Guessing Lemma, if $O^f(s, \cdot)$ has no random inputs, or $Adv$ controls the oracle $O^f(s, \cdot)$ (i.e., can choose all inputs, the $q$ observations become queries), then $Adv$ can do pre-computed dictionary guessing:

$$\frac{Adv \blacktriangleleft_{b_1} O^f(s, \blacktriangleright) \wedge Adv\ ctl_{b_2} O^f(\cdot, \cdot)}{Adv \rightsquigarrow_D s} \quad \frac{Adv\ ctl_{b_1} O^f(s, \cdot) \wedge Adv\ ctl_{b_2} O^f(\cdot, \cdot)}{Adv \rightsquigarrow_D s} \quad \begin{array}{l} b_1 \geq q \\ b_2 \geq q \cdot 2^k \end{array} \quad (5)$$

**Example 3.** Let $E$ be deterministic encryption and $H$ a hash function. Then, $E_k(\cdot), H(\cdot)$, and $E.(\alpha)$ are injective, and thus *strongly distinguishing* in one query. By Corollary 1:
$$\frac{Adv \blacktriangleleft_1 O^g(s) \wedge Adv\ ctl^{nb}\ O^g(\cdot)}{Adv \rightsquigarrow_D s}$$

$$\frac{Adv \blacktriangleleft_1 O^{E_k}(s) \wedge Adv\ ctl^{nb}\ O^{E_k}(\cdot)}{Adv \rightsquigarrow_D s} \quad \frac{Adv \blacktriangleleft_1 O^{E(\alpha)}(s) \wedge Adv\ ctl^{nb}\ O^{E(\alpha)}(\cdot)}{Adv \rightsquigarrow_D s} \quad (6)$$

We have used $ctl^{nb}$ which is weaker than $ctl^{off}$ but sufficient to verify a guess.

## 4   Case Studies

### 4.1   Anderson-Lomas Protocol

We focus on protocols which expose verifiers that fit more than one secret, a case not previously addressed using guessing rules. The Anderson-Lomas protocol [10] is relevant for the ingenuity in constructing password verifiers by using a *collisionful* hash function, i.e., a function $q(k, x)$ for which given $x$ one can find $k' \neq k$ such that $q(k, x) = q(k', x)$. The protocol description is as follows:

(1) $A \to B : g^{r_A}$

(2) $B \to A : g^{r_B}$

(3) $A \to B : H(MAC_{pw}g^{r_A r_B} \bmod 2^m, g^{r_A r_B})$

(4) $B \to A : H(MAC_{H(pw)}g^{r_A r_B} \bmod 2^m, g^{r_A r_B})$

Here, $g^{r_A r_B}$ is the regular key from the Diffie-Hellman-Merkle key exchange protocol and $pw$ is the password shared between $A$ and $B$ while $m$ is a fixed constant suggested to be $n/2$ if the size of the password space is $2^n$.

Let $Adv$ play the role of $B$ and consider the oracle $O^f(\cdot, \cdot)$, with $f(x, y) = H(MAC_x(y) \bmod 2^m, y)$. If the secret has $k$ bits then $f$ is *strongly distinguishing* in $k/m$ queries. By choosing $g^{r_A}$, $Adv$ can compute $g^{r_A r_B}$ and therefore knows both the input and output of $O^f$. Then, we have

$$\frac{Adv \blacktriangleleft_{b_1} O^f(s, \blacktriangleright) \wedge Adv \ \text{ctl}_{b_2} O^f(\cdot, \cdot)}{Adv \rightsquigarrow pw} \qquad \begin{array}{l} b_1 \geq \frac{k}{m} \\ b_2 \geq \frac{k}{m} \cdot 2^k \end{array} \qquad (7)$$

according to case (i) of the guessing lemma which allows us to formalize the attack originally presented in [10] and explain it using a general guessing rule.

## 4.2 The Norwegian ATM

With our calculus we formalize attacks in a Norwegian ATM system, shown to be flawed in [11]. The system attempts to increase password security by hiding the verifier. Cards store the PIN encrypted with a bank key $BK$, truncated to 16 bits: $\lfloor DES_{BK}(PIN) \rfloor_{16}$ (simplified, since the PIN is not encrypted directly).

To find the PIN of a stolen card, $Adv$ cannot guess the PIN off-line without $BK$, since for each PIN about $2^{40}$ of $2^{56}$ DES keys match. However, [11] presents a more subtle attack. $Adv$ gets *several honest* cards from the same bank. Each known PIN reduces the number of candidate keys by a factor of $2^{16}$. On average, 4 honest cards suffice to find $BK$, and then guess the PIN of the stolen card.

The services provided by the bank and ATM to a user are summarized below:

*Card issuing stage:* $\quad Bank \to User : \lfloor DES_{BK}(PIN) \rfloor_{16}, PIN$

*PIN change procedure:* $User \to ATM : \lfloor DES_{BK}(PIN_{old}) \rfloor_{16}, PIN_{old}, PIN_{new}$

$\qquad\qquad\qquad\quad ATM \to User : \lfloor DES_{BK}(PIN_{new}) \rfloor_{16}$

We assume PIN and card (holding $\lfloor DES_{BK}(PIN) \rfloor_{16}$) are issued securely, otherwise a Dolev-Yao adversary could get the PIN directly from the protocol.

Since $\log_2 |PIN| < 16$, $\lfloor DES_{BK}(\cdot) \rfloor_{16}$ is *strongly distinguishing* in one query and $Adv$ can guess the PIN using the PIN change procedure as oracle:

$$\frac{Adv \text{ knows } \lfloor DES_{BK}(PIN) \rfloor_{16}}{Adv \blacktriangleleft_1 O^{\lfloor DES_{BK}(\cdot) \rfloor_{16}}(PIN)} \quad \frac{Adv \to PIN_{new} \quad ATM \to \lfloor DES_{BK}(PIN_{new}) \rfloor_{16}}{Adv \ \text{ctl}^{nb} O^{\lfloor DES_{BK}(\cdot) \rfloor_{16}}(\cdot)}$$
$$\overline{\qquad\qquad\qquad\qquad\qquad Adv \rightsquigarrow_D PIN \qquad\qquad\qquad\qquad\qquad} \qquad (8)$$

In reality, the PIN is encrypted concatenated with a card-specific value $CV$. Thus, changing the PIN no longer controls the encryption oracle. To find the PIN, $Adv$ must simulate the oracle himself, and for this, $BK$ must be known:

$$\frac{Adv \text{ knows } \lfloor DES_{BK}(PIN.CV) \rfloor_{16}}{Adv \blacktriangleleft_1 O^{\lfloor DES_{BK}(\cdot) \rfloor_{16}}(PIN.CV)} \quad \frac{Adv \text{ knows } BK}{Adv \ \text{ctl}^{off} O^{\lfloor DES_{BK}(\cdot) \rfloor_{16}}}$$
$$\overline{\qquad\qquad\qquad\qquad\qquad Adv \rightsquigarrow PIN \qquad\qquad\qquad\qquad\qquad} \qquad (9)$$

For this goal, *Adv* needs to *observe* an oracle output on *BK*. One possibility is in the card issuing stage. Let $f(\sigma, x) = \lfloor DES_\sigma(x) \rfloor_{16}$. Then, $f$ is *strongly distinguishing* in 4 queries since the *DES* key has 56 bits, and we have:

$$\frac{\dfrac{Adv \text{ knows } PIN.CV_{1..4}, \lfloor DES_{BK}(PIN.CV_{1..4}) \rfloor_{16}}{Adv \blacktriangleleft_4 O^f(BK, \cdot)} \qquad Adv \text{ ctl}^{\text{off}} O^f(\cdot, \cdot)}{Adv \rightsquigarrow BK} \qquad (10)$$

Another option comes again from controlling the PIN change procedure. Let $g(\sigma, x) = \lfloor DES_\sigma(CV.x) \rfloor_{16}$ for a card of the adversary with value *CV*. Then,

$$\frac{Adv \text{ ctl}^{nb} O^g(BK, \cdot) \qquad Adv \text{ ctl}^{\text{off}} O^g(\cdot, \cdot)}{Adv \rightsquigarrow BK} \qquad (11)$$

This attack fits the real-world situation where the adversary can change his own PIN and is new to the best of our knowledge. In [11], only the attack using several cards from the bank (to guess all DES key bits) is given. Moreover, our calculus distinguishes two ways to find *BK*. The attacks illustrate both dictionary and and pre-computed dictionary guessing.

## 5   Conclusions

We have introduced a calculus based on oracles for detecting guessing attacks, with rules that supplement the deductions of a Dolev-Yao intruder. The rules are based on *observes* and *controls* relations between the adversary and oracles. Conceptually separating generating the verifier from verifying the guess justifies consideration of on-line guessing attacks which may not be detected and blocked. The calculus can be used in a mixed on-line/off-line setting. Our guessing rules also handle protocols with verifiers that match more than one secret. In this case, guessing has a probabilistic meaning, and our rules give sufficient bounds on the number of observations for successful attacks. We formalize the known flaws in the Anderson-Lomas and Norwegian ATM protocols in this framework. For the ATM protocol, our calculus finds new attacks based on a PIN change procedure and on the use of the ATM as encryption oracle.

**Acknowledgments.** Thanks to Cas Cremers who helped clarify a first writeup of our approach and to the anonymous reviewers for their valuable comments.

## References

1. Ding, Y., Horster, P.: Undetectable on-line password guessing attacks. Operating Systems Review 29(4), 77–86 (1995)
2. Lowe, G.: Analysing protocols subject to guessing attacks. Journal of Computer Security 12(1), 83–98 (2004)
3. Corin, R., Malladi, S., Alves-Foss, J., Etalle, S.: Guess what? Here is a new tool that finds some new guessing attacks. In: Proc. Workshop on Issues in the Theory of Security, pp. 62–71 (2003)

4. Delaune, S., Jacquemard, F.: A theory of dictionary attacks and its complexity. In: Proc. 17th IEEE Computer Security Foundations Workshop, pp. 2–15 (2004)
5. Drielsma, P.H., Mödersheim, S., Viganò, L.: A formalization of off-line guessing for security protocol analysis. In: Baader, F., Voronkov, A. (eds.) LPAR 2004. LNCS (LNAI), vol. 3452, pp. 363–379. Springer, Heidelberg (2005)
6. Corin, R., Doumen, J.M., Etalle, S.: Analysing password protocol security against off-line dictionary attacks. In: Proc. 2nd Int'l. Workshop on Security Issues with Petri Nets and other Computational Models (WISP), pp. 47–63 (2004)
7. Abadi, M., Baudet, M., Warinschi, B.: Guessing attacks and the computational soundness of static equivalence. In: Aceto, L., Ingólfsdóttir, A. (eds.) FOSSACS 2006. LNCS, vol. 3921, pp. 398–412. Springer, Heidelberg (2006)
8. Baudet, M.: Deciding security of protocols against off-line guessing attacks. In: Proc. 12th ACM Conf. on Computer and Communications Security, pp. 16–25 (2005)
9. Blanchet, B.: An Efficient Cryptographic Protocol Verifier Based on Prolog Rules. In: 14th IEEE Computer Security Foundations Workshop, pp. 82–96 (2001)
10. Anderson, R.J., Lomas, T.M.A.: Fortifying key negotiation schemes with poorly chosen passwords. Electronics Letters 30(13), 1040–1041 (1994)
11. Hole, K.J., Moen, V., Klingsheim, A.N., Tande, K.M.: Lessons from the Norwegian ATM system. IEEE Security and Privacy 5(6), 25–31 (2007)

# Structural Attacks on Two SHA-3 Candidates: Blender-$n$ and DCH-$n$

Mario Lamberger and Florian Mendel

Institute for Applied Information Processing and Communications (IAIK),
Graz University of Technology, Inffeldgasse 16a, A-8010 Graz, Austria
mario.lamberger@iaik.tugraz.at

**Abstract.** The recently started SHA-3 competition in order to find a new secure hash standard and thus a replacement for SHA-1/SHA-2 has attracted a lot of interest in the academic world as well as in industry. There are 51 round one candidates building on sometimes very different principles.

In this paper, we show how to attack two of the 51 round one hash functions. The attacks have in common that they exploit structural weaknesses in the design of the hash function and are independent of the underlying compression function. First, we present a preimage attack on the hash function Blender-$n$. It has a complexity of about $n \cdot 2^{n/2}$ and negligible memory requirements. Secondly, we show practical collision and preimage attacks on DCH-$n$. To be more precise, we can trivially construct a $(2^8 + 2)$-block collision for DCH-$n$ and a 1297-block preimage with only 521 compression function evaluations. The attacks on both hash functions work for all output sizes and render the hash functions broken.

**Keywords:** Hash functions, collision attacks, preimage attacks, SHA-3, Blender, DCH.

## 1 Introduction

Until 2005, the number of papers on the cryptanalysis of hash functions was quite easy to overlook. This changed significantly with the dawn of the work of Wang *et al.* [1,2]. The weaknesses discovered in MD5 and SHA-1 had wide reaching consequences and were a wake-up call for both academia and industry. The SHA-2 family [3] was only considered to be a temporary solution. Although no full attacks on a member of SHA-2 have been found to date, the fact that the design and security principles are very close to those of SHA-1 raised doubts about the long term security of the SHA-2 family.

As a consequence, the National Institute of Standards and Technology (NIST) has launched a similar competition [4] as it has done for the Advanced Encryption Standard (AES) to replace DES. This time, the goal is to find a new secure hash standard SHA-3. As of now, 51 submissions have advanced to the first round of the SHA-3 competition. Supported by the cryptographic community, the task

P. Samarati et al. (Eds.): ISC 2009, LNCS 5735, pp. 68–78, 2009.

of NIST is now to find the best hash function in terms of a wide spectrum of requirements, such as speed and security.

The proposals of round one are based on a great variety of security considerations and design principles. Many of them are block cipher based, using especially AES or parts of AES as building blocks. Some hash functions are based on asymmetric primitives and again others are mere curiosities. From the design perspective, there are Merkle-Damgård [5,6] constructions and variations thereof, sponge constructions [7], HAIFA constructions [8], wide pipe constructions [9], etc.

In this paper, we want to demonstrate vulnerabilities of two designs, namely the hash functions Blender-$n$ [10] and DCH-$n$ [11]. Although they are both based on quite different design principles, they have in common that an attacker can omit the tedious task of going into the details of the respective round transformations. We have identified weaknesses in both design principles.

We present a structural preimage attack on the hash function Blender-$n$ that has a complexity of about $n \cdot 2^{n/2}$ and negligible memory requirements. Furthermore, we show practical collision and preimage attacks on DCH-$n$ and show that we can trivially construct a $2^8 + 2$-block (multi)-collision for DCH-$n$ and 1297-block preimages with only 521 compression function evaluations.

Our paper underlines the importance of a well founded design principle for a hash function since a bad choice of the iteration mode renders the efforts put in the compression function design ineffectual.

The paper is organized as follows. Section 2 describes our preimage attack on Blender-$n$. Then, Section 3 demonstrates a practical collision and a practical (second) preimage attack on DCH-$n$. In Section 4 we conclude.

## 2   A Preimage Attack on Blender-$n$

In this section, we present a preimage attack on Blender-$n$ with a complexity of about $n \cdot 2^{n/2}$ and negligible memory requirements. The attack is independent of the compression function of Blender-$n$ and works for all output sizes. A very similar preimage attack for Blender-$n$ was independently proposed in [12]. It has a slightly higher attack complexity of about $n \cdot 2^{(n+w)/2}$, where $w$ is 32 or 64 bits depending on the word size of the hash function. We are well aware of the attacks on Blender-$n$ which concentrate on the internal structure of the compression function presented in [13,14,15]. Even though our attack is less efficient compared to the attack [14], it is superior in the sense that it isn't affected by any tweak to the compression function. Furthermore, due to the generic nature of our attack, it may be applicable to a wider range of hash function designs. For example, the SHA-3 candidate AURORA has been recently broken by similar principles [16,17,18].

### 2.1   Description of Blender-$n$

The hash function Blender-$n$ is an iterated hash function. It processes message blocks of 32 (or 64) bits and produces a hash value of 224, 256 (or 384, 512)

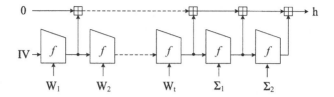

**Fig. 1.** Structure of the hash function Blender-$n$

bits. If the message length is not a multiple of 32 (or 64) bits, an unambiguous padding method is applied. For the description of the padding method we refer to [10]. Let $W = W_1 \| W_2 \| \cdots \| W_t$ be a t-block message (after padding).

In the following $\neg$ denotes the bitwise complement and $\Sigma$ denotes summation modulo $2^w$ where $w$ is the wordsize (32 or 64-bit). The hash value $h$ is computed from the chaining values $A_i$ as follows (see Figure 1):

$$h = \Sigma_{i=1}^{t+2} A_i .$$

The chaining values $A_i$ are computed as follows:

$$A_0 = IV \tag{1}$$
$$A_i = f(A_{i-1}, W_i) \quad \text{for } 0 < i \le t \tag{2}$$
$$A_{t+1} = f(A_t, \Sigma_1) \tag{3}$$
$$A_{t+2} = f(A_{t+1}, \Sigma_2) , \tag{4}$$

where $\Sigma_1 = \neg \Sigma_{i=1}^t W_i$, $\Sigma_2 = \Sigma_{i=1}^t \neg W_i$ and $IV$ is a predefined initial value.

As can be seen in (3) and (4), Blender-$n$ specifies two checksums ($\Sigma_1$ and $\Sigma_2$) consisting of the modular addition of all message blocks, which are then input to the two final application of the compression function $f$. Computing this checksum is not part of most commonly used hash functions such as MD5 and SHA-1.

The compression function $f$ basically consist of 4 steps:

1. Compute the preliminary intermediate values using add-with-carry.
2. Compute the rotation factor $r$.
3. Rotate the intermediate values.
4. Compute the next state $A_i$.

For a detailed description of the Blender-$n$ compression function we refer to [10], since we do not need it for our analysis.

## 2.2   A Preimage Attack on Blender-$n$

In this section, we present a preimage attack on the hash function Blender-$n$. It has a complexity of about $n \cdot 2^{n/2}$ and negligible memory requirements. It is based on the following two observations.

**Observation 1.** *The checksums $\Sigma_1$ and $\Sigma_2$ are strongly related.*

In other words, the second checksum does not increase the security of Blender-$n$. This will be very useful for our attack. Let $X = \Sigma_{i=1}^{t} W_i$ then:

$$\Sigma_1 = \neg \Sigma_{i=1}^{t} W_i = \neg X$$
$$\Sigma_2 = \Sigma_{i=1}^{t} \neg W_i = \Sigma_{i=1}^{t}(-W_i - 1) = -t - \Sigma_{i=1}^{t} W_i = -t - X$$

Note that $-W_i = \neg W_i + 1$ and hence $\neg W_i = -W_i - 1$.

**Observation 2.** *The final hash value $h$ of Blender-$n$ is computed from the chaining values $A_i$ by modular additions.*

In other words, the computation of $h$ is invertible. This will be very useful for our attack. Assume, that we can find $2^n$ messages $w^*$ (and hence chaining values $A_i^*$ for $0 < i \leq t$), such that all produce the same value $A_t$ and $X$, then we have constructed a preimage for $h$. This is similar to recent attacks on GOST [19] and Damgård-Merkle hash functions with linear or additive checksums [20].

Based on this short description, we will now show how to find messages $w^*$ which all produce the same value $A_t$ and lead to the same checksum value with a complexity of about $n \cdot 2^{n/2}$ and negligible memory requirements. For the sake of simplicity let $n = 512$ for the remainder of this section. Note that the attack works similar for the other output sizes of Blender-$n$.

Assume we want to construct a preimage for Blender-512 consisting of 2561 message blocks, *i.e.* $m = W_1 \| W_2 \| \cdots \| W_{2561}$. The attack basically consists of two steps and uses multicollisions. It can be summarized as follows.

**STEP 1: Constructing the multicollision.** A multicollision is a set of messages of equal length that all lead to the same hash value. As shown in [21], constructing a $2^t$ collision, *i.e.* $2^t$ messages consisting of $t$ message blocks which all lead to the same chaining value, can be done with a complexity of about $t \cdot 2^{n/2}$ for any iterated hash function.

In the attack we want to construct a $2^{512}$ collision for the iterative part (chaining values), to get $2^{512}$ messages $w^*$ (and hence chaining values $A_i^*$) leading to the same value $A_t$ and $X$. This has a complexity of about $512 \cdot 2^{288} = 2^{297}$.

However, in the case of Blender-$n$ constructing a multicollision is slightly more complicated. First, due to the small size of the message blocks (64 bits) we need several blocks to construct a collision in the chaining values. Second, to ensure that $\Sigma_1$ and $\Sigma_2$ (respectively, $X = \Sigma_{i=1}^{k} W_i$) are equal we need one additional block. In detail, by using 5 message blocks we can construct a collision in the iterative part (chaining values) and the checksums. Since for Blender-512 the chaining value has 512 bits and $X$ has 64 bits, this has a complexity of about $2^{288}$ using a generic birthday attack.

However, due to the simple structure of the checksum value $X$, we can easily guarantee that $X$ collides by choosing the message blocks carefully in the attack. It can be summarized as follows:

1. Choose an arbitrary value for $d$.
2. For all $2^{4 \cdot 64} = 2^{256}$ choices of $W_i, \ldots, W_{i+3}$ adjust $W_{i+4}$ accordingly such that $\Sigma_{j=i}^{i+4} W_j = d$ is fulfilled and compute $A_{i+4}$ with $i > 0$.
3. After computing all $2^{256}$ candidates for $A_{i+4}$ we expect to find a collision due to the birthday paradox.

In other words, we can find a collision for the iterative part (chaining values) and $X$ with a complexity of about $2^{256}$ instead of $2^{288}$. Furthermore, the memory requirements can be significantly reduced by applying a memory-less variant of the birthday attack [22].

Hence, we can construct a $2^{512}$ collision with a complexity of about $512 \cdot 2^{256} = 2^{265}$ and negligible memory requirements.

**STEP 2: Constructing the preimage for $h$.** In the previous step we constructed a $2^{512}$ collision in the first $5 \cdot 512 = 2560$ iterations of the hash function. Hence, we have $2^{512}$ messages $w^*$ leading to the same chaining value $A_{2560}$ and to a collision in $X$ (and hence in the two checksums $\Sigma_1$ and $\Sigma_2$).

Next we append an additional message block $W_{2561}$ to $w^*$ such that the padding of each of the messages $m^* = w^* \| W_{2561}$ is correct. It is easy to see that appending one message block has no effect on the multicollision in the iterative part and the checksums.

From this set of $2^{512}$ messages $m^*$ that all lead to the same chaining value $A_{2560}$ and $X$, we now have to find a message $m^*$ having $h$ as hash value. We write $h = h^* + A_{2561} + A_{2562} + A_{2563}$ where $h^*$ is one of the $2^{512}$ values:

$$h^* = \Sigma_{i=1}^{512} (A_{5i-4}^{r_i} + A_{5i-3}^{r_i} + \cdots + A_{5i}^{r_i}),$$

with $r_i \in \{0, 1\}$. Here, $(A_{5i-4}^0, A_{5i-3}^0, \ldots, A_{5i}^0)$ and $(A_{5i-4}^1, A_{5i-3}^1, \ldots, A_{5i}^1)$ are the corresponding 5-block chaining values constituting the multicollision. To find the correct $h^*$ and hence the message leading to the preimage of $h$ we make use of a meet-in-the-middle attack.

First, we save all values for

$$S_1 = \Sigma_{i=1}^{256} (A_{5i-4}^{r_i} + A_{5i-3}^{r_i} + \cdots + A_{5i}^{r_i})$$

in the list $L$. Note that we have in total $2^{256}$ values for $S_1$ in $L$. Second, we compute

$$S_2 = \Sigma_{i=257}^{512} (A_{5i-4}^{r_i} + A_{5i-3}^{r_i} + \cdots + A_{5i}^{r_i})$$

and check if $h^* - S_2$ is in the list $L$. After testing all $2^{256}$ values for $S_2$, we expect to find a matching entry in the list $L$ and hence a message $w^*$ that leads to $h^* = S_1 + S_2$. This step of the attack has a complexity of $2^{256}$ and memory requirements of $2^{256}$. Once we have found $w^*$, we have found a preimage for Blender-512 consisting of 2560+1 message blocks, namely $m^* = w^* \| W_{2561}$. Note that the memory requirements of the attack can significantly be reduced by applying a memory-less variant of the meet-in-the-middle attack introduced by Quisquater and Delescaille in [22].

Hence, a preimage can be constructed for Blender-512 with a complexity of $2^{265}$ and negligible memory requirements. Note that in a similar way one can construct preimages for all output sizes of Blender-$n$ with a complexity of about $n \cdot 2^{n/2}$.

# 3   Practical Collision and Preimage Attacks on DCH-$n$

## 3.1   Description of DCH-$n$

The hash function DCH-$n$ [11], proposed by Wilson, is an iterated hash function based on the Merkle-Damgård design principle and produces a hash value of $n = 224, 256, 384$ or $512$ bits. It processes message blocks of 504 bits and preprocesses the input blocks by adding 8 bits of dithering input. At the end, standard MD strengthening is applied.

In each iteration the compression function $f$ is used to update the chaining value of 512 bits as follows:

$$H_{i+1} = f(H_i, M_i) = H_i \oplus M_i \oplus g(M_i) , \qquad (5)$$

where $g(M)$ is some non-linear transformation. The author of DCH-$n$ claims that the hash function makes use of the Miyaguchi-Preneel mode of operation for block cipher based hash constructions [23]. Nevertheless, a quick look at equation (5) shows that the chaining value $H_i$ is not introduced to the non-linear function $g$. This fact will be exploited by our attack. For a detailed description of DCH-$n$ we refer to [11].

## 3.2   Cryptanalysis

In this section, we will present our collision and preimage attack on DCH-$n$. The attack is an extension of the attack of Khovratovich and Nikolic [24] and is based on similar principles as the attacks on SMASH [25].

A 512-bit block $M_i$ in iteration $i$ consists of $m_i \| M_i'$, where $m_i$ is the 8-bit dithering input and $M_i'$ is the original message block. The 8-bit dithering $m_i$ consists of two parts. The 5 least significant bits are a simple counter, that increments with every iteration. The 3 most significant bits are determined by an encoding of the optimal moves in the "Towers of Hanoi"-sequence, where a new step is generated whenever the 5-bit counter is reset. This sequence is a square-free sequence and therefore assumed to be a good choice for dithering. For closer details, we refer to [11]. At this point, we also want to refer to the work of Andreeva *et al.* [26] that studies the limits on dithering based designs in general.

Let us introduce the function $\gamma_i(M_i') := g(m_i \| M_i') \oplus (m_i \| M_i')$, that is, we combine the XOR from the definition and the dithering corresponding to block $i$ into one function. Then (5) can be rewritten as:

$$H_{i+1} = H_0 \oplus \gamma_0(M_0') \oplus \gamma_1(M_1') \oplus \cdots \oplus \gamma_i(M_i') \qquad (6)$$

**Collision Attack.** We now describe the collision attack. We start with a message consisting of $N + 1$ message blocks, $m = M_0 \| M_1 \| \dots \| M_N$. Each $M_i = m_i \| M_i'$, where $m_i$ is computed according to the dithering rule, for $i = 0, \dots, N$. For an 8-bit dithering, there are only $2^8$ possible dithering blocks, so if $N$ is large enough, there must be $0 \leq i, j \leq 2^8$ with $i \neq j$ such that $m_i = m_j$ and thus, $\gamma_i = \gamma_j$. Based on (6), we have

$$H_{N+1} = H_0 \oplus \gamma_0(M_0') \oplus \gamma_1(M_1') \oplus \cdots \oplus \gamma_N(M_N').$$

So setting $M_i' = M_j' = a \in \{0,1\}^{504}$ for the above $i \neq j$ implies that these blocks don't contribute to the value $H_{N+1}$ and can thus be freely chosen.

Without looking on the dithering rule for DCH-$n$, we simply could set $N = 2^8$ to get colliding messages having $2^8 + 2$ blocks (one final block for the padding). Note, that since the "Towers of Hanoi"-sequence only has 6 valid states, a smaller choice for $N$ would be $N = 6 \cdot 2^5 = 192$. The bottom line is that we can trivially construct collisions for DCH-$n$, independently of the concrete dithering method. The messages in the colliding message pair consist of $2^8 + 2$ message blocks.

Every choice of $a \in \{0,1\}^{504}$ leads to a collision. Hence, we can trivially construct $t$-collisions (for $0 < t < 2^{504}$) for DCH-$n$. Note that these attacks apply to DCH-$n$ for all output sizes. Due to size considerations, we don't include an actual colliding message pair.

**Preimage Attack.** The core observation for the preimage attack is that the outputs of DCH-$n$ form a vector space of dimension $n$ over $GF(2)$. This can be easily seen when looking at the alternative description of DCH-$n$ in (6). A similar approach was used in the attack on the hash function family SMASH-$n$ in [25]. Therefore, the task is to compute a basis of the vector space generated by the DCH-$n$ outputs in order to construct preimages for DCH-$n$. Again, the only technicality we have to take care of is the dithering of the message blocks.

In the following we assume $n = 512$ since the other output lengths of DCH-$n$ result from truncations of DCH-512. To describe our preimage attack, we will use he following two technical lemmas. As in the collision case we will need to find different indices $(i, j)$ for which the dithering blocks $m_i$ and $m_j$, and thus, $\gamma_i$ and $\gamma_j$, are the same. For the collision attack we needed only one such index pair whereas for the preimage case this won't suffice. The first lemma will tell us how many message blocks our preimage needs to have to guarantee a certain number of such index pairs.

**Lemma 1.** *For a message having $N = 2 \cdot \ell + 2^8$ or more message blocks, we can be certain to have at least $\ell$ index pairs $(i_0, j_0), \dots, (i_{\ell-1}, j_{\ell-1})$ that satisfy $\gamma_{i_k} = \gamma_{j_k}$ for all $k$ and where all occurring indices are unique.*

*Proof.* We need to guarantee that among all indices from $0, \dots, N - 1$ we can find $\ell$ pairs as described above. If we take a look at the 8-bit dithering strings $m_i$ for $i = 0, \dots, N - 1$ we know, that the 3 non-counter bits can only have 8 different values $0, 1, \dots 7$ (actually 6 for the "Towers of Hanoi"-sequence). Let $n_0, \dots, n_7$ denote the frequencies with which the values $0, \dots, 7$ occur in the non-counter part of the first $N$ dithering messages. To every non-counter block, there correspond $2^5$ counter blocks. Thus, $N = 32 \cdot \sum_{i=0}^{7} n_i$. From this, the number of sought pairs $(i_k, j_k)$ is

$$32 \cdot \sum_{i=0}^{7} \left\lfloor \frac{n_i}{2} \right\rfloor = 32 \cdot \sum_{i=0}^{7} \left( \frac{n_i}{2} - \left\{ \frac{n_i}{2} \right\} \right) \geq \frac{N}{2} - 2^7.$$

Therefore, $N = 2 \cdot \ell + 2^8$ is a valid choice of $N$.       □

The second lemma is concerned with the probability that random vectors from $GF(2)^n$ contain a basis and is a well known result (*cf.* [27]).

**Lemma 2.** *The probability for $\ell \geq n$ vectors drawn uniformly at random from $GF(2)^n$, to span a space of dimension $n$ is*

$$\prod_{i=0}^{n-1} \frac{2^{\ell} - 2^i}{2^{\ell}} = \prod_{i=0}^{n-1} (1 - 2^{i-\ell}).$$

Now, the attack can be summarized as follows:

1. Assume we want to construct a preimage for $h$ consisting of $N + 1$ message blocks. Thus, we have to find a message $M$ such that:

$$h = H_0 \oplus \bigoplus_{i=0}^{N} \gamma_i(M_i) \, .$$

2. We choose the last message block $M_N$ such that the padding is correct.
3. Once we have fixed the last message block, we have to find the remaining message blocks $M_i'$ for $0 \leq i < N$ such that:

$$\bigoplus_{i=0}^{N-1} \gamma_i(M_i') = h \oplus H_0 \oplus \gamma_N(M_N') \tag{7}$$

4. According to Lemma 1 we choose $N = 2 \cdot \ell + 2^8$ in order to have $\ell \geq 512$ index pairs $(i_0, j_0), \ldots, (i_{\ell-1}, j_{\ell-1})$ satisfying $\gamma_{i_k} = \gamma_{j_k}$ (where every $i_k, j_k$ is unique).
5. Next, we compute $\ell$ vectors $a^k = \gamma_{i_k}(M_0^{k'}) \oplus \gamma_{j_k}(M_1^{k'})$ for $k = 0, \ldots, \ell - 1$ with random $M_0^{k'}$ and $M_1^{k'}$ and save the triples $(a^k, M_0^{k'}, M_1^{k'})$ in a list $L$.
6. From the set of $\ell \geq 512$ vectors $a^k$ we try to compute a basis of the output vector space of DCH-$n$. If we succeed, this means that we can basically construct such a basis with a complexity of $2 \cdot \ell$ compression function evaluations. This can be reduced to $\ell + 1$ evaluations of the compression function by fixing the block $M_0^{k'}$ and letting only the block $M_1^{k'}$ vary when generating the vectors $a^k$ in the previous step.

   Lemma 2 implies that for a choice of $\ell = 520$ we already have a probability of 0.9961 for finding a basis among the $a^k$ and thus need 521 compression function evaluations. Note, that constructing the basis is a one time effort.

Let now $\mathcal{B} = \{a^{k_0}, \ldots, a^{k_{511}}\}$ denote the basis for the output vector space and let $\mathcal{I} = \bigcup_{k=0}^{511} i_k \cup j_k$ be the union of all the indices contributing to these basis vectors. (For simplicity we assume that the first $n$ pairs correspond to the basis vectors.)

7. We divide the indices $\mathcal{N} = \{0, \ldots, N-1\}$ into $\mathcal{I}$ and $\mathcal{N} \setminus \mathcal{I}$. For every index $i$ in $\mathcal{N} \setminus \mathcal{I}$ we set $M_i' = 0 \ldots 0$. These message blocks correspond to the indices not contributing to the basis. From (7) we thus get

$$\bigoplus_{\mathcal{I}} \gamma_i(M_i') = h \oplus H_0 \oplus \gamma_N(M_N') \bigoplus_{\mathcal{N} \setminus \mathcal{I}} \gamma_i(0 \ldots 0).$$

Once a basis $\mathcal{B}$ and the indices $\mathcal{I}$ are computed, the right side of the equation is completely known and thus we have

$$\bigoplus_{\mathcal{I}} \gamma_i(M_i') = c$$

8. An arbitrary $c$ can be represented with respect to this basis $c = x_0 a^{k_0} + \cdots + x_{511} a^{k_{511}}$ by solving the linear system over $GF(2)$. Now we choose the blocks $M_i'$ for $i \in \mathcal{I}$ as follows:
   - If $x_k = 0$ for $0 \leq k < n$, we set $M_{i_k}' = \alpha$ and $M_{j_k}' = \alpha$ for some arbitrary value of $\alpha \in \{0,1\}^{504}$ (as in the collision attack). In this case, $\gamma_{i_k}$ and $\gamma_{j_k}$ are equal, these two values cancel out and don't contribute to the result.
   - If $x_k = 1$ for $0 \leq k < n$, we set $M_{i_k}' = M_0^{k'}$ and $M_{j_k}' = M_1^{k'}$ such that $\gamma_{i_k}(M_0^{k'}) \oplus \gamma_{j_k}(M_1^{k'}) = a^k$ for $0 \leq k < n$.

Hence we can construct a preimage by solving a linear system of equations of dimension $512 \times 512$ over $GF(2)$. Constructing the basis has a complexity of $\ell + 1$ compression function evaluations and is a one time effort.

Furthermore, the preimage attack can be used to construct second preimages for DCH-$n$ with the same complexity. Note that by using the above described method, preimages (or second preimages) always consist of $N + 1 = 2\ell + 2^8 + 1$ message blocks.

## 4   Conclusion

In this paper, we were investigating two round one candidates of the SHA-3 hash function competition of NIST. Namely, we were interested in a cryptanalysis of DCH-$n$ and Blender-$n$ by solely investigating the iteration mode.

We showed a preimage attack on the hash function Blender-$n$ for all output sizes. The attack has a complexity of about $n \cdot 2^{n/2}$ compression function evaluations and negligible memory requirements. It is based on structural weaknesses in the design of the hash function and is independent of the compression function $f$. Furthermore, we also presented that it is trivial to construct collisions and (second) preimages for DCH-$n$. The presented attack applies to all similar constructions not introducing the chaining variable into the compression function.

We want to emphasize once more that the main target of the underlying paper was to identify weak design philosophies of hash functions and to learn our lessons from the attacks. It has to be noted that the vulnerabilities pinpointed in this paper are not isolated cases. Our attack on Blender-$n$ has quite some resemblance to the attack on the Russian hash function standard GOST [19] and the recent attack on the SHA-3 candidate AURORA [16,17,18]. The attacks on DCH-$n$ are relying on similar principles as the attacks on the hash function SMASH [25].

## Acknowledgements

The authors wish to thank David Wilson and the anonymous referees for useful comments and discussions. The work in this paper has been supported in part by the European Commission under contract ICT-2007-216646 (ECRYPT II) and in part by the IAP Programme P6/26 BCRYPT of the Belgian State (Belgian Science Policy).

## References

1. Wang, X., Yin, Y.L., Yu, H.: Finding Collisions in the Full SHA-1. In: Shoup, V. (ed.) CRYPTO 2005. LNCS, vol. 3621, pp. 17–36. Springer, Heidelberg (2005)
2. Wang, X., Yu, H.: How to Break MD5 and Other Hash Functions. In: Cramer, R. (ed.) EUROCRYPT 2005. LNCS, vol. 3494, pp. 19–35. Springer, Heidelberg (2005)
3. National Institute of Standards and Technology: FIPS 180-3, Secure Hash Standard, Federal Information Processing Standard (FIPS), Publication 180-3. Federal Information Processing Standard (October 2008),
   http://csrc.nist.gov/publications/PubsFIPS.html
4. National Institute of Standards and Technology: Announcing Request for Candidate Algorithm Nominations for a New Cryptographic Hash Algorithm (SHA-3) Family. Federal Register Notice (November 2007), http://csrc.nist.gov
5. Damgård, I.: A design principle for hash functions. In: Brassard, G. (ed.) CRYPTO 1989. LNCS, vol. 435, pp. 416–427. Springer, Heidelberg (1990)
6. Merkle, R.C.: One way hash functions and DES. In: Brassard, G. (ed.) CRYPTO 1989. LNCS, vol. 435, pp. 428–446. Springer, Heidelberg (1990)
7. Bertoni, G., Daemen, J., Assche, G.V., Peeters, M.: Sponge Functions. In: ECRYPT Hash Workshop 2007, Barcelona, May 24-25 (2007)
8. Biham, E., Dunkelman, O.: A Framework for Iterative Hash Functions - HAIFA. Cryptology ePrint Archive, Report 2007/278 (2007), http://eprint.iacr.org
9. Lucks, S.: A Failure-Friendly Design Principle for Hash Functions. In: Roy, B. (ed.) ASIACRYPT 2005. LNCS, vol. 3788, pp. 474–494. Springer, Heidelberg (2005)
10. Bradbury, C.: BLENDER: A Proposed New Family of Cryptographic Hash Algorithms. Submission to NIST (2008),
    http://ehash.iaik.tugraz.at/uploads/5/5e/Blender.pdf
11. Wilson, D.A.: The DCH Hash Function. Submission to NIST (2008),
    http://web.mit.edu/dwilson/www/hash/dch/Supporting_Documentation/dch.pdf

12. Newbold, C.: Observations and Attacks on the SHA-3 Candidate Blender (2008), http://ehash.iaik.tugraz.at/uploads/2/20/Observations_on_Blender.pdf
13. Klima, V.: A near-collision attack on Blender-256 (2008), http://cryptography.hyperlink.cz/BMW/near_collision_blender.pdf
14. Klima, V.: Huge Multicollisions and Multipreimages of Hash Functions BLENDER-n. Cryptology ePrint Archive, Report 2009/006 (2009), http://eprint.iacr.org/
15. Liangyu, X., Ji, L.: Semi-free start collision attack on Blender. Cryptology ePrint Archive, Report 2008/532 (2008), http://eprint.iacr.org/
16. Ferguson, N., Lucks, S.: Attacks on AURORA-512 and the Double-Mix Merkle-Damgaard Transform. Cryptology ePrint Archive, Report 2009/113 (2009), http://eprint.iacr.org/
17. Sasaki, Y.: A 2nd-Preimage Attack on AURORA-512. Cryptology ePrint Archive, Report 2009/112 (2009), http://eprint.iacr.org/
18. Sasaki, Y.: A Collision Attack on AURORA-512. Cryptology ePrint Archive, Report 2009/106 (2009), http://eprint.iacr.org/
19. Mendel, F., Pramstaller, N., Rechberger, C., Kontak, M., Szmidt, J.: Cryptanalysis of the GOST Hash Function. In: Wagner, D. (ed.) CRYPTO 2008. LNCS, vol. 5157, pp. 162–178. Springer, Heidelberg (2008)
20. Gauravaram, P., Kelsey, J.: Linear-XOR and Additive Checksums Don't Protect Damgård-Merkle Hashes from Generic Attacks. In: Malkin, T.G. (ed.) CT-RSA 2008. LNCS, vol. 4964, pp. 36–51. Springer, Heidelberg (2008)
21. Joux, A.: Multicollisions in Iterated Hash Functions. Application to Cascaded Constructions. In: Franklin, M. (ed.) CRYPTO 2004. LNCS, vol. 3152, pp. 306–316. Springer, Heidelberg (2004)
22. Quisquater, J.J., Delescaille, J.P.: How Easy is Collision Search. New Results and Applications to DES. In: Brassard, G. (ed.) CRYPTO 1989. LNCS, vol. 435, pp. 408–413. Springer, Heidelberg (1990)
23. Preneel, B., Govaerts, R., Vandewalle, J.: Hash functions based on block ciphers: A synthetic approach. In: Stinson, D.R. (ed.) CRYPTO 1993. LNCS, vol. 773, pp. 368–378. Springer, Heidelberg (1994)
24. Khovratovich, D., Nikolic, I.: Cryptanalysis of DCH-n (2008), http://lj.streamclub.ru/papers/hash/dch.pdf
25. Lamberger, M., Pramstaller, N., Rechberger, C., Rijmen, V.: Analysis of the Hash Function Design Strategy Called SMASH. IEEE Transactions on Information Theory 54(8), 3647–3655 (2008)
26. Andreeva, E., Bouillaguet, C., Fouque, P.A., Hoch, J.J., Kelsey, J., Shamir, A., Zimmer, S.: Second preimage attacks on dithered hash functions. In: Smart, N.P. (ed.) EUROCRYPT 2008. LNCS, vol. 4965, pp. 270–288. Springer, Heidelberg (2008)
27. Lidl, R., Niederreiter, H.: Finite fields, 2nd edn. Encyclopedia of Mathematics and its Applications, vol. 20. Cambridge University Press, Cambridge (1997); With a foreword by P. M. Cohn

# Meet-in-the-Middle Attacks Using Output Truncation in 3-Pass HAVAL

Yu Sasaki[1,2]

[1] NTT Information Sharing Platform Laboratories, NTT Corporation,
3-9-11 Midoricho, Musashino-shi, Tokyo, 180-8585 Japan
sasaki.yu@lab.ntt.co.jp
[2] The University of Electro-Communications,
1-5-1 Choufugaoka, Choufu-shi, Tokyo, 182-8585 Japan

**Abstract.** We propose preimage and pseudo-preimage attacks on short output lengths of the hash function 3-pass HAVAL, which is designed to be able to output various hash lengths by one algorithm. HAVAL executes a truncate function at the end of the hash computation in order to produce various output lengths. If the hash value is truncated, the internal state size becomes larger than the hash length. Hence, it appears that finding attacks faster than the exhaustive search becomes relatively hard. In this paper, we propose two types of preimage and pseudo-preimage attacks based on the meet-in-the-middle attack. A key point of our attack is how to deal with input information for truncate functions. The first approach works for various types of truncate functions. The second approach uses a property particular to the truncate function of HAVAL. As far as we know, these are the first preimage and pseudo-preimage attacks that work for short output lengths of HAVAL.

**Keywords:** HAVAL, hash, truncate, wide pipe, meet-in-the-middle, preimage, pseudo-preimage.

## 1  Introduction

Due to a widespread use of cryptographic hash functions in many applications such as digital signatures, password-based authentication, random number generators, they deserve a proper level of cryptanalytical attention. Hash functions are required to satisfy security properties such as collision resistance, 2nd preimage resistance, and preimage resistance. When the length of the hash value is $n$ bits, collisions, 2nd preimages, and preimages should not be computed with less than $2^{n/2}$, $2^n$, and $2^n$ operations, respectively.

Designing secure hash functions is a challenging task. In fact, several hash functions designed in the past such as MD5 [1] are now known to be vulnerable against a collision attack [2]. Moreover, a preimage attack has also been discovered recently [3]. To solve vulnerability of recent hash functions, NIST is conducting SHA-3 competition [4] to determine a new hash function standard.

Several hash functions designed so far have a structure where the size of the internal state is larger than that of the final output, and at the end of the computation, the internal state is truncated to produce a desired output size. For

P. Samarati et al. (Eds.): ISC 2009, LNCS 5735, pp. 79–94, 2009.
© Springer-Verlag Berlin Heidelberg 2009

example, HAVAL for short output lengths [5], SHA-224, and SHA-384 [6] have such a structure. Intuitively, this structure contributes to prevent shortcut attacks on hash functions because attackers need to control larger internal state than the final output. Recently, several researches have shown that possessing the larger internal state results in the strong or provable security of modes of operation of hash functions. For example, Coron et al. proposed the chopMD construction and proved its indifferentiability [7], Chang et al. improved its security bound [8], and Lucks proposed wide-pipe construction and showed its security [9]. Note that many hash proposals to the SHA-3 competition have the wide-pipe structure. Finally, we can say that it is important to analyze hash functions whose internal state is larger than the output.

HAVAL [5] is a hash functions designed by Zheng, Pieprzyk, and Seberry in 1992. It was designed so that various hash lengths could be produced by one algorithm. HAVAL has three variants with different security levels called $x$-pass HAVAL ($x = 3, 4, 5$). In this paper, we analyze 3-pass HAVAL. HAVAL is based on Merkle-Damgård construction, and its compression function is similar to MD5. The internal state of HAVAL is 256 bits and the hash length can be chosen from 128, 160, 192, 224, or 256 bits. When HAVAL produces a 128-, 160-, 192-, or 224-bit value, the last internal state is tailored by the truncate function.

Due to the simple structure, several attacks on 3-pass HAVAL are known [10,11,12,2,13]. The first preimage attack on 3-pass HAVAL was proposed by Aumasson et al. [14], where the attack complexity is $2^{230}$, and this was later improved by Sasaki and Aoki [15] into $2^{225}$. (In this paper, the unit of the complexity is one 3-pass HAVAL compression function operation.) Both preimage attacks are targeting only 256-bit output. Since their complexities are more than $2^{224}$, they cannot be directly applied to 128-, 160-, 192-, and 224-bit output.

Another interesting concern is the importance of pseudo-preimage attacks. A pseudo-preimage is a pair of a chaining variable $x$ and a message $M$ such that $CF(x, M) = y$, where $x$ may not be the initial value, $y$ is a given hash value, and $CF$ is a compression function. It is known that if the size of $x$ is the same as the hash size[1], a pseudo-preimage attack can be converted to a preimage attack [16, Fact9.99]. In addition, we point out that regardless of the size of the internal state, finding pseudo-preimages has stronger meaning than breaking the Enhanced Target Collision Resistance (eTCR) property [17] as a keyed hash function, where keys are used in the Key-via-IV approach. Hence, we can say that hash functions should be strong against pseudo-preimage attacks.

One should note a preimage attack by Mendel et al. [18] on hash function HAS-V [19] that can produce various hash lengths. Their attack and ours are partially identical in terms of targeting short output variants. However, Mendel et al. use almost the same technique to attack the short output variants.

## 1.1   Our Contributions

We present preimage and pseudo-preimage attacks on short output lengths of 3-pass HAVAL. We present two meet-in-the-middle based approaches.

---

[1] Ref. [9] calls such a structure "narrow-pipe."

1. The first approach uses the property where the output length is shorter than the internal state. This can be applied to not only the truncate function of HAVAL, but also other ones as long as they can be inverted at a reasonable cost. This approach finds pseudo-preimages of 3-pass HAVAL for 224-, 192-, and 160-bit output.
2. The second approach performs the meet-in-the-middle attack at an appropriate position so that results of two independent searches can be compared on variables of truncated size. This approach uses the property particular to the truncate function of HAVAL. This approach finds pseudo-preimages of 224-, 192-, 160-, and 128-bit output. Furthermore, the pseudo-preimage attack on 224-bit output can be converted to a preimage attack.

The complexity of our attacks are summarized in Table 1. In this paper, due to the limited space, we mainly discuss the attack framework and detailed procedure for 224-bit output.

**Table 1.** Comparison of previous and our attacks

| Reference | | Attack type | 256 bits | 224 bits | 192 bits | 160 bits | 128 bits |
|---|---|---|---|---|---|---|---|
| Aumasson et al. [14] | | Pseudo-preimage | $2^{224}$ | - | - | - | - |
| | | Preimage | $2^{230}$ | - | - | - | - |
| Sasaki et al. [15] | | Pseudo-preimage | $2^{192}$ | - | - | - | - |
| | | Preimage | $2^{225}$ | - | - | - | - |
| Ours | Approach 1 | Pseudo-preimage | not target | $2^{192}$ | $2^{160}$ | $2^{144}$ | - |
| | | Preimage | not target | - | - | - | - |
| | Approach 2 | Pseudo-preimage | not target | $2^{160}$ | $2^{128}$ | $2^{106}$ | $2^{84}$ |
| | | Preimage | not target | $2^{209}$ | - | - | - |

**Organization.** In section 2, we describe HAVAL specification and related work. In section 3, we explain new attacks based on the first approach. In section 4, we explain new attacks based on the second approach. In section 5, we conclude this paper.

## 2 Related Works

### 2.1 Description of 3-Pass HAVAL

HAVAL [5] is a hash function which compresses a message up to $(2^{64}-1)$ bits into 128, 160, 192, 224, or 256 bits. HAVAL has the Merkle-Damgård structure, which uses a 256-bit (8-word) chaining variable and an 1024-bit (32-word) message block to compute a compression function. After the last iteration of the Merkle-Damgård, a 256-bit chaining variable is processed by a truncate function to obtain a desired length. We describe only part of the specification of 3-pass HAVAL related to this paper. Please refer to Ref. [5] for details.

**Table 2.** Message expansion $\pi_j$

| 0 | 1 | 2 | 3 | 4 | 5 | 6 | 7 | 8 | 9 | 10 | 11 | 12 | 13 | 14 | 15 | 16 | 17 | 18 | 19 | 20 | 21 | 22 | 23 | 24 | 25 | 26 | 27 | 28 | 29 | 30 | 31 |
|---|---|---|---|---|---|---|---|---|---|----|----|----|----|----|----|----|----|----|----|----|----|----|----|----|----|----|----|----|----|----|----|
| 5 | 14 | 26 | 18 | 11 | 28 | 7 | 16 | 0 | 23 | 20 | 22 | 1 | 10 | 4 | 8 | 30 | 3 | 21 | 9 | 17 | 24 | 29 | 6 | 19 | 12 | 15 | 13 | 2 | 25 | 31 | 27 |
| 19 | 9 | 4 | 20 | 28 | 17 | 8 | 22 | 29 | 14 | 25 | 12 | 24 | 30 | 16 | 26 | 31 | 15 | 7 | 3 | 1 | 0 | 18 | 27 | 13 | 6 | 21 | 10 | 23 | 11 | 5 | 2 |

An input message $M$ is padded to be a multiple of 1024 bits. A single bit '1' is appended followed by '0's until the length becomes 944 modulo 1024. At the end, 3 bits representing the version number of HAVAL, 3 bits representing the number of the pass, 10 bits representing the output length, and 64 bits representing the unpadded message length are appended.

A padded message $M^*$ is separated into 1024-bit message blocks $(M_0, M_1, \ldots, M_{n-1})$. Let $\mathrm{CF} : \{0,1\}^{256} \times \{0,1\}^{1024} \to \{0,1\}^{256}$ be the compression function. A hash value is computed as follows.

1. $H_0 \leftarrow IV$,
2. $H_{i+1} \leftarrow \mathrm{CF}(H_i, M_i)$ for $i = 0, 1, \ldots, n-1$,

where $H_i$ is a 256-bit chaining variable and $IV$ is the initial value defined in the specification. Finally, $H_n$ is tailored by the truncate function explained later, and is output as a hash value of $M$.

**Compression Function.** The compression function iteratively computes a step function 96 times to compute a hash value. Let $p_j$ be a 256-bit value. $H_{i+1} \leftarrow \mathrm{CF}(H_i, M_i)$ is computed as follows.

1. $M_i$ is divided into 32-bit message words $m_j$ $(j = 0, 1, \ldots, 31)$.
2. $p_0 \leftarrow H_i$.
3. $p_{j+1} \leftarrow R_j(p_j, m_{\pi(j)})$ for $j = 0, 1, \ldots, 95$.
4. Output $H_{i+1}(= p_{96} + H_i)$, where "+" denotes a 32-bit word-wise addition. In this paper, we similarly use "−" to denote a 32-bit word-wise subtraction.

$R_j$ is a step function for Step $j$. Let $Q_j$ be a 32-bit value that satisfies $p_j = (Q_{j-7}\|Q_{j-6}\|Q_{j-5}\|Q_{j-4}\|Q_{j-3}\|Q_{j-2}\|Q_{j-1}\|Q_j)$. $R_j$ is defined as follows:

$$
\begin{cases}
T = f_j \circ \phi_j(Q_{j-6}, Q_{j-5}, Q_{j-4}, Q_{j-3}, Q_{j-2}, Q_{j-1}, Q_j), \\
Q_{j+1} = (Q_{j-7} \ggg 11) + (T \ggg 7) + m_{\pi(j)} + K_j, \\
R_j(p_j, m_{\pi(j)}) = (Q_{j-6}\|Q_{j-5}\|Q_{j-4}\|Q_{j-3}\|Q_{j-2}\|Q_{j-1}\|Q_j\|Q_{j+1}),
\end{cases}
$$

where $f_j$, $\phi_j$, $\ggg n$, and $K_j$ are a bitwise Boolean function, word-wise permutation, $n$-bit right rotation, and constant number defined in the specification, respectively. $\pi_j$ is a message expansion shown in Table 2. In our attacks, we do not consider the values of $f_j$, $\phi_j$, $\ggg n$, and $K_j$, but $\pi_j$ is heavily related. We graphically show the step function in Fig. 1. Note that $R_j^{-1}(\cdot, m_{\pi(j)})$ can be computed in almost the same complexity as $R_j$.

**Truncate function.** Let the 256-bit chaining variable after the last iteration of the Merkle-Damgård be $H_n = D_7\|D_6\|D_5\|D_4\|D_3\|D_2\|D_1\|D_0$. Let $L$ represent

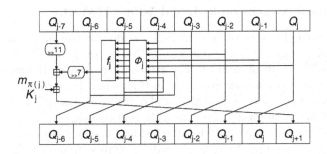

**Fig. 1.** Step function $R_j$

the output length. If $L = 256$, $H_n$ is directly output. If $L \in \{128, 160, 192, 224\}$, the hash value is computed with a truncate function $Trunc : \{0,1\}^{256} \rightarrow \{0,1\}^{L}$. Here, if a variable $X$ is $t$ bits, we use a notion $X^{[t]}$ to explicitly indicate the length of $X$. We show only the case for $L = 224$. For other case, please refer to [5].

$L=224$: Divide $D_7$ in the following way.

$$D_7 = X_{7,6}^{[5]} \| X_{7,5}^{[5]} \| X_{7,4}^{[4]} \| X_{7,3}^{[5]} \| X_{7,2}^{[4]} \| X_{7,1}^{[5]} \| X_{7,0}^{[4]}.$$

The 224-bit hash value $Y_6 \| Y_5 \| Y_4 \| Y_3 \| Y_2 \| Y_1 \| Y_0$ is computed as follows.

$$Y_6 = D_6 + X_{7,0}^{[4]}, \quad Y_5 = D_5 + X_{7,1}^{[5]}, \quad Y_4 = D_4 + X_{7,2}^{[4]}, \quad Y_3 = D_3 + X_{7,3}^{[5]},$$
$$Y_2 = D_2 + X_{7,4}^{[4]}, \quad Y_1 = D_1 + X_{7,5}^{[5]}, \quad Y_0 = D_0 + X_{7,6}^{[5]}.$$

### 2.2   Converting Pseudo-preimages to a Preimage

In $x$-bit narrow-pipe hash functions, a pseudo-preimage attack whose complexity is $2^y$, $y < x-2$ can be converted to a preimage attack with a complexity of $2^{\frac{x+y}{2}+1}$ [16, Fact9.99]. Note this algorithm cannot always be applied if the internal state is larger than the hash length. This algorithm has been used in previous preimage attacks on narrow-pipe hash functions [20,14,21,15,3].

### 2.3   Preimage Attacks on 3-Pass HAVAL

Aumasson et al. proposed two attacks that find pseudo-preimages with a complexity of $2^{224}$ and preimages with a complexity of $2^{230}$ [14]. The attacks require $16 \times 2^{64}$ words of memory. Both attacks are the meet-in-the-middle attack.

In these meet-in-the-middle attacks, results of two independent searches are compared on a 256-bit intermediate chaining variable. Therefore, even if the hash length is truncated, comparison of two independent searches needs to be performed on a 256-bit variable. Therefore, the attack complexity for 256-bit output and shorter outputs are identical. Since the attacks require $2^{224}$ to find pseudo-preimages, the attacks cannot be applied to shorter hash lengths directly.

## 2.4   Preimage Attacks on 3, 4, and 5-Pass HAVAL

Sasaki and Aoki [15] showed how to find pseudo-preimages of 3-pass HAVAL with a complexity of $2^{192}$ and preimages with a complexity of $2^{225}$. The attack requires $13 \times 2^{64}$ words of memory. The attack is the meet-in-the-middle attack, which is a base of our attack explained in this paper. In particular, techniques called splice-and-cut, partial-matching, and partial-fixing in [15], which were first proposed by Aoki and Sasaki [21], are utilized in this paper.

*Splice-and-cut* technique considers the last and first steps as consecutive steps and divide the attack target into two *chunks* of steps so that each chunk includes at least one message word that is independent of the other chunk. These message words are called *neutral words*. Then, pseudo-preimages are computed by the meet-in-the-middle attack. *Partial-matching* technique skips several steps of the attack target when the meet-in-the-middle attack is performed. Assume that one of chunks provides the value of $p_i$, where $p_i = (Q_{i-7} \| Q_{i-6} \| \cdots \| Q_i)$ and the other chunk provides the value of $p_{i-7}$, where $p_{i-7} = (Q_{i-14} \| Q_{i-13} \| \cdots \| Q_{i-7})$. $p_i$ and $p_{i-7}$ cannot be directly compared, however, part of these values, *i.e.* $Q_{i-7}$ can be compared. Therefore, results of two independent computations can be compared without a knowledge of $m_{\pi(i-1)}, m_{\pi(i-2)}, \cdots m_{\pi(i-7)}$. *Partial-fixing* technique enables attackers to skip more steps. It fixes part of the neutral words so that attackers can partially compute the step function even if a neutral word for the other chunk appears. For example, consider the step function $R_j$. Assume the lower $n$ bits of $m_{\pi(j)}$ are fixed, the upper $32 - n$ bits of $m_{\pi(j)}$ are unknown, and other variables are fully fixed. We can still compute the lower $n$ bits of $Q_{j+1}$ with a probability of 1.

Since the attack of Ref. [15] costs $2^{225}$, it can attack 256-bit output but cannot attack 224-, 192-, 160-, and 128-bit output.

# 3   Approach 1: Increasing Neutral Words Using $Trunc^{-1}$

In this section, we propose new attacks on 3-pass HAVAL that find pseudo-preimages of 224-, 192-, and 160-bit hash values. In this attack, the attacker essentially uses the property where the hash value is shorter than the internal chaining variables. The attack works for not only the truncate function of HAVAL but also other truncate functions as long as they can be inverted.

## 3.1   Attack Outline

When we divide 96 steps of 3-pass HAVAL into two chunks, the followings frequently occur.

*96 steps are separated so that one chunk includes one neutral word ($2^{32}$ freedom) but the other chunk includes several neutral words.*

Since one of chunks includes only one neutral word, previous attacks can find a pseudo-preimage faster than the brute force attack by the factor of at most $2^{32}$ even though the other chunk includes more than one neutral words. Therefore, if

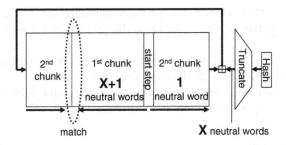

**Fig. 2.** Approach 1: increasing neutral words by inverting truncate function

---

**Input:** 224-bit hash value $Y_6\|Y_5\|Y_4\|Y_3\|Y_2\|Y_1\|Y_0$
**Output:** $2^{32}$ 256-bit chaining variables $D_7\|D_6\|D_5\|D_4\|D_3\|D_2\|D_1\|D_0$

1. `for` $D_7 = 0$ `to` `0xffffffff` {
2.     Separate $D_7$ to obtain $X_{7,6}^{[5]}\|X_{7,5}^{[5]}\|X_{7,4}^{[4]}\|X_{7,3}^{[5]}\|X_{7,2}^{[4]}\|X_{7,1}^{[5]}\|X_{7,0}^{[4]}$.
3.     Compute $D_k = Y_k - X_{7,6-k}, \quad k = 0, 1, \ldots, 6$.
4. }

---

**Fig. 3.** Inverse computation of truncate function for HAVAL 224-bit output

the hash length is shorter by 32 bits or more, the attacks cannot be faster than the brute force attack.

In our approach, we focus attention on the fact that there are many values of 256-bit intermediate chaining variables that reach a given hash value of the truncated size. If we can find those 256-bit values by inversely computing the truncate function, the number of neutral words in a chunk increases and the meet-in-the-middle attack can be performed more efficiently. We summarize this strategy for $X$-word truncating hash functions. This is also illustrated in Fig. 2.

*Separate the attack target so that one chunk that does not include IV has at least $1 + X$ neutral words and the other chunk that includes IV has at least 1 neutral word. Then, compute inversion of the truncate function to find all inverse images of the given hash value so that both chunks include at least $1 + X$ neutral words. Finally, the meet-in-the-middle attack with $1+X$ freewords is performed to find pseudo-preimages.*

### 3.2 Inverse Computation of Truncate Function

Our attack works for various truncate functions as long as all inverse images of a given hash value can be found at a reasonable cost. The truncate function of HAVAL is an example of this case. We explain how to find all inverse images in the truncate function of HAVAL.

For a given output $Y$ of the truncate function $Trunc$, the goal of the inverse computation is finding all $D$s such that $Trunc(D) = Y$. Therefore, if $t$ bits are truncated, we find $2^t$ values of $D$. Inversion of the truncate function of HAVAL can be easily computed. In Fig. 3, we show the procedure using the 224-bit case as an example.

By the similar way, the inverse computation for other output lengths can be computed. The complexity is $2^t$ truncate function operations. Note that this is negligible compared to the computation of chunks, which requires the complexity of at least $2^{t+32}$ compression function operations.

### 3.3   Attack Description

**Attack on 224-bit output.** The chunk separation for 224-bit output is shown in Table 3. The attack procedure for a given 224-bit hash value $H_n$ is as follows.

**Table 3.** Chunk separation for 224-bit output

| Step | 0 | 1 | 2 | 3 | 4 | 5 | 6 | 7 | 8 | 9 | 10 | 11 | 12 | 13 | $\cdots$ | 21 | 22 | 23 | 24 | 25 | 26 | 27 | 28 | 29 | 30 | 31 |
|------|---|---|---|---|---|---|---|---|---|---|----|----|----|----|----------|----|----|----|----|----|----|----|----|----|----|----|
| index | 0 | 1 | 2 | 3 | 4 | ⑤ | 6 | 7 | 8 | 9 | 10 | 11 | 12 | 13 | $\cdots$ | 21 | 22 | 23 | 24 | 25 | 26 | ㉗ | ㉘ | 29 | 30 | 31 |

first chunk | skip

| Step | 32 | 33 | 34 | 35 | 36 | 37 | 38 | 39 | 40 | 41 | 42 | 43 | 44 | 45 | $\cdots$ | 53 | 54 | 55 | 56 | 57 | 58 | 59 | 60 | 61 | 62 | 63 |
|------|----|----|----|----|----|----|----|----|----|----|----|----|----|----|----------|----|----|----|----|----|----|----|----|----|----|----|
| index | ⑤ | 14 | 26 | 18 | 11 | ㉘ | 7 | 16 | 0 | 23 | 20 | 22 | 1 | 10 | $\cdots$ | 24 | 29 | 6 | 19 | 12 | 15 | 13 | 2 | 25 | 31 | ㉗ |

second chunk

| Step | 64 | 65 | 66 | 67 | 68 | 69 | 70 | 71 | 72 | 73 | 74 | 75 | $\cdots$ | 83 | 84 | 85 | 86 | 87 | 88 | 89 | 90 | 91 | 92 | 93 | 94 | 95 |
|------|----|----|----|----|----|----|----|----|----|----|----|----|----------|----|----|----|----|----|----|----|----|----|----|----|----|----|
| index | 19 | 9 | 4 | 20 | ㉘ | 17 | 8 | 22 | 29 | 14 | 25 | 12 | $\cdots$ | 3 | 1 | 0 | 18 | ㉗ | 13 | 6 | 21 | 10 | 23 | 11 | ⑤ | 2 |

second chunk                    $\leftarrow|\rightarrow$     first chunk

**Attack procedure**
1. Fix $m_{29}, m_{30}$, and $m_{31}$ to satisfy the padding for a 1-block message.
2. Fix $m_i$ ($i \notin \{5, 27, 28, 29, 30, 31\}$) and $p_{88}$ to randomly chosen values.
3. For all $(m_{27}, m_{28})$, do:

$$p_j \leftarrow R_j^{-1}(p_{j+1}, m_{\pi(j)}) \qquad \text{for } j = 87, 86, \ldots, 33.$$

4. Make a table of $(m_{27}, m_{28}, p_{33})$s which are computed in the last step, where $p_{33} = (Q_{26}\|Q_{27}\|Q_{28}\|Q_{29}\|Q_{30}\|Q_{31}\|Q_{32}\|Q_{33})$.
5. For a given $H_n$, with the algorithm shown in Fig. 3, inversely compute the truncate function to find $2^{32}$ values of 256-bit $D$s s.t. $Trunc(D) = H_n$.
6. For all $(m_5, D)$,
   (a) do the following:

$$\begin{cases} p_{j+1} \leftarrow R_j(p_j, m_{\pi(j)}) & \text{for } j = 88, 89, \ldots, 95, \\ p_0 \leftarrow D - p_{96}, \\ p_{j+1} \leftarrow R_j(p_j, m_{\pi(j)}) & \text{for } j = 0, 1, \ldots, 26, \end{cases}$$

where, $p_{27} = (Q_{20}\|Q_{21}\|Q_{22}\|Q_{23}\|Q_{24}\|Q_{25}\|Q_{26}\|Q_{27})$.
   (b) Check whether $Q_{27}$ and $Q_{26}$ match those in the table.

(c) If they match, compute $p_{j+1} \leftarrow R_j(p_j, m_{\pi(j)})$ for $j = 27, 28, \ldots, 32$ by using the matched messages, and check whether all of them match or not.

(d) If match, the corresponding message and $p_0$ is a pseudo-preimage of $H_n$.

Let the complexity of 1 step function operation be $\frac{1}{96}$ compression function operation. In the above procedure, Step 3 and Step 6a cost $2^{64} \cdot \frac{55}{96}$ and $2^{64} \cdot \frac{35}{96}$, respectively. At Step 6b, $2^{128}(= 2^{64} \cdot 2^{64})$ pairs are compared and $2^{64}(= 2^{128} \cdot 2^{-64})$ pairs will remain after 64-bit matching. At Step 6c, we need the complexity of $2^{64} \cdot \frac{6}{96}$ to compute $p_{20}$ to $p_{25}$. Since $p_{20}$ to $p_{25}$ can match with a probability of $2^{-192}$, we expect $2^{-128}(= 2^{64} \cdot 2^{-192})$ pair will remain. Therefore, we repeat the above procedure $2^{128}$ times and obtain a pair that will match in all bits. Therefore, the complexity to find a pseudo-preimage is $2^{128}$ iteration of $2^{64}(= 2^{64} \cdot \frac{55}{96} + 2^{64} \cdot \frac{35}{96} + 2^{64} \cdot \frac{6}{96})$ computations, which is $2^{192}$ computations. In this attack, we use a memory to store $(2^{64} \times 10)$ words at Step 4, so the memory complexity is the order of $2^{64}$. Note that we assume the memory access is performed in negligible time compared to 1 step operation. Also note that the attack can be memoryless using the technique in [16, Remark 9.93], which requires $2^{193}$ computation and negligible memory.

**Attacks on other output lengths.** Attacks on 192-bit and 160-bit output are also possible. The chunk separation and attack procedure are almost the same as those of 224-bit. Hence, we explain details of the attacks in Appendices. Note that to attack 192-bit output, one of two chunks needs to have at least three neutral words. Similarly, to attack 160-bit output, one of two chunks needs to have at least four neutral words. Also note that the more neutral words we use, the more memory we need to store the result of chunk computations. This attack cannot be applied to 128-bit output because meet-in-the-middle attack on 256-bit internal chaining variables will cost at least $2^{128}$. In such a case, we need another attack explained in the next section.

# 4    Approach 2: Meet-in-the-Middle on Efficient Place

## 4.1    Attack Outline

In this approach, we focus attention on the following property[2].

> *In the previous meet-in-the-middle attack [15], results of two independent searches are compared on a part of bits. For other bits, attackers just wait until all of those bits happen to match.*

Based on the above observation, we explain how to attack the truncated output lengths by using the example shown in Fig. 4.

The case (1) in Fig. 4 shows the meet-in-the-middle attack for 256-bit output that checks the match of the 96 bits of $(Q_{j-5}, Q_{j-4}, Q_{j-3})$. In order to find a pair where all bits match, this meet-in-the-middle attack must be repeated $2^{160}$

---

[2] This property is also summarized by Isobe and Shibutani [22].

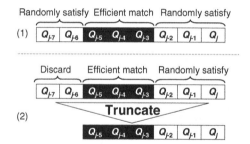

**Fig. 4.** Approach 2: meet-in-the-middle attack on truncated size

**Table 4.** Chunk separation for approach 2

| Step | 0 1 | 2 3 4 5 6 7 8 9 10 11 12 13 ⋯ 21 22 23 24 25 26 27 28 29 30 31 |
|---|---|---|
| index | ⓪ ① | 2 3 4 ⑤ 6 7 8 9 10 ① 12 13 ⋯ 21 22 23 24 25 26 27 28 29 30 31 |
| | skip | first chunk |

| Step | 32 33 34 35 36 37 38 39 | 40 41 42 43 44 45 ⋯ 53 54 55 56 57 58 59 60 61 62 63 |
|---|---|---|
| index | ⑤ 14 26 18 ① 28 7 16 | ⓪ 23 20 22 ① 10 ⋯ 24 29 6 19 12 15 13 2 25 31 27 |
| | first chunk　　←\| → | second chunk |

| Step | 64 65 66 67 68 69 70 71 72 73 74 75 ⋯ 83 84 85 86 87 88 89 90 91 92 | 93 94 95 |
|---|---|---|
| index | 19 9 4 20 28 17 8 22 29 14 25 12 ⋯ 3 ① ⓪ 18 27 13 6 21 10 23 | ① ⑤ 2 |
| | second chunk | skip |

times so that 160 bits of $Q_{j-7}, Q_{j-6}, Q_{j-2}, Q_{j-1}$, and $Q_j$ will randomly match. The case (2) in Fig. 4 shows the similar case but the truncate function discards the left 64 bits of the last 256-bit chaining variables and outputs only the right 192 bits[3]. In this case, the number of the repetition that the attacker needs is only $2^{96}$ times. This means that the efficiency of the meet-in-the-middle attack is the same and the complexity for the brute force part is reduced. Hence, the attack works at the same efficiency even if the output length is truncated.

### 4.2 Chunk Separation

In the approach 2, we use the same chunk separation in common for all hash lengths. We separate 96 steps into two chunks so that results from two chunks are compared on the input of the truncate function. The chunk separation is shown in Table 4. Note this chunk separation is the same as the one used in [15]. We exhaustively searched for the best chunks and found this was the best.

### 4.3 Attack Description

The above chunk separation can perform the efficient match on input chaining variables of the truncate function as shown in (2) of Fig. 4. In fact, the first

---

[3] This truncate function is different from that of HAVAL, but is useful to understand the attack concept.

chunk produces $p_2 = (Q_{-5}\|Q_{-4}\| \cdots \|Q_2)$ and the second chunk produces $p_{93} = (Q_{86}\|Q_{87}\| \cdots \|Q_{93})$. We thus can compute three words of $(p_0 + p_{96})$, namely, $(Q_{-3} + Q_{93}), (Q_{-4} + Q_{92})$, and $(Q_{-5} + Q_{91})$. Therefore, if the truncate function is just discarding several words, the attack is possible as explained in Fig. 4. However, since the truncate function of HAVAL is more complicated, we need more effort to attack it.

**Attack on 224-bit output.** Let $H_n = Y_6\|Y_5\| \cdots \|Y_0$ be a given 224-bit hash value. We rewrite the truncate function shown in Section 2.1 with variables $Q_j$.

$$Q_{89} + Q_{-7} = X_{7,6}^{[5]}\|X_{7,5}^{[5]}\|X_{7,4}^{[4]}\|X_{7,3}^{[5]}\|X_{7,2}^{[4]}\|X_{7,1}^{[5]}\|X_{7,0}^{[4]}.$$

$$
\begin{aligned}
Y_6 &= Q_{90} + Q_{-6} + X_{7,0}^{[4]} &\rightarrow& \qquad Y_6 - Q_{90} = Q_{-6} + X_{7,0}^{[4]}, \\
Y_5 &= Q_{91} + Q_{-5} + X_{7,1}^{[5]} &\rightarrow& \qquad Y_5 - Q_{91} = Q_{-5} + X_{7,1}^{[5]}, \\
Y_4 &= Q_{92} + Q_{-4} + X_{7,2}^{[4]} &\rightarrow& \qquad Y_4 - Q_{92} = Q_{-4} + X_{7,2}^{[4]}, \\
Y_3 &= Q_{93} + Q_{-3} + X_{7,3}^{[5]} &\rightarrow& \qquad Y_3 - Q_{93} = Q_{-3} + X_{7,3}^{[5]}, \\
Y_2 &= Q_{94} + Q_{-2} + X_{7,4}^{[4]} &\rightarrow& \qquad Y_2 - Q_{94} = Q_{-2} + X_{7,4}^{[4]}, \\
Y_1 &= Q_{95} + Q_{-1} + X_{7,5}^{[5]} &\rightarrow& \qquad Y_1 - Q_{95} = Q_{-1} + X_{7,5}^{[5]}, \\
Y_0 &= Q_{96} + Q_0 \;+ X_{7,6}^{[5]} &\rightarrow& \qquad Y_0 - Q_{96} = Q_0 \;+ X_{7,6}^{[5]}.
\end{aligned}
$$

The attack procedure is as follows. Hereafter, we use a notation $A^{U[n]}$ to explicitly denote the upper $n$ bits of a variable $A$.

**Attack procedure**
1. Fix $m_{29}, m_{30}$, and $m_{31}$ to satisfy the padding for a 1-block message for the pseudo-preimage attack or 2-block message for the preimage attack.
2. Fix $m_i$ ($i \notin \{0, 1, 5, 11, 29, 30, 31\}$) and $p_{40}$ to randomly chosen values.
3. For all $(m_0, m_1)$, do: $p_{j+1} \leftarrow R_j(p_j, m_{\pi(j)})$ for $j = 40, 41, \ldots, 92$, where $p_{93} = (Q_{86}\|Q_{87}\|Q_{89}\|Q_{89}\|Q_{90}\|Q_{91}\|Q_{92}\|Q_{93})$, and compute $(Y_6 - Q_{90}), (Y_5 - Q_{91})$, $(Y_4 - Q_{92})$, and $(Y_3 - Q_{93})$.
4. Make a table of $(m_0, m_1, Q_{86}, Q_{87}, Q_{88}, Q_{89}, (Y_6 - Q_{90}), (Y_5 - Q_{91}), (Y_4 - Q_{92}), (Y_3 - Q_{93}))$.
5. For all $(m_5, m_{11})$,
    (a) do the following: $p_j \leftarrow R_j^{-1}(p_{j+1}, m_{\pi(j)})$ for $j = 39, 38, \ldots, 2$, where, $p_2 = (Q_{-5}\|Q_{-4}\|Q_{-3}\|Q_{-2}\|Q_{-1}\|Q_0\|Q_1\|Q_2)$.
    (b) Compute possible values of $(Q_{-5} + X_{7,1}^{[5]})^{U[27]}$. Since $X_{7,1}^{[5]}$ is unknown, we cannot compute it with a probability of 1. However, there are only two possibilities of the carry from bit position 4 to 5. Let this carry number be $C_5 (\in \{0, 1\})$. For each $C_5$, we compute corresponding $(Q_{-5} + X_{7,1}^{[5]})^{U[27]}$, and store the tuple of $(C_5, (Q_{-5} + X_{7,1}^{[5]})^{U[27]})$. Similarly, for each $C_4$, we compute $(Q_{-4} + X_{7,2}^{[4]})^{U[28]}$ and for each $C_3$, we compute $(Q_{-3} + X_{7,3}^{[5]})^{U[27]}$.

(c) Check whether $(Q_{-5}+X_{7,1}^{[5]})^{U[27]}, (Q_{-4}+X_{7,2}^{[4]})^{U[28]}$, and $(Q_{-3}+X_{7,3}^{[5]})^{U[27]}$ match $(Y_5-Q_{91})^{U[27]}, (Y_4-Q_{92})^{U[28]}$, and $(Y_3-Q_{93})^{U[27]}$ stored in Step 4.

(d) If match, compute $X_{7,1}^{[5]}$ using the equation $Y_5 - Q_{91} = Q_{-5} + X_{7,1}^{[5]}$ and check whether the guess of carry $C_5$ is correct or not. Similarly, compute $X_{7,2}^{[4]}$ and $X_{7,3}^{[5]}$, and check whether $C_4$ and $C_3$ are correct or not.

(e) If $C_5, C_4$ and $C_3$ are correct, compute $p_{j+1} \leftarrow R_j(p_j, m_{\pi(j)})$ for $j = 93, 94, 95$ and $p_j \leftarrow R_j^{-1}(p_{j+1}, m_{\pi(j)})$ for $j = 1$. Then, with the similar manner to Steps 5b and 5c, check the match of $(Y_2 - Q_{94})^{[U28]}$ and $(Q_{-2} + X_{7,4}^{[4]})^{[U28]}, (Y_1 - Q_{95})^{[U27]}$ and $(Q_{-1} + X_{7,5}^{[5]})^{[U27]}, (Y_0 - Q_{96})^{[U27]}$ and $(Q_0 + X_{7,6}^{[5]})^{[U27]}$, and $(Y_6 - Q_{90})^{[U28]}$ and $(Q_{-6} + X_{7,0}^{[4]})^{[U28]}$.

(f) If match, in the same way as Steps 5d, compute $X_{7,2}^{[4]}, X_{7,1}^{[5]}, X_{7,0}^{[5]}$, and $X_{7,6}^{[4]}$, and check whether corresponding $C_2, C_1, C_0$ and $C_6$ are correct.

(g) If they are correct, $X_7 = X_{7,6}\|X_{7,5}\|\cdots\|X_{7,0}$ have already been determined. Compute $p_0$ and check whether $Q_{89} + Q_{-7} = X_7$ is satisfied.

(h) If satisfied, the corresponding message and $p_0$ is a pseudo-preimage.

In the above procedure, the complexity of Step 3 and Step 5a is approximately $2^{64}(= 2^{64} \cdot \frac{53}{96} + 2^{64} \cdot \frac{38}{96})$. Each of Step 3 and Step 5a produces $2^{64}$ items. In Step 5b, the number of items for the first chunk becomes $2^{67}(= 2^3 \cdot 2^{64})$. In Step 5c, 82-bit matching is performed for $2^{131}(= 2^{64} \cdot 2^{67})$ pairs. Hence, $2^{49}(= 2^{131} \cdot 2^{-82})$ pairs are expected to remain. Step 5d requires the complexity of $2^{49} \cdot \frac{3}{96}$ operations, and the number of remaining pair will be $2^{46}(= 2^{49} \cdot 2^{-3})$. In Step 5e, we need the complexity of $2^{42}(< 2^{46} \cdot \frac{4}{96})$. Then, because we consider two carry number patterns for 4 addition operations, the number of remaining pair will be $2^{50}(= 2^{46} \cdot 2^4)$. Finally, after the 110-bit matching, $2^{-60}(= 2^{50} \cdot 2^{-110})$ pair is expected to remain. In Step 5f, the complexity is negligible. By checking the correctness of $C_2, C_1, C_0$, and $C_6$, the number of remaining pair will be $2^{-64}(= 2^{-60} \cdot 2^{-4})$. In Step 5g, the complexity is negligible. The equation is satisfied with a probability of $2^{-32}$, hence, the number of remaining pair will be $2^{-96}(= 2^{-64} \cdot 2^{-32})$.

So far, the dominant complexity is $2^{64}$ of Steps 3 and 5a. Finally, by repeating the above procedure $2^{96}$ times, we can obtain a pseudo-preimage. The complexity of this pseudo-preimage attack is $2^{160}(= 2^{64} \cdot 2^{96})$. This is converted to the preimage attack whose complexity is $2^{209}(= 2^{1+(160+256)/2})$ by the conversion algorithm explained in Section 2.2. This attack needs to store $2^{64}$ items in Step 4. Therefore, the memory complexity is $(2^{64} \times 10)$ words.

**Attack on 192-bit output.** The attack is almost the same as that of 224-bit. Different from the attack on 224-bit, the attack on 192-bit can compare only 64 bits at the Step 5c of the attack procedure for 224-bit output. However, checking the match of 64 bits are enough to efficiently reduce the candidates of matching pairs since each chunk produces only $2^{64}$ candidates. The efficiency of the attack is the same as that of 224-bit, therefore, pseudo-preimages are computed faster than the brute force attack by the factor of $2^{64}$, which is $2^{128}$.

**Attack on 160-bit output.** The attack strategy is the same as those of 224- and 192-bit output. However, due to the structure of the truncate function, the number of bits that we can compare at the Step 5c of the attack procedure for 224-bit is only 38 bits. (Lower 7 bits and upper 12 bits of $Y_4 = D_4 + X_{7,4}^{[7]} \| X_{6,3}^{[6]} \| X_{5,2}^{[7]}$ and lower 6 bits and upper 13 bits of $Y_3 = D_3 + X_{7,3}^{[6]} \| X_{6,2}^{[7]} \| X_{5,1}^{[6]}$.) Comparing 38 bits is not enough for chunks with 64 free bits.

Our idea to solve this problem is using the partial-fixing technique, which can increase the number of matching bits by reducing the free bits of neutral words. We fix neutral words so that the number of matching bits and free bits of neutral words are balanced and the attack efficiency is maximized. We found, by fixing the lower 6 bits of $m_{11}$, we can additionally perform 6-bit matching of $Y_2 = D_2 + X_{7,2}^{[7]} \| X_{6,1}^{[6]} \| X_{5,0}^{[6]}$. Similarly, by fixing bit positions 1-10 of $m_1$, additional 7 bits of $Y_3 = D_3 + X_{7,3}^{[6]} \| X_{6,2}^{[7]} \| X_{5,1}^{[6]}$ and additional 3 bits of $Y_4 = D_4 + X_{7,4}^{[7]} \| X_{6,3}^{[6]} \| X_{5,2}^{[7]}$ can be compared. Therefore, we can perform 54-bit matching with keeping 54 free bits in $(m_0, m_1)$ and 58 free bits in $(m_5, m_{11})$. Finally, pseudo-preimages are found faster than the brute force attack by the factor of $2^{54}$, which is $2^{106}$.

**Attack on 128-bit output.** The attack is almost the same as 160-bit output. We found, by fixing the lower 20 bits of $m_{11}$ and bit positions 29, 30, 31, and 0-12 of $m_1$, we can perform 44-bit matching with keeping 48 free bits in $(m_0, m_1)$ and 44 free bits in $(m_5, m_{11})$. Hence, the attack finds pseudo-preimages faster than the brute force attack by the factor of $2^{44}$, which is $2^{84}$.

## 5 Conclusions

We proposed two types of preimage and pseudo-preimage attacks on short output lengths of 3-pass HAVAL. The first approach uses the property where the output length is shorter than the intermediate chaining variables, so it can work for various truncate functions. This approach finds pseudo-preimages of 224-, 192-, and 160-bit output with a complexity of $2^{192}, 2^{160}$, and $2^{144}$, respectively. In the second approach, we apply the meet-in-the-middle attack so that two independent results can match at a variable of truncated size. This approach finds pseudo-preimages of 224-, 192-, 160-, and 128-bit output with a complexity of $2^{160}, 2^{128}, 2^{106}$ and $2^{84}$, respectively. Considering the low complexity for 224-bit output, we also can find preimages of 224-bit output with a complexity of $2^{209}$.

We tried to apply our techniques to 4-pass and 5-pass HAVAL and found that they appeared to be enough strong against our attacks. This is because they have more steps than 3-pass HAVAL and good chunks does not exist. Hence, attacks on 4-pass and 5-pass HAVAL for short output sizes are open problems.

# References

1. Rivest, R.L.: Request for Comments 1321: The MD5 Message Digest Algorithm. The Internet Engineering Task Force (1992)
2. Wang, X., Yu, H.: How to break MD5 and other hash functions. In: Cramer, R. (ed.) EUROCRYPT 2005. LNCS, vol. 3494, pp. 19–35. Springer, Heidelberg (2005)
3. Sasaki, Y., Aoki, K.: Finding preimages in full MD5 faster than exhaustive search. In: Joux, A. (ed.) EUROCRYPT 2009. LNCS, vol. 5479, pp. 134–152. Springer, Heidelberg (2009)
4. U.S. Department of Commerce, National Institute of Standards and Technology: Federal Register /vol. 72(212)/Friday, November 2, 2007/Notices (2007)
5. Zheng, Y., Pieprzyk, J., Seberry, J.: HAVAL — one-way hashing algorithm with variable length of output. In: Seberry, J., Zheng, Y. (eds.) AUSCRYPT 1992. LNCS, vol. 718, pp. 83–104. Springer, Heidelberg (1993)
6. U.S. Department of Commerce, National Institute of Standards and Technology: Secure Hash Standard (SHS) (Federal Information Processing Standards Publication 180-3) (2008)
7. Coron, J.S., Dodis, Y., Malinaud, C., Puniya, P.: Merkle-damgård revisited: How to construct a hash function. In: Shoup, V. (ed.) CRYPTO 2005. LNCS, vol. 3621, pp. 430–448. Springer, Heidelberg (2005)
8. Chang, D., Nandi, M.: Improved indifferentiability security analysis of chopMD hash function. In: Nyberg, K. (ed.) FSE 2008. LNCS, vol. 5086, pp. 429–443. Springer, Heidelberg (2008)
9. Lucks, S.: A failure-friendly design principle for hash functions. In: Roy, B. (ed.) ASIACRYPT 2005. LNCS, vol. 3788, pp. 474–494. Springer, Heidelberg (2005)
10. van Rompay, B., Biryukov, A., Preneel, B., Vandewalle, J.: Cryptanalysis of 3-pass HAVAL. In: Laih, C.S. (ed.) ASIACRYPT 2003. LNCS, vol. 2894, pp. 228–245. Springer, Heidelberg (2003)
11. Suzuki, K., Kurosawa, K.: How to find many collisions of 3-pass HAVAL. In: Miyaji, A., Kikuchi, H., Rannenberg, K. (eds.) IWSEC 2007. LNCS, vol. 4752, pp. 428–443. Springer, Heidelberg (2007)
12. Wang, X., Feng, D., Yu, X.: An attack on hash function HAVAL-128. Science in China (Information Sciences) 48(5), 545–556 (2005)
13. Lee, E., Chang, D., Kim, J.-S., Sung, J., Hong, S.H.: Second preimage attack on 3-pass HAVAL and partial key-recovery attacks on NMAC/HMAC-3-pass HAVAL. In: Nyberg, K. (ed.) FSE 2008. LNCS, vol. 5086, pp. 189–206. Springer, Heidelberg (2008)
14. Aumasson, J.P., Meier, W., Mendel, F.: Preimage attacks on 3-pass HAVAL and step-reduced MD5. In: Workshop Records of SAC 2008, pp. 99–114 (2008)
15. Sasaki, Y., Aoki, K.: Preimage attacks on 3, 4, and 5-pass HAVAL. In: Pieprzyk, J. (ed.) ASIACRYPT 2008. LNCS, vol. 5350, pp. 253–271. Springer, Heidelberg (2008)
16. Menezes, A.J., van Oorschot, P.C., Vanstone, S.A.: Handbook of applied cryptography. CRC Press, Boca Raton (1997)
17. Halevi, S., Krawczyk, H.: Strengthening digital signatures via randomized hashing. In: Dwork, C. (ed.) CRYPTO 2006. LNCS, vol. 4117, pp. 41–59. Springer, Heidelberg (2006)

18. Mendel, F., Rijmen, V.: Weaknesses in the HAS-V compression function. In: Nam, K.H., Rhee, G. (eds.) ICISC 2007. LNCS, vol. 4817, pp. 335–345. Springer, Heidelberg (2007)
19. Park, N.K., Hwang, J.H., Lee, P.J.: HAS-V: A new hash function with variable output length. In: Stinson, D.R., Tavares, S. (eds.) SAC 2000. LNCS, vol. 2012, pp. 202–216. Springer, Heidelberg (2001)
20. Leurent, G.: MD4 is not one-way. In: Nyberg, K. (ed.) FSE 2008. LNCS, vol. 5086, pp. 412–428. Springer, Heidelberg (2008)
21. Aoki, K., Sasaki, Y.: Preimage attacks on one-block MD4, 63-step MD5 and more. In: Workshop Records of SAC 2008, pp. 82–98 (2008)
22. Isobe, T., Shibutani, K.: Preimage attacks on reduced Tiger and SHA-2. In: Fast Software Encryption 2009 Preproceedings, pp. 141–158 (2009)

# A    Attack on 192-Bit Output by Approach 1

The chunk separation for 192-bit output is shown in Table 5. The attack procedure is almost the same as that of 224-bit output. Therefore, we explain only differences of attack procedures.

**Table 5.** Chunk separation for 192-bit output

| Step | 0 | 1 | 2 | 3 | 4 | 5 | 6 | 7 | 8 | 9 | 10 | 11 | 12 | ··· | 20 | 21 | 22 | 23 | 24 | 25 | 26 | 27 | 28 | 29 | 30 | 31 |
|---|---|---|---|---|---|---|---|---|---|---|---|---|---|---|---|---|---|---|---|---|---|---|---|---|---|---|
| index | 0 | 1 | 2 | 3 | 4 | ⑤ | 6 | 7 | 8 | 9 | 10 | 11 | 12 | ··· | 20 | 21 | 22 | 23 | 24 | 25 | 26 | 27 | ㉘ | ㉙ | ㉚ | 31 |
| | | | | | | | | first chunk | | | | | | | | | | | | | | | | skip | | |

| Step | 32 | 33 | 34 | 35 | 36 | 37 | 38 | 39 | 40 | ··· | 48 | 49 | 50 | 51 | 52 | 53 | 54 | 55 | 56 | 57 | 58 | 59 | 60 | 61 | 62 | 63 |
|---|---|---|---|---|---|---|---|---|---|---|---|---|---|---|---|---|---|---|---|---|---|---|---|---|---|---|
| index | ⑤ | 14 | 26 | 18 | 11 | ㉘ | 7 | 16 | 0 | ··· | ㉚ | 3 | 21 | 9 | 17 | 24 | ㉙ | 6 | 19 | 12 | 15 | 13 | 2 | 25 | 31 | 27 |
| | | | | | | | | | second chunk | | | | | | | | | | | | | | | | | |

| Step | 64 | 65 | 66 | 67 | 68 | 69 | 70 | 71 | 72 | 73 | 74 | 75 | 76 | 77 | 78 | ··· | 86 | 87 | 88 | 89 | 90 | 91 | 92 | 93 | 94 | 95 |
|---|---|---|---|---|---|---|---|---|---|---|---|---|---|---|---|---|---|---|---|---|---|---|---|---|---|---|
| index | 19 | 9 | 4 | 20 | ㉘ | 17 | 8 | 22 | ㉙ | 14 | 25 | 12 | 24 | ㉚ | 16 | ··· | 18 | 27 | 13 | 6 | 21 | 10 | 23 | 11 | ⑤ | 2 |
| | | | | | | second chunk | | | | | | | | | | ←\|→ | | | | first chunk | | | | | | |

1. We find $2^{64}$ inputs of $Trunc^{-1}$ that result in the given output.
2. $m_{29}$ is selected as neutral words. However, at least 17 bits of $m_{29}$ are fixed by the message padding. Therefore, when all bits of $m_{29}$ are used to compute a chunk, the attack is considered as the attack on the compression function. Remember, the pseudo-preimage attack on HAVAL is still possible by using the remaining free-bits of $m_{29}$ though the attack efficiency decreases.
3. In terms of the attack on the compression function, pseudo-preimages can be discovered faster than the brute force attack by the factor of $2^{32}$. Therefore, the time complexity is $2^{160}$. Since both chunks include three neutral words, the memory complexity is approximately $2^{96}$.

# B    Attack on 160-Bit Output by Approach 1

The chunk separation for 160-bit output is shown in Table 6. Both chunks include 4 neutral words but the partial-matching technique allows the comparison of only 3 words, hence, meet-in-the-middle attack cannot be efficiently applied. This problem can be solved by increasing the number of bits compared with the partial-fixing technique proposed by [21]. We fix the lower 16-bits of $m_{28}$, and when we compute the first chunk, we partially compute $p_{29} \leftarrow R_{28}(p_{28}, m_{28})$. Consequently, both chunks have 3.5 neutral words and 3.5 words are compared in the matching part. The attack is faster than the brute force attack by the factor of $2^{16}$, which costs $2^{144}$. Because both chunks include 3.5 neutral words, the memory complexity is approximately $2^{112}$.

**Table 6.** Chunk separation for 160-bit output

| Step | 0 | 1 | 2 | 3 | 4 | 5 | 6 | 7 | 8 | 9 | 10 | 11 | 12 | $\cdots$ | 20 | 21 | 22 | 23 | 24 | 25 | 26 | 27 | 28 | 29 | 30 | 31 |
|------|---|---|---|---|---|---|---|---|---|---|----|----|----|----------|----|----|----|----|----|----|----|----|----|----|----|----|
| index | 0 | 1 | 2 | 3 | 4 | ⑤ | 6 | 7 | 8 | 9 | 10 | 11 | 12 | $\cdots$ | 20 | 21 | 22 | 23 | 24 | 25 | 26 | 27 | ㉘ | ㉙ | ㉚ | ㉛ |
| | | | | | | | | | | first chunk | | | | | | | | | | | | | skip | | | |

| Step | 32 | 33 | 34 | 35 | 36 | 37 | 38 | 39 | 40 | $\cdots$ | 48 | 49 | 50 | 51 | 52 | 53 | 54 | 55 | 56 | 57 | 58 | 59 | 60 | 61 | 62 | 63 |
|------|----|----|----|----|----|----|----|----|----|----------|----|----|----|----|----|----|----|----|----|----|----|----|----|----|----|----|
| index | ⑤ | 14 | 26 | 18 | 11 | ㉘ | 7 | 16 | 0 | $\cdots$ | ㉚ | 3 | 21 | 9 | 17 | 24 | ㉙ | 6 | 19 | 12 | 15 | 13 | 2 | 25 | ㉛ | 27 |
| | | | | | | | | | | second chunk | | | | | | | | | | | | | | | | |

| Step | 64 | 65 | 66 | 67 | 68 | 69 | 70 | 71 | 72 | 73 | 74 | 75 | 76 | 77 | 78 | 79 | 80 | $\cdots$ | 88 | 89 | 90 | 91 | 92 | 93 | 94 | 95 |
|------|----|----|----|----|----|----|----|----|----|----|----|----|----|----|----|----|----|----------|----|----|----|----|----|----|----|----|
| index | 19 | 9 | 4 | 20 | ㉘ | 17 | 8 | 22 | ㉙ | 14 | 25 | 12 | 24 | ㉚ | 16 | 26 | ㉛ | $\cdots$ | 13 | 6 | 21 | 10 | 23 | 11 | ⑤ | 2 |
| | | | | | | second chunk | | | | | | | | | | | | | $\leftarrow \mid \rightarrow$ | | | first chunk | | | | |

# On Free-Start Collisions and Collisions for TIB3

Florian Mendel and Martin Schläffer

Institute for Applied Information Processing and Communications (IAIK)
Graz University of Technology, Inffeldgasse 16a, A-8010 Graz, Austria
martin.schlaeffer@iaik.tugraz.at

**Abstract.** In this paper, we present free-start collisions for the TIB3 hash functions with a complexity of about $2^{32}$ compression function evaluations. By using message modification techniques the complexity can be further reduced to $2^{24}$. Furthermore, we show how to construct collisions for TIB3 slightly faster than brute force search using the fact that we can construct several (different) free-start collisions for the compression function. The complexity to construct collisions is about $2^{122.5}$ for TIB3-256 and $2^{242}$ for TIB3-512 with memory requirements of $2^{53}$ and $2^{100}$ respectively. The attack shows that compression function attacks have been underestimated in the design of TIB3. Although the practicality of the proposed attacks might be debatable, they nevertheless exhibit non-random properties that are not present in the SHA-2 family.

**Keywords:** Hash function, SHA-3 competition, TIB3, free-start collision, collision attack.

## 1 Introduction

A hash function maps an input of arbitrary finite length to an output of a fixed length. An important basic security requirement for a cryptographic hash function is its collision resistance – it should be computationally infeasible to find two different inputs, which hash to the same output. Recently, the collision resistance of many commonly used hash functions has been broken or doubted. Therefore, NIST has started the SHA-3 competition [1] to find a successor of the SHA-1 and SHA-2 hash functions. The cryptanalysis of the proposed SHA-3 candidates is of high importance to find a valuable hash function which is fast but still secure within the next few decades.

Many new and interesting hash functions have been proposed and some of these algorithms have a remarkable speed on certain platforms. The SHA-3 candidate TIB3 [2] is one of the fastest submissions with a speed of about 6-8 cycles/byte for all output sizes on 64-bit platforms [3]. The main design idea behind TIB3 is to use extensive parallelism by designing a "shorter" but "wider" compression function. To strengthen this short but fast compression function and to counter differential attacks, each message block is used in two subsequent compression function calls. However, in this paper we show that it is still possible to construct collisions for the hash function TIB3 below the generic complexity.

P. Samarati et al. (Eds.): ISC 2009, LNCS 5735, pp. 95–106, 2009.

**Table 1.** Summary of results on TIB3

| type of attack | target | hash size | complexity | memory |
|---|---|---|---|---|
| free-start collision | compression function | all | $2^{24}$ | - |
| collision | hash function | 256 | $2^{122.5}$ | $2^{53}$ |
| collision | hash function | 512 | $2^{242}$ | $2^{100}$ |

Using high-probability iterative characteristics, we can construct many practical free-start collisions for the compression function of TIB3 (Sect. 3). These free-start collisions are then used for the collision attacks on both TIB3-256 (Sect. 4) and TIB3-512 (Sect. 5) with a complexity slightly below the birthday bound. The results of our work are summarized in Table 1. In the following section, we first give a short description of the hash function TIB3.

## 2   Description of TIB3

The hash function TIB3 is an iterated hash function based on the Merkle-Damgård design principle [4,5]. The two main instances of TIB3 are called TIB3-256 and TIB3-512. TIB3-256 processes message blocks of 512 bits and produces hash values of 224 or 256 bits, while TIB3-512 processes message blocks of 1024 bits and produces hash values of 384 or 512 bits. If the message length is not a multiple of the block size, an unambiguous padding method is applied. For the description of the padding method we refer to [2]. Let $m = M_1 \| M_2 \| \cdots \| M_t$ be a t-block message (after padding). Then, the hash value $h = H(m)$ is computed as follows:

$$H_0 = IV_H, \ M_0 = IV_M$$
$$H_i = f(H_{i-1}, M_i \| M_{i-1}) \quad \text{for } 1 \leq i \leq t$$
$$H_{t+1} = f(H_t, 0 \| H_t \| M_t) = h$$

where $IV_H$ and $IV_M$ are predefined initial values. Note that each message block is used in two compression function calls (see Fig. 1). The compression function $f$ is used in Davies-Meyer mode [6] and consist of 2 parts: the key schedule and the state update transformation.

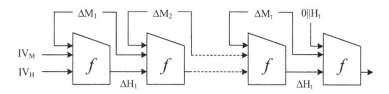

**Fig. 1.** The iteration mode of TIB3 uses the previous and current message block in each compression function call

The key schedule (or message expansion) of TIB3 takes as input the current and previous message block to compute a 4096-bit key for TIB3-256 and a 8192-bit key for TIB3-512. This key is split into 16 roundkeys $k_j$, where each roundkey is used in round $j$ of the state update transformation. For a detailed description of the key schedule function we refer to [2], since we do not need it in our analysis. In the following, we describe the state update transformation for TIB3-256 and TIB3-512 in more detail.

## 2.1    State Update Transformation for TIB3-256

The state update transformation of TIB3-256 starts from a (fixed) initial value $IV_H$ of four 64-bit words and updates them in 16 rounds each. In each round one 256-bit roundkey $k_j$ is used to update the four state variables $A$, $C$, $E$ and $G$ as follows:

$$G = G \oplus C$$
$$(A, C, E, G) = (A, C, E, G) \oplus k_j$$
$$(A, C, E) = Sbox(A, C, E)$$
$$G = PHTX(G)$$
$$C = PHTX(C)$$
$$A = A \boxplus^{32} G$$
$$G = E \boxplus^{32} G$$
$$(A, C, E, G) = (C, E, G, A),$$

where $Sbox$ is a 3-bit S-box, $PHTX$ is a bit-mixing function and $\boxplus^{32}$ denotes two 32-bit modular additions in parallel. One round of the TIB3-256 compression function is shown in Fig. 2. For the definition of the S-boxes we refer to [2]. The function $O = PHTX(I)$ is defined as follows:

$$T = I + (I \ll 32) + (I \ll 47)$$
$$O = T \oplus (T \gg 32) \oplus (T \gg 43)$$

After the last round of the state update transformation, the chaining values $A_0$, $C_0$, $E_0$, $G_0$ are XORed with the output values of the last round $A_{16}$, $C_{16}$, $E_{16}$, $F_{16}$ (feed-forward), resulting in the final value of one compression function $f$. For a detailed description of the hash function we refer to [2].

## 2.2    State Update Transformation for TIB3-512

In TIB3-512, two instances of the TIB-256 compression function are computed in parallel. The two parallel instances are mixed by two $PHTXD$ functions with inputs $C$, $D$ and $G$, $H$. A short description of the state update of TIB3-512 is given below, for more details we refer to [2].

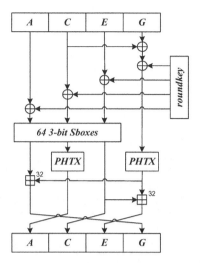

**Fig. 2.** One round of the TIB3-256 compression function

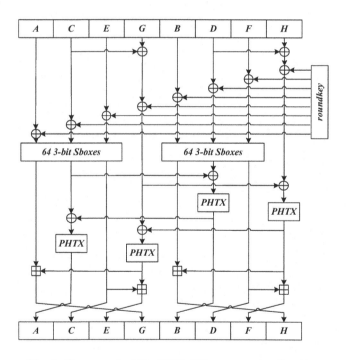

**Fig. 3.** One round of the TIB3-512 compression function

The state update transformation of TIB3-512 updates eight 64-bit words $A$, $B$, $C$, $D$, $E$, $F$, $G$ and $H$ in 16 rounds. One round of TIB3-512 is shown in Fig. 3 and defined as follows:

$$G = G \oplus C$$
$$H = H \oplus D$$
$$(A, B, C, D, E, F, G, H) = (A, B, C, D, E, F, G, H) \oplus k_j$$
$$(A, C, E) = Sbox(A, C, E)$$
$$(B, D, F) = Sbox(B, D, F)$$
$$(G, H) = PHTXD(G, H)$$
$$(C, D) = PHTXD(C, D)$$
$$A = A \boxplus G$$
$$B = B \boxplus H$$
$$G = E \boxplus G$$
$$H = F \boxplus H$$
$$(A, B, C, D, E, F, G, H) = (C, D, E, F, G, H, A, B),$$

where $k_j$ is the current 512-bit roundkey, *Sbox* is the same 3-bit S-box as in TIB3-256 and $\boxplus$ denotes a 64-bit modular addition. The function $(O, P) = PHTXD(I, J)$ is a "double" version of $PHTX$ and is defined as follows:

$$P = I \oplus J$$
$$P = PHTX(P)$$
$$O = I \oplus P$$
$$O = PHTX(O)$$

## 3   Free-Start Collisions for TIB3-256

In this section, we present a free-start collision attack on the compression function of TIB3-256 with a complexity of about $2^{24}$ compression function evaluations. Note that we use only differences in the chaining inputs and no differences in the message inputs are allowed. This is similar to the attack of den Boer and Bosselaers on MD5 [7]. However, in the case of TIB3 the complexity of the attack is much better due to its short compression function.

The attack is based on the fact that we can construct several 1-round iterative characteristics for the compression function of TIB3-256 with a probability between $2^{-2}$ and $2^{-4}$, depending on the bit position of the differences. The 1-round characteristic is shown below:

$$(-, \Delta[i], \Delta[i], \Delta[i]) \rightarrow (-, \Delta[i], \Delta[i], \Delta[i]) \tag{1}$$

where $\Delta[i]$ denotes a difference at bit position $i$. By subsequently using this 1-round characteristic 16 times, we will get a free-start collision for the whole 16-round compression function of TIB3-256. Note that the differences of the last round in $C_{16}$, $E_{16}$ and $G_{16}$ will be canceled due to the feed-forward, *i.e.* $A_0 \oplus A_{16}$, $C_0 \oplus C_{16}$, $E_0 \oplus E_{16}$, and $G_0 \oplus G_{16}$.

**Table 2.** Differential distribution table for the 3-bit S-box of TIB3 (*cf.* [2, page 15]) with input difference $S_i = (A, C, E)$ and corresponding output difference $S_o$. Probabilities are given in base 2 logarithms.

| $S_i \setminus S_o$ | 001 | 010 | 011 | 100 | 101 | 110 | 111 |
|---|---|---|---|---|---|---|---|
| 001 | | -2 | -2 | | | -2 | -2 |
| 010 | -2 | | -2 | | -2 | | -2 |
| 011 | -2 | -2 | | | -2 | -2 | |
| 100 | | | | -2 | -2 | -2 | -2 |
| 101 | | -2 | -2 | -2 | -2 | | |
| 110 | -2 | | -2 | -2 | | -2 | |
| 111 | -2 | -2 | | -2 | | | -2 |

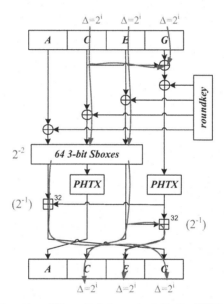

**Fig. 4.** The 1-round iterative differential characteristic for TIB3-256 which completely avoids the PHTX functions. We get a differential probability of $2^{-2}$ for the S-box, and $2^{-1}$ for the modular addition if $i \neq \{32, 64\}$.

### 3.1    On the Probability of the Characteristic

Now, lets take a closer look at the probability of the characteristic for each round $j$ which is shown in Fig. 4. Note that the xor of the roundkey in each round never changes the difference. The probabilities for all input/output differences of the 3-bit S-box of TIB3 are shown in Table 2.

– We start with the differences $\Delta[i]$ in $C$, $E$ and $G$. After the first xor operation, the difference in $G$ is canceled. In order to guarantee that the characteristic holds, we need that the differences $\Delta[i]$ in $C$ and $E$ at the input of the S-box propagate to the differences $\Delta[i]$ in $A$ and $E$ after the S-box. This holds with a probability of $2^{-2}$, see Table 2.

- Note that there are no differences in the $PHTX$ functions.
- In the case of $i = \{32, 64\}$, no carry occurs in the four 32-bit modular additions and the differences $\Delta[i]$ in $A$ and $E$ propagate to $\Delta[i]$ in $G$, $C$ and $E$ with a probability of 1. In the case of $i \neq \{32, 64\}$ no carry occurs in the two additions with a probability of $2^{-2}$.
- Hence, the resulting difference $\Delta[i]$ in $C$, $E$ and $G$ after one round is the same as in the input to this round.

The characteristic holds for one round with a probability of $2^{-2}$ for $i = \{32, 64\}$ and $2^{-4}$ for $i \neq \{32, 64\}$ and we get a characteristic for all 16 rounds with a probability of $2^{-32}$ and $2^{-64}$, respectively. Thus, we can construct a free-start collision for the compression function of TIB3-256 with a complexity of about $2^{32}$ for $i = \{32, 64\}$ and $2^{64}$ for $i \neq \{32, 64\}$ instead of $2^{128}$ as expected for a compression function with 256 bits. An example for a free-start collision for TIB3-256 with $i = 64$ is given in Table 3.

## 3.2   Improving the Attack Complexity

The complexity of the free-start collision attack can be significantly improved by using message modification techniques. Message modification was introduced by Wang et al. in the cryptanalysis of MD5 and SHA-1 [8,9]. The idea of message modification is to use the degrees of freedom one has in the choice of the message words to fulfill conditions on the chaining variables.

**Table 3.** A free-start collision for TIB3-256 with differences at bit position 64

| $H_1'$ | | $H_1''$ | | $\Delta H_1$ | |
|---|---|---|---|---|---|
| 00000000 | 00000000 | 00000000 | 00000000 | 00000000 | 00000000 |
| 00000000 | 00000000 | 80000000 | 00000000 | 80000000 | 00000000 |
| 00000000 | 00000000 | 80000000 | 00000000 | 80000000 | 00000000 |
| 00000000 | 00000000 | 80000000 | 00000000 | 80000000 | 00000000 |
| $M_1$ | | $M_2$ | | $\Delta M_1, \Delta M_2$ | |
| 90BDD5C0 | 451CE787 | E75BFF16 | FACB4B84 | 00000000 | 00000000 |
| 6BB03ABE | 8141141B | 6D6A0C85 | 52A79F37 | 00000000 | 00000000 |
| F45283B2 | 4019E54C | AECE5E32 | A5F07508 | 00000000 | 00000000 |
| 68D47A8C | EC658400 | A64F3E2B | E51D1923 | 00000000 | 00000000 |
| 20AC1B8D | 5C4F42F0 | E5079CCA | 5CC28EBE | 00000000 | 00000000 |
| B239522C | 8BF26045 | 1E7E2827 | 4E8C6B37 | 00000000 | 00000000 |
| E0EC45C2 | 3ACE0DE7 | 808C0A2F | B5E1F9AA | 00000000 | 00000000 |
| 2FB7DEBD | 84DDCF10 | 3BBF29A5 | FAB148DF | 00000000 | 00000000 |
| $H_2'$ | | $H_2''$ | | $\Delta H_2$ | |
| 55F5547C | 6AA5CC12 | 55F5547C | 6AA5CC12 | 00000000 | 00000000 |
| 40831045 | 5CC5F776 | 40831045 | 5CC5F776 | 00000000 | 00000000 |
| 43E53C0C | 4C64F862 | 43E53C0C | 4C64F862 | 00000000 | 00000000 |
| DD750B01 | DA7AD37F | DD750B01 | DA7AD37F | 00000000 | 00000000 |

In the case of TIB3-256 we can at least use the current 1024 bit message block for message modification. Using these 1024 degrees of freedom, the 16 conditions in the first 4 rounds can be fulfilled using basic message modification. In other words, we do not care about the probability of the characteristic in this part, since a message following the characteristic in the first 4 rounds can be found deterministically. Hence, the complexity of the attack can be reduced to $2^{24}$ for $i = \{32, 64\}$ and to $2^{48}$ for $i \neq \{32, 64\}$. We expect that the complexity can be further improved by using more sophisticated message modification techniques.

# 4    Collision Attack for TIB3-256

In this section, we show how the free-start collision attack on the compression function can be extended to a collision attack on the hash function. Even though the complexity of the attack is only slightly faster than a generic birthday attack, it exhibits some non-random properties that are not present in SHA-256. The attack uses the fact, that we can find several high-probability free-start collision producing characteristics for the compression function of TIB3-256.

## 4.1    Increasing the Number of Free-Start Collisions

In the previous section, we have constructed 64 different free-start collisions for $i = 1, \ldots, 64$. To increase the number of characteristics, we can fit two high probability characteristics with bit position $i \neq j$ into the compression function:

$$(-, \Delta[i, j], \Delta[i, j], \Delta[i, j]) \rightarrow (-, \Delta[i, j], \Delta[i, j], \Delta[i, j])$$

In the case of $i, j \neq \{32, 64\}$, we get a total probability of $2^{-128}$ which can be reduced to $2^{-96}$ by message modification. Note that we can further increase the number of characteristics by allowing carries at the beginning (first rounds) and end (last rounds). Hence, we can construct at least $\binom{64}{2} \sim 2^{11}$ different free-start collision.

## 4.2    From Free-Start Collisions to Collisions

In this section, we show how to use $2^x$ free-start collisions of the compression function to find collisions for the full hash function with a complexity of $2^{\frac{n-x}{2}}$. In the case of TIB3-256 we have constructed $2^{11}$ free-start collisions characteristics. Hence, the collision attack on TIB3-256 has a complexity of about $2^{122.5}$ compression function calls.

The collision attack uses 3 message blocks $M_1$, $M_2$ and $M_3$ (see Fig. 5). The main idea of the attack is to find two different first message blocks $M_1'$ and $M_1''$ which result in one of the $2^{11}$ differences of the free-start collision in $H_2$. Then, the according free-start collision is used to get a collision in $H_3$ after the second

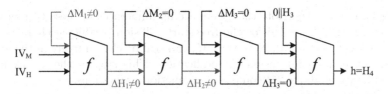

**Fig. 5.** In the collision attack on TIB3, three message blocks are used with differences only in $M_1$. The resulting near-collision in $H_2$ is transformed into a collision in $H_3$. The last message block $M_3$ is used for message modification.

compression function call. Note that we need a third message block $M_3$ for the message modification of the free-start collision:

$$H_1 = f(IV_H, M_1\|IV_M)$$
$$H_2 = f(H_1, M_2\|M_1)$$
$$H_3 = f(H_2, M_3\|M_2)$$
$$H_4 = f(H_3, 0\|H_3\|M_3) = h$$

The collision attack on TIB3-256 can then be summarized as follows:

1. Choose an arbitrary value for the message block $M_2$.
2. Use a birthday attack to find a $\Delta H_2$ (near-collision) which matches one of the $2^{11}$ free-start collision producing characteristics. Note that $M_2$ is fixed in the attack and only $M_1$ can be modified. This is important, since we do not allow any differences in $M_2$. The birthday phase has a complexity of about $2^{\frac{256-11}{2}} = 2^{122.5}$ compression function evaluations and can be implemented as follows:
   (a) Choose $2^{122.5}$ different random message blocks $M_1$ and store the messages and resulting chaining values $H_2$ in a set $S$. Then, we need to find those two messages $M_1'$, $M_1''$ with $M_1' \oplus M_1'' = \Delta M_1$, such that $H_2' \oplus H_2'' = \Delta H_2$ belongs to the set of $2^{11}$ free-start collision producing characteristics:
   (b) From the set $S$ find all pairs $(M_1', M_1'')$ with $H_2' = A'\|C'\|E'\|G'$ and $H_2'' = A''\|C''\|E''\|G''$ such that $A' = A''$, $(C' \oplus E') = (C'' \oplus E'')$ and $(C' \oplus G') = (C'' \oplus G'')$. Note that this can be done with a standard birthday attack (using a hash table) and we get $2^{122.5} \times 2^{122.5} \times 2^{-192} = 2^{53}$ pairs.
   (c) For each pair, compute $C' \oplus C''$ and check if the Hamming weight $HW(C' \oplus C'') \le 2$. In other words, check if $C' \oplus C''$ (and hence, $H_2' \oplus H_2''$) belongs to the set of $2^{11}$ free-start collision producing characteristics. This step of the attack has a complexity of about $2^{64-11} = 2^{53}$ XOR operations which is negligible compared to $2^{122.5}$ compression function evaluations.
3. Finally, we use the according free-start collision producing characteristic to turn the near-collision of $\Delta H_2$ into a collision in $H_3$ by using the message

blocks $M_2$ and $M_3$. Note that there are no differences in these two message blocks, which is needed for the free-start collision producing characteristic to work (*cf.* Section 3). Note that $M_3$ can still be chosen freely in the attack and hence, used for message modification. This step of the attack has a complexity of about $2^{96}$ compression function evaluations.

Alltogether, we can construct collisions in TIB3-256 with a complexity of about $2^{122.5}$ compression function evaluations and similar memory requirements. The complexity of this attack can be improved as soon as more than $2^{11}$ free-start collision characteristics have been constructed. One possibility to increase the number of characteristics is by allowing carries at the beginning (first rounds) and end (last rounds) of the compression function. Furthermore, the memory requirements of the attack can be significantly reduced. By using distinguished points [10,11] the first part of the birthday attack, *i.e.* 2.(a)-2.(b) can be implemented with memory requirements of $2^{53}$ instead of $2^{122.5}$.

# 5   Collision Attack for TIB3-512

The collision attack on TIB-256 can be extended to the hash function TIB3-512 as well. Since TIB3-512 uses two instances of TIB3-256 in parallel, we can reuse the free-start collision producing characteristic of TIB3-256 (Sect. 3).

## 5.1   Free-Start Collisions for TIB3-512

Since we do not have differences at the input of any PHTXD (mixing) function, the same free-start collisions of TIB3-256 can indeed be used without modification for TIB3-512. Hence, we can construct two different and independent 1-round characteristic for TIB-512 with differences in $C$, $E$, $G$:

$$(-,-,\Delta[i],-,\Delta[i],-,\Delta[i],-) \rightarrow (-,-,\Delta[i],-,\Delta[i],-,\Delta[i],-)$$

and/or with differences in $D$, $F$, $H$:

$$(-,-,-,\Delta[i],-,\Delta[i],-,\Delta[i]) \rightarrow (-,-,-,\Delta[i],-,\Delta[i],-,\Delta[i])$$

where $\Delta[i]$ denotes a difference at bit position $i$. The probability is $2^{-4}$ for $i \neq 64$ and $2^{-2}$ for $i = 64$ since one 64-bit addition is used instead of two 32-bit additions. TIB-512 has 16 rounds as well and we can construct a free-start collision for the compression function of TIB3-512 with a complexity of $2^{32}$ for $i = 64$ and $2^{64}$ for $i \neq 64$ again. Using basic message modification we can reduce the complexity to $2^{24}$ and $2^{48}$ respectively. An example for a free-start collision with a difference in bit position 64 is given in Table 4.

## 5.2   From Free-Start Collisions to Collisions

In the case of TIB3-512, the generic complexity for a collision attack is $2^{256}$. Therefore, we can easily fit up to 5 high probability characteristics next to each

**Table 4.** A free-start collision for TIB3-512 with differences at bit position 64

| $H_1'$ | $H_1''$ | $\Delta H_1$ |
|---|---|---|
| 00000000 00000000 | 00000000 00000000 | 00000000 00000000 |
| 00000000 00000000 | 00000000 00000000 | 00000000 00000000 |
| 00000000 00000000 | 00000000 00000000 | 00000000 00000000 |
| 00000000 00000000 | 80000000 00000000 | 80000000 00000000 |
| 00000000 00000000 | 00000000 00000000 | 00000000 00000000 |
| 00000000 00000000 | 80000000 00000000 | 80000000 00000000 |
| 00000000 00000000 | 00000000 00000000 | 00000000 00000000 |
| 00000000 00000000 | 80000000 00000000 | 80000000 00000000 |

| $M_1$ | $M_2$ | $\Delta M_1, \Delta M_2$ |
|---|---|---|
| 246B6D96 2C90A727 | 240E562C 5C5D4627 | 00000000 00000000 |
| 6139BD35 C099E9CC | 31C0A3B0 B3CC94A5 | 00000000 00000000 |
| 5533B6BF D6B80FB1 | 94E6BEBD 91BC6264 | 00000000 00000000 |
| 099868E2 8C9A5821 | BB665DC4 B5C3E598 | 00000000 00000000 |
| 08ED963E A808F1E6 | 7AEFABF8 3DF12657 | 00000000 00000000 |
| 1658D8E1 94925F32 | A4D3961F 2C8BFCF8 | 00000000 00000000 |
| AF7DE86F 4013CAD4 | 626DED61 3B3BE4F7 | 00000000 00000000 |
| 24573C4C 867D59A2 | 873613B2 C1F4B14A | 00000000 00000000 |

| $H_2'$ | $H_2''$ | $\Delta H_2$ |
|---|---|---|
| 8011137D 30451AA0 | 8011137D 30451AA0 | 00000000 00000000 |
| 5791600A B98C1C4A | 5791600A B98C1C4A | 00000000 00000000 |
| 60570740 31EEA496 | 60570740 31EEA496 | 00000000 00000000 |
| 31FB13D0 8A58960D | 31FB13D0 8A58960D | 00000000 00000000 |
| 15C9B361 99054AB7 | 15C9B361 99054AB7 | 00000000 00000000 |
| B6312CAB 57CF73AE | B6312CAB 57CF73AE | 00000000 00000000 |
| C7055809 B6B3BB6A | C7055809 B6B3BB6A | 00000000 00000000 |
| 422F8F0B 9DCCC9A4 | 422F8F0B 9DCCC9A4 | 00000000 00000000 |

other (complexity $2^{5\cdot48} = 2^{240}$). Hence, we can construct at least $\binom{128}{5} \sim 2^{28}$ different free-start collision producing characteristics for TIB3-512. The resulting collision attack has a complexity of about $2^{\frac{512-28}{2}} = 2^{242}$ and memory requirements of about $2^{128-28} = 2^{100}$ using distinguished points.

## 6    Conclusion

In this paper, we have presented free-start collisions for TIB3 with a complexity of about $2^{32}$ compression function evaluations. By using message modification techniques the complexity can be reduced to $2^{24}$. Furthermore, we can construct at least $2^{11}$ free-start collision producing characteristics for TIB3-256 and $2^{28}$ for TIB3-512. We show how to use these free-start collisions to construct collisions for TIB3 slightly faster than brute force search, and get a complexity of about $2^{122.5}$ compression function calls for TIB3-256 and $2^{242}$ for TIB3-512 with memory requirements of $2^{53}$ and $2^{100}$, respectively.

TIB3 is one of the fastest submissions due to its parallelism but short compression function. In the design of TIB3, compression function attacks have been underestimated. In this paper, we have shown how to find high-probability free-start collisions and turn them into an attack on the hash function. Although the practicality of the proposed attacks might be debatable, they nevertheless exhibit non-random properties that are not present in the SHA-2 family. Since there is still room for improvements, this analysis can be a starting point for future attacks on TIB3.

## Acknowledgements

We thank the designers of TIB3 for useful discussions and pointing out a small error in a previous version of this paper. The work in this paper has been supported by the European Commission under contract ICT-2007-216646 (ECRYPT II).

## References

1. NIST: Announcing Request for Candidate Algorithm Nominations for a New Cryptographic Hash Algorithm (SHA-3) Family. Federal Register Notice (November 2007), http://csrc.nist.gov
2. Montes, M., Penazzi, D.: The TIB3 Hash. Submission to NIST (2008)
3. Fleischmann, E., Forler, C., Gorski, M.: Classification of the SHA-3 Candidates. Cryptology ePrint Archive, Report 2008/511 (2008), http://eprint.iacr.org
4. Damgård, I.: A Design Principle for Hash Functions. In: Brassard, G. (ed.) CRYPTO 1989. LNCS, vol. 435, pp. 416–427. Springer, Heidelberg (1990)
5. Merkle, R.C.: One Way Hash Functions and DES. In: Brassard, G. (ed.) CRYPTO 1989. LNCS, vol. 435, pp. 428–446. Springer, Heidelberg (1990)
6. Matyas, S.M., Meyer, C.H., Oseas, J.: Generating strong one-way functions with crypographic algorithm. IBM Technical Disclosure Bulletin 27(10A), 5658–5659 (1985)
7. den Boer, B., Bosselaers, A.: Collisions for the Compression Function of MD-5. In: Helleseth, T. (ed.) EUROCRYPT 1993. LNCS, vol. 765, pp. 293–304. Springer, Heidelberg (1994)
8. Wang, X., Yin, Y.L., Yu, H.: Finding Collisions in the Full SHA-1. In: Shoup, V. (ed.) CRYPTO 2005. LNCS, vol. 3621, pp. 17–36. Springer, Heidelberg (2005)
9. Wang, X., Yu, H.: How to Break MD5 and Other Hash Functions. In: Cramer, R. (ed.) EUROCRYPT 2005. LNCS, vol. 3494, pp. 19–35. Springer, Heidelberg (2005)
10. van Oorschot, P.C., Wiener, M.J.: Parallel Collision Search with Cryptanalytic Applications. J. Cryptology 12(1), 1–28 (1999)
11. Quisquater, J.J., Delescaille, J.P.: How Easy is Collision Search. New Results and Applications to DES. In: Brassard, G. (ed.) CRYPTO 1989. LNCS, vol. 435, pp. 408–413. Springer, Heidelberg (1990)

# Detection of Database Intrusion Using a Two-Stage Fuzzy System

Suvasini Panigrahi and Shamik Sural

School of Information Technology,
Indian Institute of Technology, Kharagpur, India
{Suvasini.Panigrahi@sit,shamik@sit}.iitkgp.ernet.in

**Abstract.** This paper presents a novel approach for detecting intrusions in databases based on fuzzy logic, which combines evidences from user's current as well as past behavior. A first-order Sugeno fuzzy model is used to compute an initial belief for each transaction. Whether the current transaction is genuine, suspicious or intrusive is first decided based on this belief. If a transaction is found to be suspicious, its posterior belief is computed using the previous suspicion score and the fuzzy evidences obtained from the history databases by applying fuzzy-Bayesian inferencing. Final decision is made about a transaction according to its current suspicion score. Evaluation of the proposed method clearly shows that the application of fuzzy logic significantly reduces the number of false alarms, which is one of the core problems of existing database intrusion detection systems.

**Keywords:** Database security, Intrusion detection, Fuzzy logic, Fuzzy-Bayesian inference, Suspicion score.

## 1 Introduction

Ubiquitous use of database systems by organizations for daily operations and due to a surge in the importance of e-commerce to the world economy, concern regarding the security of databases has become crucial. Some of the data contained in these databases are quite sensitive in nature. Organizations manage access to such data meticulously with respect to internal users as well as external perpetrators. Murray [1] has found that the primary security threats come from internal misuse rather than external attacks.

Standard database security mechanisms such as authentication, authorization, access control and data encryption are often limited in their ability to protect database management systems (DBMS) from insiders. It is the presence of such inadvertent threats that have made intrusion detection systems (IDS) one of the fundamental security strategies for protecting databases.

Most of the existing IDSs are designed for networks and operating systems and hence, are not capable of detecting intrusions in databases. Also, every database user has a certain access pattern which is captured by almost all the database intrusion detection systems (DIDSs) as rules. Any violation of the

P. Samarati et al. (Eds.): ISC 2009, LNCS 5735, pp. 107–120, 2009.

stored patterns is reported as anomalous. These rules are static in nature and result in large number of false alarms when the user develops new patterns of behavior that are not yet known to the DIDS. Moreover, attackers can learn the static rules to a certain extent and, thus, avoid detection. Therefore, there is a need for developing database intrusion detection systems which can integrate multiple evidences [2] including patterns of genuine users as well as intruders and learn the behavior of users dynamically so as to minimize overall damage to the database.

It is well known that the real world is pervasively imprecise and uncertain, and hence, requires tolerance for imprecision and uncertainty to achieve better consonance with reality − features that can be exploited by fuzzy logic [3]. We feel that the use of fuzzy logic is appropriate for database intrusion detection due to two primary reasons. Firstly, several quantitative parameters that are used in the context of this problem can potentially be viewed as fuzzy variables. Secondly, the boundary between normal and anomalous behavior in databases is fuzzy.

In this paper, a novel two-stage fuzzy database intrusion detection system (TSFDIDS) is proposed, which utilizes fuzzy logic for integrating evidences from different sources to achieve flexibility and for smoothing the abrupt separation of genuine from intrusive behavior. Besides combining evidences, learning is also incorporated in TSFDIDS through application of prior knowledge and observed data on suspicious users by using fuzzy Bayesian decision method [4].

The rest of the paper is organized as follows. We present related work in database intrusion detection in Section 2. The components of our proposed system is discussed in Section 3 along with a description of the methodology. In Section 4, we discuss the results obtained from various experiments. Finally, we conclude in Section 5 of the paper.

## 2   Related Work

Research on intrusion detection has been ongoing for more than two decades and several host-based intrusion detection systems (HIDSs) and network intrusion detection systems (NIDSs) have been developed [5][6]. However, in spite of the significant role of databases in information systems, very limited research has been carried out in the field of intrusion detection in databases.

The approaches used in detecting database intrusions mainly include data mining and Hidden Markov Model (HMM). Chung et al. [7] present DEMIDS, a misuse detection system for relational database systems. This method assumes some consistency in database usage by the legitimate users. However, if this assumption does not hold, it results in a large number of false positives. Lee et al. [8] have described a method to discover intrusions in real-time databases. They have employed the time semantics of temporal data objects to detect intrusions. Barbara et al. [9] use HMM and time series to find malicious corruption of data by building database behavioral models that capture the changing behavior over time.

Lee et al. [10] designed a signature-based DIDS which matches new SQL statements against a set of legitimate transaction fingerprints to detect database intrusions. Nevertheless, if any of the legitimate transaction fingerprints are missing due to incomplete training data, it can cause many false alarms. Another relevant DIDS was proposed by Hu et al. [11]. In this approach, the data dependency relationships among the transactions are mined and this information is used to detect anomalies. Bertino et al. [12] have developed a DIDS which mines database log files to generate user profiles that model normal user actions and is used to identify intrusions. Kamra et al. [13] have proposed another method for detecting anomalous user requests by learning profiles of users and applications interacting with a database. Srivastava et al. [14] propose a data mining based IDS that considers sensitivity of attributes while mining the dependency rules.

Majority of the DIDSs as discussed above show a lot of variation in their accuracy. The main challenge identified by most of them is that any attempt to improve the rate of correct detection of intrusion, usually causes a rise in the false alarms as well. One of the motivations of our current research is to address this challenge.

We present a unique fuzzy database intrusion detection system (TSFDIDS) with learning and adaptation capabilities, which ensures high assurance and security. As mentioned previously, fuzzy logic has several important characteristics, which make it suitable for intrusion detection in databases. However, fuzzy logic has found only limited application in intrusion detection so far and that too in host-based IDSs and network-based IDSs. Dickerson et al. [15] developed a fuzzy intrusion recognition engine (FIRE) using fuzzy sets and fuzzy rules. FIRE uses simple data mining techniques to process the network input data and generate fuzzy sets for every observed feature. Seo et al. [16] have used fuzzy logic in distributed intrusion detection for network protection.

To the best of our knowledge, this is the first ever attempt to develop a database IDS using fuzzy information fusion and fuzzy-Bayesian inferencing.

## 3   Proposed Approach

The proposed approach uses a number of rules to analyze the deviation of each incoming transaction from the normal profile of users by assigning beliefs to it. The belief values from each rule are combined to obtain an initial belief. This initial belief is further strengthened or weakened according to its similarity with intrusive or genuine transaction history.

To meet the functionality as identified above, a comprehensive architecture as shown in Fig. 1, which integrates an Input Pattern Matching Component (IPMC), Fuzzy Combination Component (FCC), Fuzzy Evidences Component (FEC) and Fuzzy-Bayesian Inferencing Component (FBIC), has been proposed. Each incoming transaction of a user is first examined by the IPMC component of the system. It employs two techniques − sequence alignment and spatio-temporal outlier detection for selecting the suitable input rules that measure the deviation of an incoming transaction from the normal patterns.

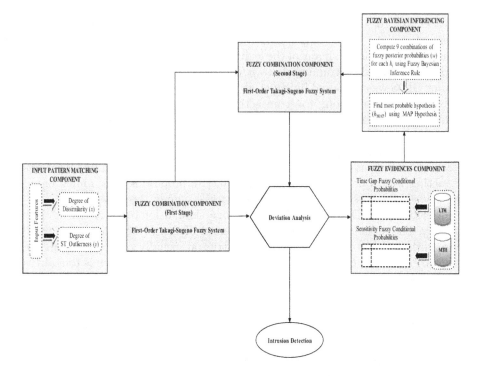

**Fig. 1.** Block Diagram of the Two-Stage Fuzzy Database Intrusion Detection System

A database user normally follows a specific sequence of database operations to accomplish a certain task. Each query in a transaction is chosen appropriately to achieve a meaningful purpose. Therefore, sequence is an effective way of representing user profiles and sequence alignment scores can be used to detect any anomalous activity. Sequence alignment is a technique used to quantify and evaluate similarity between two or more sequences. Kundu et al. [17] have introduced a novel way of applying sequence alignment for two-stage credit card fraud detection.

We use sequence alignment as a tool for comparing database access patterns of genuine users and intruders. The basic idea of our approach is that intruders are not entirely familiar with the normal database access patterns of legitimate users and they show some inter-transactional as well as intra-transactional deviation in their database access. Basic Local Alignment Search Tool (BLAST) [18] is one of the most popular heuristic approaches for sequence alignment, which is used in our database IDS for comparing sequence information.

Each new transaction is passed through the Input Pattern Matching Component (IPMC) and the new attribute sequence is aligned with each of the normal profile sequences. The *degree of dissimilarity* ($x$) is determined based on the dissimilarity between the new sequence and the user's normal profile (good) sequences. We use a simple scoring system to evaluate the *degree of dissimilarity*. A unit match score $\delta$ ($0 < \delta \leq 1$) is assigned to each matched element and a unit mismatch score $\delta'$ $(0 < \delta' \leq 1)$ to each mismatched element. Let $L$ be the length of the new sequence and $M$ be the number of matches with the aligned good sequence. The *degree of dissimilarity* $(x)$, where $0 \leq x \leq 1$, is given by the following expression:

$$x = \begin{cases} \dfrac{\delta'(L-M) - \delta M}{L} & \text{if } \delta'(L-M) > \delta M \\ 0 & otherwise \end{cases} \qquad (1)$$

In the current work, we use access sequence of table attributes for sequence alignment based deviation detection. The algorithm can be extended to include other transactional features as well.

It is also seen that the analysis of the spatio-temporal characteristics of a user's current behavior gives useful information on abnormal behavior in terms of his position (physical location of the user) and time of accessing the database. Thus, the normal spatio-temporal profile associated with each user is mined and used for the detection of intrusive activities in databases. Since an intruder is not likely to have complete knowledge regarding the normal spatio-temporal access patterns of users, some deviation from the user's profile is usually observed in its transactions, which are detected as spatio-temporal outliers. A spatio-temporal outlier (ST-outlier) can be defined as a spatio-temporal referenced object whose thematic attribute values are significantly different from those of other spatially and temporally referenced objects in its spatial and temporal neighborhood.

An approach based on the distance-based outlier (DB-outlier) detection technique [19] is utilized to filter out the ST-outliers. Other existing methods for outlier detection can efficiently handle only two dimensions or attributes of a dataset. However, the concept of DB-outlier can be applied to detect outliers effectively for any dimensional dataset. Let $N$ be the number of objects in the input dataset $T$ and let $DF$ be the underlying distance function that gives the distance between any pair of objects in $T$. An object $O$ in a dataset $T$ is considered to be a $DB(p, d)$ outlier if at least a fraction $p$ of the objects in $T$ lie at a distance greater than $d$ from $O$ ($d$-neighborhood denoted by $d_N$). Let $M$ represent the maximum number of data points within an outlier's $d_N$ (i.e., $M = N(1 - p)$). It means that an outlier needs to have less than $M$ objects within its $d_N$.

The clusters can be formed by using different attributes, although in the current work, we use the attributes $\langle loc\_ID, time\_slot, table\_ID\_seq \rangle$ for generating $ST$-outliers where:

- $loc\_ID$: location where a transaction was carried out
- $time\_slot$: time slot in which a transaction occurs. We have used the 24 hour clock and partitioned a day into 48 time slots, each of thirty minute duration. For example, at time = 6 pm ($\equiv 18 hour$), $time\_slot = 36$
- $table\_ID\_seq$: table access sequence in a transaction

The distance function $(DF)$ can be expressed as follows:

$$DF = \sqrt{(loc\_diff)^2 + (time\_diff)^2 + (tdist\_diff)^2} \qquad (2)$$

where *loc_diff*: distance between current transaction location and the user's normal profile transaction location, *time_diff*: distance between current transaction time slot and the user's normal profile time slot, *tdist_diff*: schema distance between current transaction *table_ID_seq* and the user's normal profile *table_ID_seq*.

For computing *tdist_diff*, we use a distance measure similar to that suggested in [7]. We assume a database schema $S$ with a set $RS$ of relation schemas. Consider two attributes $A_i \in R_1$ and $A_j \in R_2$ where $R_1, R_2 \in RS$ and $i, j \in \{1, \ldots, n\}$ where $n$ is the number of attributes in $S$. The pairwise schema distance between $A_i$ and $A_j$, denoted by $PS\_dist$, is defined as:

$$PS\_Dist(A_i, A_j) = \frac{SD(R_1, R_2)}{max\{SD(R_k, R_l)|R_k, R_l \in RS\}} \quad (3)$$

where $SD(R_1, R_2)$ computes the shortest distance between the two relations $R_1$ and $R_2$ based on the primary and foreign keys by which they can be related. Given a set of attributes $A = \{A_1, A_2, ..., A_n\} \subseteq attributes(S)$, the schema distance function denoted by *tdist_diff*, is defined as:

$$tdist\_diff(A_1, ..., A_n) = avg\{PS\_dist(A_i, A_j)\} \quad (4)$$

We measure the extent of deviation of an incoming transaction by its *degree of ST_outlierness*. Suppose $DF_{avg}(T_{j,\rho}^{U_k})$ and $DF_{max}(T_{j,\rho}^{U_k})$ respectively denote the average distance and maximum distance of an outlier transaction $T_{j,\rho}^{U_k}$ from the set of existing clusters. The *degree of ST_outlierness* $(y)$, where $0 \leq y \leq 1$ is then given by:

$$y = \begin{cases} \dfrac{DF_{avg}(T_{j,\rho}^{U_k})}{DF_{max}(T_{j,\rho}^{U_k})} & \text{if } |d_N| \leq M \\ 0 & otherwise \end{cases} \quad (5)$$

Three trapezoidal fuzzy sets − *low_x* $(Lx)$, *medium_x* $(Mx)$ and *high_x* $(Hx)$ are defined for the input $x$ having membership functions (MFs) $\mu_{Lx}(x)$, $\mu_{Mx}(x)$ and $\mu_{Hx}(x)$ respectively. Three similar trapezoidal fuzzy sets − *low_y* $(Ly)$, *medium_y* $(My)$ and *high_y* $(Hy)$ are also defined for the input $y$ with MFs $\mu_{Ly}(y)$, $\mu_{My}(y)$ and $\mu_{Hy}(y)$ respectively. The MFs can be expressed as follows:

$$\mu_{Lt}(t) = max\left(min\left(1, \frac{0.4 - t}{0.2}\right), 0\right) \quad (6)$$

$$\mu_{Mt}(t) = max\left(min\left(\frac{t - 0.2}{0.2}, 1, \frac{0.8 - t}{0.2}\right), 0\right) \quad (7)$$

$$\mu_{Ht}(t) = max\left(min\left(\frac{t - 0.6}{0.2}, 1\right), 0\right) \quad (8)$$

where $t \in \{x, y\}$. Here the trapezoidal MF parameters (0.2, 0.4, 0.6 and 0.8) were chosen by fitting experimental data with initially assumed function definitions.

The fuzzy inference module FCC determines whether the current transaction is normal or intrusive, based on the inputs $x$ and $y$. For each incoming transaction, it utilizes a two-input, single-output first-order *Sugeno fuzzy model* [20] to combine the inputs $x$ and $y$ for computing an *initial belief* $(z)$. The Sugeno model is used since it uses a simple inference procedure without involving the computationally expensive defuzzification operation. Based on the input MFs, FCC uses the following nine fuzzy if-then rules:

- **Rule 1:** if $x$ is *low_x* and $y$ is *low_y* then $z_1 = 0.5x + 0.5y$.
- **Rule 2:** if $x$ is *low_x* and $y$ is *medium_y* then $z_2 = 0.5x + 0.5y$.
- **Rule 3:** if $x$ is *low_x* and $y$ is *high_y* then $z_3 = 0.5x + 0.5y$.
- **Rule 4:** if $x$ is *medium_x* and $y$ is *low_y* then $z_4 = 0.5x + 0.5y$.
- **Rule 5:** if $x$ is *medium_x* and $y$ is *medium_y* then $z_5 = 0.5x + 0.5y$.
- **Rule 6:** if $x$ is *medium_x* and $y$ is *high_y* then $z_6 = 0.5x + 0.5y$.
- **Rule 7:** if $x$ is *high_x* and $y$ is *low_y* then $z_7 = 0.5x + 0.5y$.
- **Rule 8:** if $x$ is *high_x* and $y$ is *medium_y* then $z_8 = 0.5x + 0.5y$.
- **Rule 9:** if $x$ is *high_x* and $y$ is *high_y* then $z_9 = 0.5x + 0.5y$.

For simplicity, a coefficient of 0.5 is used in each of the fuzzy rules. Other values may be suitably chosen based on the specific requirements of the application. The consequents are aggregated to produce an inferred global crisp output initial belief $(z)$ by using the following expression:

$$z = \frac{w_1 z_1 + w_2 z_2 + \ldots + w_9 z_9}{w_1 + w_2 + \ldots + w_9} \quad \text{where, } 0 \le z \le 1 \tag{9}$$

The membership values on the premise part are combined using *min* operator to get the firing strength or weight $(w_i)$ of each rule. Three output fuzzy sets – *genuine_z* $(Gz)$, *suspicious_z* $(Sz)$ and *intrusion_z* $(Iz)$ are defined for the initial belief $(z)$ having MFs $\mu_{Gz}(z)$, $\mu_{Sz}(z)$ and $\mu_{Iz}(z)$ respectively. The membership grade of $z$ is computed in all the three output fuzzy sets and the incoming transaction is initially categorized as genuine, suspicious or intrusive depending on the corresponding fuzzy set in which the membership value is maximum. If $\mu_{Sz}(z)$ is maximum, the transaction is allowed but the user is labeled as suspicious and any further transaction carried out by the user is investigated by the TSFDIDS for possibility of intrusion.

For the database intrusion detection problem, there can be two possible values (hypothesis) for any suspected transaction: $h_1 = $ *intrusion*, $h_2 = \neg intrusion$. When the next transaction occurs by the same user, it is again passed through the TSFDIDS. IPMC examines the new transaction, assigns the basic probabilities, and then FCC computes the initial belief for this transaction. In case the transaction is again found to be suspicious, more information (evidence) regarding the user's database access behavior is obtained prior to deciding. The new information is expressed in the form of conditional probabilities. For accomplishing this, we have built a legitimate transactions history (LTH) for individual users from their past behavior and a generic malicious transactions history (MTH) from different types of past intrusive data.

It is observed that, in every database, there are a few attributes that are more important to be tracked for malicious modifications or leakage as compared to other attributes. We categorize all the attributes into the following three sensitivity levels – High Sensitivity ($HS$), Medium Sensitivity ($MS$) and Low Sensitivity ($LS$) by introducing weights for each attribute based on its sensitivity group. Also, modification (write) of an attribute of a particular sensitivity level is considered more important than accessing (read) the same attribute, from a database integrity point of view.

For a suspicious transaction, the initial observation done by the input pattern matching component is further strengthened by monitoring the frequency of transactions in terms of time gap from the previous transaction by the same user along with the most sensitive attribute operation in the current transaction. Thus, the parameters (evidences) considered during inspection are – time gap ($t$) and sensitivity ($s$). However, the quality of these parameters is inherently fuzzy. Hence, *fuzzy events* [4] and associated MFs are defined for each parameter.

Three *orthogonal fuzzy events* – *low_t* ($Lt$), *medium_t* ($Mt$) and *high_t* ($Ht$) are defined for time gap ($t$) that are characterized by trapezoidal MFs $\mu_{Lt}(t)$, $\mu_{Mt}(t)$ and $\mu_{Ht}(t)$ respectively. The conditional probabilities of the fuzzy events known as *fuzzy conditional probabilities* are determined in a manner similar to that suggested in [4] from the history databases. Table 1 shows the sample *time gap* fuzzy conditional probabilities for each hypothesis. The other parameter, namely, sensitivity is also dealt in a similar manner. Table 2 shows the sample *sensitivity* fuzzy conditional probabilities for each hypothesis.

**Table 1.** Fuzzy conditional probabilities for time gap

|       | $P(Lt|h_i)$ | $P(Mt|h_i)$ | $P(Ht|h_i)$ |
|-------|-------------|-------------|-------------|
| $h_1$ | 0.502       | 0.457       | 0.041       |
| $h_2$ | 0.124       | 0.182       | 0.693       |

**Table 2.** Fuzzy conditional probabilities for sensitivity

|       | $P(Ls|h_i)$ | $P(Ms|h_i)$ | $P(Hs|h_i)$ |
|-------|-------------|-------------|-------------|
| $h_1$ | 0.062       | 0.301       | 0.637       |
| $h_2$ | 0.671       | 0.272       | 0.057       |

Other than combining evidences, learning is also incorporated in TSFDIDS through application of prior knowledge and observed data on suspicious users by using the fuzzy-Bayesian decision method [4]. Bayesian learning usually fails to adequately address the uncertainties of the subjective parameters that are associated with intrusion detection. With the introduction of fuzzy set theory, it is possible to quantify the qualitative evaluation of the various subjective parameters and incorporate it into the database security problem. The two theories are therefore combined, which is popularly known as Fuzzy-Bayesian inferencing.

The fuzzy-Bayesian inferencing component employs the fuzzy-Bayesian decision method to update the suspicion score $(v)$ of the current transaction in light of the new evidence from the FEC. The *fuzzy posterior probabilities* $(w)$ are calculated for each hypothesis $(h_i)$ by using the fuzzy-Bayesian inference rule given as follows:

$$w = P(h_i|F_1F_2) = \frac{P(F_1|h_i)P(F_2|h_i)P(h_i)}{\sum_{j=1}^{2} P(F_1|h_j)P(F_2|h_j)P(h_j)} \tag{10}$$

where, $F_1$ and $F_2$ are fuzzy information. Since there are two evidence parameters and each parameter has been defined to have three possible quality ratings as discussed above, there are a total of $3^2 = 9$ combinations of inspection results for each hypothesis, namely, $P(h_i|LtLs)$, $P(h_i|LtMs)$, $P(h_i|LtHs)$, $P(h_i|MtLs)$, $P(h_i|MtMs)$, $P(h_i|MtHs)$, $P(h_i|HtLs)$, $P(h_i|HtMs)$ and $P(h_i|HtHs)$. When $F_1 = Lt$ and $F_2 = Ls$, fuzzy posterior probability of $h_1$ denoted by $P(h_1|LtLs)$ is given by the following expression:

$$P(h_1|LtLs) = \frac{P(Lt|h_1)P(Ls|h_1)P(h_1)}{\sum_{j=1}^{2} P(Lt|h_j)P(Ls|h_j)P(h_j)} \tag{11}$$

where, $P(Lt|h_1)$ and $P(Ls|h_1)$ are obtained as discussed in [4]. Derivation of the remaining posterior probabilities follows from the above Eq. (10).

*Maximum A Posteriori* (MAP) *hypothesis* is then applied, which yields the hypothesis with the highest fuzzy posterior probability as the output and is expressed as follows:

$$h_{MAP} = \max_{h_i \in H} P(h_i|e) \tag{12}$$

The fuzzy posterior probabilities are the updated beliefs about the last transaction by user based on the evidence from FEC and previous round suspicion score $v$ (last round). Since for the second suspicious transaction on a user, there is no $v$ (last round), the initial belief of the first round is itself taken as $v$ (last round) and the fuzzy posterior probabilities are computed based on this value in the second round.

The FCC applies a first-order Sugeno fuzzy model to combine the inputs — current round initial belief $z$ and highest posterior belief $w$ in the antecedent to get the output suspicion score (current round) $v$ in the same way as discussed earlier. Three output fuzzy sets — *genuine_v* ($Gv$), *suspicious_v* ($Sv$) and *intrusion_v* ($Iv$) are defined for the suspicion score, characterized by trapezoidal MFs $\mu_{Gv}(v)$, $\mu_{Sv}(v)$ and $\mu_{Iv}(v)$ respectively. The membership values of $v$ are computed in all the three output fuzzy sets and the FCC then decides whether the incoming transaction is legitimate, suspicious or intrusive depending on the corresponding fuzzy set in which the highest membership value is acquired.

## 4   Experimental Evaluation

In this section, we outline the results from an experimental evaluation of the proposed approach and illustrate its performance. The transactional web benchmark (TPC-W) [21] schema has been used for large scale simulation.

A extensive simulator has been developed which can handle various real life scenarios that are normally experienced in actual database applications. Firstly, any real-world database application contains malicious transactions interspersed with regular genuine transactions. Secondly, the genuine transactions are mostly consistent for a given user profile and are dependent on space and time. Thirdly, the genuine transactions and malicious transactions are independent events generated by two different parties and they have possibly different arrival rates. We capture such real life situations using a Markov Modulated Poisson Process (MMPP). The MMPP consists of a legitimate state $L$ and malicious state $M$ with arrival rates $\lambda_L$ and $\lambda_M$ respectively. Mixing of legitimate and malicious transactions is controlled by the transition between $L$ and $M$ states. Transition from $L$ to $M$ takes place with probability $q_{LM}$ and from $M$ to $L$ with probability $q_{ML}$.

We have used eight different simulation parameter settings (SS1 to SS8) as shown in Table 3 to show the efficacy of the proposed method. The parameter values in each setting SSi are chosen in such a way that the occurrence of intrusive transactions gradually reduces from SS1 to SS8. Standard performance metrics, namely, true positives (TP) and false positives (FP) are used to analyze the performance of the system under different test cases.

**Table 3.** Simulator Settings for arrival rate variations

| Simulator Setting | $q_{LM}$ | $q_{ML}$ | $\lambda_L$ | $\lambda_M$ |
|:---:|:---:|:---:|:---:|:---:|
| SS1 | 50 | 50 | 1 | 4 |
| SS2 | 15 | 50 | 1 | 4 |
| SS3 | 15 | 70 | 1 | 4 |
| SS4 | 10 | 80 | 1 | 4 |
| SS5 | 10 | 80 | 1 | 2 |
| SS6 | 10 | 90 | 3 | 1 |
| SS7 | 5 | 96 | 4 | 1 |
| SS8 | 5 | 99 | 8 | 1 |

### 4.1    Performance Analysis

We first study the performance of the system with variation in the percentage of overlap between the intrusive query set and the genuine query set. It is evident from the plot shown in Fig. 2 that TSFDIDS yields up to 92% TP and less than 5% FP. However, the TP rate gradually decreases with increase in the percentage of overlapping queries and the performance is worst (lowest TP) at the point of 100% overlap. With increase in the percentage of overlap, similarity among intrusive query set and genuine query set increases making it difficult to distinguish them, thus leading to degraded performance. It is seen that FP rate also reduces with rise in the percentage of overlapping queries, but the reduction is comparatively slower.

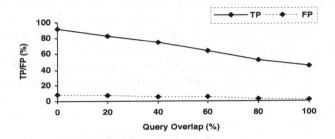

**Fig. 2.** Variation of TP/FP with intrusive and genuine query set overlap

In Fig. 3, we show the performance of the proposed system over various rounds. The first round commences with the first suspect transaction of a particular user. TSFDIDS is able to update the belief values over successive rounds and the process continues as long as the suspicion score is within the two threshold limits. It is seen that with each successive round the number of intrusive transaction detected as well as the false alarm rate rises. As a result the cumulative TP and cumulative FP increases at the end of each round.

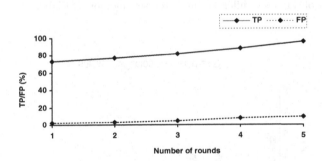

**Fig. 3.** Variation of TP/FP over Successive Rounds

## 4.2 Comparative Performance

We next compare the performance of TSFDIDS with two other systems proposed respectively by Hu et al. [11], which uses data dependency relationships, and Srivastava et al. [14], which uses weighted sequence mining. DDIDS represent the DIDS in [11] and WDIDS the DIDS in [14]. We compute TP and FP at each SSi for all the three DIDSs as mentioned above. It is evident from the plot shown in Fig. 4 that the TP rate gradually decreases from SS1 to SS8 with decrease in the percentage on intrusive transactions for all the three DIDSs. However, TSFDIDS is able to detect intrusive transactions more correctly (TP) as compared to DDIDS and WDIDS.

It is found that choosing support and confidence values is a problem in DDIDS as well as WDIDS. The TP rate in these two systems is highly dependent on the number of attribute dependency rules mined. Even if a low support value is chosen, the number of rules mined is quite less, which results in degraded performance for these systems. Moreover, the performance of TSFDIDS is also better in case of FP (lower value of FP) than the other two approaches as shown in Fig. 5.

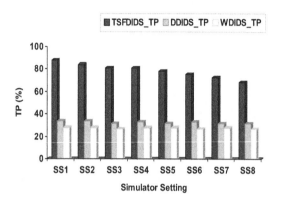

**Fig. 4.** Variation of TP with different simulator settings for TSFDIDS, DDIDS and WDIDS

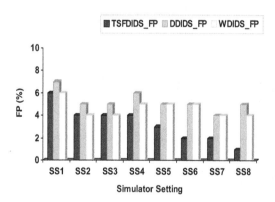

**Fig. 5.** Variation of FP with different simulator settings for TSFDIDS, DDIDS and WDIDS

## 5    Conclusions

In this paper, we have developed an innovative approach for database intrusion detection by combining evidences using fuzzy rules. In addition, belief update also takes place by means of fuzzy-Bayesian inferencing. A DIDS consisting of four components, namely, input pattern matching component, fuzzy combination component, fuzzy evidences component and fuzzy-Bayesian inference component, has been proposed. Sugeno fuzzy model is applied to combine inputs from the IPMC for computation of initial belief about each incoming transaction. Belief is updated by means of fuzzy-Bayesian inferencing using history database of both genuine users as well as intruders. MTH is built from history data about past malicious behaviors detected by any organization. The experiments yielded up to 92% TP and less than 5% FP. Combining rules using fuzzy logic accelerates detection rate while keeping the FP rate low and the use of fuzzy-Bayesian inference method further stimulates the system accuracy.

# References

1. Murray, A.C.: "The Threat From Within", Network Computing (August 2005), http://www.networkcomputing.com/showArticle.jhtml?articleID=166400792
2. Sentz, K.: Combination of Evidence in Dempster-Shafer Theory, Sandia National Laboratories, US Department of Energy (July 11, 2008), http://www.sandia.gov/epistemic/Reports/SAND2002-0835.pdf
3. Zadeh, L.A.: Soft Computing and Fuzzy Logic. IEEE Software 11(6), 48–56 (1994)
4. Ross, T.J.: Fuzzy Logic with Engineering Applications, 2nd edn. Wiley International Edition (2007)
5. Hoglund, A.J., Hatonen, K., Sorvari, A.S.: A Computer Host-Based User Anomaly Detection Using the Self-Organizing Map. In: Proceedings of the IEEE-INNS-ENNS International Joint Conference on Neural Networks, IJCNN, July 2000, vol. 5, pp. 411–416 (2000)
6. Hu, W., Hu, W., Maybank, S.: AdaBoost-Based Algorithm for Network Intrusion Detection. IEEE Transactions on Systems, Man, and Cybernetics, Part B 38(2), 577–583 (2008)
7. Chung, C.Y., Gertz, M., Levitt, K.: DEMIDS: A Misuse Detection System for Database Systems. In: Proceedings of the Integrity and Internal Control in Information System, pp. 159–178 (1999)
8. Lee, V., Stankovic, J., Son, S.: Intrusion Detection in Realtime Databases via Time Signatures. In: Proceedings of the 6th IEEE Real-Time Technology and Applications Symposium, RTAS, pp. 124–133 (2000)
9. Barbara, D., Goel, R., Jajodia, S.: Mining Malicious Data Corruption with Hidden Markov Models. In: Proceedings of the 16th Annual IFIP WG 11.3 Working Conference on Data and Application Security, July 2002, pp. 175–189 (2002)
10. Lee, S.Y., Low, W.L., Wong, P.Y.: Learning Fingerprints for a Database Intrusion Detection System. In: Gollmann, D., Karjoth, G., Waidner, M. (eds.) ESORICS 2002. LNCS, vol. 2502, pp. 264–280. Springer, Heidelberg (2002)
11. Hu, Y., Panda, B.: A Data Mining Approach for Database Intrusion Detection. In: Proceedings of the ACM Symposium on Applied Computing, pp. 711–716 (2004)
12. Bertino, E., Terzi, E., Kamra, A., Vakali, A.: Intrusion Detection in RBAC-Administered Databases. In: Proceedings of the 21st Annual Computer Security Applications Conference, ACSAC, December 2005, pp. 170–182 (2005)
13. Kamra, A., Bertino, E., Lebanon, G.: Mechanisms for Database Intrusion Detection and Response. In: Proceedings of the 2nd SIGMOD PhD Workshop on Innovative Database Research, IDAR 2008, June 2008, pp. 31–36 (2008)

14. Srivastava, A., Sural, S., Majumdar, A.K.: Weighted Intra-transactional Rule Mining for Database Intrusion Detection. In: Ng, W.-K., Kitsuregawa, M., Li, J., Chang, K. (eds.) PAKDD 2006. LNCS (LNAI), vol. 3918, pp. 611–620. Springer, Heidelberg (2006)

15. Dickerson, J.E., Juslin, J., Koukousoula, O., Dickerson, J.A.: Fuzzy Intrusion Detection. In: Proceedings of the IFSA World Congress and 20th NAFIPS International Conference, pp. 1506–1510 (2001)

16. Seo, H.S., Cho, T.H.: Application of Fuzzy Logic for Distributed Intrusion Detection. In: Hao, Y., Liu, J., Wang, Y.-P., Cheung, Y.-m., Yin, H., Jiao, L., Ma, J., Jiao, Y.-C. (eds.) CIS 2005. LNCS (LNAI), vol. 3802, pp. 340–347. Springer, Heidelberg (2005)

17. Kundu, A., Sural, S., Majumdar, A.K.: Two-Stage Credit Card Fraud Detection Using Sequence Alignment. In: Bagchi, A., Atluri, V. (eds.) ICISS 2006. LNCS, vol. 4332, pp. 260–275. Springer, Heidelberg (2006)

18. Altschul, S.F., Gish, W., Miller, W., Myers, W., Lipman, J.: Basic Local Alignment Search Tool. Journal of Molecular Biology 215, 403–410 (1990)

19. Knorr, E.M., Ng, R.T., Tucakov, V.: Distance-Based Outliers: Algorithms and Applications. The International Journal on Very Large Data Bases 8(3-4), 237–253 (2000)

20. Jang, J.S., Sun, C.T., Mizutani, E.: Neuro-Fuzzy and Soft Computing: A Computational Approach to Learning and Machine Intelligence. Prentice-Hall India, Englewood Cliffs (1997)

21. Transaction Processing Performance Council, TPC Benchmark$^{TM}$ W (Web Commerce), Specification, Version 1.8 (February 2002), http://www.tpc.org/tpcw/default.asp

# Combining Consistency and Confidentiality Requirements in First-Order Databases

Joachim Biskup and Lena Wiese

Technische Universität Dortmund, 44221 Dortmund, Germany
{biskup,wiese}@ls6.cs.uni-dortmund.de
http://ls6-www.cs.tu-dortmund.de/issi/

**Abstract.** In a logical setting, consistency of a database instance with constraints is a fundamental requirement. We show how satisfaction of a set of constraints guarantees confidentiality of some information declared secret by a security policy – albeit at the cost of some modified database entries. We identify a very general class of constraints for which this problem is effectively and in many cases efficiently solvable by means of an automatic procedure. A distance minimization ensures maximal availability of correct database entries.

## 1   Introduction and Related Work

Intelligent database systems can react in a personalized way depending on the user profile of an interacting user. Hence the protection of confidential and private information can be achieved based on a personalized *security policy* and a user's *a priori knowledge*.

In addition to denial of access, *modification* of database entries has been in use for some time to achieve protection of secrets. *Cover stories* [1,2,3,4] in MLS databases ensure a consistent database view to a low-level user without revealing high-level information; this is basically achieved by adding harmless tuples that cover up for confidential tuples. It is argued in [1,3] that without cover stories (that is, by just refusing to answer), the existence of sensitive information can be disclosed. The belief-based approach [5] presents differing views of the world according to a user's clearance; their partial databases can thus be seen as containing cover stories for users of insufficient clearance. The "provable data privacy" approach for incomplete $\mathcal{ALC}$ knowledge bases [6] also deceives a user about the correct evaluation of some queries: They use a view that omits some answers and hence makes them undefined instead of denying them. All of these approaches however address management and evaluation of cover stories but not their automatic generation. The approach in [3] restores consistency of an MLS database after an update (which we do not consider) by inserting or modifying cover stories. In contrast to our approach they do not minimize the amount of cover stories with respect to the original instance.

As for consistency, it is also of paramount importance in other related research areas: *Data exchange* aims at adding as much data as possible from source to target instances; Fagin et al. [7] allow an infinite set of null values – hence their

P. Samarati et al. (Eds.): ISC 2009, LNCS 5735, pp. 121–134, 2009.

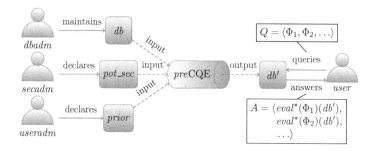

**Fig. 1.** Concept of preprocessing

chase procedure can potentially be infinite. *Consistent query answering* identifies those answers that are common to all consistent "repairs" of an inconsistent database instance; Chomicki [8] assumes an infinite domain of constant symbols.

This article regards the problem of finding a database instance that first of all is consistent with a user profile (including a set of database constraints) and second guarantees confidentiality of data that are specified in a security policy. We present a model generation procedure called *pre*CQE that achieves both goals at the same time. In a strict logical setting, *pre*CQE is hence a fully automatic procedure that generates a solution instance containing cover stories. As a tertiary goal, the procedure ensures maximal availability of correct data by optimization with respect to the input instance; in other words, the amount of cover stories is kept minimum. Figure 1 illustrates the general concept of the transformation of an input instance *db* into a solution instance *db'* that ensures consistency with constraints in a user profile *prior* and confidentiality of secrets in a security policy *pot_sec*. The work contained in this article extends and improves upon its propositional predecessor in [9] as follows. The incorporation of first-order logic significantly adds to expressiveness and user-friendliness of the system: first-order logic enables us to handle relational databases and allows for an easy declaration of database constraints, user profiles and security policies. The assumption of an infinite set of constants (combined with usage of quantifiers) enforces a purely first-order procedure and hence we avoided a direct propositionalization of the problem at hand. In contrast to the propositional case, termination as well as soundness and completeness results have been established.

We briefly outline the contributions of this article. Section 2 formalizes the logical background. Section 3 identifies a very general condition for the existence of a solution instance. Section 4 presents the *pre*CQE algorithm that uses additions and deletions of tuples to achieve consistency and confidentiality. Lastly, Section 5 puts the achievements in context.

## 2    System Settings

In this section we provide formal definitions of the system settings and all the components visualized in Figure 1. A comprehensive example at the end of this section illustrates the settings.

**Data Model:** To specify the vocabulary of an application, the database administrator *dbadm* fixes an infinite (albeit recursive) domain *dom* of constant symbols and a finite set $\mathcal{P}$ of predicate symbols. Each predicate symbol $P \in \mathcal{P}$ has an assigned arity *arity*$(P)$. Hence, *dbadm* implicitly defines a first-order language $\mathcal{L}$ (without function symbols and equality) that is based on *dom* and $\mathcal{P}$; $\mathcal{L}$ further includes an infinite set $\mathcal{V}$ of variables, the quantifiers $\exists$ and $\forall$ and the connectives $\neg$ (negation), $\vee$ (disjunction) and $\wedge$ (conjunction). On occasion, material implication $\rightarrow$ is used as an abbreviation (for a negation and a disjunction). Next, to express the invariant part of the application, *dbadm* declares a finite set $\mathcal{D}$ of closed formulas of $\mathcal{L}$ as database constraints. The predicate symbols and the constraints together form the database schema $DS = \langle \mathcal{P}, \mathcal{D} \rangle$; in Definition 2, $DS$ will also denote the set of all the instances that are compatible with the schema.

Lastly, *dbadm* maintains the data in an instance *db* of the database. From a logical perspective, all data tuples in *db* can be represented as a finite set of ground atoms of $\mathcal{L}$. By employing a closed world assumption, the database becomes a "complete" database (that returns a definite answer to any query). A database instance then induces a Herbrand-like "DB-interpretation" for $\mathcal{L}$; its characteristics are that (1) the universe of discourse uniformly coincides with *dom*, (2) predicates have a finite positive part and (3) each constant is interpreted by itself. Accordingly, we can speak of "DB-satisfiability" of a set of formulas of $\mathcal{L}$ (using the model operator $\models$ to denote satisfaction); these notions originate from [10].

**Definition 1 (DB-interpretation, DB-satisfiability).** *Consider a logical structure $I = \langle \mathcal{U}, i, j \rangle$ where $\mathcal{U}$ is the universe of discourse, and $i$ and $j$ give meaning to predicate symbols and constant symbols, respectively. $I$ is a DB-interpretation for $\mathcal{L}$ iff*

1. $\mathcal{U} = dom$
2. $i(P) \subset_{finite} \underbrace{dom \times \ldots \times dom}_{arity(P) \; times}$ for every $P$ in $\mathcal{P}$
3. $j(a) = a$ for every $a$ in $dom$

*A set $S$ of closed formulas is DB-satisfiable iff there is a DB-interpretation $I$ such that for all formulas $\Phi \in S$, $I \models \Phi$.*

To denote the interpretation that is induced by a specific database instance *db* we write $I^{db}$.

**User Model:** The user administrator models each user by a finite and consistent set *prior* of closed formulas as his assumed a priori knowledge before interacting with *db*. We assume that the user is aware of dependencies between the data as modeled by *dbadm* in the database constraints $\mathcal{D}$ – in this case, $\mathcal{D} \subseteq prior$. He may command over other (for example, instance-dependent) knowledge not explicitly modeled by *dbadm*; it is thus *useradm*'s task to identify user profiles and make such knowledge explicit in the user model *prior*.

Over the time, the user's knowledge is augmented with the responses that the database returns to his queries. Moreover, the system must be designed to protect sensitive information against a user who possesses some meta-knowledge – for example, knowledge about the data model. In particular, we need a query evaluation function that explicitly accounts for the closed world assumption. That is why we will formally define that open queries always return an infinite answer set (although for safe and domain-independent queries it is finitely representable by its positive part and an appropriate closure statement; see [10]).

**Definition 2 (Query evaluation for complete** $db$**).** *An open formula $\Phi$ is evaluated in a complete database instance db according to the following function that returns a set of closed formulas (where $\wp$ is the power set operator, $\boldsymbol{x}$ a vector of free variables and $\boldsymbol{a}$ a vector of constants):*

$$eval^*(\Phi(\boldsymbol{x})) : DS \rightarrow \wp\{\,\Psi \mid \Psi \text{ is a closed formula of } \mathscr{L}\}$$

*with*

$$eval^*(\Phi(\boldsymbol{x}))(db) := \{\Phi(\boldsymbol{a}) \mid \boldsymbol{a} \subset dom \text{ and } I^{db} \models \Phi(\boldsymbol{a})\}$$
$$\cup \ \{\neg\Phi(\boldsymbol{a}) \mid \boldsymbol{a} \subset dom \text{ and } I^{db} \not\models \Phi(\boldsymbol{a})\}$$

*For a closed formula $\Phi$, ordinary query evaluation reduces to a singleton set (for which curly braces are skipped):*

$$eval^*(\Phi)(db) := \begin{cases} \Phi & \text{if } I^{db} \models \Phi \\ \neg\Phi & \text{else} \end{cases}$$

Next, we want to analyze what the user is able to do with his knowledge; we will formalize his inference capabilities with a special form of model-theoretic implication. Again we assume that the user is aware of the data model and knows that he is querying a complete database; we encode this in the notion of "DB-implication".

**Definition 3 (DB-implication).** *For a set $S$ of closed formulas and a closed formula $\Phi$, $S$ implies $\Phi$ by DB-implication (written as $S \models_{DB} \Phi$) iff for all DB-interpretations $I$ such that $I \models S$ holds also $I \models \Phi$ holds.*

That is, the user's inferences comprise the closure of *prior* and the query responses under $\models_{DB}$.

**Policy Model:** To capture the application-specific security requirements, the security administrator *secadm* declares a security policy (more precisely confidentiality policy) *pot_sec* of "potential secrets". *pot_sec* is a finite set of closed formulas specifying sensitive information that is not to be revealed to the user if it holds (is *true*) in the instance *db*; the user may however assume that a potential secret does not hold (is *false*).

Notably, consistency of the database responses with the user profile (including the database constraints) is crucial for the enforcement of the confidentiality policy in this strict logical setting: once the user knowledge contains contradicting information, DB-implication reveals any logical sentence (including the sensitive information). Hence, enforcement of confidentiality amounts to finding a DB-interpretation that satisfies the a priori knowledge *prior* but does not satisfy any of the potential secrets. We name this property "inference-proofness".

**Definition 4 (Inference-proofness for complete database).** *Given a set prior and a set pot_sec, a complete database instance db′ is called* inference-proof *(with respect to prior and pot_sec) if and only if*

*(i)* $I^{db'} \models prior$
*(ii)* $I^{db'} \not\models \Psi$ *for every* $\Psi \in pot\_sec$

Note that an inference-proof database instance is also complete; it has to return a definite answer to each query – even one implying a potential secret. A well-informed database user might know (or at least suspect) what kind of information is sensitive. We thus assume that the user is aware of the policy specification *pot_sec* and knows that the database system will never tell him that a potential secret is *true*. He expects that instead the database system will insist that each potential secret is *false*. Hence, dealing with complete instances and a known policy specification, Item (ii) is equivalent to requiring $I^{db'} \models \neg\Psi$. In sum, we have to find a DB-interpretation $I^{db'}$ that satisfies the following "constraint set" C:

**Definition 5 (Constraint set).** *For a set prior and a set pot_sec, the constraint set is*

$$C := prior \cup Neg(pot\_sec)$$

*where* $Neg(pot\_sec) := \{\neg\Psi | \Psi \in pot\_sec\}$.

We employ a data modification technique to achieve this, but refrain from using data restriction ("refusal" or "denial of access"). More precisely, we allow additions and deletions of data tuples to present the user with a view of the data that is consistent with $C$. In this way, our kind of data modification coincides with the one used in cover story management.

While in general a constraint set $C$ might contain contradictions and hence is unsatisfiable, we identify a general condition that guarantees the existence of an inference-proof instance. Let $pot\_sec\_disj := \bigvee_{\Psi \in pot\_sec} \Psi$.

**Theorem 1 (Satisfiability of constraint set).** *Under the condition that* $prior \not\models_{DB} pot\_sec\_disj$, *the constraint set $C$ is DB-satisfiable.*

*Proof.* The assumption ensures that *pot_sec_disj* is not a tautology (otherwise it would be implied by *prior*) such that $Neg(pot\_sec)$ is indeed satisfiable. *prior* itself is satisfiable, too, as we require the user knowledge to be consistent. More precisely, applying the Definition 3 of DB-implication in contraposition, there is a DB-interpretation $I$ such that $I \models prior$ and $I \not\models pot\_sec\_disj$. But then, for all $\Psi \in pot\_sec$ also $I \not\models \Psi$ holds and thus (for complete $db$) $I \models \neg\Psi$ holds. In other words, $I \models Neg(pot\_sec)$. This ensures that indeed $I \models C$.

In addition to inference-proofness that is intended to ensure both consistency and confidentiality, "distortion minimality" responds to the issue of availability of correct data: only a minimum of the original database entries should be modified. We define the "distortion distance" of a database instance $db'$ with respect to the original input instance $db$ as the number of ground atoms that have a different evaluation in $db'$ than in $db$. In essence, we calculate the cardinality of the symmetric difference $db \oplus db' := (db \setminus db') \cup (db' \setminus db)$ of the two instances. The symmetric difference is the standard cardinality-based distance, which is widely used in belief revision and related fields; there it is often called "Dalal's distance" (see [11] for a comparison of several minimal change semantics).

**Definition 6 (Distortion distance).** *The* distortion distance *of an instance $db'$ with respect to the input instance $db$ is*

$$db\_dist(db') := card(db \oplus db').$$

A distortion minimal database instance is one that minimizes the distortion distance over all inference-proof instances.

**Definition 7 (Distortion minimality).** *An inference-proof instance $db'$ is* distortion-minimal, *iff there is no other inference-proof instance $db''$ such that $db\_dist(db') > db\_dist(db'')$.*

We illustrate the system settings with an example. We consider a database with medical data consisting of two relations: the one called "Ill" relates a person with a diagnosis, the one called "Treat" relates a person with a medical treatment.

*Example 1.* The database administrator *dbadm* specifies the language $\mathscr{L}$ as having the finite set of predicate symbols

$$\mathcal{P} = \{ Ill(Name, Diagnosis), Treat(Name, Treatment) \}.$$

In this example, we assume the predicates to be sorted and have the following infinite sort subsets of *dom*

$$Name = \{ \text{Pete, Mary, Lisa, Paul}, \ldots \}$$
$$Diagnosis = \{ \text{Aids, Flu, Cancer, Migraine}, \ldots \}$$
$$Treatment = \{ \text{MedA, MedB, MedC}, \ldots \}$$

such that $dom = Name \cup Diagnosis \cup Treatment$.
As the original database instance $db$, we have

| Ill | Name | Diagnosis |
|-----|------|-----------|
| | Pete | Aids |
| | Mary | Cancer |

| Treat | Name | Treatment |
|-------|------|-----------|
| | Pete | MedA |
| | Mary | MedB |

The user's a priori knowledge is declared in *prior* by the user administrator *useradm*; we assume in this example, that if the user knows the treatment of a patient, then he can narrow down the set of possible diagnoses (these formulas could also be present in the set $\mathcal{D}$ of database constraints):

$$prior = \{ \forall x(\, Treat(x, \text{MedA}) \rightarrow Ill(x, \text{Aids}) \vee Ill(x, \text{Cancer})),$$
$$\forall x(\, Treat(x, \text{MedB}) \rightarrow Ill(x, \text{Cancer}) \vee Ill(x, \text{Flu})) \}$$

The security administrator *secadm* specifies the potential secrets; in our example, the modeled user should neither be able to infer that there is a patient with the diagnosis Cancer nor a patient with the diagnosis Aids:

$$pot\_sec = \{\exists x\, Ill(x, \mathsf{Aids}), \exists x\, Ill(x, \mathsf{Cancer})\}$$

Unfortunately, some ordinary query responses enable the user (as modeled by *prior*) to infer a potential secret; for instance:

$$eval^*(Ill(\mathsf{Pete}, \mathsf{Aids}))(db) \models_{DB} \exists x\, Ill(x, \mathsf{Aids})$$

$$prior \cup eval^*(\exists x(\,Treat(x, \mathsf{MedB}) \wedge \neg Ill(x, \mathsf{Flu})))(db) \models_{DB} \exists x\, Ill(x, \mathsf{Cancer})$$

We see that the precondition of Theorem 1 is satisfied in this example, that is (after standardizing variables apart in *pot_sec_disj*) it holds that $prior \not\models_{DB}$ $\exists x\, Ill(x, \mathsf{Aids}) \vee \exists y\, Ill(y, \mathsf{Cancer})$. From Theorem 1 we know that the constraint set $C = prior \cup Neg(pot\_sec)$ is DB-satisfiable. Indeed we notice that the empty database instance $db'_1 = \emptyset$ is a solution candidate: $I^{db'_1}$ is a model of *prior*, but it's not a model of any of the potential secrets. We can calculate that its distortion distance is $db\_dist(db'_1) = 4$. A second inference-proof instance is the following $db'_2$:

| Ill | Name | Diagnosis |   | Treat | Name | Treatment |
|-----|------|-----------|---|-------|------|-----------|
|     | Mary | Flu       |   |       | Mary | MedB      |

The distortion distance is $db\_dist(db'_2) = 4$. Both instances are distortion-minimal: all other inference-proof instances have greater distances.

## 3    Conditions for DB-Satisfiability

Deciding whether a set of formulas is satisfiable is an undecidable problem for general first-order logic (see [12]). We are interested in establishing syntactic conditions for DB-satisfiability of input constraints $C$; in other words, by looking at the syntax of formulas in $C$ we can say whether there is a model with a finite positive part in the infinite domain *dom*. We adopt well-known results from investigations surrounding "safe queries" to our problem – that is, queries that have a finite response based on a fixed domain. As safety of queries is undecidable, syntactic restrictions for queries that define a decidable subclass of safe queries were sought. Several characterizations of differing complexity have been proposed; see for example [13] for an overview. We singled out the definitions of van Gelder and Topor (see [14]) for their "allowed" formulas as appropriate for our case. Intuitively, the allowed property ensures that each subformula that has to be evaluated returns a finite result; in other words, variables that, in principle, could be bound to infinitely many values when evaluating one subformula are actually bound to only finitely many values when evaluating another subformula and additionally this second subformula can be evaluated first. Van Gelder and Topor define the allowed property by stating a set of rules and taking the closure of these rules after a finite number of applications. This includes ground formulas of any kind.

We will apply the definition of the allowed property to formulas with only universal quantification. More precisely, we require that the formulas be in *prenex literal normal form* (PLNF, a term also borrowed from [14]): all quantifiers are moved into a prenex such that the remaining matrix is quantifier-free; additionally, the matrix is in *negation normal form* (NNF) – that is, negation signs only appear in front of atoms. In contrast to conjunctive normal form (CNF), a transformation to PLNF does not cause any loss of structural information and it does not increase the length of a formula; it is thus a very general class of formulas.

The following theoretical result will be of great use in the upcoming Section 4; it establishes the following: if an allowed universal formula is negated, transformed into PLNF and stripped of its prenex, then the resulting formula also has the allowed property (we implicitly introduce the function *dropprenex* here that returns the matrix of its input formula).

**Lemma 1 (Negations of allowed universal formulas).** *If $\Phi$ is an allowed universal formula, $dropprenex(plnf(\neg\Phi))$ is an allowed formula.*

*Proof.* (Sketch) As $\Phi$ is allowed universal of the form $\Phi = \forall \boldsymbol{x}\, \Psi(\boldsymbol{x})$, the PLNF of its negation has the form $plnf(\neg\Phi) = \exists \boldsymbol{x}\, nnf(\neg\Psi(\boldsymbol{x}))$. We drop the prenex such that $dropprenex(plnf(\neg\Phi)) = nnf(\neg\Psi(\boldsymbol{x}))$. We have not listed the definition of allowed formulas, but negation signs can be pushed inwards with it and thus the allowed property carries over to $nnf(\neg\Psi(\boldsymbol{x}))$.

# 4   Automating Inference-Proofness

Based on the theoretical background of the previous section, we present the *pre*CQE algorithm for the case of allowed universal constraints. We find the solution instance $db'$ by searching along branches in a binary "search tree". Some leaf in the search tree is then chosen as an instruction how to transform the original instance $db$ into the solution instance $db'$.

Branches are constructed by a "splitting" operation that creates two child nodes. It assigns a ground atom the value *false* in the left child node; in other words, the ground atom is either removed from the database instance (if it was included in $db$) or left out of the instance (if it was not included). In the right child node, the ground atom is assigned *true*; in other words, the ground atom is either added to or kept in the instance.

Yet, it may occur that there is actually no need for a splitting operation: only one of the two truth values (either *true* or *false*) for a ground atom promises the opportunity to satisfy a constraint. Then no new nodes are created but instead the unique truth value is assigned to the ground atom in the current node. In particular, this strategy applies to "unit constraints" – the ones containing only a single literal. Before starting with the technical details we give an example of an input in Example 2 and show its associated search tree in Figure 2.

*Example 2.* We continue Example 1. That is, we have the constraint set

$$C = \{\forall x(\neg \mathit{Treat}(x, \mathsf{MedA}) \vee \mathit{Ill}(x, \mathsf{Aids}) \vee \mathit{Ill}(x, \mathsf{Cancer})),$$
$$\forall x(\neg \mathit{Treat}(x, \mathsf{MedB}) \vee \mathit{Ill}(x, \mathsf{Cancer}) \vee \mathit{Ill}(x, \mathsf{Flu})),$$
$$\forall x \neg \mathit{Ill}(x, \mathsf{Aids}),$$
$$\forall x \neg \mathit{Ill}(x, \mathsf{Cancer}) \quad \}$$

Figure 2 shows how a solution for this input could be found; this procedure results exactly in the solution candidates $db_1'$ and $db_2'$ from Example 1. At first, we satisfy the two constraints $\forall x \neg \mathit{Ill}(x, \mathsf{Aids})$ and $\forall x \neg \mathit{Ill}(x, \mathsf{Cancer})$: the removal of the two entries $\mathit{Ill}(\mathsf{Pete}, \mathsf{Aids})$ and $\mathit{Ill}(\mathsf{Mary}, \mathsf{Cancer})$ in the root node is unequivocal. The first splitting step then corresponds to the decision whether to keep the atom $\mathit{Treat}(\mathsf{Pete}, \mathsf{MedA})$ or not. The splitting gives rise to a further splitting step in the left child node $v_1$ where ultimately in node $v_2$ the solution instance $db_1'$ is found.

Node $v_3$ also yields a solution instance – $db_2'$ – in the following manner: after the splitting operation, we have to add $\mathit{Ill}(\mathsf{Mary}, \mathsf{Flu})$ (in order to satisfy the second constraint formula). In this situation we do not have to split anymore because the other atoms in the formula (in this case, $\mathit{Treat}(\mathsf{Mary}, \mathsf{MedB})$ and $\mathit{Ill}(\mathsf{Mary}, \mathsf{Cancer})$) have already been treated.

Lastly, the root's right child $v_4$ fails (its branch is "pruned"): with the splitting operation we decided to keep the database entry $\mathit{Treat}(x, \mathsf{MedA})$ but then the atom $\mathit{Ill}(\mathsf{Pete}, \mathsf{Cancer})$ has to be added (in order to satisfy the first constraint formula). This indeed constitutes an unresolvable conflict with the constraint $\forall x \neg \mathit{Ill}(x, \mathsf{Cancer})$.

We now move on to a technical description of the *preCQE* algorithm. It assigns truth values by "marking" ground atoms in the original database instance: a

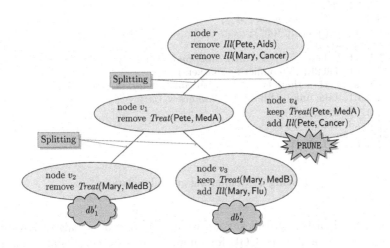

**Fig. 2.** Example of search tree

special value is temporarily appended to the ground atom that designates the intended truth valuation for the ground atom in the resulting instance $db'$. In the following, we will speak of a "marked database instance" $db_v$ and mean by it the database instance with marked ground atoms in a node $v$ of the search tree. More precisely, a marked database instance $db_v$ is a *finite* set of ground atoms some of which are marked. The following markers are used:

1. "keep" ($\mathbf{k}$): the according ground atom of $db$ should be retained in $db'$
2. "add" ($\mathbf{a}$): the according ground atom is not contained in $db$ but should be added to $db'$
3. "remove" ($\mathbf{r}$): the according ground atom of $db$ should not occur in $db'$
4. "leave" ($\mathbf{l}$): the according ground atom is not contained in $db$ and should also be left out of $db'$

As a more general notation, we use the function $marker_v(\gamma)$ to access and set the marker for a ground atom $\gamma$ in a marked database instance $db_v$. We can for example write $marker_v(P(\boldsymbol{a})) := \mathbf{a}$ to set a marker $\mathbf{a}$ for a ground atom $P(\boldsymbol{a})$; its meaning is that $P(\boldsymbol{a})$ was *false* in the input instance $db$, but that it has to be *true* in the solution instance $db'$. Markers can be implemented by an additional attribute in each relation. Yet, when using a ground atom $\gamma$ in the theoretical exposition we will *not* include the marker as an additional attribute; instead, we will access the marker only with the help of the *marker*-function.

```
1. GROUND(v): Determine ground violations in node v
   1.1. C_v^{vio} := {Φ ∈ C | eval_v(Φ) = ¬Φ};
   1.2. if (C_v^{vio} = ∅)
       1.2.1. db_best := db_v;
       1.2.2. min_lies_best := min_lies_v;
   1.3. else
       1.3.1. foreach Φ_i ∈ C_v^{vio}
           1.3.1.1. Φ'_i := plnf(¬Φ_i);
           1.3.1.2. Φ''_i := dropprenex(Φ'_i);
       1.3.2. V_v := ⋃_i eval_v^{pos}(Φ''_i);
       1.3.3. SIMP(v);
       1.3.4. if ( there is ψ ∈ V_v with card(unmarked_v(ψ)) = 0 )
           1.3.4.1. PRUNE; //(conflicting markers)
       1.3.5. else if ( there is ψ ∈ V_v with card(unmarked_v(ψ)) = 1 )
           1.3.5.1. take unique literal λ ∈ unmarked_v(ψ);
           1.3.5.2. MARK(v,λ̄);
           1.3.5.3. GROUND(v);
       1.3.6. else
           1.3.6.1. SPLIT(v);
```

**Listing 1.** Computing ground violations

As mentioned earlier, in this section we concentrate on an algorithm for allowed universal formulas. *preCQE* for allowed universal constraints comprises some procedures operating on marked database instances; the main procedures are given in pseudo code in Listings 1 to 3 with numbered lines. It starts with

2. SPLIT($v$): **Splitting** on a ground atom in node $v$
   2.1. choose $\psi \in V_v$;
   2.2. choose $\lambda \in unmarked_v(\psi)$;
   2.3. generate two child nodes $v_{left}$ and $v_{right}$;
   2.4. $db_{v_{left}} := db_{v_{right}} := db_v$;
   2.5. $min\_lies_{v_{left}} := min\_lies_{v_{right}} := min\_lies_v$;
   2.6. MARK($v_{left}, \neg|\lambda|$);
   2.7. GROUND($v_{left}$);
   2.8. MARK($v_{right}, |\lambda|$);
   2.9. GROUND($v_{right}$);

**Listing 2.** Splitting on a ground atom

3. MARK($v,\lambda$): **Marking** an unmarked ground atom $\gamma$ in $db_v$
   3.1. $\gamma := |\lambda|$;
   3.2. **if** ($\lambda = \gamma$ **and** $eval_v(\gamma) = \gamma$ )
     3.2.1. $marker_v(\gamma) :=$k;
   3.3. **else if** ($\lambda = \gamma$ **and** $eval_v(\gamma) = \neg\gamma$ )
     3.3.1. $marker_v(\gamma) :=$a;
     3.3.2. $min\_lies_v$++;
   3.4. **else if** ($\lambda = \neg\gamma$ **and** $eval_v(\gamma) = \gamma$ )
     3.4.1. $marker_v(\gamma) :=$r;
     3.4.2. $min\_lies_v$++;
   3.5. **else if** ($\lambda = \neg\gamma$ **and** $eval_v(\gamma) = \neg\gamma$ )
     3.5.1. $marker_v(\gamma) :=$l;
   3.6. **if** ($min\_lies_v \geq min\_lies_{best}$) PRUNE; //(bad bound)

**Listing 3.** Marking a ground atom

some initialization in the root node of the search tree which is not explicitly listed here; then the GROUND procedure is called.

The GROUND procedure computes in the set $C_v^{vio}$ those constraints that are violated in the current marked database $db_v$ (Line 1.1.); violated constraints are identified by evaluating them in $db_v$ respecting the markers with $eval_v$: ground atoms marked with r or l (or not contained in $db_v$) are evaluated to *false*; ground atoms marked with k or a (or contained but unmarked in $db_v$) are evaluated to *true*. If there are no violated constraints, a new optimum is found and saved in $db_{best}$ (Line 1.2.1.). Otherwise a "violation set" $V_v$ is computed (Line 1.3.2.): according to Lemma 1, the evaluation of the positive part (with $eval_v^{pos}$) returns a finite set due to the allowed property. The SIMP procedure (called in Line 1.3.3.) is responsible for simplifying ground violations according to already marked literals but is not explicitly adduced here. The algorithm tries marking ground literals in violation formulas in order to satisfy the violated constraints. Hence, if there is no unmarked ground literal left (as determined by the function $unmarked_v$ in Line 1.3.4.) we encountered a conflict in the current interpretation according to $db_v$ and can stop ("prune") exploration of the current branch; the PRUNE procedure (which is not listed here either) is responsible for backtracking in the search tree. If there is only one unmarked ground literal, we try marking it in the only way that could potentially satisfy the original constraint (1.3.5.2.). Lastly, if none of the previous cases holds, we know that there is a violation formula with more than one unmarked literal; we try both truth values for one of them with a splitting step (Line 1.3.6.1.).

In the SPLIT procedure two new child nodes are created in the search tree. In the left child node, the value *false* is tried for the chosen ground atom (Line 2.6.); in the right child node *true* is tried (Line 2.8.).

In the MARK procedure, a local lower bound $min\_lies_v$ of the distortion distance (that is, the amount of distortion in comparison to $db$) is increased whenever the new marker causes a change in the interpretation (Lines 3.3.2. and 3.4.2.); a branch is also pruned whenever a better solution was already found and saved in $db_{best}$ in a previous branch (Line 3.6.).

A marked instance corresponds to a "normal" unmarked instance when all the ground atoms that are marked with r or l are ignored; that is, the marked database instance is restricted to the "positive" ground atoms:

**Definition 8 (Positive restriction of $db_v$).** *The* positive restriction *of a marked database $db_v$ is:*

$$db_v^{pos} := \{ \, \gamma \mid \gamma \in db_v \text{ and } marker_v(\gamma) \notin \{r, l\} \, \}$$

That is, if a marked database instance was saved in $db_{best}$, we take its positive restriction as an inference-proof and distortion minimal solution.

We have proven the *preCQE* algorithm to have several nice properties we are interested in as surveyed in the remainder of this section.

1. The algorithm achieves both *consistency* and *confidentiality* (by means of inference-proofness) as well as *availability* (by means of distortion minimality). Note that the *preCQE* algorithm is guaranteed to find a distortion-minimal solution due to the Branch-and-Bound approach. If there is more than one optimal solution candidate, the first of them is chosen as the solution instance. This makes the algorithm dependent on the order of chosen violated constraints and unmarked literals therein. Yet, the security administrator *secadm* can additional state an explicit *availability policy*: the availability policy indicates preferences of (non-)distortion and can hence guide the choice of one of the optimal solution instances; see [9] for details. We also studied the use of a preference hierarchy of alternating confidentiality and availability policies in [15]. The advantage of such a preference declaration is that the algorithm remains non-interactive and the security administrator must not be asked to manually choose a solution from a set of candidates each time the algorithm is executed.

2. For allowed universal constraints we proved that the algorithm is guaranteed to terminate. We sketch the *termination* proof as follows. As already mentioned, the violation sets $V_v$ are finite and contain only constants of the "active domain" – that is, constants that either occur in the input instance $db$ or the input constraint set $C$. As the active domain is finite and markers are only set for ground atoms occuring in some formula in $V_v$, each branch has to terminate.

3. We investigated *satisfiability soundness* and *satisfiability completeness* – the algorithm finds a marked database instance if and only if there exists an inference-proof solution. Based on the termination result, we have equivalently shown the algorithm's *refutation soundness* and *refutation completeness* – the algorithm does not return a marked database instance if and only if there does not exist a solution.
4. The *runtime* of the algorithm is exponential in the cardinality of the active domain but it can be stopped prematurely with the first inference-proof solution when distortion minimality can be neglected. A prototypical implementation for propositional input showed *favorable performance* for a magnitude of ten thousands of propositional variables; a comprehensive study of the prototype is given in [16].

The somewhat lengthy proofs cannot be adduced here; especially the proof of refutation soundness requires an intricate construction using a "semantic tree" (see [17]) and Herbrand's theorem.

# 5    Discussion and Conclusion

We presented a terminating, sound and complete algorithm for allowed universal formulas that introduces a minimum of lies (cover stories) into a database instance. The algorithm has been extended to formulas with existential quantifiers: purely existential formulas (that can describe knowledge about the existence of some database entries) as well as tuple-generating dependencies (TGDs; see [7]) can be used in the constraint set. Existentially quantified variables are handled with "finite invention" (see [18]). This handling of existentially quantified variables differs from the one in [7,8] as it does not introduce null values and hence the output database instance is still complete. The most comprehensive class of constraints we have shown termination, soundness and completeness for are sets of TGDs plus denial (see [8]) and existential constraints.

The purpose of this work was to present a fully automatic procedure that runs without interaction from the administrators' side. The procedure is also more flexible than using the chase procedure (as in [7]) or just using tuple deletions (as in [8]): our algorithm incorporates both addition and deletion of tuples. In comparison to [8] we also handle a much more general class of formulas with only minor syntactic restrictions. Respecting the minimal change semantics, we achieve high availability assurance.

Future work can integrate built-in predicates (as in [14]) and examine their influence on data modification. A major open question is how to regain consistency after an update (by the user or the administrator).

We also consider it an interesting new field to combine our approach with fragmentation and encryption techniques as in [19].

# References

1. Galinovic, A., Antoncic, V.: Polyinstantiation in relational databases with multilevel security. In: Proceedings of 29th International Conference on Information Technology Interfaces, pp. 127–132. IEEE, Los Alamitos (2007)

2. Jukic, N., Nestorov, S., Vrbsky, S.V., Parrish, A.S.: Enhancing database access control by facilitating non-key related cover stories. Journal of Database Management 16(3), 1–20 (2005)
3. Cuppens, F., Gabillon, A.: Cover story management. Data & Knowledge Engineering 37(2), 177–201 (2001)
4. Sandhu, R.S., Jajodia, S.: Polyinstantation for cover stories. In: Second ESORICS. In: Deswarte, Y., Quisquater, J.-J., Eizenberg, G. (eds.) ESORICS 1992. LNCS, vol. 648, pp. 307–328. Springer, Heidelberg (1992)
5. Smith, K., Winslett, M.: Entity modeling in the MLS relational model. In: Proceedings of 18th International Conference on Very Large Data Bases, pp. 199–210. Morgan Kaufmann, San Francisco (1992)
6. Stouppa, P., Studer, T.: Data privacy for knowledge bases. In: Artemov, S., Nerode, A. (eds.) LFCS 2009. LNCS, vol. 5407, pp. 409–421. Springer, Heidelberg (2009)
7. Fagin, R., Kolaitis, P.G., Miller, R.J., Popa, L.: Data exchange: semantics and query answering. Theoretical Computer Science 336(1), 89–124 (2005)
8. Chomicki, J.: Consistent query answering: Five easy pieces. In: Schwentick, T., Suciu, D. (eds.) ICDT 2007. LNCS, vol. 4353, pp. 1–17. Springer, Heidelberg (2006)
9. Biskup, J., Wiese, L.: Preprocessing for controlled query evaluation with availability policy. Journal of Computer Security 16(4), 477–494 (2008)
10. Biskup, J., Bonatti, P.A.: Controlled query evaluation with open queries for a decidable relational submodel. Annals of Mathematics and Artificial Intelligence 50(1-2), 39–77 (2007)
11. Winslett, M.: Updating Logical Databases. Cambridge University Press, Cambridge (1990)
12. Börger, E., Grädel, E., Gurevich, Y.: The Classical Decision Problem. Springer, Heidelberg (2001)
13. Abiteboul, S., Hull, R., Vianu, V.: Foundations of Databases. Addison-Wesley, Reading (1995)
14. Van Gelder, A., Topor, R.W.: Safety and translation of relational calculus queries. ACM Transactions on Database Systems 16, 235–278 (1991)
15. Biskup, J., Burgard, D.M., Weibert, T., Wiese, L.: Inference control in logic databases as a constraint satisfaction problem. In: McDaniel, P., Gupta, S.K. (eds.) ICISS 2007. LNCS, vol. 4812, pp. 128–142. Springer, Heidelberg (2007)
16. Tadros, C., Wiese, L.: Using SAT solvers to compute inference-proof database instances (submitted, 2009)
17. Chang, C.L., Lee, R.C.T.: Symbolic Logic and Mechanical Theorem Proving. Academic Press, London (1973)
18. Hull, R., Su, J.: Domain independence and the relational calculus. Acta Informatica 31(6), 513–524 (1994)
19. Ciriani, V., De Capitani di Vimercati, S., Foresti, S., Jajodia, S., Paraboschi, S., Samarati, P.: Fragmentation and encryption to enforce privacy in data storage. In: Biskup, J., López, J. (eds.) ESORICS 2007. LNCS, vol. 4734, pp. 171–186. Springer, Heidelberg (2007)

# Cancelable Iris Biometrics Using Block Re-mapping and Image Warping*

Jutta Hämmerle-Uhl, Elias Pschernig, and Andreas Uhl**

Department of Computer Sciences, Salzburg University, Austria
uhl@cosy.sbg.ac.at

**Abstract.** The concept of *cancelable biometrics* provides a way to protect biometric templates. A possible technique to achieve such protection for iris images is to apply a repeatable, non-reversible transformation in the image domain prior to feature extraction. We applied two classical transformations, *block re-mapping* and *texture warping*, to iris textures obtained from the CASIA V3 Iris database and collected experimental results on the matching performance and key sensitivity of a popular iris recognition method.

## 1 Introduction

The use of biometrics comes with different problems as compared to conventional authentication systems. As biometric features are specific to an individual person, they cannot be changed (or not often, one person for example has only ten fingerprints and two iris patterns available). So where a password can simply be changed or an e-card invalidated, this is not possible with biometrics. In the same way, it is not possible to use different keys for different applications - for example if one wants to use a different key for the bank account and for access to the workplace computer.

This gets problematic further as the biometrics of a person are not even very secret - for example low-quality fingerprints are left everywhere, and eye images could be captured by hidden cameras. This does not only open the possibility to forge biometrics (like showing a picture of a person to a face recognition system), but also can raise privacy concerns. Large databases with biometric data would gain the potential of misuse, for example by cross-matching.

There exist several approaches in the area of biometric cryptosystems to cope with these issues - one of them is cancelable biometrics [1], which implements security by applying a key-dependent transformation to the captured biometric signals. The transformation must be non-invertible so that the original data cannot be reconstructed from the stored transformed version. At the same time matching still has to be possible with the distorted version.

---

* This work has been partially supported by the Austrian Science Fund, project no. L554-N15.
** Corresponding author.

P. Samarati et al. (Eds.): ISC 2009, LNCS 5735, pp. 135–142, 2009.

The concept of cancelable biometrics has been applied to various kinds of biometric features, like fingerprints, palmprints, face recognition and iris recognition [2,3,4].

The classical approach for cancelable biometrics [1] which we examined in this work for the case of iris recognition is to apply transformations in the image domain prior to feature extraction. This has the additional advantage that existing iris recognition algorithms can be used unmodified for the later feature extraction and matching stages. In section 2 we describe the iris recognition system we used for our tests, and in section 3 the transformations we applied - one which permutates blocks, and one which distorts images along translated grid points. Section 4 presents our experimental results.

## 2   Iris Recognition

Many iris recognition methods follow a quite common scheme [5], close to the well known and commercially most successful approach by Daugman [6]. After image acquisition, in a first step the iris texture is localised and extracted. From this texture, discriminative features are derived, which then can be used for comparison.

Since only the iris part of captured eye images is used for later feature extraction, it makes sense to consider only the iris textures in our image transformation experiments. We therefore always extract an iris texture from eye images as a first step. In our approach (following e.g. Ma et al. [7]) we assume the texture to be the area between the two almost concentric circles of the pupil and the outer iris. These two circles are found by contrast adjustment, followed by Canny edge detection and Hough transformation. After the circles are detected, unwrapping along polar coordinates is done to obtain a rectangular texture of the iris (as shown in Fig. 1.a). In our case, we always resample the texture to a size of 512x64 pixels.

Working now only on such textures, we employ a wavelet-based approach proposed by Ma et al. [7] to extract a bit-code. The texture is divided into $N$ stripes to obtain $N$ one-dimensional signals, each one averaged from the pixels of $M$ adjacent rows. We used $N = 10$ and $M = 5$ for our 512x64 pixel textures (only the 50 rows close to the pupil are used from the 64 rows, as suggested in [7]). A dyadic wavelet transform is then performed on each of the resulting 10 signals, and two fixed subbands are selected from each transform. This leads to a total of 20 subbands. In each subband we then locate all local minima and maxima above some threshold, and write a bitcode alternating between 0 and 1 at each extreme point. Using 512 bits per signal, the final code is then 512x20 bit.

Once bitcodes are obtained, matching can be performed on them and Hamming distance lends itself as a very simple distance measure. For matching to work well, we compensate for eye tilt by shifting the bit-masks during matching and use the concept of a "noise mask" to take care of hidden or distorted parts of the iris [6].

# 3   Cancelable Iris Templates

Using a key to seed a pseudo-random-number generator, we construct transformed versions of our textures. In the first transformation, each block of the
target texture is mapped to a block from the source texture. An actual example
with a block size of 16x16 pixel is shown in Fig. 1.b.

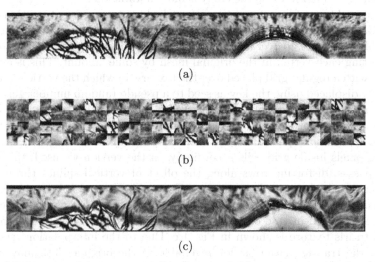

(a)

(b)

(c)

**Fig. 1.** (a) An example iris texture; (b) after randomly re-mapping blocks; (c) after
warping along a grid with randomly offset vertices

As stated in [1], using a re-mapping of blocks instead of a permutation should be
preferred for the application of cancelable biometrics, as it is not reversible. Source
blocks which are not part of the mapping are not contained in the transformed
texture at all, and therefore it is impossible to reconstruct the complete original.

The possible key space for $n$ blocks are the $n^n$ different mappings - however
many mappings are not suitable as keys. They can be close to the original -
for example if the key corresponds to the identity mapping. Or if the mapping
is a permutation, as mentioned above, it still contains all information from the
source. The other kind of unwanted key is when it corresponds to a mapping
which will not include enough information from the original texture for reasonable matching performance to be achievable - for example a mapping which maps
all blocks to the same source block.

We could therefore require some properties for mappings generated from a
key to be allowed. The number of used source blocks should be within a reasonable choice of $[a; b], 1 < a \leq b < n$, and the re-mapped texture also should be
sufficiently different from the original.

Using 30 blocks which was a reasonable number according to our experiments,
there would be $30^{30}$ or about $2^{147}$ different mappings. When requiring a number
of blocks to be present, the number will be reduced, down to 30! or about $2^{107}$ with
only permutations. When for example using only derangements out of those, so no

block retains its original position, it would be further reduced to about $2^{106}$ possibilities. Treating them all as keys still has the problem that many keys are close to each other - so this cannot directly be seen as possible key strength. Our experimental results in Section 4 use randomly generated keys out of all possibilities, so they include a degradation in matching performance from such similar keys. At the same time results from settings with fewer overall blocks have a greater chance of containing similar keys, so the effect is more notable there.

The second type of transformation we applied to textures is a distortion called mesh warping [8]. In this approach the texture is re-mapped according to a distorted grid mesh laid over it. A key is used to specify one particular distortion, by offsetting each vertex in the original mesh by some amount. This is done by starting with a regular grid placed over the texture, in which the vertices are then randomly displaced using the key as seed to a pseudo random number generator.

The transformation distorts the texture by sampling each pixel in the target texture from the corresponding area in the source texture, so that each vertex of the source mesh is placed to its translated position in the target mesh, interpolating pixels inside grid cells accordingly. In the version we used, this works in two passes, distorting rows along the offset of vertical splines through the mesh vertices, and then columns along offsets of horizontal splines. In the case of miniaturisation, a box-filter is applied to rows and columns, and linear interpolation is used in case of magnification. An example of the transformation applied to iris texture is shown in Fig. 1.c. Due to the interpolation strategies applied, the transformation is non-reversible as the original data may not be exactly recovered even if the warping parameters are known. The effect of non-revertability is more pronounced of course in the case of miniaturisation.

The maximum theoretical key space is the number of different meshes we can generate by offsetting vertices. It is $m^n$ if $n$ is the number of vertices, and $m$ is the number of possible offsets for one vertex. When choosing these parameters, the grid should not be distorted so much that large areas of the source texture are compressed to a single pixel in the target or are overlapped by other parts, as there may not be enough information left for matching. On the other hand, if the distorted picture too closely resembles the original, it is not of good use as a key-dependent signal either.

A realistic size used in our experiments would be a grid of 32x32 pixel cells - for our 512x64 pixel texture a regular grid with 16x2 vertices inside the texture can then be fit over it. If each vertex can be moved by 8 pixels horizontally and vertically, there are $17^2$ possibilities and a total of $289^{32}$ different transformations. Like in the case of re-mapped blocks, many transformations are very similar though. Using only 8 vectors with maximum offset as possible translations for each vertex, the total number of possible transformations would be $8^{32} = 2^{96}$. Like in the case of re-mapped blocks, we use a pseudo random number generator to generate one of the possible transformations for each key and therefore the experimental results use random offsets out of all possibilities. The effect of similar keys is therefore again reflected in the results, especially when there is a small parameter space.

# 4   Experiments

## 4.1   Experimental Setup

For testing, we used the *Interval* dataset out of the *CASIA Iris V3 database*, consisting of 2653 images in 396 classes (i.e. persons). In a first test, we assigned a random key to each class, then calculated the Hamming distance of resulting bit-codes between any two images (3517878 iris comparisons, 9008 of which are intra-class comparisons). If irises from the same class match worse after transformation, or irises from different classes match better after transformation, this shows in the match results as increased false non-match rate (FNMR) and increased false match rate (FMR), respectively. We plot the resulting FNMR against FMR as receiver operating characteristics (ROC) curves, and also indicate the equal error rate (EER) where FNMR and FMR are closest to each other. This is then used as indication of how usable transformed textures remain for the used feature extraction.

However, even when we can see no degradation in matching performance, this does not mean the key-dependent transformations increase security. For example applying the same key to each class could lead to good results in the first test. Therefore we performed another test to evaluate how discernible one transformation is from another, when they result from different keys (i.e. key-sensitivity is investigated). For this purpose, an iris class is copied multiple times, and each such class is then assigned a random key as before. If the key-dependent transformations don't lead to sufficiently distinct features, in this case it will shows up as high FMR because features from different classes will match. For this second test, we used the first 20 classes with at least 10 samples out of the *Interval* dataset, and created 50 random keys for each to have a roughly similar number of comparisons to the first test (2495000 iris comparisons, 45000 of which are intra-class).

## 4.2   Experimental Results

Figure 2.a shows the matching results using different block sizes for block re-mapping. For comparison, also the ROC curve for matching without any transformation applied is included. It is the lowest curve, obtaining an EER of about 1.1% with the used data and our implementation. Using random re-mappings of 32x32 pixel blocks (only 32 such blocks fit into the used 512x64 pixel textures), matching performance decreases, with a resulting EER of 1.6%. Matching performance decreases further using 16x16, 8x8 and 4x4 pixel sized blocks - which corresponds to 128, 512 and 2048 blocks respectively. The resulting EERs are 7.0%, 10.3% and 17.6%. Surprisingly, for the case of 2x2 and 1x1 blocks, where the latter amounts to a random re-mapping of single pixels, EER values decrease again to 5.6% and 7.3%. This means that even textures looking like random noise can be classified by the matching algorithm to some extent. Instead of quadratic blocks, blocks also can be rectangles. Table 1 compares the matching results when using different rectangular grid sizes for the block re-mapping, from fitting 81 blocks of 56x7 pixels to the 512x64 texture, down to fitting 16 blocks of

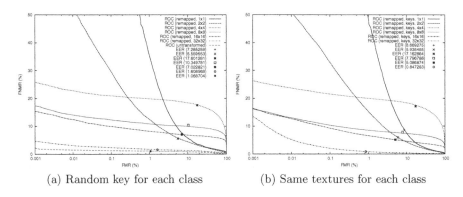

(a) Random key for each class          (b) Same textures for each class

**Fig. 2.** ROC curves for block re-mapping with different block sizes

128x16 pixel. The used sizes all obtain EERs below 1.6%. We notice, that we are able to obtain almost identical behaviour as compared to the original algorithm when choosing appropriate block re-mapping parameters.

Figure 2.b shows the result of using the same block sizes as figure 2.a, but now using the same images and only different keys for each class, to get an indication on the security of the keys, as described earlier. The big 32x32 pixel blocks obtain an EER of 0.8%, which in this case is an indicator for how distinct keys are from each other. A FMR of 1% means that 1% of comparisons of images from different classes resulted in a wrong match. As all classes use the same images, it means despite having different keys and therefore different block re-mappings, they were still close enough to match. As in the case when using actual eye classes instead of only copies with different keys, a block size of 4x4 pixels yields the worst results. The rectangular block sizes in Table 2 all result in an EER below 1%. Overall, the best results from the compared sizes are obtained when using 73x9 or 85x10 pixel sized blocks, with an EER of 1.2% for normal matching and 0.2% when only comparing different keys.

When using the mesh warping transformation, the size of the used mesh as well as the range of random offsets available can be adjusted. Figure 3.a compares

**Table 1.** EER with block re-mapping for different rectangular block sizes, using a random key for each class

| size(pixel) | 56x7 | 64x8 | 73x9 | 85x10 | 102x12 | 128x16 |
|---|---|---|---|---|---|---|
| blocks | 81 | 64 | 49 | 36 | 25 | 16 |
| EER (%) | 1.3 | 1.6 | 1.2 | 1.2 | 1.2 | 1.6 |

**Table 2.** EER with block re-mapping for different block sizes, same textures for each class

| size(pixel) | 56x7 | 64x8 | 73x9 | 85x10 | 102x12 | 128x16 |
|---|---|---|---|---|---|---|
| blocks | 81 | 64 | 49 | 36 | 25 | 16 |
| EER (%) | 0.2 | 0.4 | 0.2 | 0.2 | 0.3 | 1.0 |

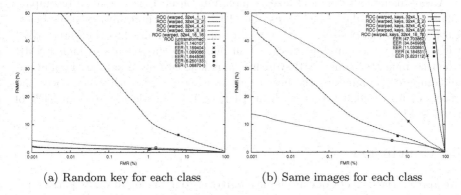

(a) Random key for each class          (b) Same images for each class

**Fig. 3.** ROC curves for warping a grid of 32x32 pixel blocks by different amounts

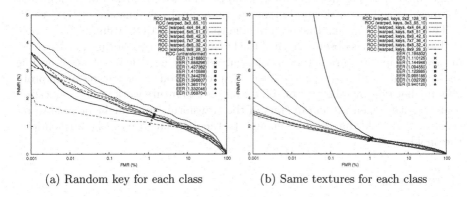

(a) Random key for each class          (b) Same textures for each class

**Fig. 4.** ROC curves for warping with different (rectangular) block sizes

matching performance when using a fixed grid with 128 vertices, consisting of 16x16 pixel sized blocks (note the expression $32 \times 4$ in the legend of the figure designates the number of 16x16 pixel blocks that can be accommodated in our texture patch). In the transformed grid, varying ranges for the horizontal and vertical pixel offsets of each mesh vertex are used. An offset in the range of 8 pixels increases the EER to 1.6%, an offset of 16 pixels to over 6%. In the case of 8 pixels offset and a 16 pixel grid, in the worst case two vertices can coincide if they are offset by the maximum of 8 pixels towards each other. In the case of 16 pixels however, the grid will overlap often, which loses even more information and also introduces additional features to the transformed texture. Figure 3.b shows the result when using the same iris textures in each class again. As expected, using only small translation offsets results in high FNMR, as transformations are too similar. The case of 8x8 pixel offsets with an EER of 4% has the best result of the compared offsets for 16x16 pixel blocks.

Always using half the block size as maximum offset, Figs. 4.a and 4.b compare different (rectangular) block sizes. The EER for matching always stays below 1.6% with the used sizes. Using a grid of of only 4 nodes has the lowest EER with 1.2% for the matching test, but in figure 4.b this results in high FNMR when keeping the same FMR, which means keys are often similar. Looking at all compared parameters, the best result is using a mesh with 9x9 vertices and offsets of up to 28x3 pixel - with an EER of 1.3% in Fig. 4.a and an EER of 0.9% in Fig. 4.b.

## 5   Conclusion

We applied the concept of cancelable biometrics to iris recognition by performing two transformations to iris textures.

The best parameters found for block re-mapping resulted in an EER of 1.2% when applying the transformation instead of 1.1% without transformations. For the mesh-warping transformation, our tests resulted 1.3% instead of 1.1% EER.

Therefore, this means that we are able to maintain ROC values (only a very slight degradation is observed) when the key-dependent transformations are used in case appropriate parameters are chosen. While we have demonstrated sensible key-sensitivity for the settings considered, a restriction of the keyspace will probably further improve the results. This is subject to future work as is the question if an attacker given the parameters of the transformations might be able to recover an iris code that is close enough to the original to pass the authenticity threshold.

## References

1. Ratha, N.K., Connell, J.H., Bolle, R.M.: Enhancing security and privacy in biometrics-based authentication systems. IBM Syst. J. 40(3), 614–634 (2001)
2. Chong, S.C., Jin, A.T.B., Ling, D.N.C.: High security iris verification system based on random secret integration. Computer Vision and Image Understanding 102(2), 169–177 (2006)
3. Chong, S.C., Jin, A.T.B., Ling, D.N.C.: Iris authentication using privatized advanced correlation filter. In: Zhang, D., Jain, A.K. (eds.) ICB 2005. LNCS, vol. 3832, pp. 382–388. Springer, Heidelberg (2005)
4. Zuo, J., Ratha, N., Cornell, J.: Cancelable iris biometric. In: Proceedings of the 19th International IAPR Conference on Pattern Recognition (ICPR 2008), pp. 1–4 (2008)
5. Bowyer, K., Hollingsworth, K., Flinn, P.: Image understanding for iris biometrics: A survey. Computer Vision and Image Understanding 110(2), 281–307 (2008)
6. Daugman, J.: How iris recognition works. IEEE Transactions on Circiuts and Systems for Video Technology 14(1), 21–30 (2004)
7. Ma, L., Tan, T., Wang, Y., Zhang, D.: Efficient iris recognition by characterizing key local variations. IEEE Transactions on Image Processing 13, 739–750 (2004)
8. Wolberg, G.: Image morphing: a survey. The Visual Computer 14(8/9), 360–372 (1998)

# Iris Recognition in Nonideal Situations

Kaushik Roy and Prabir Bhattacharya

Concordia Institute for Information Systems Engineering,
Concordia University, Montreal, Canada
{kaush_ro,prabir}@ciise.concordia.ca

**Abstract.** Most of the state-of-the-art iris recognition algorithms focus on processing and recognition of the ideal iris images which are captured in a controlled environment. In this paper, we process the nonideal iris images which are acquired in an unconstrained situation and are affected severely by gaze deviation, eyelids and eyelashes occlusion, non uniform intensity, motion blur, reflections, etc. To segment the nonideal iris images accurately, we deploy a variational level set based curve evolution scheme, which uses significantly larger time step for numerically solving the evolution partial differential equation (PDE), and therefore, speeds up the curve evolution process drastically. Genetic Algorithms (GAs) are deployed to select the subset of informative features by combining the valuable outcomes from the multiple feature selection criteria without compromising the recognition accuracy. The verification performance of the proposed scheme is validated using three nonideal iris datasets, namely, UBIRIS Version 2, ICE 2005, and WVU datasets.

**Keywords:** Iris recognition, variational level set method, curve evolution, genetic algorithms, nonideal situations.

## 1 Introduction

Iris recognition has been in the limelight for high security biometrics applications. Most state-of-the-art literatures on iris biometrics focused on processing of frontal view iris image of an eye [1,2]; however, a few new dimensions have been identified in iris biometric research, including processing and recognition of 'non frontal irises' and 'iris at a distance' [3]. For iris segmentation, most of the researchers assume that iris is circular or elliptical. However, in the cases of nonideal iris images, which are affected by gaze direction, motion blur, pupil dilation, nonlinear deformations, eyelids and eyelashes occlusions, reflections, etc, iris may appear as noncircular or nonelliptical [4,5]. In this paper, we use the methodologies to account for the nonideal irises to develop a nonideal iris recognition scheme. Previous techniques for the nonideal iris recognition do not adjust specifically for the nonideal situation [3]. Recently, several researchers have proposed different nonideal iris segmentation algorithms [3,4,5]. Most of the current nonideal iris segmentation schemes based on active contours take huge computational time due to expensive curve evolution approach [3,4,5]. In this research effort, we apply a variational level set based curve evolution approach to find the inner and outer boundaries accurately [6]. The proposed segmentation scheme with the variational level set approach uses larger time step to numerically solve the evolution partial differential

P. Samarati et al. (Eds.): ISC 2009, LNCS 5735, pp. 143–150, 2009.

equation (PDE), and thereby, speeding up the curve evolution process drastically [6]. The applied variational level set evolution could be developed using a simple finite difference scheme, and the level set function could be initialized as more efficient function than the traditional signed distance function. Also, the contours represented by the level set may break and merge naturally during evolution, and thus, the topological changes are handled automatically. Prior to applying the curve evolution approach using the active contours, we deploy an elliptical model to approximate the pupil and the iris boundaries. The iris biometrics template with the huge number of features increases the computational time. Hence, the optimal features set selection from a feature sequence with a relative high dimension has become an important issue in the field of iris recognition. The conventional feature selection techniques require sufficient number of samples per subjects to select the most representative features sequence. The Genetic Algorithms (GAs) suggest a particularly attractive approach to solve this kind of problem since they are generally quite effective in rapid global search of large, non-linear and poorly understood spaces [7]. Furthermore, different feature selection algorithms based on various theoretical arguments may produce different results on the same data set [7]. This makes selecting the optimal features subset from the original data set difficult. Therefore, we propose GAs to select the significant features subset by combining the valuable outcomes from the multiple feature selection criteria, and the proposed approach provides a convenient way of selecting a better feature subset based on the performance of the different feature selection schemes. To evaluate the proposed scheme, Support Vector Machine (SVM)-Recursive Feature Elimination (RFE), k-Nearest Neighbor (k-NN), T-statistics, and entropy-based methods are used to provide the candidate features for the selection of optimal features subset using GAs [8,9].

## 2    Nonideal Iris Segmentation

The segmentation of the nonideal iris image is a difficult task because of the noncircular shape of the pupil and the iris, and the shape differs depending on the image acquisition techniques [3]. First, we segment the iris and pupil boundaries from the eye image and then unwrap the localized iris region into a rectangular block of fixed dimension. We divide the iris segmentation process into two steps. In the first step, we use an elliptical model to approximate the inner (pupil) and outer (iris) boundaries of the iris, and then, we apply the geometric active contours with the variational formulation to find the exact inner and outer boundaries of the iris based on the estimated boundaries obtained in the previous step.

To find an approximation of the inner boundary, we select an elliptical region with the five parameters $(p_1, p_2, r_1, r_2, \phi_1)$:horizontal and vertical coordinates of the pupil center $(p_1, p_2)$ , length of the major and minor axes $(r_1, r_2)$, and the orientation of the ellipse $\phi_1$ and measure the intensity values for a fixed number of points on the pupil circumference. We vary the ellipse parameters with a small step size of three pixels to increase the ellipse size, and choose a fixed number of points randomly on the circumference to calculate the total intensity value. We repeat this process to find the boundary with a maximum variation in luminance and the center of the pupil [4]. The approximate estimate, $(I_1, I_2, R_1, R_2, \phi_2)$ for the outer boundary, on the other hand, can be found in the similar way. Based on the approximation of the inner and outer boundaries, the curve is evolved by using the level set with a variational formulation technique for accurate segmentation of the pupil and the iris regions. In the following paragraph, we briefly discuss the segmentation process based on variational level set approach.

In the level set formulation, the active contours, denoted by $C$, can be represented by the zero level set $C(t) = \{(x, y)|\phi(t, x, y) = 0\}$ of a level set function $\phi(t, x, y)$. To evolve the curve towards the inner and outer boundaries, we define the following total energy functional [6]

$$\varepsilon(\phi) = \mu\rho(\phi) + \varepsilon_{g,\lambda,v}(\phi) \tag{1}$$

where $\varepsilon_{g,\lambda,v}(\phi)$ denotes the external energy, which depends on the image data and drives the zero level set toward the object boundaries, and $\mu\rho(\phi)(\mu > 0)$ denotes the internal energy, which penalizes the deviation of $\phi$ from the signed distance function during evolution and is defined as below

$$\rho(\phi) = \int_\Omega \frac{1}{2}(|\nabla\phi| - 1)^2 dxdy \tag{2}$$

where $\Omega$ is the image domain. In (1), $g$ is the edge detector function and defined by

$$g = \frac{1}{1 + |\nabla G_{\sigma*I}|^2} \tag{3}$$

where $G_\sigma$ is the Gaussian kernel with standard deviation $\sigma$, and $I$ denotes an input image. We can further define the external energy term $\varepsilon_{g,\lambda,v}(\phi)$ of (1) as follows

$$\varepsilon_{g,\lambda,v}(\phi) = \lambda L_g(\phi) + vA_g(\phi) \tag{4}$$

where $\lambda > 0$ and $v$ are constants, and the terms $L_g(\phi)$ and $A_g(\phi)$ in (4) are defined by

$$L_g(\phi) = \int_\Omega g\delta(\phi)|\nabla_\phi| dxdy \tag{5}$$

and

$$A_g(\phi) = \int_\Omega gH(-\phi)dxdy \tag{6}$$

respectively, where $\delta$ is the univariate Dirac function, and $H$ is the Heaviside function. The energy functional $L_g(\phi)$ measures the length of the zero level set curve of $\phi$, and $A_g(\phi)$ is used to speed up curve evolution. From the calculus of variations, the Gateaux derivative of the functional $\varepsilon$ in (1) can be written as

$$\frac{\delta\varepsilon}{\delta\phi} = -\mu[\Delta\phi - div(\frac{\nabla\phi}{|\nabla\phi|})] - \lambda\delta(\phi)div(g\frac{\nabla\phi}{|\nabla\phi|}) - vg\delta(\phi) \tag{7}$$

where $\Delta$ is the Laplacian operator. The function $\phi$ that minimizes this functional satisfies the Euler-Lagrange equation $\frac{\delta\varepsilon}{\delta\phi}=0$. Now, the desired evolution equation of the level set function is defined as

$$\frac{\delta\phi}{\delta t} = \mu[\Delta\phi - div(\frac{\nabla\phi}{|\nabla\phi|})] + \lambda\delta(\phi)div(g\frac{\nabla\phi}{|\nabla\phi|}) + vg\delta(\phi) \tag{8}$$

The second and third terms in the right hand side of (8) represent the gradient flows of the energy functional and are responsible of driving the zero level set curve towards the object boundaries. The Dirac function $\delta(x)$ in (8) is defined by [6]

$$\delta_\varepsilon(x) = \begin{cases} 0, & |x| > \epsilon \\ \frac{1}{2\epsilon}[1 + \cos(\frac{\pi x}{\epsilon})], & |x| \le \epsilon \end{cases} \tag{9}$$

In order to estimate the exact boundary of the pupil, we initialize the active contour $\phi$ to the approximated pupil boundary, and evolve the curve in the narrow band of 10 pixels. We evolve the curve from outside the approximated inner boundary to suppress the effect of reflections. Similarly, for computing outer boundary, the active contour $\phi$ is initialized to the estimated iris boundary, and the optimal estimation of the iris boundary is computed by evolving the curve in a narrow band of 20 pixels. In this case, the curve is evolved from inside the approximated iris boundary to reduce the effect of the eyelids and the eyelashes. Fig. 1(b, c) shows the segmentation results.

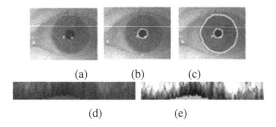

(a)          (b)          (c)

(d)                    (e)

**Fig. 1.** (a) Original image from WVU dataset (b) Pupil detection (c) Iris detection (d) Normalized image (e) Enhanced image

Besides reflections, eyelid occlusion, and camera noise, the iris image data may be corrupted by the occlusion of the eyelashes [8,9]. We deploy one dimensional Gabor filters and variance of intensity to isolate the eyelashes. We unwrap the iris region to a normalized rectangular block with a fixed dimension of size $64 \times 512$ [9]. Since the normalized iris image has relatively low contrast and may have non-uniform intensity values due to the position of the light sources, a local intensity based histogram equalization technique is applied to enhance the contrast of the normalized iris image within a small image block of size $10 \times 10$. Fig. 1(d, e) shows the unwrapped image and the effect of contrast enhancement.In this paper, Daubechies Wavelet Transform (DBWT) is used to extract the distinctive features set from normalized and enhanced iris image block of size $64 \times 512$ pixels [8]. We first divide the normalized image block into four sub images of size $32 \times 256$, and then apply the Daubechies four coefficient wavelet transform to each sub image [9]. Therefore, the normalized image is represented by a distinctive features set of $(2 \times 16 + 3) \times 4 = 140$ components.

## 3   Best Features Selection

In this paper, we apply GAs to select the prominent features based on the outcomes of the four feature selection algorithms as discussed in our previous work[8]. In order to choose the sets of feature selected by several feature selection algorithms, we deploy four existing feature selection algorithms, two filters (entropy-based, T-statistics) approaches and two wrapper (SVM-RFE, k-NN) approaches to form the *feature pool*. We apply each algorithm to the extracted features sequence and generate a ranking of those features. Given a ranking of features, we pick a number of top ranked features from each algorithm and provide these top-ranked features into the feature pool. In this paper, we propose the following fitness function

$$Fitness = W_1.(1-RR) + W_2.FAR + W_3.FRR + W_4.\frac{(Feature\,Size)}{(Total\,Number\,of\,Features)} \quad (10)$$

where $W_1, W_2, W_3$ and $W_4$ are constant weighting parameters which reflect the relative importance between *Recognition Rate* (RR), *False Accept Rate* (FAR), *False Reject Rate* (FRR) and *Feature Size*. In this paper, we use asymmetrical SVMs classifier as an induction algorithm in the experiments to separate the cases of false accepts and false rejects [8]. We use Roulette wheel selection to probabilistically select individuals from a population for latter breeding.

## 4  Experimental Results

We conduct the experimentation on three iris datasets namely, the ICE (Iris Challenge Evaluation) dataset, [12], the WVU (West Virginia University) dataset [13], and the UBIRIS version 2 dataset [14]. The ICE 2005 [12] contains 2953 images corresponding to 244 classes. The ICE database consists of left and right iris images for experimentation (1528 left iris images from 120 classes and 1425 right iris images from 124 classes). We evaluated the performance of the proposed iris recognition scheme on the WVU dataset [13]. The WVU iris dataset has a total of 1852 iris images from 380 different persons. The performance is also evaluated using UBIRIS version 2 (session 1 and session 2) dataset [14] which contains 2410 iris images from 241 different persons. In order to perform an extensive experimentation and to validate our proposed scheme, we generate a non- homogeneous dataset by combining the above three datasets, and this dataset contains 865 classes and 7215 images. We select a common set of curve evolution parameters based on variational level set approach to segment the nonideal iris

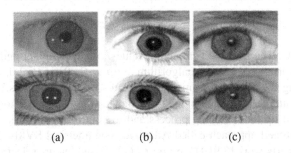

(a)               (b)               (c)

**Fig. 2.** Segmentation results on datasets (a) WVU, (b) ICE, and (c) UBIRIS

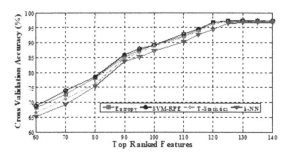

**Fig. 3.** Cross validation accuracy vs. top ranked features on combined dataset

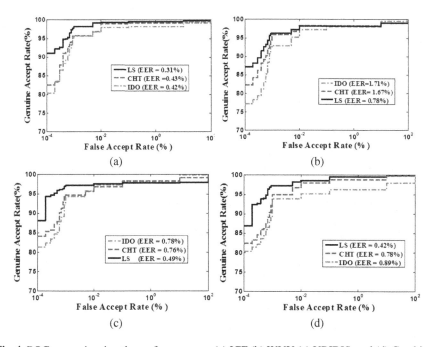

**Fig. 4.** ROC curve showing the performance on (a) ICE (b) WVU (c) UBIRIS, and (d) Combined datasets

images accurately. The selected parameters values to find the inner and outer boundaries using the variational level set algorithm are $\mu = 0.001$, $v = 2.0$, $\lambda = 5.0$, time step,$\tau = 3.0$. Fig 2 shows the segmentation results on three datasets.We deploy the SVMs for iris pattern classification due to its outstanding generalization performance [10,11]. To speed up the matching process and to control the misclassification error, we apply a combined approach called Adaptive Asymmetrical SVMs (AASVMs) as proposed in our early work [8,9].The proposed GAs based feature selection approach is used to reduce the feature dimension without compromising the recognition rate based on the multiple outcomes of four feature selection algorithms. Since the number of samples from most iris research is limited, cross-validation procedure is commonly used to evaluate the performance of a classifier. Leave-One-Out Cross Validation (LOOCV) is used for ICE dataset, and for WVU and UBIRIS datasets, we use 3-fold cross validation to obtain the training accuracy for GAs. Fig. 3 shows the accuracy of the selected feature subsets with a different number of top-ranked features from the four feature selection algorithms on the combined data set. Fig. 3 demonstrates the performance of four feature selection algorithms for the first 130 top ranked features. For the combined dataset, we can see that SVM-RFE finds the better accuracy than the other the algorithms with the 130-top ranked features. Therefore, after obtaining the 130 top-ranked features from different feature reduction algorithms, we input them to the feature pool used by the GAs. The proposed GAs based scheme reduces the features dimension from 140 to 105.In order to provide a comparative analysis, we apply the proposed level set approach (LS), integro-differential operator (IDO) method [1], and the Canny edge detection and Hough transform (CHT) based traditional approach for segmentation on all the datasets. ROC curves in Fig. 4 show that the matching performance is improved when the variational level set approach is used for segmentation on all the datasets. The Genuine Accept Rate (GAR) at a fixed FAR of 0.001% is (a) 96.24% in WVU, (b) 98.16% in ICE, and (c) 97.17% in UBIRIS datasets. However, the overall GAR on the combined dataset at the fixed FAR of 0.001% is 97.30%.

## 5 Conclusions

The accurate segmentation of the iris plays an important role in iris recognition. In this paper, we present a nonideal iris segmentation scheme using the variational level set based curve evolution approaches. The GAs are used to find the subset of informative texture features that can improve the analysis of iris data. The experimental results show that the proposed method is capable of finding feature subsets with better classification accuracy and/or smaller size than each single individual feature selection algorithm does. We validate the proposed iris recognition scheme on the ICE, the WVU, the UBIRIS, and the nonhomogeneous combined datasets with an encouraging performance.

## References

1. Daugman, J.: How iris recognition works. IEEE Transaction on Circuits, Systems and Video Technology 14(1), 1–17 (2003)
2. Daugman, J.: New methods in iris recognition. IEEE Transactions on Systems, Man, and Cybernetics-Part B 37(5), 1167–1175 (2007)
3. Schuckers, S.A.C., Schmid, N.A., Abhyankar, A., Dorairaj, V., Boyce, C.K., Hornak, L.A.: On techniques for angle compensation in nonideal iris recognition. IEEE Transactions on Systems, Man, and Cybernetics-Part B 37(5), 1176–1190 (2007)
4. Vatsa, M., Singh, R., Noore, A.: Improving iris recognition performance using segmentation, quality enhancement, match score fusion, and indexing. IEEE Transactions on Systems, Man, and Cybernetics-Part B 38(4), 1021–1035 (2008)

5. Ross, A., Shah, S.: Segmenting non-ideal irises using geodesic active contours. In: Biometric Consortium Conference, IEEE Biometrics symposium, pp. 1–6 (2006)
6. Li, C., Xu, C., Gui, C., Fox, M.D.: Level set evolution without re-initialization: a new variational formulation. In: Int. Conf. on Comp. Vis. and Pattern Recog., vol. 1, pp. 430–436 (2005)
7. Deb, K.: Multi-objective Optimization using Evolutionary Algorithms. J. Wiley, West Sussex (2004)
8. Roy, K., Bhattacharya, P.: Adaptive asymmetrical SVM and genetic algorithms based iris recognition. In: Int. Conf. on Pattern Recog., pp. 1–4 (2008)
9. Roy, K., Bhattacharya, P.: Level set approaches and adaptive asymmetrical SVMs applied for nonideal iris recognition. In: Kamel, M., Campilho, A. (eds.) ICIAR 2009. LNCS, vol. 5627, pp. 418–428. Springer, Heidelberg (2009)
10. Vapnik, V.N.: Statistical Learning Theory. J. Wiley, New York (1998)
11. Li, Q., Jiao, L., Hao, Y.: Adaptive simplification of solution for support vector machine. Pattern Recognition 40, 972–980 (2007)
12. Iris Challenge Evaluation (ICE) dataset found, http://iris.nist.gov/ICE/
13. Iris Dataset obtained from West Virginia University (WVU), http://www.wvu.edu/
14. UBIRIS dataset obtained from department of computer science, University of Beira Interior, Portugal, http://iris.di.ubi.pt/

# Efficient Conditional Proxy Re-encryption with Chosen-Ciphertext Security

Jian Weng[1,2], Yanjiang Yang[3], Qiang Tang[4], Robert H. Deng[1], and Feng Bao[3]

[1] School of Information Systems,
Singapore Management University, Singapore 178902
[2] Department of Computer Science, Jinan University, Guangzhou 510632, P.R. China
cryptjweng@gmail.com, robertdeng@smu.edu.sg
[3] Institute for Infocomm Research (I2R), Singapore, 119613
yyang@i2r.a-star.edu.sg, baofeng@i2r.a-star.edu.sg
[4] DIES, Faculty of EEMCS, University of Twente, The Netherlands
q.tang@utwente.nl

**Abstract.** Recently, a variant of proxy re-encryption, named conditional proxy re-encryption (C-PRE), has been introduced. Compared with traditional proxy re-encryption, C-PRE enables the delegator to implement fine-grained delegation of decryption rights, and thus is more useful in many applications. In this paper, based on a careful observation on the existing definitions and security notions for C-PRE, we re-formalize more rigorous definition and security notions for C-PRE. We further propose a more efficient C-PRE scheme, and prove its chosen-ciphertext security under the decisional bilinear Diffie-Hellman (DBDH) assumption in the random oracle model. In addition, we point out that a recent C-PRE scheme fails to achieve the chosen-ciphertext security.

**Keywords:** Conditional proxy re-encryption, chosen-ciphertext security, random oracle.

## 1 Introduction

In 1998, Blaze, Bleumer and Strauss [1] introduced the notion of proxy re-encryption (PRE). In a PRE scheme, a proxy is given a re-encryption key, and thus can translate ciphertexts under Alice's public key into ciphertexts under Bob's public key[1]. The proxy, however, cannot learn anything about the messages encrypted under either key. PRE turns out to be a useful primitive, and has found many applications requiring delegation of decryption right, such as encrypted email forwarding, secure distributed file systems, and outsourced filtering of encrypted spam.

Nevertheless, there exist some situations which are hard for traditional PRE to tackle. For example, suppose some of Alice's second level ciphertexts are *highly*

---

[1] In [2,3,4], the original ciphertext is called *second level ciphertext*, and the transformed ciphertext is named *first level ciphertext*. Through out this paper, we will follow these notations.

P. Samarati et al. (Eds.): ISC 2009, LNCS 5735, pp. 151–166, 2009.

*secret*, and she wants to decrypt these ciphertexts *only* by herself. Unfortunately, traditional PRE enables the proxy to convert *all* of Alice's second level ciphertexts, without any discrimination. To address this issue, two variants of PRE were independently introduced: one is named type-based proxy re-encryption (TB-PRE) introduced by Tang [5], and the other is named conditional proxy re-encryption (C-PRE) introduced by Weng *et al.* [6]. Although different in naming, C-PRE and TB-PRE are the same in spirit (for consistency, in the rest of the paper, we use C-PRE to denote the two variants.). In such systems, ciphertexts are generated with respect to a certain condition, and the proxy can translate a ciphertext only if the associated condition is satisfied. Compared with traditional PRE, C-PRE enables the delegator to implement fine-grained delegation of decryption rights, thereby more useful in many applications.

## 1.1   Our Motivations and Results

We first investigate the definitions and security notions for C-PRE defined in [6,5]. Both have their respective pros and cons: (i) In Weng *et al.*'s definition, the proxy needs *two* key pairs (i.e., the partial re-encryption key and the condition key) to perform the transformation, while the proxy in Tang *et al.*'s definition has only one key pair; (ii) In Tang's definition, the delegators and the delegatees have to be in different systems, which means that the user in a given system can *only* act as either (not both) a delegator or a delegatee. In contrast, in Weng *et al.*'s definition, a user can be the delegator for any other users, and can also be the delegatee for any other users. (iii) Both of the security notions in [5,6] only consider the second level ciphertext security, and do not address the first level ciphertext security.

In this paper, we re-formalize the definition for C-PRE by incorporating the advantages in [6,5]. More specifically, in our formalization the proxy holds only one key (re-encryption key) for performing transformations, and a user can act as the delegator or the delegatee for any other users. We also define the first level ciphertext security for C-PRE. We then propose a new C-PRE scheme, and prove its CCA-security under the well-studied decisional bilinear Diffie-Hellman (DBDH) assumption in the random oracle model. Our scheme has better overall efficiency in terms of both computation and communication than Tang's and Weng *et al.*'s schemes. In addition, we show that Weng *et al.*'s C-PRE scheme fails to achieve the CCA-security.

## 1.2   Related Work

Mambo and Okamoto [7] firstly introduced the concept of delegation of decryption rights, as a better-performance alternative to the trivial approach of decrypting-then-encrypting of ciphertexts. Blaze, Bleumer and Strauss [1] formalized the concept of proxy re-encryption, and proposed the first bidirectional PRE scheme (in which the delegation from Alice to Bob also allows re-encryption from Bob to Alice). In 2005, Ateniese *et al.* [2,3] presented unidirectional PRE schemes based on bilinear pairings.

The schemes in [1,2,3] are only secure against chosen-plaintext attacks (CPA). However, applications often require the CCA-security. In ACM CCS'07, Canetti and Hohenberger [8] presented a CCA-secure bidirectional PRE scheme from bilinear pairings. Later, Libert and Vergnaud [4] gave a unidirectional PRE scheme secure against replayable chosen-ciphertext attacks (RCCA) [9]. In their extended version, Libert and Vergnaud [10] further consider the the problem of conditional proxy re-encryption, and suggested a RCCA-secure C-PRE scheme in the standard model without assuming registered public keys[2].

Previous PRE schemes rely on the costly bilinear pairings. Thus Canetti and Hohenberger [8] left an open question to construct CCA-secure PRE without pairings. In CANS'08, Deng et al. [11] proposed a CCA-secure bidirectional PRE scheme without pairings. In PKC'09, Shao and Cao [12] proposed a unidirectional PRE scheme without pairings, and claimed that their scheme is CCA-secure. However, Weng et al. [13] pointed out that Shao and Cao's PRE scheme is not CCA-secure by presenting a concrete attack. Weng et al. [13] further presented an efficient CCA-secure unidirectional PRE scheme without pairings.

Traceable proxy re-encryption, introduced by Libert and Vergnaud [14], attempts to solve the problem of disclosing re-encryption keys, by tracing the proxies who have done so. Proxy re-encryption has also been studied in identity-based scenarios, such as [15,16,17]. Recently, Chu et al. [18] introduced a generalized version of C-PRE named conditional proxy broadcast re-encryption (CPBRE), in which the proxy can re-encrypt the ciphertexts for a set of users at a time.

# 2    Model of Conditional Proxy Re-encryption

Before re-formalizing the definition and security notions for C-PRE, we first explain some notations used in the rest of this paper. For a finite set $S$, $x \in_R S$ means choosing an element $x$ from $S$ with a uniform distribution. For a string $x$, $|x|$ denotes its bit-length. We use $\mathcal{A}(x, y, \cdots)$ to indicate that $\mathcal{A}$ is an algorithm with the input $(x, y, \cdots)$. By $z \leftarrow \mathcal{A}(x, y, \cdots)$, we indicate the running of $\mathcal{A}(x, y, \cdots)$ and letting $z$ be the output. We use $\mathcal{A}^{\mathcal{O}_1, \mathcal{O}_2, \cdots}(x, y, \cdots)$ to denote that $\mathcal{A}$ is an algorithm with the input $(x, y, \cdots)$ and can access to oracles $\mathcal{O}_1, \mathcal{O}_2, \cdots$. By $z \leftarrow \mathcal{A}^{\mathcal{O}_1, \mathcal{O}_2, \cdots}(x, y, \cdots)$, we denote the running of $\mathcal{A}^{\mathcal{O}_1, \mathcal{O}_2, \cdots}(x, y, \cdots)$, and letting $z$ be the output.

## 2.1    Definition of C-PRE Systems

Weng et al.'s definition differentiates between partial re-encryption key and condition key. A more standard model should combine them into an integral entity. Our definition is standard in this regard, having only re-encryption key; and we allow the delegators and the delegatees to share the same systems, unlike Tang's model. Formally, a C-PRE scheme consists of the following algorithms:

---

[2] We sincerely thank one of the anonymous reviewers for pointing out that, Libert and Vergnaud [10] also suggested a C-PRE scheme in the standard model without assuming registered public keys.

**Setup**$(1^\kappa)$: On input a security parameter $1^\kappa$, this algorithm outputs a global parameter $param$, which includes the message space $\mathcal{M}$. For brevity, we assume that $param$ is implicitly included in the input of the rest algorithms.

**KeyGen**$(1^\kappa)$: all parties use this randomize key generation algorithm to generate a public/private key pair $(pk_i, sk_i)$.

**ReKeyGen**$(sk_i, w, pk_j)$: On input the delegator's private key $sk_i$, a condition $w$ and the delegatee's public key $pk_j$, the re-encryption key generation algorithm outputs a re-encryption key $rk_{i \xrightarrow{w} j}$.

**Enc₂**$(pk, m, w)$: On input a public key $pk$, a plaintext $m \in \mathcal{M}$ and a condition $w$, the second encryption algorithm outputs a second level ciphertext CT, which can be re-encrypted into a first level one (intended for a possibly different receiver) using the suitable re-encryption key.

**Enc₁**$(pk, m)$: On input a public key $pk$ and a plaintext $m \in \mathcal{M}$, this first encryption algorithm outputs a first level ciphertext CT that cannot be re-encrypted for another party.

**ReEnc**$(\mathbf{CT}_i, rk_{i \xrightarrow{w} j})$: On input a second level ciphertext $\mathrm{CT}_i$ associated with $w$ under public key $pk_i$, and a re-encryption key $rk_{i \xrightarrow{w} j}$, this re-encryption algorithm, run by the proxy, outputs a first level ciphertext $\mathrm{CT}_j$ under public key $pk_j$.

**Dec₂**$(\mathbf{CT}, sk)$: On input a second level cipertext CT and a private key $sk$, this second decryption algorithm outputs a message $m$ or the error symbol $\perp$.

**Dec₁**$(\mathbf{CT}, sk)$: On input a first level cipertext CT and a private key $sk$, this first decryption algorithm outputs a message $m$ or the error symbol $\perp$.

The correctness of C-PRE means that, for any condition $w$, any $m \in \mathcal{M}$, and any couple of private/public key pairs $(pk_i, sk_i)$, $(pk_j, sk_j)$, it holds that

$$\mathsf{Dec}_2(\mathsf{Enc}_2(pk_i, m, w), sk_i) = m, \quad \mathsf{Dec}_1(\mathsf{Enc}_1(pk_i, m), sk_i) = m,$$
$$\mathsf{Dec}_1\left(\mathsf{ReEnc}(\mathsf{Enc}_2(pk_i, m, w), \mathsf{ReKeyGen}(sk_i, w, pk_j)), sk_j\right) = m.$$

### 2.2 Security Notions

In this subsection, we will define the security notions for C-PRE systems. Before giving these security notions, we first consider the following oracles which together model the ability of an adversary. These oracles are provided for the adversary $\mathcal{A}$ by a challenger $\mathcal{C}$ who simulates an environment running C-PRE.

- Uncorrupted key generation oracle $\mathcal{O}_u(i)$: $\mathcal{C}$ runs algorithm KeyGen to generate a public/private key pair $(pk_i, sk_i)$, and returns $pk_i$ to $\mathcal{A}$.
- Corrupted key generation oracle $\mathcal{O}_c(i)$: $\mathcal{C}$ runs algorithm KeyGen to generate a public/private key pair $(pk_j, sk_j)$,, and returns $(pk_j, sk_j)$ to $\mathcal{A}$.
- Re-encryption key oracle $\mathcal{O}_{rk}(pk_i, w, pk_j)$: Challenger $\mathcal{C}$ first runs $rk_{i \xrightarrow{w} j} \leftarrow$ ReKeyGen$(sk_i, w, pk_j)$, and then returns $rk_{i \xrightarrow{w} j}$ to $\mathcal{A}$.
- Re-encryption oracle $\mathcal{O}_{re}(pk_i, pk_j, (w, \mathrm{CT}_i))$: Challenger $\mathcal{C}$ first runs $\mathrm{CT}_j \leftarrow$ ReEnc$(\mathrm{CT}_i, rk_{i \xrightarrow{w} j})$, where $rk_{i \xrightarrow{w} j} = $ ReKeyGen$(sk_i, w, pk_j)$, and then returns $\mathrm{CT}_j$ to $\mathcal{A}$.

– First level decryption oracle $\mathcal{O}_{1d}(pk, \text{CT})$: Here CT is a first level ciphertext. $\mathcal{C}$ runs $\text{Dec}_1(\text{CT}, sk)$, and returns the corresponding result to $\mathcal{A}$.

Note that for the last three oracles, it is required that $pk_i$, $pk_j$ and $pk$ were generated beforehand by either $\mathcal{O}_c$ or $\mathcal{O}_u$.

We are now ready to define the semantic security for C-PRE under chose-ciphertext attacks. Libert and Vergnaud [4]differentiated two kinds of semantic security for traditional (single-hop) unidirectional PRE systems: *first level ciphertext security* and *second level ciphertext security*. We here follow Libert and Vergnaud's definitions, and define these two kinds security notions for C-PREs.

**Second level ciphertext security.** Intuitively speaking, second level ciphertext security models the scenario that the adversary $\mathcal{A}$ is challenged with a second level ciphertext $\text{CT}^*$ encrypted under a target public key $pk_{i^*}$ and a target condition $w^*$. $\mathcal{A}$ can issue a series of queries to the above five oracles. These queries are allowed as long as they would not allow $\mathcal{A}$ to decrypt trivially. For examples, $\mathcal{A}$ should not query on $\mathcal{O}_{rk}(pk_{i^*}, w^*, pk_j)$ to obtain an re-encryption key $rk_{i^* \xrightarrow{w^*} j}$ where $pk_j$ came from oracle $\mathcal{O}_c$. Otherwise, $\mathcal{A}$ can trivially decrypt the challenge ciphertext by first re-encrypting it into a first level ciphertext and then decrypting it with $sk_j$. Similarly, $\mathcal{A}$ cannot query on $\mathcal{O}_{re}(pk_{i^*}, pk_j, (w^*, \text{CT}^*))$ where $pk_j$ came from oracle $\mathcal{O}_c$. Also, for a first level ciphertext $\text{CT}' = \text{ReEnc}(\text{CT}^*, rk_{i^* \xrightarrow{w^*} j})$, $\mathcal{A}$ is disallowed to query on $\mathcal{O}_{1d}(pk_j, \text{CT}')$. One might wonder that why we do not provide the second level decryption oracle for $\mathcal{A}$. In fact, explicitly providing adversary $\mathcal{A}$ with this oracle is useless, since (i). for the challenge ciphertext $\text{CT}^*$, $\mathcal{A}$ is obviously not allowed to ask the second level decryption oracle to decrypt it; (ii). while for any other second level ciphertext $\text{CT}_t$ encrypted under public key $pk_t$ and condition $w$ such that $(pk_t, w, \text{CT}_t) \neq (pk_{i^*}, w^*, \text{CT}^*)$, adversary $\mathcal{A}$ can first issue a re-encryption query $\mathcal{O}_{re}(pk_t, pk_j, (w, \text{CT}_t))$ to obtain a first level ciphertext $\text{CT}_j$, and then issue a first level decryption query $\mathcal{O}_{1d}(pk_j, \text{CT}_j)$ to obtain the underlying plaintext. Below gives the formal definition for second level ciphertext's sematic security under adaptive chosen ciphertext attack (IND-2CPRE-CCA).

**Definition 1.** *For a C-PRE scheme $\mathcal{E}$ and a probabilistic polynomial time adversary $\mathcal{A}$ running in two stages* find *and* guess, *we define $\mathcal{A}$'s advantage against the* IND-2CPRE-CCA *security of $\mathcal{E}$ as*

$$\text{Adv}_{\mathcal{E},\mathcal{A}}^{\text{IND-2CPRE-CCA}}(1^\kappa) = \left| \Pr \left[ \delta' = \delta \left| \begin{array}{c} param \leftarrow \text{Setup}(1^\kappa) \\ (pk_{i^*}, w^*, (m_0, m_1), \text{st}) \leftarrow \mathcal{A}_{\text{find}}^{\mathcal{O}_u, \mathcal{O}_c, \mathcal{O}_{rk}, \mathcal{O}_{re}, \mathcal{O}_{1d}}(param) \\ \delta \in_R \{0, 1\}, \text{CT}^* \leftarrow \text{Enc}_2(pk_{i^*}, m_\delta, w^*) \\ \delta' \leftarrow \mathcal{A}_{\text{guess}}^{\mathcal{O}_u, \mathcal{O}_c, \mathcal{O}_{rk}, \mathcal{O}_{re}, \mathcal{O}_{1d}}(param, \text{CT}^*, \text{st}) \end{array} \right. \right] - \frac{1}{2} \right|,$$

*where* st *is some internal state information of adversary $\mathcal{A}$. Here it is mandated that $|m_0| = |m_1|$, and the following requirements are simultaneously satisfied: (i). $pk_{i^*}$ is generated by oracle $\mathcal{O}_u$; (ii). For a public key $pk_j$ generated by oracle $\mathcal{O}_c$, $\mathcal{A}$ cannot issue the query $\mathcal{O}_{rk}(pk_{i^*}, w^*, pk_j)$; (iii) For a public key $pk_j$ generated* by oracle $\mathcal{O}_c$, $\mathcal{A}$ cannot issue the query $\mathcal{O}_{re}(pk_{i^*}, pk_j, (w^*, \mathrm{CT}^*))$; (iv). For a public key $pk_j$ and the first level ciphertext $\mathrm{CT}' = \mathsf{ReEnc}(\mathrm{CT}^*, rk_{i^* \xrightarrow{w^*} j})$, $\mathcal{A}$ cannot issue the query $\mathcal{O}_{1d}(pk_j, \mathrm{CT}')$.

We refer to adversary $\mathcal{A}$ as an IND-2CPRE-CCA adversary. A C-PRE scheme $\mathcal{E}$ is said to be $(t, q_u, q_c, q_{rk}, q_{re}, q_{1d}, \epsilon)$-IND-2CPRE-CCA secure, if for any $t$-time IND-2CPRE-CCA adversary $\mathcal{A}$, who makes at most $q_u, q_c, q_{rk}, q_{re}$ and $q_d$ queries to $\mathcal{O}_u, \mathcal{O}_c, \mathcal{O}_{rk}, \mathcal{O}_{re}$ and $\mathcal{O}_{1d}$, respectively, we have $\mathrm{Adv}_{\mathcal{E},\mathcal{A}}^{\mathsf{IND\text{-}2CPRE\text{-}CCA}}(1^\kappa) \leq \epsilon$.

**First Level Ciphertext Security.** The above definition provides the adversary with a second level ciphertext in the challenge phase. Next, we define a complementary definition of security (denote by IND-1CPRE-CCA) by providing the adversary with a first level ciphertext in the challenge phase. Note that, since the first level ciphertext cannot be re-encrypted in a single hop C-PRE scheme, $\mathcal{A}$ is allowed to obtain *any* re-encryption keys. Furthermore, given these re-encryption keys, $\mathcal{A}$ can re-encrypt ciphertexts by himself, and hence there is no need to provide the re-encryption oracle $\mathcal{O}_{re}$ for him. As argued before, the second level decryption oracle is also unnecessary.

**Definition 2.** *For a C-PRE scheme $\mathcal{E}$ and a probabilistic polynomial time adversary $\mathcal{A}$ running in two stages* find *and* guess, *we define $\mathcal{A}$'s advantage against the* IND-1CPRE-CCA *security of $\mathcal{E}$ as*

$$\mathrm{Adv}_{\mathcal{E},\mathcal{A}}^{\mathsf{IND\text{-}1CPRE\text{-}CCA}}(1^\kappa) = \left| \Pr \left[ \delta' = \delta \middle| \begin{array}{c} param \leftarrow \mathsf{Setup}(1^\kappa) \\ (pk_{i^*}, (m_0, m_1), \mathtt{st}) \leftarrow \mathcal{A}_{\mathtt{find}}^{\mathcal{O}_u, \mathcal{O}_c, \mathcal{O}_{rk}, \mathcal{O}_{1d}}(param) \\ \delta \in_R \{0,1\}, \mathrm{CT}^* \leftarrow \mathsf{Enc}_1(pk_{i^*}, m_\delta) \\ \delta' \leftarrow \mathcal{A}_{\mathtt{guess}}^{\mathcal{O}_u, \mathcal{O}_c, \mathcal{O}_{rk}, \mathcal{O}_{1d}}(param, \mathrm{CT}^*, \mathtt{st}) \end{array} \right] - \frac{1}{2} \right|,$$

where $\mathtt{st}$ is some internal state information of adversary $\mathcal{A}$. Here it is mandated that, $|m_0| = |m_1|$, $pk_{i^*}$ is generated by $\mathcal{O}_u$, and $\mathcal{A}$ cannot issue the query $\mathcal{O}_{1d}(pk_{i^*}, \mathrm{CT}^*)$.

We refer to the above adversary $\mathcal{A}$ as an IND-1CPRE-CCA adversary. We say that a C-PRE scheme $\mathcal{E}$ is $(t, q_u, q_c, q_{rk}, q_{1d}, \epsilon)$-IND-1CPRE-CCA secure, if for any $t$-time IND-1CPRE-CCA adversary $\mathcal{A}$ that makes at most $q_u, q_c, q_{rk}$ and $q_d$ queries to oracles $\mathcal{O}_u, \mathcal{O}_c, \mathcal{O}_{rk}$ and $\mathcal{O}_{1d}$, respectively, we have $\mathrm{Adv}_{\mathcal{E},\mathcal{A}}^{\mathsf{IND\text{-}1CPRE\text{-}CCA}}(1^\kappa) \leq \epsilon$.

**Remark.** In [2], Ateniese *et al.* defined the notion *master secret security*, for unidirectional proxy re-encryption. This security notion catches the intuition that, even if the dishonest proxy colludes with the delegatee, it is still impossible for them to derive the delegator's private key. Note that for C-PREs, there is no need to define master secret security, since this security is implied by the first level ciphertext security. This is due to the fact that, if the dishonest proxy and the delegatee can collude to derive the delegator's private key, they can certainly use this private key to decrypt the challenge ciphertext, and thus break the first level ciphertext security.

# 3  Proposed CCA-Secure C-PRE Scheme

In this section, we propose a new C-PRE scheme with CCA-security. Before presenting our scheme, we list three important and *necessary* principles for designing CCA-secure C-PRE schemes: (i) the validity of the second level ciphertexts should be *publicly verifiable*; otherwise, it will suffer from a similar attack as illustrated in [11, 19]; (ii) the second level ciphertexts should be able to resist the adversary's malicious manipulating; (iii) it should also be impossible for the adversary to maliciously manipulate the first level ciphertext. We remark that it is non-trivial to design a C-PRE scheme satisfying these three requirements, especially the last one. To help understand our scheme, we first present an insecure attempt, and then improve it to obtain our final CCA-secure scheme.

## 3.1  A First Attempt

We denote this first attempt by S1, which is specified as below:

**Setup**($1^\kappa$): On input a security parameter $1^\kappa$, the setup algorithm first determines $(q, \mathbb{G}, \mathbb{G}_T, e)$, where $q$ is a $\kappa$-bit prime, $\mathbb{G}$ and $\mathbb{G}_T$ are two cyclic groups with prime order $q$, and $e$ is the bilinear pairing $e : \mathbb{G} \times \mathbb{G} \to \mathbb{G}_T$. Next, it chooses $g \in_R \mathbb{G}$, and five hash functions $H_1, H_2, H_3, H_4$ and $H_5$ such that $H_1 : \{0,1\}^* \to \mathbb{Z}_q, H_2 : \{0,1\}^* \to \mathbb{G}, H_3 : \mathbb{G} \to \{0,1\}^n, H_4 : \{0,1\}^* \to \mathbb{G}$ and $H_5 : \mathbb{G} \to \mathbb{Z}_q$, where $n$ is polynomial in $\kappa$ and the message space is $\mathcal{M} = \{0,1\}^n$. The global parameter is $param = ((q, \mathbb{G}, \mathbb{G}_T, e), g, n, H_1, \cdots, H_5)$.

**KeyGen**($1^\kappa$): To generate the public/private key pair for user $U_i$, it picks $x_i \in_R \mathbb{Z}_q$, and sets the public key and private key to be $pk_i = g^{x_i}$ and $sk_i = x_i$, respectively.

**ReKeyGen**($sk_i, w, pk_j$): On input a private key $sk_i$, a condition $w$ and a public key $pk_j$, this algorithm randomly picks $s \in_R \mathbb{Z}_q$, and outputs the re-encryption key as

$$rk_{i \overset{w}{\to} j} = (rk_1, rk_2) = \left( \left(H_2(pk_i, w)pk_j^s\right)^{-sk_i}, pk_i^s \right). \tag{1}$$

**Enc$_2$**($pk, m, w$): On input a public key $pk$, a condition $w$ and a message $m \in \mathcal{M}$, the sender first picks $R \in_R \mathbb{G}_T$. Then he computes $r = H_1(m, R)$, and outputs the second level ciphertext $CT = (C_1, C_2, C_3, C_4)$ as

$$\left( g^r, \ R \cdot e(pk, H_2(pk, w))^r, \ m \oplus H_3(R), \ H_4(C_1, C_2, C_3)^r \right). \tag{2}$$

Note that the last ciphertext component, $C_4$, is used to ensure the public verifiability of the ciphertext, while the first three components, $(C_1, C_2, C_3)$, are in fact the ciphertext of the CCA-secure ElGamal encryption scheme [20] applying the Fujisaki-Okamoto transformation [21].

**Enc$_1$**($pk, m$): On input a public key $pk$ and a message $m \in \mathcal{M}$, the sender first picks $R \in_R \mathbb{G}_T$ and $\bar{s} \in_R \mathbb{Z}_q^*$. Then he computes $r = H_1(m, R)$, and outputs the first level ciphertext $CT$ as

$$CT = (\overline{C}_1, \overline{C}_2, \overline{C}_3, \overline{C}_4) = \left( g^r, \ R \cdot e(g, pk)^{-r \cdot \bar{s}}, \ m \oplus H_3(R), \ g^{\bar{s}} \right). \tag{3}$$

**ReEnc($\mathbf{CT}_i, rk_{i \xrightarrow{w} j}$):** On input a second level ciphertext $CT_i = (C_1, C_2, C_3, C_4)$ associated with condition $w$ under public key $pk_i$, and a re-encryption key $rk_{i \xrightarrow{w} j} = (rk_1, rk_2)$, it generates the first level ciphertext under public key $pk_j$ as follows: Check whether the following equality holds:

$$e(C_1, H_4(C_1, C_2, C_3)) = e(g, C_4). \tag{4}$$

If not, output $\perp$; else output $CT_j = (\overline{C}_1, \overline{C}_2, \overline{C}_3, \overline{C}_4)$ as

$$\overline{C}_1 = C_1, \quad \overline{C}_2 = C_2 \cdot e(C_1, rk_1), \quad \overline{C}_3 = C_3, \quad \overline{C}_4 = rk_2. \tag{5}$$

Observe that $CT_j = (\overline{C}_1, \overline{C}_2, \overline{C}_3, \overline{C}_4)$ is indeed of the following form:

$$\overline{C}_1 = g^r, \quad \overline{C}_3 = m \oplus H_3(R), \quad \overline{C}_4 = pk_i^s = g^{s \cdot sk_i},$$

$$\overline{C}_2 = R \cdot e(pk_i, H_2(pk_i, w))^r \cdot e\left(g^r, \left(H_2(pk_i, w)pk_j^s\right)^{-sk_i}\right) = R \cdot e\left(g, pk_j\right)^{-r \cdot s \cdot sk_i}.$$

Letting $\overline{s} = s \cdot sk_i$, it can be seen that the above first level ciphertext has the same form as Eq. (3).

**Dec$_2$(CT, $sk$):** On input a private key $sk$ and a second level ciphertext $CT = (C_1, C_2, C_3, C_4)$, it first checks whether Eq. (4) holds. If not, it returns $\perp$. Otherwise, it computes $R = \dfrac{C_2}{e(C_1, H_2(pk, w))^{sk}}$, $m = C_3 \oplus H_3(R)$, and check whether $g^{H_1(m,R)} = C_1$ holds. If yes, it returns $m$; else it returns $\perp$.

**Dec$_1$(CT, $sk$):** On input a private key $sk$ and a first level ciphertext $CT = (\overline{C}_1, \overline{C}_2, \overline{C}_3, \overline{C}_4)$ under public key $pk$, it computes $R = \overline{C}_2 \cdot e(\overline{C}_1, \overline{C}_4)^{sk}$ and $m = \overline{C}_3 \oplus H_3(R)$. Return $m$ if $g^{H_1(m,R)} = \overline{C}_1$ holds and $\perp$ otherwise:

**Analysis.** At first glance, it seems that scheme S1 is CCA-secure. Unfortunately, this is not true, since the adversary can maliciously manipulate the first level ciphertext to get a *new* yet *valid* one. Concretely, given the first level ciphertext as in Eq. (3), the adversary can pick $\ell \in_R \mathbb{Z}_q$ and produces another first level ciphertext $CT' = (\overline{C}'_1, \overline{C}'_2, \overline{C}'_3, \overline{C}'_4)$ such that:

$$\overline{C}'_1 = \overline{C}_1 = g^r, \quad \overline{C}'_2 = \overline{C}_2 \cdot e(\overline{C}_1, pk)^{-\ell} = R \cdot e(g, pk)^{-r \cdot (\overline{s}+\ell)}.$$

$$\overline{C}'_3 = \overline{C}_3 = m_\delta \oplus H_3(R), \quad \overline{C}'_4 = \overline{C}_4 \cdot g^\ell = g^{\overline{s}+\ell}.$$

Letting $\overline{s} =' \overline{s} + \ell$, we can easily see that $CT'$ is another new and valid ciphertext as Eq. (3). Thus the CCA-security can be trivially broken.

## 3.2   CCA-Secure C-PRE Scheme

Indeed, the insecurity of S1 lies in the construction of the re-encryption key, i.e., $rk_2$ is loosely integrated with $rk_1$. This enables the adversary to maliciously manipulate the resulting first level ciphertext and obtain another *valid* first level ciphertext. So, to design a CCA-secure C-PRE scheme, we should carefully design the re-encryption key, so that the resulting first level ciphertext cannot be maliciously manipulated by the adversary. Based on this observation, we present our CCA-secure C-PRE scheme (denoted by S2) as below:

**Setup**($1^\kappa$) **and KeyGen**($1^\kappa$): The same as in S1.

**ReKeyGen**($sk_i, w, pk_j$): On input a private key $sk_i$, a condition $w$ and a public key $pk_j$, this algorithm picks $s \in_R \mathbb{Z}_q$, and outputs $rk_{i \xrightarrow{w} j} = (rk_1, rk_2)$ as

$$rk_2 = pk_i^s, \quad rk_1 = \left( H_2(pk_i, w)pk_j^{s \cdot H_5(pk_j^{s \cdot sk_i})} \right)^{-sk_i}.$$

Observe that in the re-encryption key $rk_{i \xrightarrow{w} j}$, $rk_2$ is now *seamlessly* integrated with $rk_1$. That is, we integrate $rk_2$ with $rk_1$ by embedding $H_5(pk_j^{s \cdot sk_i}) = H_5(rk_2^{sk_j})$ in $rk_1$. This is an important trick for scheme S2 to achieve the CCA-security.

**Enc$_2$**($pk, m, w$): The same as in S1.

**Enc$_1$**($pk, w$): On input a public key $pk$ and a message $m \in \mathcal{M}$, the sender first picks $R \in_R \mathbb{G}_T$ and $\bar{s} \in_R \mathbb{Z}_q^*$. Then he computes $r = H_1(m, R)$, and outputs the first level ciphertext $\mathrm{CT} = (\overline{C}_1, \overline{C}_2, \overline{C}_3, \overline{C}_4)$ as

$$\left( g^r, \ R \cdot e(g, pk)^{-r \cdot \bar{s} \cdot H_5(pk^{\bar{s}})}, \ m \oplus H_3(R), \ g^{\bar{s}} \right). \tag{6}$$

**ReEnc**($\mathrm{CT}_i, rk_{i \xrightarrow{w} j}$): The same as in S1. Note that, since the re-encryption key is different from that in S1, the resulting first level ciphertext $\mathrm{CT}_j = (\overline{C}_1, \overline{C}_2, \overline{C}_3, \overline{C}_4)$ is of the following forms:

$$\left( g^r, \ R \cdot e(g, pk_j)^{-r \cdot s \cdot sk_i \cdot H_5(pk_j^{s \cdot sk_i})}, \ m \oplus H_3(R), \ g^{s \cdot sk_i} \right),$$

where $r = H_1(m, R)$ and $R \in_R \mathbb{G}_T$. Letting $\bar{s} = s \cdot sk_i$, it can be seen that the above first level ciphertext has the same form as Eq. (6).

Note also that, now $\overline{C}_4$ is tightly integrated with $\overline{C}_2$ by embedding $\overline{C}_4$ in $H_5(\overline{C}_4^{sk_j}) = H_5(pk_j^{s \cdot sk_i})$, and hence it is unable for the adversary to modify the first level ciphertext to obtain a new and *valid* one. Therefore, the attack against scheme S1 does not apply to scheme S2.

**Dec$_2$**($\mathbf{CT}, sk$): The same as in S1.

**Dec$_1$**($\mathbf{CT}, sk$): On input a private key $sk$ and a first level ciphertext $\mathrm{CT} = (\overline{C}_1, \overline{C}_2, \overline{C}_3, \overline{C}_4)$ under public key $pk$, this algorithm first computes $R = \overline{C}_2 \cdot e(\overline{C}_1, \overline{C}_4)^{sk \cdot H_5(\overline{C}_4^{sk})}$ and $m = \overline{C}_3 \oplus H_3(R)$. Next, it returns $m$ if $g^{H_1(m,R)} = \overline{C}_1$ holds and $\bot$ otherwise.

### 3.3 Security Analysis

The CCA-security of our schemes S2 is based on a complexity assumption called decisional Bilinear Diffie-Hellman (DBDH) assumption. The DBDH problem in groups $(\mathbb{G}, \mathbb{G}_T)$ is, given a tuple $(g, g^a, g^b, g^c, Z) \in \mathbb{G}^4 \times \mathbb{G}_T$ with unknown $a, b, c \in_R \mathbb{Z}_q$, to decide whether $Z = e(g, g)^{abc}$. A polynomial-time algorithm $\mathcal{B}$ has *advantage* $\epsilon$ in solving the DBDH problem in groups $(\mathbb{G}, \mathbb{G}_T)$, if

$$\left| \Pr[\mathcal{B}\left(g, g^a, g^b, g^c, Z = e(g, g)^{abc}\right) = 1] - \Pr[\mathcal{B}\left(g, g^a, g^b, g^c, Z = e(g, g)^d\right) = 1] \right| \geq \epsilon,$$

where the probability is taken over the random choices of $a, b, c, d$ in $\mathbb{Z}_q$, the random choice of $g$ in $\mathbb{G}$, and the random bits consumed by $\mathcal{B}$.

**Definition 3.** *We say that the* $(t, \epsilon)$*-DBDH assumption holds in groups* $(\mathbb{G}, \mathbb{G}_T)$, *if there exists no t-time algorithm* $\mathcal{B}$ *that has advantage* $\epsilon$ *in solving the DBDH problem in* $(\mathbb{G}, \mathbb{G}_T)$.

For our scheme's CCA-security at the second level, we have the following theorem, whose detailed proof can be found in Appendix B.

**Theorem 1.** *Our scheme* S2 *is* IND-2CPRE-CCA *secure in the random oracle model, assuming the DBDH assumption holds in groups* $(\mathbb{G}, \mathbb{G}_T)$. *More specifically, if there exists an* IND-2CPRE-CCA *adversary* $\mathcal{A}$, *who asks at most* $q_{H_i}$ *random oracle queries to* $H_i$ *with* $i \in \{1, \cdots, 5\}$ *and breaks the* $(t, q_u, q_c, q_{rk}, q_{re}, q_d, \epsilon)$-IND-2CPRE-CCA *security of scheme* S2, *then there exists an algorithm* $\mathcal{B}$ *that can break the* $(t', \epsilon')$-DBDH *assumption in groups* $(\mathbb{G}, \mathbb{G}_T)$ *with*

$$\epsilon' \geq \frac{\epsilon}{\dot{e}(1 + q_{rk})} - \frac{q_{H_1} + q_{H_5} + q_{re} + q_d}{q},$$

$$t' \leq t + \mathcal{O}(\tau(q_{H_2} + q_{H_4} + q_u + q_c + 3q_{rk} + q_{H_1}q_{re} + (q_{H_1} + q_{H_5})q_d)),$$

*where* $\tau$ *is the maximum over the time to compute an exponentiation in* $\mathbb{G}, \mathbb{G}_T$, *and the time to compute a pairing;* $\dot{e}$ *denotes the base of the natural logarithm.*

The first level ciphertext security of S2 is ensured by the following theorem.

**Theorem 2.** *Our scheme* S2 *is* IND-1CPRE-CCA *secure in the random oracle model, assuming the DBDH assumption holds in groups* $(\mathbb{G}, \mathbb{G}_T)$. *More specifically, if there exists an* IND-1CPRE-CCA *adversary* $\mathcal{A}$, *who asks at most* $q_{H_i}$ *random oracle queries to* $H_i$ *with* $i \in \{1, \cdots, 5\}$ *and can break the* $(t, q_u, q_c, q_{rk}, q_d, \epsilon)$-IND-1CPRE-CCA *security of scheme* S2, *then there exists an algorithm* $\mathcal{B}$ *that can break the* $(t', \epsilon')$-DBDH *assumption in groups* $(\mathbb{G}, \mathbb{G}_T)$ *with*

$$\epsilon' \geq \epsilon - \frac{q_{H_1} + q_{H_5} + q_d}{q},$$

$$t' \leq t + \mathcal{O}(\tau(q_{H_2} + q_{H_4} + q_u + q_c + 3q_{rk} + (q_{H_1} + q_{H_5})q_d)),$$

*where* $\tau$ *and* $\dot{e}$ *have the same meaning as in Theorem 1.*

The proof for Theorem 2 is similar to that of Theorem 1 with some modifications. For example, the simulation for the random oracle $H_2$ no longer need to flip a biased coin, and the simulation for oracle $\mathcal{O}_{rk}$ has to successfully answer all the re-encryption key queries without aborting. Due to the space limit, we give the detailed proof in the full paper.

### 3.4    Comparisons

In Table 1, we compare our scheme with Tang's scheme [5] [3], Weng et al.'s scheme [6] and Livert-Vergnaud's scheme [10]. We first explain some notations used in

---

[3] Tang presented two schemes: one is CPA-secure, and the other is CCA-secure. To be fair, we here choose Tang's CCA-secure scheme for comparison.

**Table 1.** Comparisons among Ours Scheme and the C-PRE Schemes in [5, 6, 4]

| | Schemes | Our Scheme S2 | Tang's Scheme [5] | Weng's Scheme [6] | Livert-Vergnaud's Scheme [10] |
|---|---|---|---|---|---|
| **Length** | 2nd-level ciphtxt | $2\|G\|+1\|G_T\|+1\|M\|$ | $2\|G\|+1\|G_T\|+1\|M\|$ | $3\|G\|+1\|M\|+l_1$ | $\|svk\|+3\|G\|+1\|G_T\|+\|\sigma\|$ |
| | 1st-level ciphtxt | $2\|G\|+1\|G_T\|+1\|M\|$ | $2\|C_{PKE}\|+1\|G\|+1\|G_T\|+1\|M\|$ | $1\|G_T\|+1\|M\|+l_1$ | $\|svk\|+7\|G\|+1\|G_T\|+1\|\sigma\|$ |
| | public key | $1\|G\|$ | $1\|G\|$ | $2\|G\|$ | $(n+2)\|G\|$ |
| | private key | $1\|Z_q\|$ | $1\|Z_q\|$ | $1\|Z_q\|$ | $1\|Z_q\|$ |
| | re-encryption key | $2\|G\|$ | $1\|C_{PKE}\|+1\|G\|$ | $2\|G\|$ | $2\|G\|$ |
| **Cost** | $Enc_2$ | $1t_p+3t_e$ | $1t_p+3t_e$ | $1t_p+5t_e$ | $1t_s+4t_e$ |
| | $Enc_1$ | $1t_p+4t_e$ | $1t_p+2t_e+2t_{Enc_{PKE}}$ | $1t_p+2t_e$ | $1t_s+8t_e$ |
| | ReEnc | $3t_p$ | $3t_p+1t_{Enc_{PKE}}$ | $3t_p+2t_e$ | $4t_p+6t_e$ |
| | $Dec_2$ | $3t_p+2t_e$ | $3t_p+2t_e$ | $4t_p+5t_e$ | $1t_p+1t_e+1t_v$ |
| | $Dec_1$ | $1t_p+3t_e$ | $2t_{Dec_{PKE}}+1t_p+1t_e$ | $2t_e$ | $9t_p+1t_e+1t_v$ |
| **Security** | | CCA | CCA | Not CCA | RCCA |
| **Without RO?** | | No | No | No | Yes |

Table 1. Here $|M|$, $|G|$, $|G_T|$, $|svk|$ and $|\sigma|$ denote the bit-length of a plaintext, an element in groups $G$ and $G_T$, the verification key and signature of one-time signature, respectively. We use $t_p, t_e, t_s, t_v$ to represent the computational cost of a bilinear pairing, an exponentiation, signing and verifying a one-time signature, respectively. $l_1$ denotes the security parameter used in Weng et al.'s scheme. Tang's scheme needs an additional public key encryption scheme PKE, which is assumed to be deterministic and one-way[4]. We here use $t_{Enc_{PKE}}$ and $t_{Dec_{PKE}}$ to represent the computational cost of an encryption and a decryption in the public key encryption(PKE) scheme used in Tang's scheme. For $|C_{PKE}|$, it denotes the ciphertext length of scheme PKE used in Tang's scheme.

The comparison results indicate that our scheme S2 outperforms Tang's scheme in terms of both computational and communicational costs. Our scheme has a better overall performance than Weng et al.'s scheme: The ciphertext length and computation cost for first level encryption and decryption in Weng et al.'s scheme lead ours, while ours beats theirs in the other metrics; most importantly, our scheme is CCA-secure, while theirs fails. Our scheme also has a better overall performance than Libert-Vergnaud's scheme. Besides, ours is CCA-secure under the well-studied DBDH assumption, while Libert-Vergnaud's scheme only satisfies the RCCA-security (which is a weaker variant of CCA-security assuming a harmless mauling of the challenge ciphertext is tolerated) under a less studied assumption, named 3-weak decisional bilinear Diffie-Hellman inversion (3-wDBDH) assumption. However, like Tang and Weng et al.'s schemes, our scheme suffers from a limitation that its security relies on the random oracle in the know secret key model, while Libert-Vergnaud's scheme can be proved without random oracles in the chosen-key model.

## 4    Conclusions

We re-formalized the definition and security notions for conditional proxy re-encryption (C-PRE), and proposed an efficient CCA-secure C-PRE scheme un-

---

[4] To the best of our knowledge, the ciphertext in such a PKE scheme needs at least two group elements, and its computational cost for encryption and decryption involves at least two exponentiations and one exponentiation respectively. Hence, we have $|C_{PKE}| \geq 2|G|$, $t_{Enc_{PKE}} \geq 2t_e$, $t_{Dec_{PKE}} \geq 1t_e$.

der our model. In addition, we gave an attack to Weng *et al.*'s C-PRE scheme, showing that it fails to achieve the CCA-security.

This work motivates some interesting open questions. One is how to construct a CCA-secure (instead of RCCA-secure) C-PRE scheme without random oracles. Another is how to construct CCA-secure C-PRE schemes supporting "OR" and "AND" gates over conditions.

## Acknowledgement

We are grateful to the anonymous reviewers for their helpful comments. This work is partially supported by the Office of Research, Singapore Management University.

## References

1. Blaze, M., Bleumer, G., Strauss, M.: Divertible Protocols and Atomic Proxy Cryptography. In: Nyberg, K. (ed.) EUROCRYPT 1998. LNCS, vol. 1403, pp. 127–144. Springer, Heidelberg (1998)
2. Ateniese, G., Fu, K., Green, M., Hohenberger, S.: Improved Proxy Re-Encryption Schemes with Applications to Secure Distributed Storage. In: NDSS, The Internet Society (2005)
3. Ateniese, G., Fu, K., Green, M., Hohenberger, S.: Improved proxy re-encryption schemes with applications to secure distributed storage. ACM Trans. Inf. Syst. Secur. 9(1), 1–30 (2006)
4. Libert, B., Vergnaud, D.: Unidirectional Chosen-Ciphertext Secure Proxy Re-encryption. In: Cramer, R. (ed.) PKC 2008. LNCS, vol. 4939, pp. 360–379. Springer, Heidelberg (2008)
5. Tang, Q.: Type-based proxy re-encryption and its construction. In: Chowdhury, D.R., Rijmen, V., Das, A. (eds.) INDOCRYPT 2008. LNCS, vol. 5365, pp. 130–144. Springer, Heidelberg (2008)
6. Weng, J., Deng, R.H., Ding, X., Chu, C.K., Lai, J.: Conditional proxy re-encryption secure against chosen-ciphertext attack. In: ASIACCS, pp. 322–332 (2009)
7. Mambo, M., Okamoto, E.: Proxy cryptosystems: delegation of the power to decrypt ciphertexts. IEICE Transactions on Fundamentals of Electronics, Communications and Computer Sciences E80-A(1), 54–63 (1997)
8. Canetti, R., Hohenberger, S.: Chosen-Siphertext Cecure Proxy Re-Encryption. In: Ning, P., di Vimercati, S.D.C., Syverson, P.F. (eds.) ACM Conference on Computer and Communications Security, pp. 185–194. ACM, New York (2007)
9. Canetti, R., Krawczyk, H., Nielsen, J.B.: Relaxing Chosen-Ciphertext Security. In: Boneh, D. (ed.) CRYPTO 2003. LNCS, vol. 2729, pp. 565–582. Springer, Heidelberg (2003)
10. Libert, B., Vergnaud, D.: Unidirectional chosen-ciphertext secure proxy re-encryption, http://hal.inria.fr/inria-00339530/en/, This is the extended version of [4]
11. Deng, R.H., Weng, J., Liu, S., Chen, K.: Chosen-Ciphertext Secure Proxy Re-encryption without Pairings. In: Franklin, M.K., Hui, L.C.K., Wong, D.S. (eds.) CANS 2008. LNCS, vol. 5339, pp. 1–17. Springer, Heidelberg (2008)

12. Shao, J., Cao, Z.: CCA-Secure Proxy Re-encryption without Pairings. In: Jarecki, S., Tsudik, G. (eds.) Public Key Cryptography. LNCS, vol. 5443, pp. 357–376. Springer, Heidelberg (2009)
13. Weng, J., Chow, S.S., Yang, Y., Deng, R.H.: Efficient unidirectional proxy re-encryption. Cryptology ePrint Archive, Report 2009/189 (2009), http://eprint.iacr.org/
14. Libert, B., Vergnaud, D.: Tracing malicious proxies in proxy re-encryption. In: Galbraith, S.D., Paterson, K.G. (eds.) Pairing 2008. LNCS, vol. 5209, pp. 332–353. Springer, Heidelberg (2008)
15. Matsuo, T.: Proxy Re-encryption Systems for Identity-Based Encryption. In: Takagi, T., Okamoto, T., Okamoto, E., Okamoto, T. (eds.) Pairing 2007. LNCS, vol. 4575, pp. 247–267. Springer, Heidelberg (2007)
16. Green, M., Ateniese, G.: Identity-Based Proxy Re-encryption. In: Katz, J., Yung, M. (eds.) ACNS 2007. LNCS, vol. 4521, pp. 288–306. Springer, Heidelberg (2007)
17. Chu, C.K., Tzeng, W.G.: Identity-Based Proxy Re-encryption Without Random Oracles. In: Garay, J.A., Lenstra, A.K., Mambo, M., Peralta, R. (eds.) ISC 2007. LNCS, vol. 4779, pp. 189–202. Springer, Heidelberg (2007)
18. Chu, C.K., Weng, J., Chow, S.S.M., Zhou, J., Deng, R.H.: Conditional proxy broadcast re-encryption. In: ACISP, pp. 327–342 (2009)
19. Weng, J., Deng, R.H., Liu, S., Chen, K., Lai, J., Wang, X.: Chosen-ciphertext secure proxy re-encryption without pairings. Cryptology ePrint Archive, Report 2008/509 (2008), http://eprint.iacr.org/, This is the full paper of [11]
20. Gamal, T.E.: A Public Key Cryptosystem and a Signature Scheme Based on Discrete Logarithms. In: Blakely, G.R., Chaum, D. (eds.) CRYPTO 1984. LNCS, vol. 196, pp. 10–18. Springer, Heidelberg (1985)
21. Fujisaki, E., Okamoto, T.: Secure Integration of Asymmetric and Symmetric Encryption Schemes. In: Wiener, M.J. (ed.) CRYPTO 1999. LNCS, vol. 1666, pp. 537–554. Springer, Heidelberg (1999)
22. Coron, J.S.: On the Exact Security of Full Domain Hash. In: Bellare, M. (ed.) CRYPTO 2000. LNCS, vol. 1880, pp. 229–235. Springer, Heidelberg (2000)

# Appendix

## A Cryptanalysis of Weng *et al.*'s C-PRE Scheme

In this section, we will explain why Weng *et al.*'s C-PRE scheme [6] fails to achieve the CCA-security. Due to the space limit, here we only give a brief review of the scheme (please refer to [6] for the detailed scheme and the corresponding security notions). In Weng *et al.*'s scheme, a user's private key for the user is $sk = x \in \mathbb{Z}_q^*$, and his public key is $pk = (g^x, g_1^{1/x})$. The re-encryption key, from one public key $pk_i = (g^{x_i}, g_1^{1/x_i})$ to another public key $pk_j = (g^{x_j}, g_1^{1/x_j})$ associated with condition $w$, consists of two parts: a partial re-encryption key $rk_{i,j} = g^{x_j/x_i}$ and a condition key $ck_{i,w} = H_3(w, pk_i)^{1/x_i}$. A second level ciphertext $CT_i = (A, B, C, D)$ under $pk_i$ is

$$\left( g_1^r, \ (g^{x_i})^r, \ H_2\left(e(g,g)^r\right) \oplus (m\|r') \oplus H_4\left(e(Q_i, H_3(w, pk_i))^r\right), \ H_5(A, B, C)^r \right),$$

while a first level ciphertext $CT_j = (B', C)$ re-encrypted from $pk_i$ to $pk_j$ is

$$\left( e(g, g^{sk_j})^r, \ H_2(e(g,g)^r) \oplus (m\|r') \right).$$

According to the security model defined in [6], for a target public key $pk_{i^*}$ and a target condition $w^*$, even if the adversary has corrupted another user's secret key $sk_j$, he is still allowed to obtain one (not *both*) of the partial re-encryption key $rk_{i^*,j}$ and the condition key $ck_{i^*,w^*}$. Now, we explain how an adversary can break the CCA-security of Weng *et al.*'s scheme: she first obtains $sk_j = x_j$ and $rk_{i^*,j} = g^{x_j/x_{i^*}}$, and then computes $g^{1/x_{i^*}} = \left(g^{x_j/x_{i^*}}\right)^{1/x_j}$. Next, she calculates $e(g,g)^r$ as $e\left((g^{x_{i^*}})^r, g^{1/x_{i^*}}\right)$, where $(g^{x_{i^*}})^r$ is exactly the second component of the second level ciphertext. Using $e(g,g)^r$, she can certainly decrypt the first level ciphertext to obtain the underlying plaintext.

## B    Security Proof for Theorem 1

*Proof.* Suppose algorithm $\mathcal{B}$ is given a DBDH instance $(g, g^a, g^b, g^c, Z) \in \mathbb{G}^4 \times \mathbb{G}_T$ with unknown $a, b, c \in_R \mathbb{Z}_q$. $\mathcal{B}$'s goal is to decide whether $Z = e(g,g)^{abc}$. $\mathcal{B}$ works by interacting with adversary $\mathcal{A}$ in the IND-2CPRE-CCA game as follows:

**Initialize Stage.** $\mathcal{B}$ gives $param = ((q, \mathbb{G}, \mathbb{G}_T, e), g, n, H_1, \cdots, H_5)$ to $\mathcal{A}$. Here $H_1, \cdots, H_5$ are the random oracles controlled by $\mathcal{B}$ and can be adaptively asked by $\mathcal{A}$ at any time. $\mathcal{B}$ maintains five hash lists $H_i^{\text{list}}$ with $i \in \{1, \cdots, 5\}$, which are initially empty, and responds the random oracle queries for $\mathcal{A}$ as shown in Figure 1.

---

- $H_1(m, R)$: If this query already appears on $H_1^{\text{list}}$ in a tuple $(m, R, r)$, return $r$. Otherwise, choose $r \in_R \mathbb{Z}_q$, add the tuple $(m, R, r)$ to the $H_1^{\text{list}}$ and respond with $H_1(m, R) = r$.
- $H_2(pk_i, w)$: If this query already appears on the $H_2^{\text{list}}$, then return the predefined value. Otherwise, choose $\mu, \mu' \in_R \mathbb{Z}_q$, and use the Coron's proof technique [22] to flip a biased coin $\text{coin}_i \in \{0, 1\}$ that yields 1 with probability $\theta$ and 0 with probability $1 - \theta$. If $\text{coin}_i = 0$, define $H_2(pk_i, w) = g^\mu \cdot (g^b)^{-\mu'}$; otherwise, define $H_2(pk_i, w) = g^{\mu + \mu'}$. Finally, add the tuple $(pk_i, w, \text{coin}_i, \mu, \mu')$ to the list $H_2^{\text{list}}$ and respond with $H_2(pk_i, w)$.
- $H_3(R)$: If this query already appears on the $H_3^{\text{list}}$, then return the predefined value. Otherwise, choose $\omega \in_R \{0, 1\}^n$, add the tuple $(R, \omega)$ to the $H_3^{\text{list}}$ and respond with $H_3(R) = \omega$.
- $H_4(C_1, C_2, C_3)$: If this query already appears on the $H_4^{\text{list}}$, then return the predefined value. Otherwise, choose $\gamma \in_R \mathbb{Z}_q$, add the tuple $(C_1, C_2, C_3, \gamma)$ to the $H_4^{\text{list}}$ and respond with $H_4(C_1, C_2, C_3) = g^\gamma$.
- $H_5(V)$: If this query already appears on the $H_5^{\text{list}}$, then return the predefined value. Otherwise, choose $\lambda \in_R \mathbb{Z}_q$, add the tuple $(V, \lambda)$ to the $H_5^{\text{list}}$ and respond with $H_5(V) = \lambda$.

---

**Fig. 1.** The Simulations for $H_i$ for $i = 1, \cdots, 5$

**Find Stage.** In this stage, adversary $\mathcal{A}$ issues a series of queries subject to the restrictions of the IND-2CPRE-CCA game. $\mathcal{B}$ maintains a list $K^{\text{list}}$ which is initially empty, and answers these queries for $\mathcal{A}$ as follows:

- Uncorrupted key generation oracle $\mathcal{O}_u(i)$: Algorithm $\mathcal{B}$ first picks $x_i \in_R \mathbb{Z}_q$, and defines $pk_i = (g^a)^{x_i}$. Next, it sets $c_i = 0$ and adds the tuple $(pk_i, x_i, c_i)$ to the $K^{\text{list}}$. Finally, it returns $pk_i$ to adversary $\mathcal{A}$.
- Corrupted key generation oracle $\mathcal{O}_c(j)$: $\mathcal{B}$ first picks $x_j \in_R \mathbb{Z}_q$ and defines $pk_j = g^{x_j}$ and $c_j = 1$. Next, it adds the tuple $(pk_j, x_j, c_j)$ to the $K^{\text{list}}$ and returns $(pk_j, x_j)$ to adversary $\mathcal{A}$.

- Re-encryption key oracle $\mathcal{O}_{rk}(pk_i, w, pk_j)$: $\mathcal{B}$ first recovers $(pk_i, w, \text{coin}_i, \mu, \mu')$ from the $H_2^{\text{list}}$ and tuples $(pk_i, x_i, c_i)$ and $(pk_j, x_j, c_j)$ from the $K^{\text{list}}$. Next, it constructs the re-encryption key $rk_{i \overset{w}{\to} j}$ for adversary $\mathcal{A}$ according to the following situations:
  - Case 1: $c_i = 1$, it means that $sk_i = x_i$. Using $sk_i$, $\mathcal{B}$ can certainly generate the re-encryption key $rk_{i \overset{w}{\to} j}$ for $\mathcal{A}$ as in algorithm ReKeyGen.
  - Case 2: $(c_i = 0 \land c_j = 1 \land \text{coin}_i = 1)$, it means that $sk_i = ax_i$, $sk_j = x_j$ and $H_2(pk_i, w) = g^{\mu+\mu'}$. $\mathcal{B}$ picks $s \in_R \mathbb{Z}_q$, computes $rk_2 = pk_i^s, rk_1 = (g^a)^{-(\mu+\mu'+x_j \cdot s \cdot H_5((g^a)^{x_i \cdot s \cdot x_j}))x_i}$ and returns $(rk_1, rk_2)$ to $\mathcal{A}$. Observe that this is indeed a valid re-encryption key, since

$$rk_1 = (g^a)^{-(\mu+\mu'+x_j \cdot s \cdot H_5((g^a)^{x_i \cdot s \cdot x_j}))x_i} = \left(g^{\mu+\mu'+sk_j \cdot s \cdot H_5(pk_j^{s \cdot sk_i})}\right)^{-a \cdot x_i}$$

$$= \left(g^{\mu+\mu'} g^{sk_j \cdot s \cdot H_5(pk_j^{s \cdot sk_i})}\right)^{-sk_i} = \left(H_2(pk_i, w) pk_j^{s \cdot H_5(pk_j^{s \cdot sk_i})}\right)^{-sk_i}.$$

  - Case 3: $(c_i = 0 \land c_j = 0 \land \text{coin}_i = 1)$, it means that $sk_i = ax_i$, $sk_j = ax_j$ and $H_2(pk_i, w) = g^{\mu+\mu'}$. $\mathcal{B}$ picks $s' \in_R \mathbb{Z}_q$, computes $rk_2 = g^{x_i s'}, rk_1 = (g^a)^{-(\mu+\mu'+x_j s' \cdot H_5(pk_j^{s' \cdot x_i}))x_i}$, and returns $(rk_1, rk_2)$ to $\mathcal{A}$. Observe that, letting $s = \frac{s'}{a}$, one can see that it is indeed a valid re-encryption key.
  - Case 4: $(c_i = 0 \land c_j = 0 \land \text{coin}_i = 0)$, it means that $sk_i = ax_i$, $sk_j = ax_j$ and $H_2(pk_i, w) = g^{\mu} \cdot (g^b)^{-\mu'}$. $\mathcal{B}$ picks $s \in_R \mathbb{Z}_q$, computes $rk_2 = pk_i^s$, $rk_1 = pk_i^{-u}$, and returns returns $rk_{i \overset{w}{\to} j} = (rk_1, rk_2)$ to $\mathcal{A}$. Observe that, if implicitly let $H_5(pk_j^{s \cdot sk_i}) = \frac{b \cdot \mu'}{s \cdot a \cdot x_j}$ (note that $pk_j^{s \cdot sk_i}$ is unknown to $\mathcal{A}$, since $sk_i, sk_j$ and $s$ are all unknown to him), we can easily see that this is indeed a valid re-encryption key as required.
  - Case 5: $(c_i = 0 \land c_j = 1 \land \text{coin}_i = 0)$, $\mathcal{B}$ outputs $\beta' \in_R \{0,1\}$ and **aborts**.
- Re-encryption oracle $\mathcal{O}_{re}(pk_i, pk_j, (w, \text{CT}_i))$: $\mathcal{B}$ parses $\text{CT}_i = (C_1, C_2, C_3, C_4)$. If Eq. (4) does not hold, it outputs $\perp$; otherwise, it works as follows:
  1. Recover $(pk_i, x_i, c_i)$ and $(pk_j, x_j, c_j)$ from the $K^{\text{list}}$ and $(pk_i, w, \text{coin}_i, \mu, \mu')$ from the $H_2^{\text{list}}$.
  2. If $(c_i = 0 \land c_j = 1 \land \text{coin}_i = 0)$ does not hold, then $\mathcal{B}$ can construct the re-encryption key $rk_{i \overset{w}{\to} j}$ as in the re-encryption key query, and then can certainly generate the first level ciphertext $\text{CT}_j$ for $\mathcal{A}$.
  3. Otherwise, it implies that $c_j = 1$, i.e., $sk_j = x_j$. In this case, $\mathcal{B}$ picks $s \in_R \mathbb{Z}_q$ and generates the first level ciphertext as follows: search whether there exists a tuple $(m, R, r) \in H_1^{\text{list}}$ such that $g_1^r = C_1, R \cdot e(pk_i, H_2(pk_i, w))^r = C_2, m \oplus H_3(R) = C_3$ and $H_4(C_1, C_2, C_3)^r = C_4$ hold. If yes, pick $s \in_R \mathbb{Z}_q$, compute $\overline{C}_4 = pk_i^s, \overline{C}_2 = R \cdot e(C_1, pk_i^{s \cdot H_5(\overline{C}_4^{x_j})})^{-x_j}$, and return $\text{CT}_j = (C_1, \overline{C}_2, C_3, \overline{C}_4)$ as the first level ciphertext to $\mathcal{A}$; otherwise return $\perp$. Note that we can store $s$ in a table to keep the consistency of $s$ for the same re-encryption queries $\mathcal{O}_{re}(pk_i, pk_j, (w, *))$.

- First level decryption oracle $\mathcal{O}_{1d}(pk_j, \text{CT})$: $\mathcal{B}$ first recovers $(pk_j, x_j, c_j)$ from the $K^{\text{list}}$. If $c_j = 1$ (meaning $sk_j = x_j$), $\mathcal{B}$ decrypts the ciphertext using $sk_j$ and returns the plaintext to $\mathcal{A}$. Otherwise, it searches $H_1^{\text{list}}$ and $H_5^{\text{list}}$ to see whether there exist a tuple $(m, R, r) \in H_1^{\text{list}}$ and a tuple $(V, \lambda) \in H_5^{\text{list}}$ such that $g^r = \overline{C}_1, R \cdot e\left(\overline{C}_4, pk_j\right)^{-r \cdot \lambda} = \overline{C}_2, m \oplus H_3(R) = C_3$ and $e(V, g) = e(\overline{C}_4, pk_j)$. If yes, return $m$ to $\mathcal{A}$; else return $\perp$.

**Challenge Stage.** When $\mathcal{A}$ decides that **Find** stage is over, it outputs a target public key $pk_{i^*}$, a condition $w^*$ and two equal-length messages $m_0, m_1 \in \{0, 1\}^n$. $\mathcal{B}$ responds as follows:

1. Recover $(pk_{i^*}, x_{i^*}, c_{i^*})$ from the $K^{\text{list}}$ and $(pk_{i^*}, w^*, \text{coin}_{i^*}, \mu, \mu')$ from the $H_2^{\text{list}}$. If $\text{coin}_{i^*} = 1$, output a random bit $\beta' \in_R \{0, 1\}$ and **aborts**. Otherwise, it means that $H_2(pk_{i^*}, w^*) = g^\mu \cdot (g^b)^{-\mu'}$.
2. Flip a random coin $\delta \in_R \{0, 1\}$ and pick $R^* \in_R \mathbb{G}_T$. Compute $C_1^* = g^c$, $C_2^* = R^* \cdot Z^{-\mu' \cdot x_{i^*}} \cdot e(g^a, g^c)^{x_{i^*} \mu}$ and $C_3^* = m_\delta \oplus H_3(R^*)$.
3. Issue an $H_4$ query on $(C_1^*, C_2^*, C_3^*)$ to obtain the tuple $(C_1^*, C_2^*, C_3^*, \gamma^*)$, and define $C_4^* = (g^c)^{\gamma^*}$.
4. Finally, give $\text{CT}^* = (C_1^*, C_2^*, C_3^*, C_4^*)$ to $\mathcal{A}$.

Note that by the above construction, if $Z = e(g, g)^{abc}$, $\text{CT}^*$ is indeed a valid ciphertext for $m_\delta$ under $pk_{i^*}$ and $w^*$. To see this, implicit letting $H_1(m_\delta, R^*) = c$, we have

$$C_2^* = R^* \cdot Z^{-\mu' \cdot x_{i^*}} \cdot e(g^a, g^c)^{x_{i^*} \mu} = R^* \cdot e(g, g)^{-\mu' \cdot abc \cdot x_{i^*}} \cdot e(g^a, g^c)^{x_{i^*} \mu}$$
$$= R^* \cdot e(g^{a \cdot x_{i^*}}, g^\mu g^{-\mu' \cdot b})^c = R^* \cdot e(pk_{i^*}, H_2(pk_{i^*}, w^*))^c,$$
$$C_1^* = g^c, \quad C_3^* = m_\delta \oplus H_3(R^*), \quad C_4^* = (g^c)^{\gamma^*} = (g^{\gamma^*})^b = H_4(C_1^*, C_2^*, C_3^*)^c.$$

On the other hand, when $Z$ is uniform and independent in $\mathbb{G}_T$, the challenge ciphertext $\text{CT}^*$ is independent of $\delta$ in the adversary's view.

**Guess Stage.** $\mathcal{A}$ continues to issue the rest of queries as in **Find** stage, with the restrictions described in the IND-2CPRE-CCA game. $\mathcal{B}$ responds to these queries as in **Find** stage.

**Output Stage.** Eventually, adversary $\mathcal{A}$ returns a guess $\delta' \in \{0, 1\}$ to $\mathcal{B}$. If $\delta' = \delta$, $\mathcal{B}$ outputs $\beta' = 1$; otherwise, $\mathcal{B}$ outputs $\beta' = 0$.

This completes the description of the simulation. Due to space limit, in the full paper, we will show that $\mathcal{B}$'s advantage against the DBDH assumption is at least $\epsilon' \geq \frac{\epsilon}{\tilde{e}(1 + q_{rk})} - \frac{q_{H_1} + q_{H_5} + q_{re} + q_d}{q}$, and $\mathcal{B}$'s running time is bounded by $t' \leq t + \mathcal{O}(\tau(q_{H_2} + q_{H_4} + q_u + q_c + 3q_{rk} + q_{H_1}q_{re} + (q_{H_1} + q_{H_5})q_d))$. $\qquad\square$

# Practical Algebraic Attacks on the Hitag2 Stream Cipher

Nicolas T. Courtois[1], Sean O'Neil[2], and Jean-Jacques Quisquater[3]

[1] University College London, UK
[2] VEST Corporation, France
[3] Université Catholique de Louvain, Belgium

**Abstract.** Hitag2 is a stream cipher that is widely used in RFID car locks in the automobile industry. It can be seen as a (much) more secure version of the [in]famous Crypto-1 cipher that is used in MiFare Classic RFID products [14,20,15]. Recently, a specification of Hitag2 was circulated on the Internet [29]. Is this cipher secure w.r.t. the recent algebraic attacks [8,17,1,25] that allowed to break with success several LFSR-based stream ciphers? After running some computer simulations we saw that the Algebraic Immunity [25] is at least 4 and we see no hope to get a very efficient attack of this type.

However, there are other algebraic attacks that rely on experimentation but nevertheless work. For example Faugère and Ars have discovered that many simple stream ciphers can be broken experimentally with Gröbner bases, given an extremely small quantity of keystream, see [17]. Similarly reduced-round versions of DES [9] and KeeLoq [11,12] were broken using SAT solvers, that actually seem to outperform Gröbner basis techniques. Thus, we have implemented a generic experimental algebraic attack with conversion and SAT solvers, [10,9]. As a result we are able to break Hitag2 quite easily, the full key can be recovered in a few hours on a PC. In addition, given the specific protocol in which Hitag2 cipher is used in cars, some of our attacks are practical.

**Keywords:** RFID tags, Hitag 2 algorithm, MiFare Crypto-1 cipher, stream ciphers, algebraic cryptanalysis, Boolean functions, Gröbner bases, SAT solvers.

## 1 Introduction

Hitag2 is a stream cipher that is primarily used in RFID transponder systems manufactured by Philips/NXP, and used by many car manufacturers for unlocking car doors remotely. According to [33] it is used for example in all Alfa Romeo 156 and 166 models, Ford Galaxy and Transit, GM Corsa and Zafira, numerous Nissan, Opel, Peugeot, Seat and Volvo models, most Honda cycles and most Iveco trucks. It is not clear whether this cipher is still used in many new cars, but it is used in numerous models still in widespread use.

This system was introduced by Philips Semiconductors in the late 90's. The security level of this cipher was quite moderate from the start. The key is only 48

P. Samarati et al. (Eds.): ISC 2009, LNCS 5735, pp. 167–176, 2009.
© Springer-Verlag Berlin Heidelberg 2009

bits. Moreover, the internal state is also only 48 bits, not bigger than the key size, which makes Hitag2 vulnerable to time/data/memory tradeoff attacks [3]. These however require both large quantities of keystream and large quantities of RAM, and therefore are neither practical to run nor it is very realistic for the attacker to get a large amount of keystream generated with the same key. However given the key length, exhaustive search is feasible and requires a tiny quantity of data. In this paper we show that in identical (or very close) conditions: given a very tiny amount of known (or chosen) plaintext, attacks substantially faster than brute force do exist.

The security of Hitag2 can be compared to the security of the famous MiFare Crypto 1 cipher that is used in hundreds of millions of smart cards worldwide [14,20,15]. In MiFare Crypto 1, the taps of the Boolean function are regular which makes it indeed breakable by SAT solver attacks [14], but these actually have very soon became obsolete, because there is an even faster direct attack that takes 0.05 seconds on a PC [20]. In Hitag 2 the taps are not regular. Yet we will show that the cipher is still breakable faster than by brute force, with this type of automated cryptanalysis with SAT solvers.

## 1.1   Algebraic Cryptanalysis

In the past, many researchers asked if ciphers such as the DES could be broken by solving a system of Boolean equations, see for example [32,22,16,23,7,28] following the idea of algebraic cryptanalysis (breaking cipher seen as solving a system of equations) that was formulated as early as by Shannon in 1949, see [31]. It turns out that so far nobody has been able to break standard cryptography such as the DES or the AES, see [9,7]. Moreover, very few block ciphers were ever broken. Many researchers have tried to crack small scale variants of AES [5] and similar toy ciphers [18]. However the authors also showed very strong limitations of these attacks, for example these attacks can break (only) 6 rounds of the DES, but this is accomplished given 1 single known plaintext [9], compared to linear and differential cryptanalysis that require larger amounts of encrypted material in order to work. So far, only one full-round block cipher that is used in practice can be broken by an algebraic attack. This cipher is KeeLoq, an industrial cipher used by hundreds of millions of people every day to unlock their cars. Here the algebraic attacks can break up to 160 rounds of the cipher, but can also break the full 528 rounds of the cipher if one uses another weakness of the cipher: its periodic structure, see [11]. In this paper we break another cipher that is used in automobile locks and alarms, Hitag2.

Algebraic cryptanalysis was much more successful for LFSR-based stream ciphers. Several stream ciphers were broken since 2003 Courtois, Meier, Krause et al, see [1,8,17]. These attacks exploit low degree I/O relations for the output filter/combiner Boolean function of the cipher (and augmented versions of it), and we have verified that these attacks cannot break Hitag2 algorithm, because the Boolean function used is quite large and has good "Algebraic Immunity" of at least 4. However, Faugère and Ars have discovered another major attack on LFSR-based stream ciphers that is not widely known: they showed that many simple stream ciphers can be broken experimentally with Gröbner bases, given an extremely small quantity of keystream, see [17], not much more than one needs to make the solution uniquely defined. This type of attack is a bit obscure and relies on experimentation. It remains quite rare, and the bottom line is that more or less any cipher can be broken provided that it is "not too complex", and this in an automated way without human intervention.

In this paper we will break Hitag2 in a similar way, showing that it is a very weak design. We do not however use Gröbner bases. In fact, it appears that Gröbner bases algorithms such as F4 or F5 are not always necessary, in several cases they either they can be replaced by a simpler attack that does not require a fixed monomial ordering and is essentially a linear algebra attack [9,11], or there is a faster attack. For example, the method of solving multivariate low-degree equations via algebraic representation and later conversion to a logical SAT problem [10,9], (see also [24,27]) is in almost all that cases we are aware of, much faster and can break much more instances than the current Gröbner basis techniques.

Algebraic representation and conversion to SAT is also the method we exploit in the present paper. Other very promising algorithms were also proposed for solving systems of equations derived from symmetric ciphers, see [30], but so far they (or at least their current implementation) still lag behind the SAT solvers. For example they allow to break only 4 rounds of DES, cf. [30].

### 1.2 Our Contribution

In this paper we show that Hitag2 is extremely weak w.r.t. algebraic attacks. This result surprised us, as nothing in the description of Hitag2 allows to believe that it will be weak. In fact, we will demonstrate that if we simply XOR two of the LFSR bits to the output of the nonlinear function used in Hitag2, it will already be much stronger against our attacks achieving a level of security close to its key size. So there is a real mystery here that remains unsolved: why is this cipher comparatively quite weak? We don't answer this question, just demonstrate the weakness experimentally and compare to the tweaked version.

The attacks described in this paper are essentially black box attacks: we try some tools, they are able to recover the key, and we have a limited understanding of why these attacks work. The method — write equations, convert and solve — is very general and applicable to any block or stream cipher, see [10,9]. It is very hard to know what exactly makes systems efficiently solvable, but it appears that sparsity alone can make systems efficiently solvable, both by SAT solvers and classical Gröbner bases methods, see [10,30]. However these attacks are far from being obscure attacks that nobody can reproduce: They are very simple to implement following [10] and combined with a public domain SAT solver MiniSat 2.0. [26]. We contend that any researcher who is also a good programmer can implement this attack in one day.

## 2  Cipher Description

Hitag2 is a very simple cipher: a filter generator with 48-bit LFSR and a non-linear function with 20 inputs, that produces 1 output bit per clock (the other components on the picture are not used for encryption, only at the initialisation stage). The reference source code together with test vectors can be downloaded from [29]. The picture below (Fig. 1) describes both the encryption (keystream generation) and the initialisation process.

First, the state is filled with 32 bits of the serial number and 16 lower-ranking bits of the key. Then for 32 steps, the LFSR feedback is not used, instead the state is filled on the right with 32 bits, each of them being a XOR of three bits: one output by the Boolean function applied to the previous state, one bit of the key, and one bit of the IV. After 32 steps the state of the cipher becomes the initial setting of the LFSR.

Then the keystream is generated as follows. At each clock, the LFSR is updated first, and then the output bit is computed by using the whole Boolean function on 20 inputs. This large Boolean function is composed of 6 instantiations of 3 smaller Boolean functions that are described on this picture by their truth tables and their Algebraic Normal Forms (ANF).

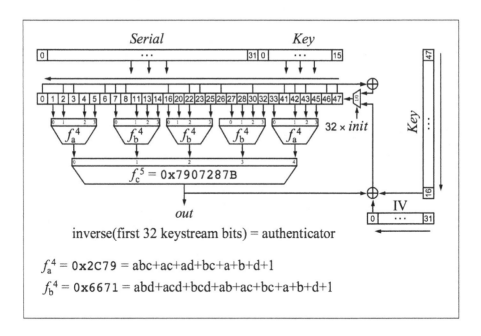

$f_a^4 = 0x2C79 = abc+ac+ad+bc+a+b+d+1$

$f_b^4 = 0x6671 = abd+acd+bcd+ab+ac+bc+a+b+d+1$

**Fig. 1.** Hitag2 Initialization and Encryption

## 3   The Hitag 2 Protocol

According to [21] HITAG2 RFID chips have 256-bit of data that is divided in 8 pages of 32-bits. They have several read-only modes (A, B and C) in which data are broadcast in cleartext. These modes are suitable for applications such as animal identification, but offer no security. When Hitag2 systems are used in automobile locks, they work in a more secure so called "Crypto mode" as described in Section 4.2.1. of [21]. The protocol works as follows:

1. The reader (embedded inside a car) sends the command 0x18 to the transponder (which is battery-powered and embedded in the portable key-fob).
2. The tag responds by 11111 followed by 32-bit serial number (SN).
3. Then the reader sends a pseudo-random IV on 32 bits and the authenticator on 32 bits. The authenticator is the first 32 bits of the keystream (but transmitted in the reverse order), that are obtained from the Hitag2 cipher as on Fig. 1 initialised with this IV and with the secret key shared both by the reader and the transponder key.
4. If the key is happy with the 32-bit authenticator that should match the one it has computed itself, it sends 11111 followed by the content of Page 3 of its memory (these are 8 bits of configuration flags and 24 bits of some personalised "transponder password" denoted PSWT) that are encrypted by bitwise XOR with the next 32 bits of the keystream.

**Important Remarks.** This isn't a classical challenge-response authentication protocol, it is the car that proves his identity to the key first, and clearly a chosen-IV attack is not possible here (this would allow us to break the system in 6 hours, see Section 7). Moreover, unless the attacker already knows the content of Page 3 of the transponder memory, the attacker obtains only 32 bits of the keystream, which makes that not all our attacks are practical, see Section 6.

## 4 Handling Algebraic Attacks

All key recovery attacks in this paper are algebraic attacks with SAT solvers [10,9] that work as follows:

1. First we write the key recovery problem as a large system of equations with a very large number of unknowns over $GF(2)$ with non-linear components of the cipher being described by a localised small system of very sparse low-degree Boolean equations. This is done by following directly the description of the cipher. Linear operations give linear equations, and nonlinear operations can be written as multivariate polynomials in their Algebraic Normal Form (ANF). There is however a better method: in order to keep equations as simple as possible, we may add additional variables. Thus, Boolean functions are actually implemented by cutting them in smaller chunks, following closely their gate-efficient (bitslice) representation that can be found directly in the Hitag2 reference code [29]. In this bitslice representation each bit is computed using negations, OR gates and AND gates with two inputs. All these gates are then directly described as quadratic equations as follows: $a\ OR\ b \equiv a + b + ab$ and $a\ AND\ b \equiv ab$ and $NOT\ a \equiv 1 + a$. With an appropriate naming scheme for variables, our equations look typically as follows:

R_905_0_4=S_938_0*S_947_0+S_938_0+S_947_0

Writing these equations present no difficulty. It requires just to follow step by step the source code of an efficient implementation of the cipher [29] that is decomposed in elementary operations on bits.

2. We will fix some $g$ key bits to their correct values (in a full attack all $2^g$ possibilities are checked until the correct key is found). Typically $g$ is between 0 and 16. There is a right number of key bits that one needs to guess. If we get too few the attack will be slower. (Quite surprisingly, also when we guess too many, the attack can frequently be slower, but when guess even more, at some moment the system will be solved instantly).

3. Then we use a conversion to a SAT problem, we use exactly the same general-purpose program that is described and used in [10,9].

4. Then the key and all the other variables are determined in seconds by an open-source program MiniSat 2.0. [26]. We assume that this takes time $T$.

5. We run our attack with the correct key and we get time $T$. The total complexity of the whole attack is estimated to be $2^g \cdot T$ for chance of success of at least 80 %, see below for additional explanations.

**Notes on the Total Complexity.** There are two basic ways of handling such an attack, that can be called serial and parallel, and these will lead to various trade-offs between the speed and success probability. In the parallel version, we run all the computation in parallel, and abort as long as one of them finds the key. The total time is $2^g \cdot T$ for 100 % success probability. In practice, this it cannot be done because this would result in much lower speeds due to the influence of the CPU cache, and the amounts of RAM needed to run processes concurrently. However it is a good strategy for a distributed attack. In the serial version, we assume in advance that $T$ lies in a certain interval, for example $T < 2 \cdot E(T)$, where $E(T)$ is the average value. Then we abort all computations after time $2 \cdot E(T)$. This an early abort strategy and the success probability will be, this is a rough estimation, about 80 %. Only in about 20 % of cases the time will exceed $2 \cdot E(T)$. With this strategy, 80 % of the time we do only $2^{g-1}$ trials on average, and in 20 % of cases we do full $2^g$ trials and fail to find the key. Here the total time is also on average about $0.8 \cdot 2^{g-1} \cdot 2T + 0.2 \cdot 2^g \cdot 2T \approx 2^g \cdot T$ but for the success probability of only about 80 %.

**Further Remarks.** In addition, we can also restart all the processes with a slightly bigger interval, this would be a mix of parallel and serial strategy. The exact complexity of such attacks simply cannot be computed without extra considerations such as number of CPUs available, memory bandwidth of the solver program that is function that depends on time, and will greatly depend on the sizes and speeds of the CPU cache, the RAM and disk drives available. So we keep a simple estimate as above.

## 5   Brute Force Attacks

We assume that in a software implementation, 1 output bit for Hitag2 cipher requires about 40 CPU clocks. The reference implementation of Hitag2 can be found here [29]. Accordingly, to output 48 keystream bits: we need abut $2^{11}$ CPU clocks. Then $2^{20}$ Keys will be checked in 0.5 s on average with 2 GHz CPU, and $2^{32}$ Keys will be checked in $2000s$ on average. The full key space can therefore be checked in about 4 years on average.

We don't consider FPGA implementations, these would considerably speed-up both the brute force and our attacks. However since no FPGA implementations of SAT solvers or other algebraic attacks are known, we cannot make comparisons. Therefore we consider and compare only software attacks.

## 6   Known-IV Attacks

The equations that we are solving include the complete initialisation process of the cipher. The results are as follows:

Given 1 single known IV and 50 known keystream bits, and if we guess correctly 16 key bits, the remaining 32 key bits are found by a SAT solver in 0.6 s on a 2 GHz Centrino CPU. This is about 4000x faster than brute force. The full attack on full 48-bit key takes about $2^{16} \cdot 0.6s$ which is about 11 hours.

In contrast, for our tweaked strengthened version of the cipher with a linear function of two state bits (we chose bits number 3 and 47) XORed to the output, the fastest attack we found takes $2^{16} \cdot 300$ seconds which is equivalent to 1 year. This is only very marginally faster than brute force (that would be about 4 years).

**Is this practical?** Following Section 3, only 32 keystream bits per IV (the authenticator) are known to the attacker, this attack with 50 output bits is not practical.

### 6.1   Practical Known-IV Attacks

In the real-life, when only 32 bits of the keystream per known IV are available, we have the following attack: We fix 14 bits of the key and write our equations for four known IVs. The solution is then found in 10 seconds on a PC. Then a full attack on a full 48-bit key takes about $2^{14} \cdot 10s$ which is less than 2 days.

## 7   Chosen-IV Attacks

In our chosen IV attacks, the IVs are chosen to be consecutive integers encoded on 32 bits in the inversed order: i.e. the least significant bit is the one that enters the cipher the last. The first IV is chosen at random.

For Hitag2, our fastest attack requires 16 chosen IVs, and takes **6 hours**.

For the tweaked cipher, our fastest attack requires 16 chosen IVs, and takes 6 months.

**Is this chosen IV attack practical?** Following Section 3, this chosen IV attack is not practical. The fastest practical attack remains the attack with 4 known IVs form Section 6.1 that takes less than 2 days.

## 8    Conclusion

In this paper we show that, it is not easy to know if a cipher is broken by an already published attack in a matter of hours, until we actually try the attack. Even though the notion Algebraic Immunity [25] suggests that Hitag2 should be secure, there is a larger variety of algebraic attacks on stream cipher, and this cipher simply isn't so secure. We also demonstrated that a minor modification of this cipher may render it secure (or secure enough) for reasons that are far from being understood. It appears that experimental cryptanalysis is now well ahead of theory and we are surprised by the very existence attacks that are practical enough to be exploited by hackers. Many older and simple industrial ciphers can be broken, and examples of Hitag2 (this paper), KeeLoq [2] and MiFare Classic [20,15] show that the common industrial practice of choosing some fast or economical cipher and using it in products for many years to come is now bankrupted. Weak ciphers should be discontinued before they are broken, and not after.

Trying to keep the specification of insufficiently secure products secret may make things even worse. Then the security of real-life products does collapse very badly in one day, when the cipher is reverse engineered [20,29].

We will not however fall into the trap of believing that 'open source cryptography' is a solution to this problem. Maybe open-source cryptography is a plausible and economical solution. Really good security however, occurs when there are several layers of security, and the confidentiality of the cryptographic algorithm can help. Especially in the embedded hardware world, where some attacks such as side channel attacks are hard and costly to avoid, a secret algorithm can be an effective barrier that prevents this type of analysis. So a cryptographic algorithm can be secret. Provided it is as good as a standard 'open source' cipher and does not mislead the customer about the security level offered.

In this paper we are dealing with automated cryptanalysis techniques, that are computer programs that recover they key from the specification of the cipher. This opens new interesting perspectives for the industry. Traditionally, the industrial firm would develop or just use a cipher and would (or rather should) ask cryptanalysts to evaluate its security. But then they have to disclose their algorithms. Now however, the firm can also ask cryptanalysts to develop software that to a large extent automates the cryptanalysis process, and therefore the industry might also be able to evaluate the security of their (confidential) designs themselves, at least against attacks such as described in this paper.

## References

1. Armknecht, F., Krause, M.: Algebraic Atacks on Combiners with Memory. In: Boneh, D. (ed.) CRYPTO 2003. LNCS, vol. 2729, pp. 162–176. Springer, Heidelberg (2003)
2. Biham, E., Dunkelman, O., Indesteege, S., Keller, N., Preneel, B.: How to Steal Cars – A Practical Attack on KeeLoq. In: Smart, N.P. (ed.) EUROCRYPT 2008. LNCS, vol. 4965, pp. 1–18. Springer, Heidelberg (2008)

3. Biryukov, A., Shamir, A.: Cryptanalytic Time/Memory/Data Tradeoffs for Stream Ciphers. In: Okamoto, T. (ed.) ASIACRYPT 2000. LNCS, vol. 1976, pp. 1–13. Springer, Heidelberg (2000)
4. Buchmann, J., Pychkine, A., Weinmann, R.-P.: Block Ciphers Sensitive to Gröbner Basis Attacks. In: Pointcheval, D. (ed.) CT-RSA 2006. LNCS, vol. 3860, pp. 313–331. Springer, Heidelberg (2006)
5. Cid, C., Murphy, S., Robshaw, M.J.B.: Small Scale Variants of the AES. In: Gilbert, H., Handschuh, H. (eds.) FSE 2005. LNCS, vol. 3557, pp. 145–162. Springer, Heidelberg (2005)
6. Courtois, N.: The security of hidden field equations (HFE). In: Naccache, D. (ed.) CT-RSA 2001. LNCS, vol. 2020, pp. 266–281. Springer, Heidelberg (2001)
7. Courtois, N., Pieprzyk, J.: Cryptanalysis of Block Ciphers with Overdefined Systems of Equations. In: Zheng, Y. (ed.) ASIACRYPT 2002. LNCS, vol. 2501, pp. 267–287. Springer, Heidelberg (2002)
8. Courtois, N., Meier, W.: Algebraic Attacks on Stream Ciphers with Linear Feedback. In: Biham, E. (ed.) EUROCRYPT 2003. LNCS, vol. 2656, pp. 345–359. Springer, Heidelberg (2003)
9. Courtois, N., Bard, G.V.: Algebraic Cryptanalysis of the Data Encryption Standard. In: Galbraith, S.D. (ed.) Cryptography and Coding 2007. LNCS, vol. 4887, pp. 152–169. Springer, Heidelberg (2007), http://eprint.iacr.org/2006/402/, Also presented at ECRYPT workshop Tools for Cryptanalysis, Krakow, September 24-25 (2007)
10. Bard, G.V., Courtois, N.T., Jefferson, C.: Efficient Methods for Conversion and Solution of Sparse Systems of Low-Degree Multivariate Polynomials over GF(2) via SAT-Solvers, http://eprint.iacr.org/2007/024/
11. Courtois, N., Bard, G.V., Wagner, D.: Algebraic and Slide Attacks on KeeLoq. In: Nyberg, K. (ed.) FSE 2008. LNCS, vol. 5086, pp. 97–115. Springer, Heidelberg (2008), http://eprint.iacr.org/2007/062/
12. Courtois, N., Bard, G.V., Bogdanov, A.: Periodic Ciphers with Small Blocks and Cryptanalysis of KeeLoq. In: Tatra Mountains Mathematic Publications, post-proceedings of Tatracrypt 2007 conference (2008) (to apperar)
13. Courtois, N., O'Neil, S.: Reverse-engineered Philips/NXP Hitag2 Cipher. Talk given at the Rump Session of Fast Sotware Encryption conference (FSE 2008), Lausanne, Switzerland, February 12 (2008), http://fse2008rump.cr.yp.to/00564f75b2f39604dc204d838da01e7a.pdf
14. Courtois, N., Nohl, K., O'Neil, S.: Algebraic Attacks on MiFare RFID Chips, http://www.nicolascourtois.com/papers/mifare_rump_ec08.pdf
15. Courtois, N.T.: The Dark Side of Security by Obscurity and Cloning MiFare Classic Rail and Building Passes Anywhere, Anytime. In: SECRYPT 2009, International Conference on Security and Cryptography, Milan, Italy, July 7-10 (2009)
16. Davio, M., Desmedt, Y., Fosseprez, M., Govaerts, R., Hulsbosch, J., Neutjens, P., Piret, P., Quisquater, J.-J., Vandewalle, J., Wouters, P.: Analytical Characteristics of the DES. In: Chaum, D. (ed.) Crypto 1983, pp. 171–202. Plenum Press, New York (1984)
17. Ars, G., Faugère, J.-C.: An Algebraic Cryptanalysis of Nonlinear Filter Generators using Gröbner Bases. INRIA research report, https://hal.ccsd.cnrs.fr/
18. Faugère, J.-C., Perret, L.: Algebraic Cryptanalysis of Curry and Flurry using Correlated Messages (September 2008), http://eprint.iacr.org/2008/402
19. Faugère, J.-C.: A new efficient algorithm for computing Gröbner bases ($F_4$). Journal of Pure and Applied Algebra 139, 61–88 (1999), http://www.elsevier.com/locate/jpaa

20. Garcia, F.D., de Koning Gans, G., Muijrers, R., van Rossum, P., Verdult, R., Schreur, R.W., Jacobs, B.: Dismantling MIFARE classic. In: Jajodia, S., Lopez, J. (eds.) ESORICS 2008. LNCS, vol. 5283, pp. 97–114. Springer, Heidelberg (2008)
21. Philips Semiconductors Data Sheet, HT2 Transponder Family, Communication Protocol, Reader ¡=¿ HITAG2(R) Transponder, Product Specification, Version 2.1 (October 1997), http://www.phreaker.ru/showthread.php?p=226
22. Hulsbosch, J.: Analyse van de zwakheden van het DES-algoritme door middel van formele codering. Master thesis, K. U. Leuven, Belgium (1982)
23. Jakobsen, T.: Cryptanalysis of Block Ciphers with Probabilistic Non-Linear Relations of Low Degree. In: Krawczyk, H. (ed.) CRYPTO 1998. LNCS, vol. 1462, pp. 212–222. Springer, Heidelberg (1998)
24. Massacci, F., Marraro, L.: Logical cryptanalysis as a SAT-problem: Encoding and analysis of the U.SS. Data Encryption Standard. Journal of Automated Reasoning 24, 165–203 (2000)
25. Meier, W., Pasalic, E., Carlet, C.: Algebraic Attacks and Decomposition of Boolean Functions. In: Cachin, C., Camenisch, J.L. (eds.) EUROCRYPT 2004. LNCS, vol. 3027, pp. 474–491. Springer, Heidelberg (2004)
26. MiniSat 2.0. An open-source SAT solver package, by Niklas Eén, Niklas Sörensson, http://www.cs.chalmers.se/Cs/Research/FormalMethods/MiniSat/
27. Mironov, I., Zhang, L.: Applications of SAT Solvers to Cryptanalysis of Hash Functions. In: Biere, A., Gomes, C.P. (eds.) SAT 2006. LNCS, vol. 4121, pp. 102–115. Springer, Heidelberg (2006), http://eprint.iacr.org/2006/254
28. Murphy, S., Robshaw, M.: Essential Algebraic Structure within the AES. In: Yung, M. (ed.) CRYPTO 2002. LNCS, vol. 2442, p. 1. Springer, Heidelberg (2002)
29. Hitag2 specification, reference implementation and test vectors, http://cryptolib.com/ciphers/hitag2/
30. Raddum, H., Semaev, I.: New Technique for Solving Sparse Equation Systems, http://eprint.iacr.org/2006/475/
31. Shannon, C.E.: Communication theory of secrecy systems. Bell System Technical Journal 28 (1949); see in particular page 704
32. Schaumuller-Bichl, I.: Cryptanalysis of the Data Encryption Standard by the Method of Formal Coding. In: Beth, T. (ed.) EUROCRYPT 1982. LNCS, vol. 149, pp. 235–255. Springer, Heidelberg (1983)
33. Transponder Table, a list of cars and transponders used in these cars. Each time the table says PH/CR, which means Philips transponder in crypto mode, we assumed that this car uses Hitag2, http://www.keeloq.boom.ru/table.pdf

# A New Construction of Boolean Functions with Maximum Algebraic Immunity

Deshuai Dong[1], Shaojing Fu[1], Longjiang Qu[1,2], and Chao Li[1,2]

[1] Department of Mathematics and System Science,
National University of Defence Technology
Changsha 410073, China
[2] State Key Laboratory of Mobile Communication
Southeast University, Nanjing 210018, China

**Abstract.** Because of the algebraic attacks, a high algebraic immunity is now an important criteria for Boolean functions used in stream ciphers. In this paper, we study the construction of Boolean functions with maximum algebraic immunity. We first present a new method to construct Boolean functions, in any number of variables, with maximum algebraic immunity(AI), and we also improve our algorithm to construct balanced functions with optimum algebraic immunity for any even number of variables. Furthermore, the enumeration and algebraic degree of the constructed Boolean functions are investigated.

**Keywords:** Boolean Function, Algebraic Attacks, Algebraic Immunity, Balancedness.

## 1 Introduction

Algebraic attack to LFSR-based stream cipher was proposed by Coutois and Meier in 2003 [1]. Its main idea is to deduce the security of a stream cipher to solve an over-defined system of multivariate nonlinear equations whose unknowns are the bits of the initialization of the LFSR. By searching low degree annihilator, some LFSR-based stream ciphers such as Toyocrypt [2], LILI-128 [1] and SFINKS [3] etc were successfully attacked.

To resist algebraic attack, a new cryptographic property of Boolean functions which is known as algebraic immunity (AI) has been proposed by Meier et al [4]. The AI of a Boolean function expresses its ability to resist standard algebraic attack. Thus the AI of Boolean function used in cryptosystem should be sufficiently high. Courtois and Meier [1,4] showed that, for any $n$-variable Boolean function, its AI is bounded by $\lceil n/2 \rceil$. If the bound is achieved, we say the Boolean function has maximum AI. Obviously, a Boolean function with maximum AI has strongest ability to resist standard algebraic attack. Therefore, the construction of Boolean functions with maximum AI is of great importance, and several classes of Boolean functions with large algebraic immunity have been investigated and constructed [5,6,7,8,9,4,10,11,12].

P. Samarati et al. (Eds.): ISC 2009, LNCS 5735, pp. 177–185, 2009.
© Springer-Verlag Berlin Heidelberg 2009

In this paper, we use a specific order on elements of $\mathbb{F}_2^n$ and divide the elements of $\mathbb{F}_2^n$ with weight no more than $\lceil n/2 \rceil - 1$ into some intervals, then we construct Boolean functions with the support from the intervals. We also improve our method to construct balanced functions with optimum algebraic immunity for any even number of variables. Then we enumerate the constructed Boolean functions, and we also study the case that the constructed Boolean functions have optimum algebraic degree.

The paper is organized as follows. Section 2 provides basic definitions and notations. In Section 3, we present our main construction of Boolean functions, in any number of variables, with maximum algebraic immunity. In Section 4, we improve our method to construct balanced maximum AI Boolean functions on even number of variables. the enumeration and algebraic degree of the constructed Boolean functions are investigated in Section 5. Section 6 concludes this paper.

## 2    Preliminaries

Let $\mathbb{F}_2$ be the binary finite field, the vector space of dimension $n$ over $\mathbb{F}_2$ is denoted by $\mathbb{F}_2^n$. A Boolean function on $n$ variables may be viewed as a mapping from $\mathbb{F}_2^n$ into $\mathbb{F}_2$. The set of all $n$-variable Boolean function is denoted by $B_n$. A Boolean function $f(x_1, x_2, \cdots, x_n)$ is also interpreted as the output column of its truth table, that is, a binary string of length $2^n$ having the form:

$$\{f(0,0,\cdots,0), \quad f(0,0,\cdots,1), \quad \cdots, \quad f(1,1,\cdots,1)\}.$$

The weight of $f$ is the number of ones in its output column, and is denoted by $wt(f)$. The support of $f$ denoted by $supp(f)$ is the set of inputs $X \in \mathbb{F}_2^n$ such that $f(X) = 1$, that is,

$$supp(f) = \{X \in F_2^n | f(X) = 1\}.$$

For vector $X = (x_1, \cdots, x_n)$ and $Y = (Y_1, \cdots, Y_n)$, we denote $X^Y = x_1^{Y_1} \cdots x_n^{Y_n}$, and the support of $X$ is denoted by $supp(X) = \{i | x_i = 1\}$.

**Definition 1.** *An $n$-variable function $f$ is balanced if and only if $wt(f) = 2^{n-1}$.*

Let us denoted the addition operator over $\mathbb{F}_2$ by $+$. An $n$-variable function $f(x_1, \cdots, x_n)$ can be seen as a multivariate polynomial over $\mathbb{F}_2$, that is,

$$f(x_1, \cdots, x_n) = a_0 + \sum_{i=1}^{n} a_i x_i + \sum_{1 \le i < j \le n} a_{i,j} x_i x_j + \cdots + a_{1,2,\cdots,n} x_1 x_2 \cdots x_n$$

where the coefficients $a_0, a_i, a_{i,j}, a_{1,2,\cdots,n}$ are constants in $\mathbb{F}_2$. This representation of $f$ is called the algebraic normal form (ANF) of $f$. The algebraic degree $\deg(f)$ of $f$ is the number of variables in the highest order term with nonzero coefficient. A Boolean function is affine if it has algebraic degree at most 1.

A nonzero $n$-variable Boolean function $g$ is called an annihilator of an $n$-variable Boolean function $f$ if $f * g = 0$. We denote the set of all annihilators of $f$ by $AN(f)$. That is,

$$AN(f) = \{g \in B_n | g * f = 0\}$$

**Definition 2.** *For $f \in B_n$, the algebraic immunity(AI) of $f$ is the minimum degree of non-zero functions $g \in B_n$ such that $g * f = 0$ or $g * (f + 1) = 0$. Namely,*

$$AI(f) = \min\{\deg(g)|0 \neq g \in AN(f) \cup AN(1 + f)\}$$

**Lemma 1.** *[1] Let $f$ be an $n$-variable boolean functions, then $AI(f) \leq \lceil n/2 \rceil$.*

If a function has maximum algebraic immunity $AI(f) = \lceil n/2 \rceil$ with $n$ odd, then it is balanced, and the algebraic immunity property takes care of three fundamental properties of a Boolean function, balancedness, algebraic degree and nonlinearity, but it does this incompletely in the case of balancedness when $n$ is even.

## 3  Construction of Boolean Functions with Maximum AI

In this section, We will present the main idea of our construction. We will use a specific order on elements of $\mathbb{F}_2^n$. This order is induced by the integer order on the set of integers $[0, 2^n - 1]$ by viewing an $n$-tuple $X$ as a binary representation of an integer. More precisely an element $X = (x_1, \ldots, x_n)$ are associated to the integer $\sum_{i=1}^{n} x_i 2^{i-1}$. This identification allows us to compare elements in $\mathbb{F}_2^n$ and to speak about intervals. For instance, for $Y_1 = (y_1^{(1)}, \ldots, y_n^{(1)}), Y_2 = (y_1^{(2)}, \ldots, y_n^{(2)}) \in \mathbb{F}_2^n$, $Y_1 < Y_2$ means that $\sum_{i=1}^{n} y_i^{(1)} 2^{i-1} < \sum_{i=1}^{n} y_i^{(2)} 2^{i-1}$, and we define

$$[Y_1, Y_2) = \{Y \in \mathbb{F}_2^n | Y_1 \leq Y < Y_2\}.$$

We index from $Y_0$ to $Y_k$ the elements $Y$ in $\mathbb{F}_2^n$ of weight $\leq \lceil n/2 \rceil - 1$ arranged in increasing order, so $k = \sum_{i=0}^{\lceil n/2 \rceil - 1} \binom{n}{i} - 1$.

The following results advance the ability to decide whether a boolean function has maximum algebraic immunity.

**Lemma 2.** *[13] Let $n$ be odd, and $f \in B_n$ be balanced. Then $AI(f) = \lceil n/2 \rceil$ if and only if $f$ does not have a nonzero annihilator of degree$\leq \lceil n/2 \rceil - 1$.*

**Lemma 3.** *[14] Let $n$ be even, $f \in B_n$, and its weight equals to $\sum_{i=0}^{\lceil n/2 \rceil - 1} \binom{n}{i}$. Then $AI(f) = \lceil n/2 \rceil$ if and only if $f$ does not have a nonzero annihilator of degree$\leq \lceil n/2 \rceil - 1$.*

**Lemma 4.** *Given a monomial $x_1^{y_1} \ldots x_n^{y_n}$ of degree $d$, the associated monomial function is 1 on $X = (x_1, \ldots, x_n) \in \mathbb{F}_2^n$ if and only if $Y = (y_1, \ldots, y_n) \subset X$ which means $supp(Y) \subset supp(X)$. Moreover, this function is equal to zero on the interval $[0, Y)$, and is equal to 1 on the interval $[Y, Y')$ where $Y'$ is the first point in $\mathbb{F}_2^n$ greater than $Y$ of weight$\leq d$.*

*Proof.* The monomial function $x_1^{y_1} \ldots x_n^{y_n}$ evaluates to 1 on an $n$-tuple $X = (x_1, \ldots, x_n)$ if and only if for all $1 \leq i \leq n$ such that $y_i = 1$ we also have $x_i = 1$, which means $supp(Y) \subset supp(X)$. We also have,

$$\sum_{i=1}^{n} y_i 2^{i-1} \leq \sum_{i=1}^{n} x_i 2^{i-1}$$

so this function is equal to zero on the interval $[0, Y)$.

Now, let us define $j$ to be the minimum of the set $\{i | y_i \neq 0\}$. Denote by $Y'$ the first point greater than $Y$ of weight less than or equal to $d$. The last assertion follows from the fact that all $X$ strictly between $Y$ and $Y'$ coincide with $Y$ for all positions greater than or equal to $j$. This implies that the monomial evaluates to 1 on the interval $[Y, Y')$.                                          □

Our new idea is mainly based on Lemma 4. Now, the new algorithm to construct Boolean functions with maximum algebraic immunity is given below.

**Algorithm 1**

Step 1: For $i = 0$ to $k - 1$, choose element $X_i$ in $[Y_i, Y_{i+1})$;
Step 2: if $i = k$, choose $X_i$ such that $Y_i \subset X_i$ ;
Step 3: Construct Boolean function $f \in B_n$ such that $supp(f) = \cup_{i=0}^{k}\{X_i\}$;
Step 4: Output $f$, then $AI(f) = \lceil n/2 \rceil$.

**Theorem 1.** *The Boolean functions constructed in Algorithm 1 have maximum algebraic immunity.*

*Proof.* Suppose the constructed Boolean function is $f(X)$, and the support of $f(X)$ is $\cup_{i=0}^{k}\{X_i\}$. If it has an annihilator $g(X)$ of degree $\leq \lceil n/2 \rceil - 1$, then we have $g(X_i) = 0$ for all $X_i(0 \leq i \leq k)$, and the ANF of $g$ is of form as follows

$$g(X) = \sum_{i=0}^{k} a_i X^{Y_i}$$

where $a_i$ is the coefficient of the monomial $X^{Y_i}$ and is from $\mathbb{F}_2$. Then we have a homogeneous linear equation for each input $X_i \in supp(f)$,

$$g(X_i) = \sum_{i=0}^{k} a_i X_i^{Y_i} = 0$$

We prove all $a_i = 0(0 \leq i \leq k)$ by induction on i.

When $i = 0$, $Y_0 = (0, 0 \cdots, 0)$ and $Y_1 = (1, 0 \cdots, 0)$, then $X_0 = (0, 0 \cdots, 0)$, By lemma 4, we get $g(X_0) = a_0$, which gives $a_0 = 0$.

Now we prove the inductive step. Assume that, for all $i < l$, the induction assumption holds. We will show it for $i = l$. Since $a_i = 0$ for all $0 \leq i < l$, then

$$g(X) = \sum_{i=l}^{k} a_i X^{Y_i}$$

By Lemma 4, we have $X_l^{Y_i} = 1$ and $X_l^{Y_i} = 0$ for all $l < i$, then we get $g(X_l) = a_l$, which implies $a_l = 0$. So $a_i = 0 (0 \le i \le k)$, which is to say that $g(X) \equiv 0$. So $f$ does not have a nonzero annihilator of degree$\le \lceil n/2 \rceil - 1$. For $n$ is odd (or even), we can obtain that $f$ has maximum algebraic immunity because of lemma 2(or 3). $\qquad\square$

## 4  Constructing Balanced Boolean Functions with Maximum AI on Even Numbers of Variables

It is obvious that when $n$ is even the weights of constructed Boolean functions in Algorithm 1 are $\sum_{i=0}^{n/2-1} \binom{n}{i}$, so the functions are not balanced. In this section we will give another algorithm for even $n$ so that the constructed Boolean functions are also balanced. Now let's begin with some existing results.

**Lemma 5.** *[8]Let $n$ be an even number, and define $F_n(X)$ as follows,*

$$F_n(X) = \begin{cases} 1, wt(X) \le \frac{n}{2}, \\ 0, wt(X) > \frac{n}{2}. \end{cases}$$

*then $AI(F_n) = \frac{n}{2}$.*

Given an even number $n$ and $x \in \mathbb{F}_2^n$, let $v(x)$ be a vector,

$$v(x) = (1, x_1, \cdots, x_n, x_1 x_2, \cdots, x_{n-1} x_n, \cdots, x_1 \cdots x_{n/2-1}, \cdots, x_{n/2+1} \cdots x_n)$$

that is, $v(x)$ is the vector generated by monomials of degree less than $n/2$ valued at $x$.

**Definition 3.** *Given $f \in B_n$, let $M(f)$ be the $wt(f) \times \sum_{i=0}^{n/2-1} \binom{n}{i}$ matrix defined as*

$$M(f) = \begin{pmatrix} v(X_1) \\ v(X_2) \\ \vdots \\ v(X_{wt(f)}) \end{pmatrix}$$

*where $\{X_1, \cdots, X_{wt(f)}\}$ are the support of $f$.*

**Lemma 6.** *[15] Let $f$ be an $n$-variable boolean functions, then $f$ does not have a nonzero annihilator of degree$\le \lceil n/2 \rceil - 1$ if and only if $M(f)$ has column full rank.*

The Algorithm is as below.

**Algorithm 2** (only for $n$ is even)

Step 1: For $i = 0$ to $k - 1$, choose element $X_i$ in $[Y_i, Y_{i+1})$ and $wt(X_i) \le n/2$;

Step 2: if $i = k$, choose $X_i$ such that $Y_i \subset X_i$ and $wt(X_i) \le n/2$;

Step 3: for $i = k + 1$ to $2^{n-1} - 1$, choose any $X_i \notin \cup_{j=0}^{i-1} \{X_j\}$ and $wt(X_i) \le n/2$;

Step 4: Construct Boolean function $f \in B_n$ such that $supp(f) = \cup_{i=0}^{2^{n-1}-1} \{X_i\}$;

Step 5: Output $f$, then $AI(f) = n/2$ and $wt(f) = 2^{n-1}$.

**Theorem 2.** *The Boolean functions constructed in Algorithm 2 is balanced and have maximum algebraic immunity.*

*Proof.* Let $f$ be the functions constructed in Algorithm 2, and it is clear that $f$ is balanced. Then $wt(f) = 2^{n-1} > k + 1 = \sum_{i=0}^{n/2-1} \binom{n}{i}$. From Lemma 6 and Theorem 1 we have

$$M(f^*) = \begin{pmatrix} v(X_0) \\ v(X_1) \\ \vdots \\ v(X_k) \end{pmatrix}$$

have column full rank. Then column rank of

$$M(f) = \begin{pmatrix} v(X_0) \\ v(X_1) \\ \vdots \\ v(X_k) \\ \vdots \\ v(X_{wt(f)-1}) \end{pmatrix}$$

is no less than $M(f^*)$, which indicates $M(f)$ has column full rank, that is, $f$ does not have a nonzero annihilator of degree less than $n/2$.

In Lemma 5, the function $F_n$ has maximum AI, then the corresponding Matrix $M(F_n + 1)$ of the function $F_n + 1$ have column full rank, and it is obvious that $M(f + 1)$ can be obtained by inserting some rows into $M(F_n + 1)$, then $M(f + 1)$ have column full rank, which indicates that $f + 1$ does not have a nonzero annihilator of degree$\leq \lceil n/2 \rceil - 1 = n/2 - 1$.                      □

## 5   The Enumeration and Property of the Constructed Maximum AI Functions

In this section, we discuss how many maximum AI functions can be constructed by using our algorithm. It is obvious that the two algorithm can construct Boolean functions. We can especially suppose $X_i = a_i (0 \leq i \leq k)$. As to Algorithm 2, for $i = k + 1$ to $2^{n-1} - 1$, we can choose any $X_i \in \mathbb{F}_2^n$ whose weight is $n/2$. We first discuss the enumeration of Algorithm 1.

**Theorem 3.** *Let $c = \lceil n/2 \rceil - 1$, then the number of $n$-variable Boolean functions with maximum AI constructed in Algorithm 1 is*

$$2^{n-c} \prod_{d=3}^{n} \prod_{t=\max\{1,c+3-d\}}^{\min\{c,n-d+1\}} 2^{(t+d-2-c)\binom{n-d}{t-1}}.$$

*Proof.* It is obvious that the accurate enumeration of Algorithm 1 is

$$2^{n-c} \prod_{i=1}^{k} |Y_i - Y_{i-1}|,$$

where $|Y_i - Y_{i-1}|$ is the choices of $X_{i-1}$ and $2^{n-c}$ denotes the choices of $X_k$. So if we want $X_{i-1}$ to have more than one choice, then $Y_i \geq Y_{i-1} + 2$, which means that $wt(Y_i - 1) > c$.

For $Y_i = (y_1, \cdots, y_n)$, let $d$ be its first position such that $y_d \neq 0$ and $t$ be the weight of $Y_i$. Then the weight of $Y_i - 1$ is $t + d - 2$. So if $t + d - 2 > c$, the choices of $X_{i-1}$ are $2^{t+d-2-c}$. The number of such $Y_i$ are $\binom{n-d}{t-1}$. Because of $t + d - 2 > c$, $d > c + 2 - t \geq 3$, and $\max\{1, c + 3 - d\} \leq t \leq \min\{c, n - d + 1\}$. So the enumeration of Algorithm 1 is

$$2^{n-c} \prod_{d=3}^{n} \prod_{t=\max\{1,c+3-d\}}^{\min\{c,n-d+1\}} 2^{(t+d-2-c)} \binom{n-d}{t-1}.$$

Different from Algorithm 1, the accurate number of the constructed functions in Algorithm 2 is hard to calculate. Note that the constructed Boolean functions have the property that the weight of any element in their support does not exceed $n/2$, so the number of this case is much fewer than that of Algorithm 1, and the bound of this case can be given as follows.

**Theorem 4.** *Let $n$ be an even integer, $c = n/2 - 1$, and Enum2 denote the number of $n$-variable Boolean functions with maximum AI constructed in Algorithm 2. Let $\Lambda(d,t) = (n-c) \prod_{d=3}^{n} \prod_{t=\max\{1,c+3-d\}}^{\min\{c,n-d+1\}} (t + d - 1 - c) \binom{n-d}{t-1}$. Then*

$$\Lambda(d,t) \leq Enum2 \leq \left( \frac{\binom{n}{n/2}}{\binom{n}{n/2}/2} \right) \Lambda(d,t).$$

*Proof.* Similar to Algorithm 1, let $d$ denote the first position $d(1 \leq d \leq n)$ of $Y_i$ where $y_d \neq 0$ and $t$ be the weight of $Y_i$. If $t + d - 2 > c$, the choices of $X_{i-1}$ are $t + d - 1 - c$. So for all $X_i(0 \leq i \leq k)$, the choices of all such $X_i$ are $\Lambda(d,t) = (n-c) \prod_{d=3}^{n} \prod_{t=\max\{1,c+3-d\}}^{\min\{c,n-d+1\}} (t + d - 1 - c) \binom{n-d}{t-1}$, and this enumeration has no repeated cases. The number of remaining vectors of weight $\leq n/2$ are $\binom{n}{n/2}$, and we need choice $\binom{n}{n/2}/2$ vectors randomly, so the choices of $X_i(k+1 \leq i \leq 2^{n-1} - 1)$ are $\left( \frac{\binom{n}{n/2}}{\binom{n}{n/2}/2} \right)$. But it is obvious that the enumeration of all $X_i(0 \leq i \leq 2^{n-1} - 1)$ has some repeated cases. From the analysis of above, we can gain the conclusion. □

In the rest of this section, we will study the case that the constructed functions have optimum algebraic degree.

**Theorem 5.** *Suppose $f \in B_n$, and its support is $\cup_{i=0}^{k}\{X_i\}$, $x_d^{(i)}$ denotes the $d$-th position of $X_i$, then $\deg(f) = n - 1$ if and only if $k$ is odd and there is a number $d(1 \leq d \leq n)$ such that $t_d = |\{X_i | x_d^{(i)} = 0, 0 \leq i \leq k\}|$ is odd.*

*Proof.* If $k$ is odd, which means $wt(f)$ is even, it is obvious $\deg(f) \leq n - 1$. If there is a number $d(1 \leq d \leq n)$ such that $t_d$ is odd, then the ANF of $f$ is

$$
\begin{aligned}
f(y_1, y_2, \cdots, y_n) = \; & (y_1 + x_1^{(1)}) \cdot (y_2 + x_2^{(1)}) \cdots (y_d + x_d^{(1)}) \cdots (y_n + x_n^{(1)}) + \cdots \\
& + (y_1 + x_1^{(k)}) \cdot (y_2 + x_2^{(k)}) \cdots (y_d + x_d^{(k)}) \cdots (y_n + x_n^{(k)}).
\end{aligned}
$$

Then the coefficient of the monomial $y_1 y_2 \cdots y_{d-1} y_{d+1} \cdots y_n$ is $\sum_{i=0}^{k} x_d^{(i)} = 1 (mod\ 2)$, so $\deg(f) = n - 1$.

If $\deg(f) = n - 1$, it is obvious $k$ is odd. If there is not a number $d(1 \leq d \leq n)$ such that $t_d$ is odd, then the ANF of $f$ does not contain the term $y_1 y_2 \cdots y_{d-1} y_{d+1} \cdots y_n$ for any $d(1 \leq d \leq n)$, so $\deg(f) < n - 1$, which is contrary. So there is a number $d(1 \leq d \leq n)$ such that $t_d$ is odd.     □

Based on theorem 5, we can modify both Algorithm 1 and Algorithm 2 so that the degree of the constructed Boolean functions is $n - 1$. As to Algorithm 1, this can be achieved by choosing $X_i(0 \leq i \leq k)$ so that there is a number $d\ (1 \leq d \leq n)$ such that $t_d$ is odd. The modify for Algorithm 2 is similar.

*Example 1.* For $n = 5$ by using Algorithm 1, $k = \sum_{i=0}^{2} \binom{5}{i} - 1 = 15$, we can choose $X_i(0 \leq i \leq 15)$ as $\cup_{i=0}^{k}\{X_i\} = \{(00000), (10000), (01000), (11000), (00100), (10100), (11100), (00010), (10010), (11010), (11110), (00001), (10001), (11001), (11101), (00011)\}$, Let $X_i(0 \leq i \leq 15)$ be the support of the Boolean function $f$, then ANF of $f$ is as below:

$f(x) = 1 + x_2 x_4 + x_2 x_5 + x_3 x_4 + x_2 x_3 + x_3 x_4 + x_1 x_2 x_3 + x_1 x_2 x_4 + x_2 x_4 x_5 + x_1 x_2 x_5 + x_1 x_4 x_5 + x_3 x_4 x_5 + x_1 x_2 x_3 x_4 + x_1 x_2 x_4 x_5 + x_1 x_2 x_3 x_5 + x_1 x_3 x_4 x_5$,
the AI of $f$ is 3, and it is obvious that the degree of $f$ is 4.

## 6   Conclusion

Possessing a high algebraic immunity is a necessary criteria for Boolean functions used in stream ciphers against algebraic attacks. In this paper, we present a new method to construct Boolean functions, in any number of variables, with maximum algebraic immunity. We also improve our method to construct balanced Boolean functions with maximum algebraic immunity for any even number of variables. However, it is still an open problem to generalize our construction to obtain more balanced Boolean functions with maximum algebraic immunity on even variables. Furthermore, there are still some problems need to be studied such as whether the constructed functions can achieve high nonlinearities and be robust against fast algebraic attacks.

## Acknowledgments

The work in this paper is supported by the National Natural Science Foundation of China (No:60803156) and the open fund of State Key Laboratory of Mobile Communication of Southeast University (W200807).

# References

1. Courtois, N., Meier, W.: Algebraic attacks on stream ciphers with linear feedback. In: Biham, E. (ed.) EUROCRYPT 2003. LNCS, vol. 2656, pp. 345–359. Springer, Heidelberg (2003)
2. Mihaljevic, W., Imai, H.: Cryptanalysis of toyocrypt-hs1 stream cipher. IEICE Transactions on Fundamentals E85-A, 66–73 (2002)
3. Courtois, N.: Cryptanalysis of sfinks. In: Won, D.H., Kim, S. (eds.) ICISC 2005. LNCS, vol. 3935, pp. 261–269. Springer, Heidelberg (2006)
4. Meier, W., Pasalic, E., Carlet, C.: Algebraic attacks and decomposition of boolean functions. In: Cachin, C., Camenisch, J.L. (eds.) EUROCRYPT 2004. LNCS, vol. 3027, pp. 474–491. Springer, Heidelberg (2004)
5. Carlet, C.: A method of construction of balanced functions with optimum algebraic immunity, http://eprint.iacr.org/2006/149
6. Carlet, C., Dalai, D.K., Gupta, K.C., Maitra, S.: Algebraic immunity for cryptographically significant boolean functions: Analysis and construction. IEEE Transactions on Information Theory 52(7) (2006)
7. Carlet, C., Feng, K.: An infinite class of balanced functions with optimal algebraic immunity, good immunity to fast algebraic attacks and good nonlinearity. In: Pieprzyk, J. (ed.) ASIACRYPT 2008. LNCS, vol. 5350, pp. 425–440. Springer, Heidelberg (2008)
8. Dalai, D.K., Maitra, S., Sarkar, S.: Basic theory in construction of boolean functions with maximum possible annihilator immunity. Des. Codes, Cryptography 40(1) (2006)
9. Li, N., Qi, W.-F.: Construction and analysis of boolean functions of 2t+1 variables with maximum algebraic immunity. In: Lai, X., Chen, K. (eds.) ASIACRYPT 2006. LNCS, vol. 4284, pp. 84–98. Springer, Heidelberg (2006)
10. Qu, L.J., Li, C., Feng, K.Q.: A note on symmetric boolean functions with maximum algebraic immunity in odd number of variables. IEEE Transactions on Information Theory 53(8) (2007)
11. Qu, L.J., Li, C.: On the $2^m$-variable symmetric boolean functions with maximum algebraic immunity. Science in China Series F-Information Sciences 51(2) (2008)
12. Qu, L., Li, C.: Weight support technique and the symmetric boolean functions with maximum algebraic immunity on even number of variables. In: Pei, D., Yung, M., Lin, D., Wu, C. (eds.) Inscrypt 2007. LNCS, vol. 4990, pp. 271–282. Springer, Heidelberg (2008)
13. Canteaut, A.: Open problems related to algebraic attacks on stream ciphers. In: Ytrehus, Ø. (ed.) WCC 2005. LNCS, vol. 3969, pp. 120–134. Springer, Heidelberg (2006); 1–10 invited talk
14. Liu, M.C., Pei, D.Y.: Construction of boolean functions with maximum algebraic immunity. In: Chinacrypt 2008, pp. 79–92 (2008)
15. Qu, L.J., Feng, G.Z., Li, C.: On the boolean functions with maximum possible algebraic immunity: Construction and a lower bound of the count, http://eprint.iacr.org/2005/449

# A²M: Access-Assured Mobile Desktop Computing

Angelos Stavrou[1], Ricardo A. Barrato[2], Angelos D. Keromytis[2], and Jason Nieh[2]

[1] Computer Science Department, George Mason University
[2] Computer Science Department, Columbia University

**Abstract.** Continued improvements in network bandwidth, cost, and ubiquitous access are enabling service providers to host desktop computing environments to address the complexity, cost, and mobility limitations of today's personal computing infrastructure. However, distributed denial of service attacks can deny use of such services to users. We present A²M, a secure and attack-resilient desktop computing hosting infrastructure. A²M combines a stateless and secure communication protocol, a single-hop Indirection-based network (IBN) and a remote display architecture to provide mobile users with continuous access to their desktop computing sessions. Our architecture protects both the hosting infrastructure and the client's connections against a wide range of service disruption attacks. Unlike any other DoS protection system, A²M takes advantage of its low-latency remote display mechanisms and asymmetric traffic characteristics by using multipath routing to send a small number of replicas of each packet transmitted from client to server. This packet replication through different paths, diversifies the client-server communication, boosting system resiliency and reducing end-to-end latency. Our analysis and experimental results on PlanetLab demonstrate that A²M significantly increases the hosting infrastructure's attack resilience even for wireless scenarios. Using conservative ISP bandwidth data, we show that we can protect against attacks involving thousands (150, 000) attackers, while providing good performance for multimedia and web applications and basic GUI interactions even when up to 30% and 50%, respectively, of indirection nodes become unresponsive.

## 1 Introduction

In today's world of commodity computers and increasing broadband network connectivity, the existing computing infrastructure imposes severe limitations on increasingly mobile users. Such users lack a common computing environment as they move between home, office, and while on the road. Mobile users have been forced to adapt by carrying around bulky laptop computers and other stateful devices with battery draining power needs. This approach is increasingly unsustainable as the management and security costs of owning and maintaining these devices grow, especially for large organizations with many users. Maintenance is particularly difficult with devices that may be roaming anywhere, on any network. Furthermore, these portable devices are inherently physically insecure and it is not uncommon for these machines to be damaged or stolen, resulting in the loss of any important data stored on them. This is a critical problem especially in health care computing, where HIPAA compliance is a requirement in supporting the clinical information access of highly mobile medical professionals. Even

P. Samarati et al. (Eds.): ISC 2009, LNCS 5735, pp. 186–201, 2009.

when such data can be recovered from backup, the time-consuming process of reconstituting the state of the lost machine on another device results in a huge disruption in critical computing service for the user.

Outsourced IT systems often utilize a thin-client computing model to decouple a user's applications and desktop computing session from any particular end-user device by moving all application logic to hosting providers. Graphical displays are virtualized and served across a network to a client device using a remote-display protocol, with application logic executed on the server. Clients transmit user input to the server, and the server returns screen updates. Examples of popular thin-client platforms include Citrix MetaFrame [1], Microsoft Terminal Services [2], AT&T Virtual Network Computing (VNC) [3]. Because all application processing is done on the server, the client only needs to be able to display and manipulate the user interface, enabling clients to be simple and stateless.

A key issue that must be addressed to ensure that users obtain reliable and secure access to hosted computing services is protection of the server infrastructure and the client's connection against denial of service attacks, particularly of the distributed kind (DDoS). DDoS attacks are an increasing occurrence in today's Internet, aiming to deny use of a service to legitimate users [4]. The same ubiquitous network connectivity that improves access to a service provider for legitimate mobile users, also increases an attacker's ability to launch a DDoS against a service provider, sometimes as part of an extortion scheme [5]. One type of DoS attack that is difficult to identify and isolate involves sending enough attack traffic which will cause the links close to the servers to be congested and eventually drop all useful traffic. The potential of such attacks to disrupt user access to applications and data is an important challenge that must be addressed before ASPs can achieve mass acceptance. Unfortunately, existing DDoS protection mechanisms either require large-scale deployment, or offer unacceptably high latency and latency variance [6, 7], especially when under attack. To be of any practical use, interactive and real-time applications such as GUI operations and multimedia streaming demand a low-latency pipe at all times.

In this context, we introduce *Access-Assured Mobile* (A$^2$M) desktop computing, a hosted computing infrastructure that combines a remote-display architecture with a stateless indirection-based network (IBN) composed of dedicated nodes. A$^2$M provides both protected and efficient access to hosted desktop computing environments, even in the presence of denial of service attacks. Nodes participating in the IBN communicate only to exchange control messages, but not to route the client's data, unlike previous overlay-based approaches [6, 7]. A$^2$M clients exploit the path diversity naturally exhibited by a distributed IBN to "spread" their traffic such that directed attacks do not cause service disruptions. To further alleviate any potential delays introduced by the IBN and reduce the latency in the end-to-end communication, A$^2$M uses a number of other optimizations at the remote display level to minimize the impact these delays may have on the user's experience. A$^2$M combines a simple low-level display protocol and a server-push model to minimize client-server synchronization and network round-trips. Atop this basic model, A$^2$M implements higher-lever mechanisms, such as client-managed cursor display, shortest-job-first display command scheduling, and a non-blocking drawing pipeline, further increasing the overall interactive response of the system. The contributions of our work are:

- We implement and evaluate $A^2M$ in the real Internet using PlanetLab. Our experiments show that $A^2M$ introduces very little latency in most scenarios.
- We are the first to conduct realistic (non-simulation) experiments to evaluate the resilience of our system against DDoS attacks using wireless nodes, and its performance under attack. Our results validate the design of $A^2M$, showing good performance for multimedia and interactive applications even with 30%–50% of the IBN nodes under attack.

## 2   $A^2M$ Architecture

As shown in Figure 1, $A^2M$'s architecture is divided in two major components: the hosting infrastructure and the access infrastructure. The hosting infrastructure provides an environment for desktop sessions where a user's session is decoupled from any particular end-user access device, by moving all application state to hosting servers. Applications run within these servers, and their display output is redirected over the network to the access device. Redirection is performed by a per-session virtual display driver that translates from application display-draw commands to $A^2M$'s display protocol commands. The protocol commands are then forwarded to the client device for display. $A^2M$ extends previous work on desktop hosting infrastructures such as MobiDesk [8] by providing mechanisms that provide continuous access to hosted desktop sessions, even in the presence of distributed denial of service attacks on the hosting servers.

$A^2M$'s access infrastructure provides the connection between users on the network and the applications running on the hosting servers. Users make use of a simple client application that merely forwards input events to the applications running on the server,

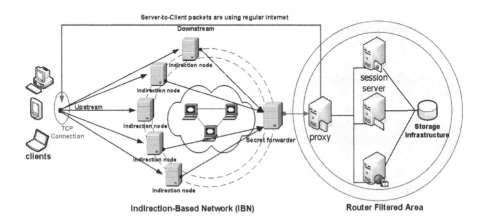

**Fig. 1.** $A^2M$ Architecture. The two directions of the client-server connection take different paths: the client-to-server direction goes over the indirection-based network, while the server-to-client direction goes directly to the client (not through the infrastructure).

and processes display updates generated in response to these events. This application model results in a highly asymmetric network traffic pattern. On one side, input events (headed uplink, or upstream toward the server) are very small pieces of information that are generated at a relatively slow, human-dependent rate. On the other hand, display updates (headed downlink, or downstream toward the client) are orders of magnitude larger and are generated as fast bursts of activity. For example, during web browsing, a single user input event (a mouse click on a link) results in a full-screen update having to be displayed (the destination web page).

The traffic asymmetry is made more pronounced when we consider the different roles and importance of input events and display updates. In an interactive system user experience is dictated by the response time, which in turn is determined by how quickly input events are processed and display updates are made visible to the user. If response time is too high, the user will become exasperated and frustrated with the system. Since a single input event triggers the generation of display updates, guaranteed delivery of each event becomes crucial for the performance of the system. On the other hand, humans are known to be more tolerant to partial updates than to longer response times, because partial updates provide feedback to their actions. Delivery of updates should then be made such that updates can begin to be displayed as soon as possible, even if the complete update takes longer to appear.

The resource centralization around the hosting infrastructure results in a threat model where denial of service attacks on the system will only affect the uplink direction, *i.e.,* the traffic **to** the hosting servers, by saturating the network links and queuing buffers close to the servers or by directly attacking the hosting infrastructure servers. Therefore, it is crucial for A$^2$M to protect this communications channel from interference, blocking unwanted traffic close to the attacker before it can reach the service providing machines. On the other hand, the downlink direction will for the most part be relatively free of noise, and without any need to be protected. Note that denial of service attacks typically affect the uplink direction, *i.e.,* the traffic **to** the server, by saturating the network links and queuing buffer close to the server. The downlink direction is relatively free of noise. Thus, we are primarily interested in protecting the client-to-server traffic from interference; the opposite direction does need typically any such protection.

Taking advantage of both the traffic asymmetry and the threat model, A$^2$M partitions bi-directional connections between the client and the server into an indirected client-to-server multi-path and a direct server-to-client path. The IBN takes care of routing input events and other client-to-server traffic and protects the hosting infrastructure. Protection is performed by acting as a distributed firewall that conceptually distinguishes between authorized client-generated traffic, and unauthorized and possibly malicious traffic. Traffic permitted to traverse through the IBN is directed to a filtering router close to the hosting servers, whereas all other traffic is dropped or rate-limited providing a distributed "shield" against both network congestion and host directed attacks. In Section 2.2, we provide an estimation of the resistance of the system, using this filtering mechanism, to denial of service attacks, in terms of the average number of machines that must participate in the attack.

The direct server-client path in turn ensures that large and bursty display updates are delivered to the client as quickly as possible, even if parts of them are lost or delayed and need to be retransmitted. $A^2M$'s approach represents a sharp departure from traditional interactive client-server architectures, where a vulnerable bi-directional direct connection provides the only means of communication between the client and the server. We should note that $A^2M$ does not preclude routing both traffic directions over the IBN, albeit at a possible increase in the end-to-end latency when no replication of packets is present. Since this mode is not necessary for our usage scenario, we do not further consider it in this paper.

## 2.1 System Operation

To provide seamless and ubiquitous connectivity, $A^2M$ encapsulates all functionality within a self-contained client application that manages communication with the indirection infrastructure, forwards user events to hosted applications, and displays application output on the local device. To access a desktop session, users must first obtain access to the IBN, which in turn allows them to authenticate with the hosting infrastructure, and then gain access to their session. Users need to be recognized as legitimate in order for the IBN to distinguish their traffic from other unauthorized, possibly malicious traffic. In contrast to traditional service providing infrastructures such as web-content distributors, $A^2M$ requires users to be authenticated and does not allow anonymous users, because only authorized users should be able to connect to the hosting infrastructure. $A^2M$ ties the authentication requirements of the IBN and the hosting infrastructure into a single, seamless process.

**Client Authentication:** Before a client is allowed to send traffic through the IBN, it must obtain a *ticket,* which is then included in all subsequent packets sent to the IBN, until it expires. The ticket is used by the IBN nodes to authenticate the user, validate the routing decisions, and prevent malicious (or subverted) clients from utilizing a disproportionate amount of bandwidth. To obtain a ticket, the client contacts an indirection node at random using a ticket establishment protocol described in detail in previous work [9]. This protocol is fully distributed and resilient to CPU exhaustion attacks. Furthermore, the ticket issuing process is protected against replay and IP spoofing attacks. At the end of the protocol, the client and the IBN have authenticated each other, and the client is in possession of a ticket. The ticket contains a session key $K_u$, a range of sequence numbers for which it is valid (more on this later), and the IP address of the client, all encrypted under $K_M$, a secret key negotiated periodically (*e.g.,* every few hours) among all indirection nodes. Note that only the indirection nodes can decrypt the ticket; clients treat the ticket as an opaque value that they must provide to the AAN with each packet they need to forward. A second copy of $K_u$ is independently encrypted under the client's public key. This ticket can only be used by the client to continue the authentication protocol (*i.e.,* prove liveness for both the IBN nodes and the client. Once the full two-party authentication is completed, the last indirection node provides the client with a ticket that is not "restricted," *i.e.,* the corresponding flag inside the ticket is cleared. As we discussed in the previous section, the tickets are periodically refreshed, to avoid situations where a malicious user distributes a valid session key and ticket to a large number of zombies that then simultaneously send attack traffic through the IBN.

The connections to the hosting infrastructure are asymmetric: the client-to-server traffic will travel through the IBN, while the server-to-client traffic will use regular Internet routing. In the case where a session does not already exist, a new session is created and populated, before the client is allowed to connect to it. The authentication and connection setup process is done transparently by the client application, and it does not require special support from the underlying devices. This simplicity allows A$^2$M users to access their sessions from almost any number of Internet-enabled devices.

Once the connection to the hosting server is established, the client will be recognized as a legitimate user, and user input events will be allowed to traverse the indirection nodes and be routed to the hosted applications. This process continues until the user disconnects from the session, at which point the client's ticket is revoked and the connections are closed. Since a disconnected client is no longer allowed to use the system, previously legitimate devices cannot be reused as attack tools on the infrastructure.

## 2.2 Assured Access Indirection Network

We have implemented the *Assured Access Network* (AAN), which significantly extends the ideas of SOS [6] and Mayday [7]. Our approach, shown in Figure 1, is to spread the packets from the client across all indirection nodes in a pseudo-random manner. This new communication mechanism protects the client-server connection establishment and provides uninterrupted connectivity to the target server throughout the client's session. The admitted packets are internally forwarded to a secret forwarder (selected at random, and changing over time), which is allowed to forward traffic to the utility server. Only authorized clients are allowed to use the IBN and contact the hosting servers and these clients are provisioned in advance (*e.g.*, at registration time) with the appropriate authentication material, such as an RSA public/private key pair and a public-key certificate [10, 11]. AAN works in conjunction with filtering routers close to the hosting infrastructure, to allow only traffic from the IBN's secret forwarders to reach A$^2$M's hosting servers. All other traffic is considered unauthorized and possibly malicious, and therefore filtered out.

Contrary to previous overlay architectures, our system achieves this filtering without the use of overlay routing to transfer the client's request to the server. In our system, legitimate packets are reflected to the secret servlet(s) generating a one-hop indirection network. As shown in Figure 1 there is no single path between the client and the server - instead packets are spread from the client to the indirection nodes creating a single-hop multi-path effect. Both the use of the single-hop indirection and the multi-path routing permit our system to scale well in terms of latency, as we shall see in Section 3. For more details on the overlay architecture itself, see [9].

## 2.3 AAN Encapsulation

When using AAN, every packet sent by a client to an indirection node contains four fields: a client identifier, the ticket, an authenticator, and a monotonically increasing sequence number. Recall that the ticket contains the session key and the maximum sequence number for which the ticket is valid, and is encrypted and authenticated under a secret key known only to the indirection nodes. Note that these indirection nodes are *not* user machines, but are hosts dedicated to offering a DoS protection service.

The sequence number is a 32-bit value that is incremented by the client for each packet transmitted through the IBN with a given session key. The client identifier is a random 32-bit value that is selected by the indirection node that authenticated the client, and is used as an index in the table of last-seen sequence number, maintained by each indirection node for each active client. The authenticator is a fast hash function, such as UMAC [12], computed over the session key and the whole packet (including the ticket, sequence number, and client identifier). Thus, the only amount of state each indirection node needs to maintain per active client are the client's identifier and the last sequence number seen from that particular client. Assuming that both the client identifier and the sequence number are 32-bit values, each indirection node needs to maintain only 64 bits of state for each client; thus, if the system has 1 million active clients, we will only need 8 MB of state — easily manageable even if it is stored in main memory, given current prices of RAM.

### 2.4   AAN Operation

A client transmitting a packet through the IBN uses the session key and the sequence number as inputs to a pseudo-random function (PRF). The output is treated as an index to a publicly available list of indirection nodes, through which the packet will be routed. The list of available indirection nodes does not need to change frequently, even if nodes become unavailable (*e.g.,* for maintenance purposes), and can be downloaded by clients independently of the protected communication. For the purposes of this paper, we assume that clients trust the IBN's entry points. Discussion and analysis of an environment where access points cannot be safely trusted can be found in [13].

The client encapsulates the original packet (addressed to the final destination) inside a packet for the indirection node, along with the information identified above (client identifier, ticket, sequence number, authenticator). The packet is then forwarded through the IBN to the secret forwarder for that particular destination, and from there to the final destination.

An indirection node that receives such a packet first verifies that the sequence number on the packet is larger than the last sequence number seen from that client, by using the client identifier to index the internal table. It then decrypts the ticket, obtaining the session key for that client, with which it verifies the authenticator. The indirection node also verifies that the sequence number is within the acceptable range of sequence numbers for this ticket. Finally, it uses the key and the sequence number along with the PRF to determine whether the client correctly routed the traffic. If all steps are successful, the indirection node updates the sequence number table and forwards the packet to the secret forwarder. Packets with lower or equal sequence numbers are considered duplicates (either accidental artifacts of the underlying network, or malicious replays by attackers) and are quietly dropped.

### 2.5   Attack Resistance

Here we attempt to give a simple analysis on the expected resiliency of our system. Additional work is needed to further refine the model and validate our assumptions.

However, this analysis should serve as a good first-order approximation on the effectiveness of the approach.

Since the Internet (and ISPs') backbones are well provisioned, the limiting factors are going to be the links close to the target of the attack. The aggregate bandwidth for most major ISP POPs is on the order of 10 to 20 Gbps, according to an informal poll of several providers. If the aggregate bandwidth of the attack plus the legitimate traffic is less than or equal to the POP capacity, legitimate traffic will not be affected, and the POP routers can drop the attack traffic (by virtue of dropping any traffic that did not arrive through the IBN). Unfortunately, there do not exist good data on DDoS attack volumes; network telescopes [14] tend to underestimate their volume, since they only detect response packets to spoofed attack packets. However, we can attempt a simple, back of the envelope calculation of the effective attack bandwidth available to an attacker that controls $X$ hosts that are (on average) connected to an aDSL or cable network, each with 256Kbps uplink capacity. Assuming an effective yield (after packet drops, self-interference, and lower capacity than the nominal link speed) of 50%, the attacker controls $128 \times X$ Kbps of attack traffic. If the POP has an OC-192 (10 Gbps) connection to the rest of the ISP, an attacker needs 78, 000 hosts to saturate the POP's links. If the POP has a capacity of 20 Gbps, the attacker needs 156, 000 hosts. Although we have seen attack clouds of that magnitude (or larger), the ones used in actual attacks are much smaller in practice. Thus, an IBN-protected system should be able to withstand the majority of DDoS attacks. If attacks of that magnitude are a concern, we can expand the scope of the filtering region to neighboring POPs of the same ISP (and their routers); this would increase the link capacity of the filtered region significantly, since each of the neighboring POPs see only a fraction of the attack traffic. Additional work is needed to determine the practical limits of the system. In Section 3 we give some experimental results on the resilience of our system against attacks that target the IBN itself.

## 3   Implementation and Experimental Results

To demonstrate the feasibility of the proposed architecture, we have implemented an A$^2$M prototype which hosts and protects Linux-based desktop sessions. We deployed the indirection nodes of our prototype in 80 PlanetLab nodes, while having the client and server reside in our local network. Our architecture spreads the packets across all indirection nodes. Perhaps the most surprising aspect of our implementation is its size: excluding cryptographic libraries and the JFK protocol, the code implementing the complete functionality of the system consists of 1,600 lines of well commented $C$ code. The JFK implementation itself adds another 2,500 lines of code. Although this is a prototype implementation and does not include management code and other facilities that would be required in a production system, we feel that the system is surprisingly lightweight and easy to comprehend.

The implementation consists of the code for the indirection nodes, as well as code running on each client that does the encapsulation and initial routing. A detailed description of MobiDesk may be found in [8]. On the client, a routing-table entry redirects all IP packets destined for the protected servers to a virtual interface, implemented using the *tun* pseudo-device driver. This device consists of a linked pairs of virtual network interfaces and character devices that a user-level process can read and write. IP packets sent to the *tun0* network interface can be read by a user process reading the device */dev/tun0*. Similarly, if the process writes 1a complete IP packet to */dev/tun0* this will appear in the kernel's IP input queue as if it were coming from the network interface *tun0*. Thus, whenever an application on the client tries to access a protected server, all outgoing traffic is intercepted by the virtual interface. A user-level proxy daemon process reading from the corresponding device captures each outgoing IP packet, encapsulates it in a UDP packet along with authentication information, and sends it to one of the indirection nodes according to the protocol. The code running on indirection nodes receives these UDP packets, authenticates and forwards them to the secret forwarder, which forwards them to the final destination. There, the packets are decapsulated and delivered to the original intended recipient (*e.g.,* web server). The decapsulation can be done by a separate box or by the end-server itself. In addition to the decapsulation code on the indirection nodes, there is also a daemon listening for connection establishment packets from the clients.

In evaluating $A^2M$, we focused on two metrics: the quality of service in terms of latency, as this is perceived by the end user, and the system's resilience when under attack *i.e.,* node failures. PlanetLab provides a realistic network environment for our experiments that stresses the performance of our system because the packets follow different, highly variant paths to reach the protected server. In our experiments, we protected the uplink traffic from the client to the server routing it through the IBN, while the return path followed normal Internet routing (outside the IBN).

Our testbed consisted of a client PC simulating a typical remote-display access device, a server where the benchmark applications executed, and 80 indirection hosts deployed across various PlanetLab locations in the US and Canada. The client computer had a 450Mhz Intel Pentium-II CPU and 128MB RAM running Debian with Linux 2.4.27. Our client PC was chosen to reflect the type of low-power, thin-client devices which we expect to become $A^2M$'s access devices. The laptop PC had a 1.5Ghz Intel Pentium M and 1GB RAM running Debian with Linux 2.6.10. The server was an Intel dual-Xeon 2.80GHz with 1GB of RAM running RedHat 9 with Linux 2.4.20.

We measured the performance of $A^2M$ in web, video, and basic interactive tasks as representative applications of typical desktop usage. Our web measurements used the Mozilla 1.6 browser to run a benchmark based on the Web Page Load test from the Ziff-Davis i-Bench benchmark suite. The benchmark consists of a sequence of 54 web pages containing a mix of text and graphics. The browser window was set to full-screen resolution for all platforms measured. Video playback performance was measured using Mplayer 1.0pre3 to play a 34.75 second video clip of original size $352x240$ pixels displayed at full-screen resolution. For our interactive tests we recorded a number of sessions where simple interactive tasks were performed. Recording the sessions allowed us to reliably play back the exact same tasks under different network conditions. The measure of performance for these tests was the latency experienced by a user performing the specific task. The primary measure of web browsing performance was the average page-download latency in response to a mouse-click on a web page link. To minimize any additional overhead from the retrieval of web pages, we used a conservative setup where the web server was directly connected to the hosting server through a LAN connection. The primary measure of video playback performance was video quality [15], which accounts for both playback delays and frame drops that degrade playback quality. For example, 100% video quality means that all video frames were displayed at real-time speed. On the other hand, 50% video quality means either that half the video frames were dropped when displayed at real-time speed or that the clip took twice as long to play even though all of the video frames were displayed.

We first examined the effects that the basic indirection network and various levels of packet replication had on the overall performance of the system. The levels of replication tested were no replication, 50% (meaning one extra copy of each packet with probability 0.5), 100% replication (one extra copy of each packet) and 200% replication (two extra copies of each packet). We also measured the impact of the IBN size by running our experiments on 8 and 80 nodes participating in the IBN. We ran a baseline test where we used a direct LAN connection between the client and the server. Since the indirection nodes were deployed over a wide area with varying network latency, this test provided us with a very conservative measurement of the indirection overhead. In a realistic A²M deployment, the client and server will typically reside at different, topologically distant locations. In that case, it is entirely possible for the indirection path to provide better connectivity characteristics than a direct connection due to the multi-path effect, which allows the packets originating from the client to follow a route with lower latency towards the end server [16, 17, 18, 19]. Although not shown in our results for ease of viewing, we also compared the performance of A²M to that of MobiDesk and found it to be the same on the direct connection case.

Figure 2 illustrates the end-to-end average web latency results as perceived by the client. We can see that even for the worst-case scenario, an 80-node IBN without packet replication, the overhead from the indirection results in a latency increase of only 2 (*i.e.*, twice the latency of the baseline direct connection). When 50% packet replication is used (*i.e.*, replicating a packet with probability 0.5), the overhead drops significantly to 40% for the 80-node IBN. The drop in the overhead is due to the variant path latency

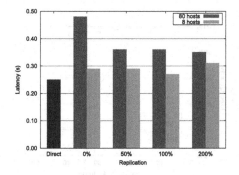

**Fig. 2.** Web latency *vs.* packet replication. The leftmost bar shows the latency when connected directly to the server using a LAN and no protection.

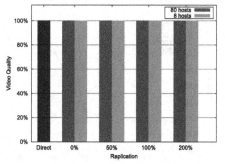

**Fig. 3.** Video quality *vs.* packet replication — video quality remains 100% under all test scenarios even for a 80-node IBN with no packet replication

of nodes participating in the IBN. TCP does not behave optimally when packets appear to have high variance when arriving at the end server out of order. Adding packet replication decreases this variance, as the same packet follows more than one paths with different latency and the end server uses the one that arrives first. Boosting the replication beyond 50% follows the law of diminishing returns, as each additional increase in replication gives us less latency improvements. Care must be taken however, as too much packet replication can cause performance degradation, since bandwidth is "wasted" on duplicate packets. This is better exemplified by the results on the 8-node network using 200% replication. The 80-node network does not exhibit the same adverse affect because its average path latency is higher, allowing the secret gateway enough time to process the encapsulated packets received by the IBN.

To measure our system with an application that could generate more upstream traffic and required the system to maintain its quality of service above a threshold for latency, we used video playback. Figure 3 shows the results for video quality as measured at the client side. We can clearly see that A$^2$M performs optimally under all test scenarios, providing the same perfect video quality as the direct LAN connection scenario, even for the worst-case scenario of the 80-node IBN deployed over a WAN with no packet replication.

The behavior of the overall system under attack was measured using a simulated denial of service attack that targeted the IBN itself. Our threat model assumes that the attacker can render a fraction of the nodes participating in the IBN unresponsive, thus inducing packet loss in the TCP connection of a user connected to the hosting server. All resilience tests were run on the 80-node IBN network. When attacked, a node stops forwarding packets from the client to the end host, acting as a mute node. Since there is no immediate feedback, clients do not know which A$^2$M nodes are operating and which are suppressed by the attacker. Figure 4 illustrates the effects on the average web page latency as we increase the percentage of node failure, and demonstrates both the resilience of A$^2$M and the advantages of packet replication. Without packet replication, latency quickly degrades to twice that of the direct connection when we have 15% of node failures, and reaches three times for 20% node failure. On the other hand,

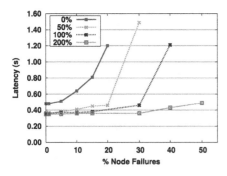
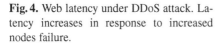
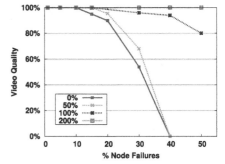

**Fig. 4.** Web latency under DDoS attack. Latency increases in response to increased nodes failure.

**Fig. 5.** Video quality under DDoS attack. Video quality drops only after a substantial percentage of nodes become unresponsive.

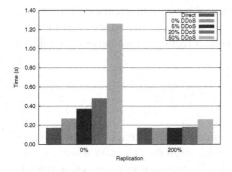

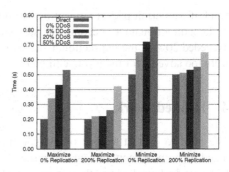

**Fig. 6.** Interactive performance for the echo test. Even without replication and with attacks affecting up to 20% of the IBN nodes, the client's end-to-end latency increases only by a factor of 2.5 when compared to the direct, non-protected case.

**Fig. 7.** Interactive performance for minimize/maximize window test. Without replication and for attacks affecting up to 20% of the IBN nodes, the client's end-to-end latency increases only by a factor of 2.

employing packet replication allows A$^2$M to maintain an almost constant latency that is very close to the direct connection, even under 50% A$^2$M node failure, in the case of 200% replication.

*Interactive Applications.* Although video streaming and web browsing are both representative and demanding applications, we felt that we needed to include another set of experiments that require a high level of synchronization between the upstream and the downstream channel. We performed four different tests, each representing typical interactive operations on a desktop environment. The tests were performed by first recording a session of a user performing the appropriate operation, and then playing back the session in a number of different experimental scenarios. Our measure of performance was the user-perceived latency in response to the interactive operations. The four tests performed were: echo, minimize/maximize window, scroll, and move window. The echo test measured the time it takes for a line of text to appear on the screen after the user has pressed and depressed a key. The minimize/maximize window tests measures the time it takes to maximize a window after the user has pressed the maximize button, and then (after the window has been maximized) to minimize it after the user has pressed the minimize button. The scroll test measures the time it takes to scroll down a full-screen web page in response to a single Page Down key-press, and then the time it takes to scroll back to the top by leaving the Arrow Up key pressed. Finally, the move window test measures the time it takes to move a window across the screen. The window's size is about one fifth of the screen's size, and it is moved by dragging the window while the left-mouse button is pressed. The window operation is opaque, *i.e.,* the contents of the window are continuously redrawn as the user performs the move operation.

The end-to-end latency the end users experience for these operations is shown in Figures 6, 7. These measurements show that without using packet replication, and for attacks up to 20% of the indirection nodes, the client's end-to-end latency increases only by a factor of 2.5 when compared to the direct, non-protected case. On the other hand, if we permit packet replication, we notice an increase in latency only after 50% of the indirection nodes become unresponsive. In some cases, for attack intensities that exceeded 20% of the indirection nodes and without replication the network conditions were too adverse for the test to complete.

## 4   Related Work

The need to protect against or mitigate the effects of DoS attacks has been recognized by both the commercial and research world, given the ease with which such attacks can be launched and their frequency [14]. $A^2M$ provides an attack-resilient utility infrastructure for mobile desktop computing. Due to space limitations, we do not discuss here the extensive work on DDoS prevention or mitigation that requires widespread deployment inside the network or the use of new protocols and end-applications [20,21,22,23,24,25,26,27,28].

$A^2M$ builds on the ideas proposed by the MobiDesk [8] desktop hosting infrastructure, and its remote display architecture, THINC [29]. In contrast to $A^2M$, MobiDesk and THINC do not address the problem of potential attacks on their infrastructure and use a direct connection to communicate between user devices and hosting servers. This makes the system defenseless and vulnerable to simple denial of service attacks that may cause hosted desktop sessions to become unavailable to users. Attacks may either target the hosting infrastructure or the communication channels providing the service, and render the MobiDesk infrastructure useless.

SOS [6]  first suggested the concept of using an overlay network to preferentially route traffic, using multi-hop overlay routing, from legitimate users to a secret node that is allowed to reach the protected server. All other traffic is restricted at the Internet Service Provider's Point-of-Presence (POP), which in most cases has enough capacity to handle all attack and legitimate traffic. The same idea is used in MayDay [7]. WebSOS [30] relaxes the requirement for a priori knowledge of the legitimate users, by adding a Graphic Turing Test to the overlay, allowing the system to differentiate between human users and attack zombies. MOVE [31] eliminates the dependency on network filtering at the ISP POP routers by keeping the current location of the server secret and using process migration to move away from targeted locations. A system similar to MOVE is described in [32]. There, it is observed that in some cases the various security properties offered by SOS can still be maintained using mechanisms that are simpler and more predictable. However, some second-order properties, such as the ability to rapidly reconfigure the architecture in anticipation of or in reaction to a breach of the filtering identity (*e.g.*, identifying the secret forwarder) are compromised.

All of these overlay-based systems impose a high latency overhead, making them unfit for time-critical applications. To route the client traffic, these systems create an overlay route whose length increases with the number of overlay nodes (usually with $O(log(n))$, where $n$ is the size of the overlay). Such an increase in the path length leads to higher (and highly varying) end-to-end latency. Moreover, these systems are vulnerable to attacks that target the connection state that is kept by each of their overlay nodes: by attacking a specific node, the attacker forces the users connected to it to detect this attack and re-establish both their connectivity and authentication credentials to another, potentially healthy, overlay node. An attacker can force the users to reset their connections repeatedly, making the system impractical.

Several remote display and thin-client architectures are widely used today, including the X-Window System [33], Citrix MetaFrame (ICA) [1], Microsoft Terminal Services (RDP) [2], VNC [3], and SunRay [34]. However, none of these systems provide resiliency against DoS attacks as A$^2$M does. None of these systems take advantage of the asymmetry in remote-display traffic to improve performance in environments with high variability of network latency. As shown on previous studies [15, 35], all of these systems suffer major performance degradations in high-latency network environments. X, ICA, and RDP use high-level display protocol primitives that can result in worse performance due to the additional synchronization required.

## 5 Conclusions

We presented A$^2$M, an attack-resilient and latency efficient desktop hosting infrastructure based on a single-hop indirection network. A$^2$M exploits multi-path routing, packet replication, and the high asymmetry inherent to interactive display traffic, to assure access to remote desktop sessions, even in the presence of high-volume DoS attacks. Contrary to the current DoS protection mechanisms, our system guarantess both availability and uninterrupted connectivity to the end server providing a truly secure end-to-end connectivity model. Furthermore, in a departure from traditional client-server systems, A$^2$M provides an asymmetric client-server connection consisting of an indirected client-to-server multi-path, and a direct server-to-client connection. A$^2$M's indirection-based overlay acts both as a first-level distributed firewall and as a redirection mechanism for performance-critical user input-events going from the client device to the hosting servers. In turn, the direct server-to-client connection provides quick delivery of display updates, to provide quick response time and good user experience.

A prototype of A$^2$M was implemented in Linux and we evaluated its performance on PlanetLab. Our experimental results show that, as opposed to existing DDoS protection mechanisms, A$^2$M has minimum latency overhead and can provide good interactive performance for web, video, and general interactive applications. Furthermore, we demonstrate that A$^2$M significantly increases the attack resilience of the hosting infrastructure, being able to provide perfect video playback and low-latency web browsing and GUI interactions even in the presence of large attacks on the infrastructure. A$^2$M maintains 100% video quality in a number of remote video display scenarios, despite the use of overlay routing. Furthermore, end-to-end latency increases by less than 5% even when 40% of nodes have been rendered unusable by an attacker. Given its performance and resilience to DoS attacks, A$^2$M represents a step forward towards realizing the vision of computer utilities that provide ubiquitous, secure, and assured-access desktop computing.

## Acknowledgements

This work was supported by NSF Grants CNS-07-14277, CNS-04-26623, and Google Inc.

# References

1. Citrix ICA Technology Brief. Technical White Paper, Boca Research (1999)
2. Cumberland, B., Carius, G., Muir, A.: Microsoft Windows NT Server 4.0, Terminal Server Edition: Technical Reference. Microsoft Press (August 1999)
3. Richardson, T., Stafford-Fraser, Q., Wood, K.R., Hopper, A.: Virtual Network Computing. IEEE Internet Computing 2(1), 33–38 (1998)
4. DoS-Resistant Internet Working Group Meetings (February 2005), http://www.communicationsresearch.net/dos-resistant
5. Hulme, G.: Extortion online. Information Week (September 13, 2004)
6. Keromytis, A.D., Misra, V., Rubenstein, D.: SOS: Secure Overlay Services. In: Proceedings of ACM SIGCOMM, August 2002, pp. 61–72 (2002)
7. Andersen, D.G.: Mayday: Distributed Filtering for Internet Services. In: Proceedings of the 4th USENIX Symposium on Internet Technologies and Systems (USITS) (March 2003)
8. Baratto, R., Potter, S., Su, G., Nieh, J.: MobiDesk: Mobile Virtual Desktop Computing. In: Proceedings of the 10th Annual ACM International Conference on Mobile Computing and Networking (MobiCom) (September 2004)
9. Stavrou, A., Keromytis, A.: Countering DoS Attacks With Stateless Multipath Overlays. In: Proceedings of the 12th ACM Conference on Computer and Communications Security (CCS), November 2005, pp. 249–259 (2005)
10. Blaze, M., Feigenbaum, J., Ioannidis, J., Keromytis, A.D.: The KeyNote Trust Management System Version 2. RFC 2704 (September 1999)
11. CCITT: X.509: The Directory Authentication Framework. International Telecommunications Union, Geneva (1989)
12. Black, J., Halevi, S., Krawczyk, H., Krovetz, T., Rogaway, P.: UMAC: Fast and Secure Message Authentication. In: Wiener, M. (ed.) CRYPTO 1999. LNCS, vol. 1666, pp. 216–233. Springer, Heidelberg (1999)
13. Xuan, D., Chellappan, S., Wang, X.: Analyzing the Secure Overlay Services Architecture under Intelligent DDoS Attacks. In: Proceedings of the 24th International Conference on Distributed Computing Systems (ICDCS), March 2004, pp. 408–417 (2004)
14. Moore, D., Voelker, G., Savage, S.: Inferring Internet Denial-of-Service Activity. In: Proceedings of the 10th USENIX Security Symposium, August 2001, pp. 9–22 (2001)
15. Nieh, J., Yang, S.J., Novik, N.: Measuring Thin-Client Performance Using Slow-Motion Benchmarking. ACM Transactions on Computer Systems (TOCS) 21(1), 87–115 (2003)
16. Gummadi, K.P., Madhyastha, H.V., Gribble, S.D., Levy, H.M., Wetherall, D.: Improving the Reliability of Internet Paths with One-hop Source Routing. In: Proceedings of the 6th Symposium on Operating Systems Design & Implementation (OSDI) (December 2004)
17. Andersen, D.G., Snoeren, A.C., Balakrishnan, H.: Best-Path vs. Multi-Path Overlay Routing. In: Proceedings of the Internet Measurement Conference (October 2003)
18. Kaella, A., Pang, J., Shaikh, A.: A Comparison of Overlay Routing and Multihoming Route Control. In: Proceedings of ACM SIGCOMM, August/September 2004, pp. 93–106 (2004)
19. Su, A., Choffnes, D.R., Kuzmanovic, A., Bustamante, F.E.: Drafting Behind Akamai (Travelocity-Based Detouring). In: Proceedings of ACM SIGCOMM, September 2006, pp. 435–446 (2006)
20. Ioannidis, J., Bellovin, S.M.: Implementing Pushback: Router-Based Defense Against DDoS Attacks. In: Proceedings of the ISOC Symposium on Network and Distributed System Security (SNDSS) (February 2002)
21. Dean, D., Franklin, M., Stubblefield, A.: An Algebraic Approach to IP Traceback. In: Proceedings of the ISOC Symposium on Network and Distributed System Security (SNDSS), February 2001, pp. 3–12 (2001)

22. Savage, S., Wetherall, D., Karlin, A., Anderson, T.: Practical Network Support for IP Trace-back. In: Proceedings of ACM SIGCOMM, August 2000, pp. 295–306 (2000)

23. Snoeren, A., Partridge, C., Sanchez, L., Jones, C., Tchakountio, F., Kent, S., Strayer, W.: Hash-Based IP Traceback. In: Proceedings of ACM SIGCOMM (August 2001)

24. Li, J., Sung, M., Xu, J., Li, L.: Large-Scale IP Traceback in High-Speed Internet: Practical Techniques and Theoretical Foundation. In: Proceedings of the IEEE Symposium on Security and Privacy (May 2004)

25. Reiher, P., Mirkovic, J., Prier, G.: Attacking DDoS at the source. In: Proceedings of the 10th IEEE International Conference on Network Protocols (November 2002)

26. Yaar, A., Perrig, A., Song, D.: An Endhost Capability Mechanism to Mitigate DDoS Flood-ing Attacks. In: Proceedings of the IEEE Symposium on Security and Privacy (May 2004)

27. Papadopoulos, C., Lindell, R., Mehringer, J., Hussain, A., Govindan, R.: COSSACK: Coor-dinated Suppression of Simultaneous Attacks. In: Proceedings of DISCEX III, April 2003, pp. 2–13 (2003)

28. Parno, B., Wendlandt, D., Shi, E., Perrig, A., Maggs, B., Hu, Y.C.: Portcullis: protecting con-nection setup from denial-of-capability attacks. SIGCOMM Comput. Commun. Rev. 37(4), 289–300 (2007)

29. Baratto, R., Kim, L., Nieh, J.: THINC: A Virtual Display Architecture for Thin-Client Com-puting. In: Proceedings of the 20th ACM Symposium on Operating Systems Principles (SOSP) (October 2005)

30. Morein, W.G., Stavrou, A., Cook, D.L., Keromytis, A.D., Misra, V., Rubenstein, D.: Us-ing Graphic Turing Tests to Counter Automated DDoS Attacks Against Web Servers. In: Proceedings of the 10th ACM International Conference on Computer and Communications Security (CCS), October 2003, pp. 8–19 (2003)

31. Stavrou, A., Keromytis, A.D., Nieh, J., Misra, V., Rubenstein, D.: MOVE: An End-to-End Solution To Network Denial of Service. In: Proceedings of the ISOC Symposium on Network and Distributed System Security (SNDSS), February 2005, pp. 81–96 (2005)

32. Khattab, S.M., Sangpachatanaruk, C., Moss, D., Melhem, R., Znati, T.: Roaming Honeypots for Mitigating Service-Level Denial-of-Service Attacks. In: Proceedings of the 24th Interna-tional Conference on Distributed Computing Systems (ICDCS), March 2004, pp. 238–337 (2004)

33. Scheifler, R.W., Gettys, J.: X Window System, 3rd edn. Digital Press (1992)

34. Schmidt, B.K., Lam, M.S., Northcutt, J.D.: The interactive performance of SLIM: a stateless, thin-client architecture. In: 17th ACM Symposium on Operating Systems Principles (SOSP), December 1999, vol. 34, pp. 32–47 (1999)

35. Lai, A., Nieh, J.: Limits of Wide-Area Thin-Client Computing. In: Proceedings of the ACM International Conference on Measurement and Modeling of Computer Systems (SIGMET-RICS), June 2002, pp. 228–239 (2002)

# Automated Spyware Collection and Analysis

Andreas Stamminger[1], Christopher Kruegel[1],
Giovanni Vigna[1], and Engin Kirda[2]

[1] University of California, Santa Barbara
{as,chris,vigna}@cs.ucsb.edu
[2] Institut Eurecom, France
kirda@eurecom.fr

**Abstract.** Various online studies on the prevalence of spyware attest overwhelming numbers (up to 80%) of infected home computers. However, the term spyware is ambiguous and can refer to anything from plug-ins that display advertisements to software that records and leaks user input. To shed light on the true nature of the spyware problem, a recent measurement paper attempted to quantify the extent of spyware on the Internet. More precisely, the authors crawled the web and analyzed the executables that were downloaded. For this analysis, only a single anti-spyware tool was used. Unfortunately, this is a major shortcoming as the results from this single tool neither capture the actual amount of the threat, nor appropriately classify the functionality of suspicious executables in many cases.

For our analysis, we developed a fully-automated infrastructure to collect and install executables from the web. We use three different techniques to analyze these programs: an online database of spyware-related identifiers, signature-based scanners, and a behavior-based malware detection technique. We present the results of a measurement study that lasted about ten months. During this time, we crawled over 15 million URLs and downloaded 35,853 executables. Almost half of the spyware samples we found were not recognized by the tool used in previous work. Moreover, a significant fraction of the analyzed programs (more than 80%) was incorrectly classified. This underlines that our measurement results are more comprehensive and precise than those of previous approaches, allowing us to draw a more accurate picture of the spyware threat.

## 1 Introduction

In general, spyware is used to describe a broad class of software that is surreptitiously installed on a user's machine to intercept or take control over the interaction between the user and her computer. This broad definition includes programs that monitor a user's Internet surfing habits, but might also apply to software that redirects browser activity, commits click fraud, or downloads additional malware. Unfortunately, over time, the term spyware has become increasingly imprecise, and different companies or researchers often label the same

P. Samarati et al. (Eds.): ISC 2009, LNCS 5735, pp. 202–217, 2009.

program differently. In this paper, we use the term spyware in a more narrow sense – as browser-based software that records privacy-sensitive information and transmits it to a third party without the user's knowledge and consent. This definition is more faithful to the "original" purpose of spyware, which is to record the activity of a user while surfing the web.

A host can become infected with spyware in various ways. For example, the spyware component might come bundled with shareware, such as a peer-to-peer client or a supposed Internet "accelerator." It is common practice that small software companies, unable to sell their products in retail, cooperate with spyware/adware distributors to fund the development of their products [1]. Most of the time, however, users have no choice to "unselect" the installation of the piggybacked nuisance without disrupting the desired software functionality.

In this paper, we are interested in the extent to which executables on the web present a spyware threat. This allows us to compare our results to a previous study [2]. For our analysis, we focus on spyware that uses Microsoft's Internet Explorer to monitor the actions of a user. Typically, this is done either by using the Browser Helper Object (BHO) interface or by acting as a browser toolbar. We feel that this focus is justified by the fact that the overwhelming majority of spyware uses a component based on one of these two technologies, a fact that is confirmed by a number of previous papers [3,4,5,6].

As mentioned above, the authors of a previous measurement study [2] attempted to assess the prevalence of spyware on the Internet. To this end, they crawled the web for executables, downloaded them, and installed the programs in an automated fashion. Then, the authors executed a single anti-spyware program, Lavasoft's Ad-Aware [7], to assess the threat level of each program.

Unfortunately, in the previous study, little attention was devoted to the fact that relying on the output and correctness of a *single* tool can significantly misrepresent the true problem, and thus, the perception of the spyware threat. Obviously, scanner-based systems cannot detect novel threats for which no signature exists. Also, such systems are often trivial to evade by using techniques such as obfuscation or code substitution. Hence, scanner-based systems may introduce false negatives, and as a result, cause the threat to be *underestimated*. However, it is also possible that a detection tool mislabels programs as being more dangerous than they actually are. Such false positives may cause an *overestimation* of the threat.

In our work, one of the aims was to show the bias that is introduced by deriving statistics from the results of a single tool. Obviously, we did not simply want to re-run the experiments with more anti-spyware tools (although we did employ a second, scanner-based application). Instead, we wanted to perform our analysis using substantially different approaches that aim to detect spyware. To this end, we checked for spyware-related identifiers in the Windows registry, using a popular, publicly-available database [8]. Moreover, we employed a behavior-based approach [3] that monitors the execution of a component in a sandbox and checks for signs of suspicious behavior. By combining multiple techniques and employing further, manual analysis in cases for which different tools disagree, we aimed to establish a level of "ground truth" for our sample set. Based on this ground truth, we identify the weaknesses of each technique when exposed to a large set of real-world, malicious code.

In summary, the contributions of this paper are the following:

- In about ten months, we crawled over 15 million URLs on the Internet and analyzed 35,853 executables for the presence of spyware components.
- We employed three different analysis techniques and devoted additional manual effort to identify the true nature of the components that we obtained. This allows us to expose the weaknesses of individual analysis approaches.
- We compare our results to a previous study that attempted to measure the spyware threat on the Internet and critically review their results.

## 2    Methodology

In this section, we describe our approach to analyze the extent of spyware on the Internet. In order to keep a consistent terminology within the rest of the paper, we first define the behavior that constitutes spyware activity. Then, we explain how we crawl the web for executables and briefly discuss our approach to automatically install these programs. Finally, we describe how we identify a program as spyware,

As mentioned previously, the term spyware is overloaded. For example, it is not uncommon that a component that displays advertisements is considered spyware, even it does not read nor leak any privacy-related information. Other examples of mislabeled spyware are toolbars that provide search fields that send input to a search engine of the user's choice. Clearly, information that is entered into the search field is forwarded to the search engine. Hence, the component is not malicious, as it informs the user where the data is sent to.

Because of the ambiguous use of the spyware term, it is possible that the actual risk of downloading a spyware-infected executable is overstated. Consequently, we need a more precise discrimination between different classes of activity. As mentioned previously, we focus in our study on browser extensions (BEs) for the Microsoft Internet Explorer (from now on, we use the term browser extension to refer to both BHOs and toolbars). To make our discussion of browser extensions more precise, we propose the following taxonomy:

**Benign.** An extension is benign when it does not perform any function that might be undesirable from a privacy-related point of view, nor exposes the user to unwanted content.

**Adware.** Adware is benign software with the purpose of advertising a certain product, e.g., via pop-up windows. These components do not leak any sensitive information, though.

We also consider a toolbar as adware when it provides a search field to the user that sends the input to a particular (often, less well-known) search engine. The reason is that the toolbar promotes, or advertises, the use of a particular search engine. Of course, the user is free to use the toolbar or not.

**Grayware.** Grayware occasionally performs actions that send sensitive data to third parties in a way that is not completely transparent to users, especially inexperienced ones.

An important class of grayware are browser hijacker components. A browser hijacker is software that modifies the default browser behavior. Depending on the resource that is controlled, we distinguish between different types of hijackers. A *homepage hijacker* modifies the default home page that is displayed when the browser is launched. A *search hijacker* modifies the default search engine of the browser. It allows the user to enter keywords directly into the browser's address bar without the need to request the website of a search engine. Typically, the user is redirected to a less popular search engine with sponsored results. Similarly, a *error page hijacker* causes the browser to display a particular web site whenever a misspelled URL is entered into the address bar. Usually, the original URL is passed as a query parameter to this web site. While a hijacker component might appear to be a useful feature, it is also profitable for the author of the landing site. This is for two reasons: First, it increases the hit count for his site (which drives up advertising revenue) and second, it reveals popular URL misspellings to facilitate domain squatting. Since a hijacker component is not triggered for regular pages that are visited, it is not a means to capture *all* of the user's surfing activity. Also, an alert user can notice the modified browser behavior and reset it accordingly. These are the two differences that distinguish grayware from spyware.

**Spyware.** Spyware, as defined in this paper, serves the purpose of secretly and comprehensively collecting data about the user, such as her surfing habits or form inputs. The data collection process is invisible, and a significant amount of user data (for example, most or all of the visited URLs) are leaked to a third, untrusted party.

**Malware.** Some components are reported to perform actions that are typically associated with "regular" malware. An example are Trojan downloaders that run in the context of the browser when accessing malicious content on the Internet so that they can bypass personal firewalls. These components do not necessarily access private information, but perform clearly undesirable activity. For such components, we use the generic term malware.

It is possible that the same component implements functionalities that fall into different categories. For example, a spyware component could also display ads. In this case, the program is assigned to the category that captures the more malicious behavior (spyware, for this example).

## 2.1   Web Crawling

To find a representative amount of programs that install spyware components, we developed a fully automated system that crawls the web for potential candidates and downloads them. To this end, we make use of the Heritrix [9] web crawler, which can be easily extended and customized. For downloading interesting web resources, we focus on binary content, such as executables or zip archives. Similar to [2], we identify such content by examining two properties for each candidate URL. If either (1) the URL's file extension is .exe, or (2) the "Content-type" HTTP header of the corresponding web resource is application/octet-stream, we download the file. We then check the first bytes of the file header and compare it with the "magic" value that denotes a Windows executable. We perform similar checks for zip, cabinet (.cab), and MS Installer files (.msi).

To determine whether Internet users with a specific field of interest are more likely to encounter spyware on the web, we defined eight categories, similar to [2]: adult, games, kids, music, desktop (office), pirate, shareware, and toolbar. For each category, we fed the Google search engine with category-specific keywords and used the fifty most relevant search results as a seed for our crawler. We consider this a reasonable approach, because these are the pages that users would most likely encounter when searching for content in the categories mentioned above. To focus our crawling to those web sites that are found by the Google search, we do a breath-first crawl only up to a depth of three links away from the seed.

## 2.2 Automatic Installation

To determine whether an executable contains spyware, we install it on a Windows guest system running on top of a Qemu virtual machine emulator [10]. Each executable is installed on a system that has been reverted to a known, clean state. Since we wish to analyze thousands of programs, the installation process has to be performed automatically. To this end, we had to find a way to simulate user interaction, which is typically necessary when navigating through Windows installation wizards that have a graphical user interface.

Once an executable is successfully installed, we have to determine whether a browser extension (BE) is present or not. Fortunately, this is quite straightforward. The reason is that, in order to be loaded on startup by the Internet Explorer, a BE must register its CLSID (i.e., Component ID) under a particular (directory) key in the Windows Registry. Thus, after each installation, we simply check for the presence of CLSIDs in this special directory. Note that it is difficult for a spyware to avoid setting this key, as the Internet Explorer would otherwise simply not load the BE at startup. We proceed with the subsequent analysis phase when any BE is identified.

## 2.3 Analysis

The purpose of the analysis phase is to determine whether a BE is malicious or not. More precisely, we attempt to classify each browser extension based on the taxonomy introduced previously. We use three different approaches to determine the type of a BE: an identifier-based mechanism, two scanner-based tools, and a behavior-based technique. They are discussed in more detail below.

**Identifier-based Detection.** The identifier-based detection relies on the value of the CLSID of the BE component. Interestingly, many spyware programs use the same CLSID to register their component (possibly because the developers were lazy or use the same code base). Thus, the value of the identifier can provide some insight into the nature of the corresponding program. Moreover, also the file name of the extension component can be revealing. Of course, both identifiers can be easily modified by miscreants.

CastleCops [8] is[1] a community of security professionals that provides a free and searchable database of BHOs and toolbars. At the time of writing, it contained 41,144 entries. For each BE, the database lists various information, including the BE's CLSID and its file name. Furthermore, a classification is provided. This classification includes $X$ for spyware and malware, $O$ for programs that are open to debate (such as grayware and adware), and $L$ for legitimate items.

To perform identifier-based detection, we use HijackThis [11], a free utility that scans a computer for installed browser extensions, reporting both the CLSIDs and path names of the identified components. Based on the information provided by HijackThis, we consult the CastleCops database. Using the classification provided by this database, we can classify the browser extension accordingly. The absence of any entry results in the BE being classified as legitimate.

**Scanner-based Detection.** Our scanner-based detection was based on two commercial anti-spyware products, Ad-Aware [7] and Spybot [12] – both popular and well-known spyware scanners.

Ad-Aware uses a number of threat categories to specify the precise nature of a sample. During our analysis, we encountered the following categories:

- *Adware*: Programs displaying advertising on the user's computer, without leaking sensitive information.
- *Data miner*: Programs designed to collect and transmit private user information to a third party. This behavior may be disclosed to the user through to the EULA. This is the equivalent to our spyware definition.
- *Malware*: A generic category for harmful programs, equivalent to our malware class.

To ensure that we had the newest signatures, we always updated Ad-Aware's signature database before launching a scan. To check for suspicious code, we perform a *full system scan*. Once the tool is finished, we check the report for the presence of any component that is recognized as being suspicious. If this is the case, we record the corresponding threat category.

Spybot is a spyware scanner that attempts to detect threats on the user's computer by comparing registry entries and files against a database with signatures of well-known malware samples. This tool allows to choose the threat categories for which a user wants to check. For our study, we chose those categories that we assumed to be most-closely related to spyware: hijackers, keyloggers, malware, potentially unwanted programs, and spyware. After we run a *system scan*, Spybot lists each detected threat, without providing any further classification.

---

[1] Unfortunately, CastleCops has recently ceased its operations, but was still active while we performed our analysis.

**Behavior-based Detection.** To perform behavior-based detection, we build upon an analysis tool that we obtained from the authors of [3]. This tool allows the identification of unknown browser components as spyware by dynamically observing their behavior. Specifically, the tool monitors the flow of sensitive information (such as the URL that a user visits or the content of the web pages that are loaded) as it is processed by the Internet Explorer and any loaded browser extension (BHOs and toolbars). Whenever it observes any leak of sensitive information on behalf of a BE, such as submitting data to a remote server, this BE is considered spyware. For its analysis, two types of sensitive data are considered:

- URLs that the browser navigates to, and
- the contents of web pages retrieved by the browser in response to browser navigation.

Whenever sensitive (tainted) information is written out on behalf of the monitored BE, this action is recorded as suspicious. Writing out information can refer to saving data in a file, but also considers the case when data is sent over a network socket. This allows one to identify two different kinds of suspicious behavior:

- *Browser hijackers (grayware)*: As mentioned previously, hijacker components modify the default browser behavior such that certain user input is redirected to particular web sites. This behavior is detected when search terms or malformed URLs are entered into the browser address bar and then leaked by the BE.
- *Spyware*: These programs are detected when URLs are secretly leaked to an entity outside the browser (such as a remote host or a local file).

The behavior-based analysis is dynamic. Hence, it is necessary to monitor the activity of a BE while the browser is used to surf the web. To perform the dynamic analysis in an automated fashion, we had to develop an additional tool that allows us to drive the browser and simulate a user surfing the web (while monitoring the activity of browser extensions). This tool interacts with the browser in three different ways: by entering data directly into the address bar, by filling out and submitting form fields, and by following links on web pages. This variety of actions should provide for the realistic simulation of a user that browses the web. Moreover, to trigger hijacker programs, the tool enters keywords directly into the browser's address bar and intentionally requests malformed web addresses.

To simulate a browsing session, we require a list of URLs that should be visited as well as a number of keywords that we can enter into form fields. Our list of URLs included various popular search engine sites, such as `google.com`, `yahoo.com`, and `altavista.com`. During our browsing session, we surfed these sites and entered numerous keywords with the aim to "trigger" the spyware program to leak information to a remote server or redirect the browser to a different site. We compiled our list of keywords using Google HotTrends, selecting the most popular search terms. Besides search engine sites, we also entered some of these keywords directly into the browser's address bar. To trigger BEs that hijack error pages, we also entered misspelled URLs.

# 3  Results

In this section, we discuss the results of our measurement study. More precisely, we show the prevalence of spyware-infected executables for a number of different "regions" on the web. Moreover, we compare the effectiveness of different detection techniques, examining their strengths and limitations. In particular, we are interested in the possible bias that Ad-Aware introduces, since this was the sole tool used in a previous attempt to quantify the extent of spyware on the Internet [2]. Finally, we compare the findings in the previous study to the results of our analysis.

## 3.1  Overall Results

We crawled the web for ten months (from January 2007 until the end of October 2007), visited over 15 million URLs, and found 35,853 executables. Table 1 shows the number of binary resources that we discovered, categorized by their file type. The vast majority of downloads were Win32 executables and Zip archives. As shown in Table 2, 9.4% of all executables were installing at least one BHO or toolbar. This underlines the popularity of these techniques. Each browser extension that we obtained during the ten month crawl period was analyzed using the three approaches described in the previous section. Then, based on the (often differing) results of the individual techniques, we performed manual analysis to obtain a "ground truth" for our data set.

To obtain ground truth, we inspected those BEs for which the analysis methods reported different results. The manual analysis was carried out by launching the sample in a virtual machine, manually monitoring its network traffic as well as other system modifications (e.g., created files or registry entries). Also, we performed more extensive web surfing. The recorded behavior was then compared

Table 1. Crawler results by file type

| Win32 exec. | Zip archive | MS install | Cabinet |
|---|---|---|---|
| Files 29,104 (72.5%) | 10,260 (25.6%) | 425 (1.1%) | 335 (0.8%) |

Table 2. Overall crawler results

| URLs crawled | executables found | executables w/ BEs | unique BEs |
|---|---|---|---|
| 15,111,548 | 35,853 | 3,356 (9.4%) | 512 |

to various, online malware descriptions, and, based on all available information, a final classification was assigned to each sample. Moreover, especially in cases where a particular BE was more popular (i.e., part of several executables), we contacted the developers of the anti-spyware products to resolve classification errors. Although small errors are clearly possible, we believe that we have established a solid data set of benign and malicious components that can meaningfully serve as ground truth for our evaluation.

**Table 3.** Overall analysis results

| executables w/ non-benign BEs | unique non-benign BEs | executables w/ spy/malware BEs | unique spy/malware BEs |
| --- | --- | --- | --- |
| 2,384 (6.6%) | 205 (40.0%) | 117 (0.3%) | 22 (4.3%) |

Table 3 shows the overall analysis results. It can be seen that about 6.6% of all executables contain non-benign BEs. However, most of these programs belong to the adware category, while the fraction of executables that contain malicious components (spyware and malware) is significantly less - only 0.3% or 117 executables. This clearly underlines that the spyware threat might appear much more dramatic when the analysis does not distinguish precisely between non-invasive adware and malicious spyware. A breakdown of the non-benign BEs according to our taxonomy is presented in Table 4.

## 3.2 Distribution of Infected Executables

In this section, we discuss in more detail the prevalence of particular, malicious browser extensions, as well as their habitat (i.e., domains and regions on the web that are primary sources for these browser extensions).

Table 5 shows those ten browser extensions that we encountered most frequently in executables. Note that, for this table, we only consider grayware, spyware, and malware extensions. The reason for not considering adware is that we want to specifically focus on the more invasive, malicious programs. It can be seen that *NewDotNet* is by far the most popular component found by our crawler, being bundled with 197 executables. Most of these executables are peer-to-peer software (e.g., Limewire, Gnutella) and download accelerators. *NewDotNet* is an error hijacker that redirects URLs that cannot be resolved via DNS to various remote hosts, such as `r404.qsrch.net`. *Webhancer* is the most popular spyware component, and it is bundled particularly often with screensavers. In our experiments, this component secretly recorded the URLs that were visited and forwarded them to `dr2.webhancer.com`.

In the next step, we analyzed the prevalence of malicious browser extensions based on the categories of the web sites that are serving them. As mentioned in Section 2.1, for finding sites to crawl, we seeded the Google searches with keywords that were chosen from eight categories (adult, games, kids, music,

**Table 5.** Top 10 BEs - counting only grayware, spyware, and malware

**Table 4.** Non-benign BEs, by class

| class | # BEs | times observed |
|---|---|---|
| adware | 162 (79.0%) | 1,985 (83.3%) |
| grayware | 21 (10.2%) | 282 (11.8%) |
| spyware | 18 (8.8%) | 91 (3.8%) |
| malware | 4 (2.0%) | 26 (1.1%) |

| name | class | times observed |
|---|---|---|
| NewDotNet | grayware | 197 |
| Webhancer | spyware | 60 |
| P2P Energy | grayware | 45 |
| TR/Agent.A | malware | 24 |
| NavExcel | grayware | 21 |
| Acez.SiteError | grayware | 6 |
| Pal.PCSpy | spyware | 6 |
| ClickSpring | spyware | 5 |
| SmartKeyLogger | spyware | 5 |
| CasinoBar | spyware | 2 |

desktop (office), pirate, shareware, and toolbar). The results for the prevalence of non-benign components on pages of these categories are shown in Table 6. As the numbers demonstrate, we encountered spyware in all categories.

Before analyzing the results in detail, we conjectured that most spyware would be found on shareware or freeware sites. This is not only because of the large amount of executables hosted on those sites, but also because shareware is often offered together with dubious adware to finance its development. Our results confirm the initial intuition: The *shareware* category is not only the richest source for executables in general, but also holds the largest number of executables that install a BE. Interestingly, although over 15% of the shareware applications come with a non-benign BE, the actual fraction causing a spyware or malware infection is comparatively low (0.4%). The categories of the sites where BEs are most likely misused for malicious purposes are adult, desktop (office), and games, as indicated by the highest fraction of spyware BEs (last row in Table 6).

### 3.3 Detection Effectiveness

This section provides a detailed comparison between the ground truth and the results delivered by each detection technique that we used for our study. This allows us to identify interesting cases in which a certain technique is particularly effective or ineffective.

**Identifier-based Detection.** Table 7 contrasts our ground truth classification with the labeling according to CastleCops. Each table entry shows the number of unique BEs and, in parenthesis, the number of corresponding executables, based on their classification by CastleCops versus their true nature.

When examining this table, the considerable number of debatable components reflects the general difficulty analysts face when they have to assign a certain category to a certain browser extension. Often, it is up to the user whether they consider the behavior of a component acceptable or not. Also, there are a quite

**Table 6.** Penetration of non-benign BEs across different web categories

|  | adult | games | kids | music | office | pirate | share | toolbar |
|---|---|---|---|---|---|---|---|---|
| URLs (in K) | 660 | 536 | 2,375 | 3,573 | 1,089 | 4,589 | 1,791 | 498 |
| domains | 790 | 1,678 | 1,821 | 1,662 | 1,911 | 3,795 | 3,298 | 2,087 |
| executables | 1,298 | 3,048 | 3,732 | 3,053 | 3,363 | 6,586 | 11,043 | 3,730 |
| executables w/ BEs | 49 (3.8%) | 85 (2.8%) | 278 (7.4%) | 273 (8.9%) | 59 (1.8%) | 143 (2.2%) | 2,270 (20.6%) | 199 (5.3%) |
| executables w/ non-ben. BEs | 30 (2.3%) | 14 (0.5%) | 158 (4.2%) | 163 (5.3%) | 31 (0.9%) | 81 (1.2%) | 1,825 (16.5%) | 82 (2.2%) |
| domains w/ non-ben. BEs | 16 (2.0%) | 9 (0.5%) | 48 (2.6%) | 56 (3.4%) | 26 (1.4%) | 44 (1.2%) | 88 (2.7%) | 39 (1.9%) |
| executables w/ spy/mal. BEs | 7 (0.5%) | 3 (0.1%) | 13 (0.3%) | 22 (0.7%) | 10 (0.3%) | 12 (0.2%) | 42 (0.4%) | 8 (0.2%) |
| domains w/ spy/mal. BEs | 5 (0.6%) | 3 (0.2%) | 10 (0.5%) | 16 (1.0%) | 7 (0.4%) | 8 (0.2%) | 15 (0.5%) | 7 (0.3%) |
| BEs | 17 | 13 | 201 | 208 | 32 | 79 | 232 | 172 |
| non-benign BEs | 6 (35.3%) | 4 (30.8%) | 120 (59.7%) | 127 (61.1%) | 16 (50.0%) | 25 (31.6%) | 110 (47.4%) | 68 (39.5%) |
| spy/malware BEs | 3 (17.6%) | 2 (15.4%) | 8 (4.0%) | 12 (5.8%) | 5 (15.6%) | 9 (11.4%) | 10 (4.3%) | 5 (2.9%) |

**Table 7.** Ground truth vs. CastleCops

|  | - | legitimate | debatable | ad-/spyware |
|---|---|---|---|---|
| benign | 62 (166) | 186 (583) | 57 (220) | 2 (3) |
| adware | 106 (278) | 4 (10) | 31 (1,641) | 21 (56) |
| grayware | 2 (2) | 1 (3) | 6 (52) | 12 (225) |
| spyware | 2 (4) | 0 (0) | 4 (15) | 12 (72) |
| malware | 0 (0) | 0 (0) | 0 (0) | 4 (26) |

large number of CLSIDs (106) used by adware BEs that we could not find in the online database. This is mainly due to *Softomate* components, discussed in the following paragraph.

In general, it can be seen that identifier-based identification works surprisingly well. Unfortunately, this kind of detection can be easily evaded, and certain spyware variants (e.g., *Win32.Stud.A*) already use randomly-generated CLSIDs.

**Scanner-based Detection.** Table 8 shows our comparison with the reports provided by Ad-Aware. When we consider the similarity of our definition of *spyware* and Ad-Aware's description of a *data miner*, our results show a surprising mismatch in the number of detected samples. During our analysis, Ad-Aware (mis)labeled 130 unique adware components as data miner. All other techniques could not confirm these threats.

Closer examination of Ad-Aware's report showed that 92% of these mislabeled components are toolbars. To determine whether these components only track

**Table 8.** Ground truth vs. Ad-Aware

|          | benign       | adware   | data miner    | malware |
|----------|--------------|----------|---------------|---------|
| benign   | 303 (963)    | 4 (9)    | 0 (0)         | 0 (0)   |
| adware   | 15 (20)      | 14 (99)  | 130 (1,863)   | 3 (3)   |
| grayware | 8 (238)      | 3 (8)    | 10 (36)       | 0 (0)   |
| spyware  | 4 (9)        | 2 (3)    | 7 (67)        | 5 (12)  |
| malware  | 4 (26)       | 0 (0)    | 0 (0)         | 0 (0)   |

user data that is entered into the toolbar, we additionally performed manual testing. Some of these toolbars provide search results for paid advertisers, but only when we use the search function of the toolbar. Clearly, this is the expected behavior, and thus, should not be classified as data miner. We also contacted Lavasoft to resolve this issue. We were told that one possible cause for their classification might be the fact that the installation routine does not clearly state the purpose of an adware program, and thus, it is labeled as data miner. Additionally, they admit that some samples were misclassified.

One particular problem was caused by the *Softomate Toolbar*, which is a developer aid for creating customized Internet Explorer components. A few malicious samples are created using this tool. However, Ad-Aware tags *all* toolbars that are developed with the help of Softomate as data miner. This is unfortunate, because we observed that over 50% of all executables with browser extensions were using a component produced by Softomate. However, only a tiny fraction is recognized as malicious by all other detection techniques. Given the significant amount of adware BEs that were tagged as data miners by Ad-Aware, we recognize a significant bias that overstates the actual threat.

On the other hand, we also found four actual spyware threats not reported by Ad-Aware. Three of these threats were revealed by the behavior-based detection technique (as we show later below), and three could also be identified using Spybot. This demonstrates the limitations of signature-based detection and the possibility to underestimate the threat because of novel, malicious code instances. However, four cases are still a relatively small number. In two additional cases, a spyware threat was misclassified as adware.

Table 9 shows our comparison with Spybot. At first glance, it appears that Spybot misses a considerable amount of adware samples. On further examination, 93% of these BEs are Softomate Toolbars, a popular type of extension. As we discussed previously, we labeled these BEs as (mildly annoying) adware, but one could also argue that they are benign. Therefore, we consider this mismatch as negligible.

**Behavior-based Detection.** Table 10 shows the performance of our taint analysis with respect to ground truth. As expected, those BEs leaking sensitive user information, such as URLs surfed by the user, could be found in the categories grayware and spyware. Since *benign* software and *adware* do not disclose private user information to a remote server, we cannot distinguish between these components.

**Table 9.** Ground truth vs. Spybot

| | not detected | detected |
|---|---|---|
| benign | 304 (965) | 3 (7) |
| adware | 131 (1,831) | 31 (154) |
| grayware | 8 (59) | 13 (223) |
| spyware | 3 (7) | 15 (84) |
| malware | 1 (2) | 3 (24) |

**Table 10.** Ground truth vs. behavior-based

| | not detected | detected |
|---|---|---|
| benign | 300 (956) | 7 (16) |
| adware | 161 (1,984) | 1 (1) |
| grayware | 6 (8) | 15 (274) |
| spyware | 4 (13) | 14 (78) |
| malware | 4 (26) | 0 (0) |

A significant advantage of behavior-based, dynamic analysis is the fact that also novel threats can be identified. Thus, we would expect that the behavior-based approach can identify more spyware components than scanner-based techniques. Table 11 lists those BEs that were detected by the behavior-based analysis but missed by Ad-Aware. For seven unique extensions, we detected redirections for keywords entered directly into the browser's address bar. Two different, unique BEs leaked all surfed URLs to a remote third party.

**Table 11.** BEs detected by behavioral analysis but not Ad-Aware

| name | # variants | class |
|---|---|---|
| 811_Toolbar | 1 | grayware |
| Camfrog Toolbar | 1 | grayware |
| CasinoDownloader | 2 | spyware |
| CyberDefender | 1 | grayware |
| NewDotNet | 4 | grayware |
| Offsurf Proxy | 1 | grayware |
| P2P Energy | 1 | grayware |
| Win32.Stud.A | 1 | spyware |

**Table 12.** False positives raised by behavior-based detection

| name | # variants |
|---|---|
| ChildWebGuardian | 3 |
| GL-AD Popup Term. | 1 |
| PCTools Browser | 1 |
| SurfLogger | 1 |
| WhereWasI | 1 |

The fact that Ad-Aware misses *NewDotNet* is problematic, as this component is the most popular grayware found by our crawler (accounting for 197 infected executables, as can be seen in Table 5). This introduces an imprecision into statistics that depend on Ad-Aware output. In addition to the seven grayware components, Ad-Aware also missed two spyware programs. Both programs transmit all the URLs that are surfed to a third party. More precisely, *Casino-Downloader* transmits all surfed URLs to `ad.outerinfoads.com` and various other affiliated severs. *Win32.Stud.A* is a BHO that is installed silently by a free picture viewer application. Interestingly, we observed that different CLSIDs are used every time the BHO is installed. This clearly indicates an attempt to evade identifier-based detection. This BHO records the URLs visited by the user and transmits them encrypted to `www.googlesyndikation.com`. When it detects certain keywords or URLs, it aggressively displays pop-up advertisements.

The behavioral analysis failed to recognize a few malicious components as spyware. One important reason was that several components attempted to connect to remote hosts that were no longer available. Thus, collected information could not be leaked. In other cases, the components were waiting for a particular trigger (a specific URL) that was not part of our set of visited URLs.

The behavior-based analysis considers a BE as spyware whenever it leaks tainted (sensitive) user information from the Internet Explorer process. However, there might be cases in which this operation is legitimate, giving raise to false positives. In the following, we discuss the samples that have been incorrectly labeled as spyware, although their behavior is (likely) legitimate. Table 12 provides an overview. For example, *ChildWebGuardian* tracks user surfing habits and is intended to give parental control over the sites visited by a child. Thus, it logs the list of URLs that a user visits to a local file, presenting it later to the parent for inspection.

It is interesting to note that all components that caused false positives write information (such as URLs) to the local file system only. Thus, the behavioral analysis could be modified so that a component is marked as spyware only when sensitive information is sent over the network (possibly via the file system or another process). For the analyzed components, this would *not* have caused additional false negatives.

Overall, the behavioral analysis captured the spyware threat most accurately. Together with Spybot, this technique correctly detected the largest fraction of malicious browser extensions. Moreover, it raised by far the smallest number of false positives (and this number could be further decreased, as discussed previously). Thus, when repeating our experiments without any manual analysis, the results of the behavioral technique can be used to classify unknown components. Adding tools such as Spybot can improve detection rates but also incorrectly inflates the number of spyware components due to false positives.

## 3.4    Comparison to Previous Work

When we compare our measurement results to findings in the previous study [2], we note certain similarities. For example, the previous study observed that between 5.5% and 13.4% of all crawled executables are spyware-infected. If we consider non-benign BEs of all categories, the fraction of infected executables we detect in our study is 6.6%. However, this number does not reflect the actual spyware threat present on the Internet. Rather than focusing on the real spyware-threat, it only provides a rough estimate of the number of programs that ship with possibly annoying, but nevertheless non-intrusive, advertising components. The reason is that only a small fraction of non-benign samples actually perform privacy-invasive operations (as shown in Table 4). A major reason for the different assessment of the threat level between our study and previous work is Ad-Aware. Ad-Aware was the only tool used in the previous study, and it mislabels a significant number of non-malicious adware programs as spyware (data miner). This leads to an overestimation of the actual number of executables that are infected with privacy-invasive components.

## 4   Related Work

As detailed in previous sections, our work was inspired by the measurement study presented in [2]. Similar to the methodology presented in that paper, we crawled the web for executables that were then automatically installed and analyzed. The major difference of our work is the way in which we perform our analysis. Instead of relying on a single tool, we use three different approaches to classify each executable. This allows us to derive a more precise assessment of the extent of the spyware threat on the Internet than was reported by the authors of the study in [2]. Moreover, we are able to identify the weaknesses of individual detection and analysis techniques. As a result, we can understand in which ways the results reported in the previous work might be biased.

Since malicious code is an important problem, a number of researchers have proposed techniques to analyze and detect malware. The details of the behavioral-based approach, which we used and extended in this paper to automatically identify spyware components, were previously presented in [4]. Other dynamic approaches [13,14] to identify more general classes of malware based on their behavior often use taint propagation to detect suspicious information flows. Complementary to dynamic techniques, there are static analysis approaches [15] to identify malicious code patterns, and techniques [16] to extract network-based signatures that capture suspicious traffic flows.

## 5   Conclusion

In this paper, we present the results of a measurement study that attempts to quantify the extent of spyware-infected executables on the Internet. Inspired by previous work, we crawled the web for executables that were then installed and analyzed. In total, our experiment lasted around ten months. We crawled over 15 million URLs and downloaded more than 35 thousand executables. An important difference to previous work is the fact that we used three different analysis techniques. By combining the views from different vantage points, we were able to identify the limitations of each individual technique. In particular, we found that Ad-Aware, the tool used for the previous study, significantly overestimates the severity of many samples. As a result, previous work might have overestimated the prevalence of privacy-invasive spyware. While we did find a non-negligible number of spyware-infested executables, the vast majority of non-benign browser extensions were not stealing private information but displaying annoying advertisements.

**Acknowledgments.** This work has been supported by the Austrian Science Foundation (FWF) and by Secure Business Austria (SBA) under grants P-18764, P-18157, and P-18368, and by the European Commission through project FP7-ICT-216026-WOMBAT.

# References

1. Good, N., Dhamija, R., Grossklags, J., Thaw, D., Aronowitz, S., Mulligan, D., Konstan, J.: Stopping Spyware at the Gate: A User Study of Privacy, Notice and Spyware. In: Symposium On Usable Privacy and Security, SOUPS (2005)
2. Moshchuk, A., Bragin, T., Gribble, S.D., Levy, H.M.: A Crawler-based Study of Spyware on the Web. In: Network and Distributed Systems Security Symposium, NDSS (2006)
3. Egele, M., Kruegel, C., Kirda, E., Yin, H., Song, D.: Dynamic Spyware Analysis. In: Usenix Annual Technical Conference (2007)
4. Kirda, E., Kruegel, C., Banks, G., Vigna, G., Kemmerer, R.: Behavior-based Spyware Detection. In: Usenix Security Symposium (2006)
5. Wang, Y., Roussev, R., Verbowski, C., Johnson, A., Wu, M., Huang, Y., Kuo, S.: Gatekeeper: Monitoring Auto-Start Extensibility Points (ASEPs) for Spyware Management. In: Large Installation System Administration Conference (2004)
6. Hackworth, A.: Spyware. US-CERT Publication (2005)
7. Lavasoft: Ad-Aware, http://www.lavasoftusa.com/software/adaware
8. Castlecops: The CLSID / BHO List / Toolbar Master List, http://www.castlecops.com/CLSID.html
9. Mohr, G., Stack, M., Rnitovic, I., Avery, D., Kimpton, M.: Introduction to Heritrix. In: 4th International Web Archiving Workshop (2004)
10. Bellard, F.: QEMU, a Fast and Portable Dynamic Translator. In: Usenix Annual Technical Conference, Freenix Track (2005)
11. Trendmicro: HijackThis, http://www.trendsecure.com/portal/en-US/tools/security_tools/hijackthis
12. Spybot: Spybot Search & Destroy, http://www.safer-networking.org/
13. Moser, A., Kruegel, C., Kirda, E.: Exploring Multiple Execution Paths for Malware Analysis. In: Symposium on Security and Privacy (2007)
14. Yin, H., Song, D., Egele, M., Kruegel, C., Kirda, E.: Panorama: Capturing System-wide Information Flow for Malware Detection and Analysis. In: ACM Conference on Computer and Communication Security, CCS (2007)
15. Christodorescu, M., Jha, S., Seshia, S., Song, D., Bryant, R.: Semantics-Aware Malware Detection. In: Symposium on Security and Privacy (2005)
16. Wang, H., Jha, S., Ganapathy, V.: NetSpy: Automatic Generation of Spyware Signatures for NIDS. In: Annual Computer Security Applications Conference, ACSAC (2006)

# Towards Unifying Vulnerability Information for Attack Graph Construction

Sebastian Roschke, Feng Cheng, Robert Schuppenies, and Christoph Meinel

Hasso Plattner Institute (HPI), University of Potsdam,
P.O. Box 900460, 14440, Potsdam, Germany
{sebastian.roschke,feng.cheng,
robert.schuppenies,christoph.meinel}@hpi.uni-potsdam.de

**Abstract.** Attack graph is used as an effective method to model, analyze, and evaluate the security of complicated computer systems or networks. The attack graph workflow consists of three parts: information gathering, attack graph construction, and visualization. To construct an attack graph, runtime information on the target system or network environment should be monitored, gathered, and later evaluated with existing descriptions of known vulnerabilities. The output will be visualized into a graph structure for further measurements. Information gatherer, vulnerability repository, and the visualization module are three important components of an attack graph constructor. However, high quality attack graph construction relies on up-to-date vulnerability information. There are already some existing databases maintained by security companies, a community, or governments. Such databases can not be directly used for generating attack graph, due to missing unification of the provided information. This paper challenged the automatic extraction of meaningful information from various existing vulnerability databases. After comparing existing vulnerability databases, a new method is proposed for automatic extraction of vulnerability information from textual descriptions. Finally, a prototype was implemented to proof the applicability of the proposed method for attack graph construction.

## 1 Introduction

Along with the rapid development and extension of IT-Technology, computer and network attacks and their countermeasures become more and more complicated. Attack Graphs have been proposed for years as a formal way to simplify the modeling of complex attacking scenarios. Based on the interconnection of single attack steps, they describe multi-step attacks [1]. Attack Graphs not only describe one possible attack, but many potential ways for an attacker to reach a goal. In an attack graph, each node represents a single attack step in a sequence of steps. Each step may require a number of previous attack steps before it can be executed, denoted by incoming edges, and on the other hand may lead to several possible next steps, denoted by outgoing edges. With the help of attack graphs most of possible ways for an attacker to reach a goal can be computed.

P. Samarati et al. (Eds.): ISC 2009, LNCS 5735, pp. 218–233, 2009.

This takes the burden from security experts to evaluate hundreds and thousands of possible options. Thus, a program can identify weak spots much faster than a human. At the same time, representing attack graphs visually allows security personal a faster understanding of the problematic pieces of a network [2,3].

As depicted in Figure 1, the attack graph workflow consists of three independent phases: Information Gathering, Attack Graph Construction, as well as Visualization and Analysis. In the information gathering phase, all necessary information to construct attack graphs is collected and unified, such as information on network structure, connected hosts, and running services. In the attack graph construction phase, a graph is computed based on the gathered system information and existing vulnerability descriptions. Finally, the attack graph is processed in the visualization and analysis phase. Attack graphs always require a certain set of input information. For one, a database of existing vulnerabilities has to be available, as without it, it would not be possible to identify or evaluate the effects of host-specific weaknesses. Also, the network structure must be known beforehand. It is necessary to identify which hosts can be reached by the attacker. Often, an host-based vulnerability analysis is performed before the attack graph is constructed.

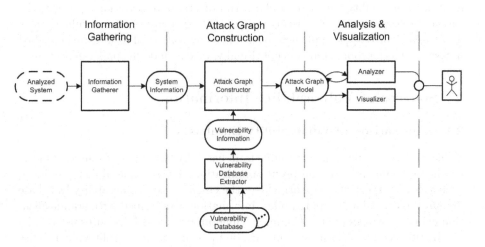

**Fig. 1.** Workflow of Attack Graph Construction and Analysis

Vulnerability information is stored in so called vulnerability databases (VDB), which collect known software vulnerabilities. Such databases comprehend large compilations of software weaknesses in a non-uniform manner. Well known databases are the VDB from SecurityFocus [9], advisories from Secunia [5], and the Open Source Vulnerability Database (OSVDB) [6], operated by the Open Security Foundation. Besides these known VDB from different providers, there is another important effort called the Common Vulnerabilities and Exposures list (CVE) [7], which is a meta vulnerability database. Its goal is to provide a common identifier for known weaknesses which can be used across various VDBs. Before 1999, each vulnerability database has its own name and it was difficult to detect when entries referred to the same weakness. With the help of CVE entries, vulnerabilities at least have a unique identifier.

For attack graph construction, up-to-date vulnerability information is crucial to provide high quality results. Automatic extraction of up-to-date vulnerability descriptions provide capabilities to build high quality attack graphs. As vulnerability descriptions are stored in semi-structured textual descriptions, automatic extraction is a challenging task. However, we address this task in this paper and present our approach and some preliminary results. Existing vulnerability databases are analyzed in terms of usability for attack graph construction. The information provided by these databases is unified by means of a data structure, which is later used for attack graph construction. Furthermore, a prototype is developed for automatic extraction of vulnerability descriptions from several databases. An analysis of the transformation shows between 70-90 percent of correctness in extraction from textual descriptions. The integration of the third party tool MulVAL [4] proofs that automatic extraction can be used to create meaningful attack graphs.

This paper is organized as follows. Section 2 provides a short analysis of existing vulnerability databases and description standards. Section 3 presents the way and design to extract existing vulnerability descriptions from vulnerability databases. A prototype implementation and some statistical results are described in Section 4 as a proof of our concept. Finally, we conclude the paper in Section 5.

## 2    Sources of Vulnerability Information

### 2.1    Comparison of Vulnerability Databases

Vulnerability information is available from basically two types of sources. On the one hand, commercial or non-profit organizations act as vulnerability providers, such as Secunia security advisories [5] or the Open Source Vulnerability Database [6]. On the other hand, vulnerability information is described with standardization efforts, for example the Common Vulnerabilities and Exposures list (CVE) [7]. In this section, the most popular commercial and non-profit vulnerability providers will be examined. A closer look is taken at the information each of them provides about a vulnerability and which of them may be useful in the context of attack graph construction. Not included are sources which focus primary on virus and exploit descriptions, because they rather describe how an attack is conducted instead of stating what the preconditions and postconditions are.

As commercial vulnerability database providers, DragonSoft (D.Soft)[1], Secunia [5], SecurityFocus (S.Focus) [9], Securiteam[2], and X-Force [10] are selected. As non-commercial providers, the Cooperative Vulnerability Database

---

[1] http://vdb.dragonsoft.com/

[2] http://www.securiteam.com/cves/

(CoopVDB)[3], the Department of Energy Cyber Incident Response Capability (DoE-CIRC)[4], the National Vulnerability Database (NVD) [11], the Open Source Vulnerability Database (OSVDB) [6], and the United States Computer Emergency Readiness Team (US-CERT) [12] have been chosen. Besides the US-CERT, many other regional computer security incident response teams, such as the Australian CERT AusCERT and the German CERT-Bund exist. The US-CERT was chosen as a representative.

| | D.Soft | S.Focus | Secunia | Securit. | X-Force | CoopVDB | DoE-CIRC | NVD | OSVDB | US-CERT |
|---|---|---|---|---|---|---|---|---|---|---|
| title | x | x | x | x | x | x | x | | x | x |
| description | x | x | x | x | x | x | x | x | x | x |
| range | x | x | x | x | x | x | | x | x | x |
| OS | x | x | x | x | x | x | x | x | x | x |
| software | | x | x | x | x | x | | x | x | x |
| critical | x | | x | x | x | | x | x | | x |
| impact | x | | x | x | x | x | x | x | x | x |
| authentication | | | | | | | | x | | |
| class | | x | | | | | | x | x | |
| access complexity | | | | | | | | x | | |
| format 2) | H | H | H | H | H | H | H | H, X | C, H, M, S, X | H |
| exploit | x | x | | | | x | | x | x | |
| solution status | x | x | x | x | x | x | x | x | x | x |
| solution | x | x | x | x | x | x | x | | x | |
| release date | x | x | x | x | x | | x | x | x | x |
| last update | | x | x | | | | | x | x | x |
| popularity | | | x | | | | | | x | |
| discovered by | | x | x | | | | | | x | x |
| vendor-specific ID | x | x | x | | x | | x | | x | x |
| CVE reference | x | x | x | x | x | x | x | x | x | x |
| CVSS reference | x | | | | x | | x | x | | |

**Fig. 2.** Comparison of Vulnerability Databases

Except for the NVD, all VDBs provide a title for a listed vulnerability as well as textual descriptions. Both are intended for human readers to grasp the context and impact of a weakness. Several VDBs have their own vendor-specific identifier for each vulnerability, but all of them also provide the corresponding CVE ID, the unified identifier. This allows to quickly identify information on the same subject provided by different vendors. All but the CoopVDB list the release date of a weakness, only half provide information on when a vulnerability was updated the last time. Especially the update information can be useful to decide whether previously gained knowledge may have become outdated. The Secunia as well as the OSVDB also give a popularity indicator, i.e., how often information on a specific weakness was requested. Four out of ten databases state who discovered a vulnerability. The range from which a vulnerability can be exploited is given by all but the DoE-CIRC database. The operating system on which a weakness may occur can be specified in all VDBs, whereas it is sometimes subsumed in a generic software section. On the other hand, Secunia and the CoopVDB are the only one having a distinct software section. Only four VDBs use Common Vulnerability Scoring System (CVSS) values to indicate how critical a weakness is, two more have their own benchmark. The impact is provided by all but SecurityFocus, where it is described in the title of an advisory. The NVD is the only one that

---

[3] https://cirdb.cerias.purdue.edu/coopvdb/public/
[4] http://doecirc.energy.gov/ciac/

also provides information about the complexity of an attack as well as whether some kind of authentication is required. Links to exploit descriptions are given by half of the here presented vulnerability databases, but all provide a status indicator to show if a weakness has been addressed with, for example, a software update. A description of mitigation measurements is given in all cases, except by the NVD and the DoE-CIRC.

Most vulnerability information of these sources are only available as HTML page descriptions. Only the NVD and the OSVDB provide this data in additional formats, such as XML. Last but not least, all vulnerability databases provide cross-references to entries in other VDBs as well as software vendors describing a weakness. HTML has the disadvantage of being more difficult for information extraction. Instead of data which is gathered in a single file, it is distributed over thousands of web pages. The references between VDB entries are helpful to find new information on vulnerabilities of HTML-based vulnerability databases. Without knowing the vendor-specific ID assigned to a weakness, it is often difficult to deduce the correct URL for it. With the provided links, these can be found easily in a web-crawling-like fashion. Also helpful in this process is the extraction of the CVE identifier which can serve as a validation that the correct entry has been found. The range is an important piece of information for attack graph construction, because it allows to determine whether network or local access to a machine is required. Also important to determine the preconditions of an attack is data on the targeted programs. Once a vulnerability scanner has identified the software configuration of a host, matching vulnerabilities can be inferred. Title and textual description of a vulnerability seem to be less useful on first sight, but it will be shown in Section 4 that this does not have to be true.

The impact of security advisories can be used to gather information on the postcondition of an attack. If it states, for example, 'gain root access' a violation at the highest level is given, providing the attacker with access to every given resource. Therefore, this means that all, confidentiality, integrity, and availability of the target system and the hosted programs and data are violated. A classification of an attack may provide useful input as well, since a class such as 'Code Injection' indicates an integrity violation. Furthermore, information whether some kind of authentication is required for an exploitation can be used to deduce precondition information. More explicit are vulnerability databases with CVSS values, such as X-Force or NVD. The CVSS provides a rating of the extend to which a vulnerability may violate CIA security goals. Instead of inferring these values from impact descriptions, CVSS values can be extracted.

Not explicitly described by any of the VDBs are affected programs and data of an attack. Currently, only programs required for an attack are defined. This information is provided only implicitly in impact and textual descriptions, such as 'root access' indicates that the operating system is affected. Exploit information are often given in the form of references to exploit implementations and therefore not regarded as useful for vulnerability information generation. Similar, the fact that a solution exists is by itself not helpful to deduce pre- or postconditions of an attack. Solution descriptions are often references to patches from the vendor of the vulnerable software. Thus, they provide merely more information than the list of affected programs. Release and popularity information as well as who discovered a weakness is of no affect to attack graph generation, but data on latest updates can be. It can be used to check for outdated vulnerability descriptions, for instance, if an additional security violation caused by a vulnerability is found. Both, access complexity and criticality of a vulnerability may provide useful input, if the corresponding attack graph tool models the attacker as well. But if the assumption is made that an attacker will exploit any given vulnerability, access complexity, and criticality can be disregarded.

The National Vulnerability Database (NVD) provides most of the presented vulnerability information. It also has the advantage of making this data available in a well-defined XML format, which alleviates the amount of work to implement a parser. Except for the OSVDB, all other vulnerability databases will require a web scraping approach to retrieve data. Another benefit of the NVD is the explicit inclusion of extensive CVSS information. This means no additional source must be parsed to extract this data. Additionally, the NVD refers to Open Vulnerability and Assessment Language (OVAL) descriptions, that is detailed characterizations of the software configuration which is vulnerable. For these reasons, the NVD should be chosen as the primary source of input for vulnerability information generation. Nevertheless, other vulnerability databases can not be disregarded. Especially for cross-verification of retrieved information an analysis of these VDBs will be helpful. Finally, although different databases provide the same type of information, it does not mean they provide them to the same extend.

## 2.2   Common Vulnerability Scoring System

The Common Vulnerability Scoring System (CVSS) [13] addresses the problem of incompatible vulnerability assessments. Based on different metrics, every vulnerability is evaluated and scored. Each vulnerability is attributed with values for base metrics, temporal metrics, and environmental metrics. Base metrics include access vector and access complexity information, the degree of Confidentiality, Integrity, and Availability (CIA) violations, and the number of required authentication steps. Temporal metrics note, for example, if a vulnerability has already been verified, and environmental metrics are specific to, for example, a company. The scoring is done by analysts who dissect a weakness. This means that different VDB vendors can possibly provide different CVSS scores for the same vulnerability.

Compared to the other information provided by vulnerability databases, the CVSS shows several similarities. The access vector is similar to the range attribute, and the CIA impacts are a formalization of the textual impact description given by other VDBs. Also, temporal metrics reflect information about the availability of exploits and solutions. Only the environmental metrics have no counterpart, since they are specific to each organization and their threat exposure. One could argue that this eliminates the need to extract the same information twice, but it will be shown in Section 4 that CVSS scores are not always conform with other facts given in security advisories.

Of the three possible metrics defined by the CVSS, only base metric information are provided by the National Vulnerability Database (NVD). The X-Force VDB additionally provides temporal information.

### 2.3   Open Vulnerability and Assessment Language

The Open Vulnerability and Assessment Language (OVAL) [8] allows to give detailed and structured description of configurations affected by vulnerabilities. In contrast to the CVSS scores, these vulnerability definitions do not contain information on the severity or locality, but on installed programs. Definitions are created and submitted by security experts. After a review by the OVAL team, these definitions are made public in the OVAL repository. An OVAL entry could be one of the following five types. *Vulnerability definitions* describe tests which determine the presence of vulnerabilities, *compliance definitions* are tests that determine whether the configuration settings of a system meets a specific security policy, *inventory definition* tests describe whether a specific piece of software is installed on the system, and *patch definition* tests determine whether a particular patch has been applied. Additionally, a *miscellaneous type* is available for any other test description not fitting in one of the first four categories. Of these types, we focus on work with vulnerability definitions. The other four can be useful to determine how a system is configured, but the former helps to describe requirements and effects of a specific vulnerability. OVAL definitions are based on XML schemes. A core schema defines a common set of XML elements which can be used by any OVAL definition. Additionally, component schemes are available for various systems such as Microsoft Windows or Sun Solaris, each containing element definitions specific to the corresponding system.

The topmost element of an OVAL definition is the definition element which has three attributes: an id, a version, and a class attribute. The id attribute gives a unique identifier for this definition in the scope of the OVAL namespace. The version indicates the number of times a definition was modified. As pointed out above, we will rely on vulnerability definitions which is reflected in the class attribute with the value 'vulnerability'. The first child element of a definition is the non-optional metadata element. It starts with a title describing the vulnerability definition. Next, the affected systems are enumerated. This enumeration may contain two different types, platform and product entries. Then, references can be given, e.g., to the CVE entry found at the MITRE website. Note that the reference element includes a source attribute and an id attribute. Since OVAL vulnerability definitions are based on CVE entries, it is possible to link a vulnerability definition to a specific CVE entry. Also required is the specification of a description element, which gives a human-readable description of a vulnerability. Not used in many definitions is the note element. It is therefore omitted from further considerations. The last element of a definition is of the type criteria, which can have three attributes: 'operator', 'negate', and 'comment'. The operator attribute defines boolean-like operators, such as $AND$ and $OR$. The negate attribute enables the unary boolean operator *NOT*. Finally, the comment attribute allows to describe the corresponding criteria.

A major advantage of OVAL definitions is their detail level. Compared to other vulnerability descriptions, system configurations specified with OVAL are structured and explicitly described. Additionally, the enumeration of affected platforms and programs allow to deduce not only software requirements for preconditions, but also the affected postcondition programs. This is an advantage over current attack graph approaches, which always assume that the attacked host will be affected by an exploit. But sometimes this assumption is not correct, and only the attacked program will suffer a loss of either confidentiality, integrity, or availability. Note that version definitions in criteria elements allow to define ranges of affected software versions. The SecurityFocus VDB, for example, explicitly lists all vulnerable versions, such as "Microsoft Windows XP Professional", "Microsoft Windows XP Professional SP1", "Microsoft Windows XP Professional SP2", and "Microsoft Windows XP Professional SP3", whereas for OVAL definitions it is sufficient to state "Microsoft Windows XP Professional less or equal SP3". A drawback of the existing vulnerability definitions is that they are usually provided by vendors and therefore sometimes too specific. The Unix-based Mail Transfer Agent sendmail[5] is vulnerable to remote code execution as specified in CVE entry 2006-0058[6]. But although sendmail can be installed and executed on any Unix-based operating system, the corresponding OVAL definition only refers to the Redhat distribution of Linux: "The operating system installed on the system is Red Hat Enterprise Linux 4 [..]." Finally, note that criteria definitions with operators need to be comparable.

# 3   Automatic Extraction of Vulnerability Information

## 3.1   A Data Model for Vulnerability Descriptions

To use vulnerability descriptions from different databases in attack graph construction, these descriptions need to be unified. We used a flexible and extensible data model to unify vulnerability descriptions of multiple vulnerability databases. The data model is capable to express vulnerability descriptions provided by vulnerability databases. The logical data model describes *system*, *influence*, and *range properties*. *System properties* describe states a system can be in, e.g., running programs, existing accounts, and existing databases. *Influence properties* describe the influence an attacker has on system properties by successful exploitation. *Range properties* describe the location from which an attacker can perform successful exploitation, e.g., local or remote. As depicted in Figure 3, a vulnerability requires a precondition and a postcondition, which can be represented by system properties. Two basic types are used for descriptions: *properties and sets*. *Properties* represent predicates and *sets* allow a grouping of properties based on boolean logic. Both types facilitate a simple evaluation

---

[5] http://www.sendmail.org/

[6] http://cve.mitre.org/cgi-bin/cvename.cgi?name=CVE-2006-0058

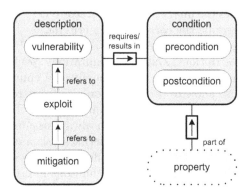

**Fig. 3.** Vulnerability Concept

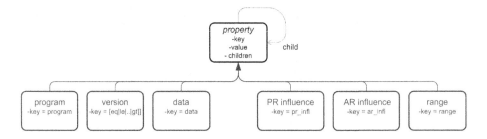

**Fig. 4.** Logical Data Model of System Properties

based on matching of True or False values. Finally, descriptions link different system states together, one as the requirement and the other as the result of an attack. Based on this properties and sets, we can flexibly describe many different system states.

*System properties* are characteristics and resources of a computer system which are considered relevant vulnerability information. Each system property describes one specific attribute of such a system, whereas properties are related to one another as depict in Figure 4. For example, the installed version of an application can be a system property. An application's version is meaningless if it cannot be linked to a certain application. Properties and their relations may change over time due to modifications, such that an application may be upgraded to a newer version. *System properties* can be found in two layers, the network layer and the software layer. The network layer describes properties of interconnected computers, such as network addresses and port numbers. The software layer describes properties of software systems, such as programs, data, and account information. We defined several different *system properties* which are useful to create attack graphs, such as network properties, host connectivity, programs, protocols, data, accounts, and others. To describe actions performed on systems, *influence properties* will be used. I*nfluence properties* describe the

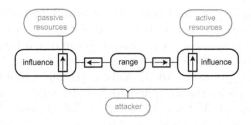

**Fig. 5.** Influence Properties

relationship between a potential attacker and system properties which represent computer resources (see Figure 5).

## 3.2   Extracting Textual Descriptions

Textual descriptions of facts cannot be neglected for an important reason. Many values of important attributes, for example the attack range and the impact, are described with a selection of English words. This is adequate for human readers who can interpret the meaning of these words, but will not be sufficient by itself for an application to put logic into it. What is needed instead is a mapping of these words to distinct, well-defined values which can be interpreted and compared by a program.

With the Common Vulnerability Scoring System, one such approach was already presented and considered to be valuable for vulnerability information generation. Whereas the CVSS targets at a common evaluation of impact a vulnerability can have, the Common Platform Enumeration (CPE)[7] creates a concerted naming convention for software applications. It is supported by the NVD, but fairly new and therefore currently not widely adopted. For this reason, current approaches have to rely on other means to extract comparable program names and versions. Nevertheless, future undertakings should consider the use of the CPE to obtain this information. We realized that names can be extracted fairly simple from OVAL criteria definition comments, because they follow a simple pattern. Take a look at the definition given in in CVE 2008-4250[14] for the Windows 2000 installation: **"Microsoft Windows 2000 SP4 or later is installed"**.

These descriptions always begin with the product name, followed by the version, and optionally a version relation stating if vulnerable versions of the described program are either *equal to, smaller than, smaller than or equal to, greater than,* or *greater or equal to* the given version. Therefore, an extraction of the program name can be done easily.

The program version can be extracted from OVAL definitions in the same fashion program names can be. But similarly, a comparable and distinct naming schema which at the same time provides unique identifiers for the corresponding

---

[7] http://cpe.mitre.org/

programs is missing. Again, the Common Platform Enumeration addresses this deficiency, but can currently not be used, because it has not been widely adopted, yet. Having a closer look at common version descriptions, it turns out that application versions are generally described with consecutive numerical, alphabetical, or alpha-numerical expressions. Additionally, major and minor versions are commonly used. Examples for this are Windows XP SP1, SP2, or SP3 or the Apache HTTP daemon version 1.3.41, 2.0.63, or 2.2.11. Taking into account that many programming languages rank the string "2.0.63" to be smaller than "2.2.11", distinct and comparable version numbers are already given in the form of textual descriptions.

The next relevant attribute that has to be retrieved in order to create attack descriptions is the range from which an attack can be conducted. If the range of a certain vulnerability cannot be retrieved from CVSS encoded data, another vulnerability database or the short textual description of a given CVE entry has to be used. As shown in section 2, almost all vulnerability databases provide a range evaluation that can be used as an alternative input. The provided values have to be mapped to values included in the data model. Additionally, most of these textual vulnerability descriptions contain phrases such as 'local user' or 'remote attacker' which can be used to identify the range from within an attack can take place. Comparing the results of this analysis to existing CVSS definitions for vulnerabilities discovered in 2008, 95% of the range values were identified correctly. The remaining 5% are due to unspecific attacker descriptions or incorrect CVSS assignments, such as the CVSS attack vector assigned to CVE-2008-0840 is 'local' although the vulnerability is described as **"Directory traversal vulnerability in Public Warehouse LightBlog 9.6 allows remote attackers to execute arbitrary local files"**.

Similar to range information, loss type data or impact descriptions are provided by most VDBs, but can also be extracted from CVSS entries. Again, a mapping for each vulnerability database has to be found which converts the given values to distinct and measurable values. It is important to create information which can be interpreted as both, preconditions and postconditions. For this reason, the notion of influence on passive and active resources is used. To convert the three CIA values confidentiality, integrity, and availability to either influence category, the following mapping is used. Confidentiality loss for passive resources is comparable to read access, integrity loss to write access, and availability loss to the deletion of data. Influence on active resources, that is influence on the input, the output or the existence of this resource is not affected by confidentiality loss. Confidentiality violations can be caused by active resources, but factually confidential are considered only passive resources. As a consequence, the disclosure of secret procedures can not be addressed by this mapping, which is justified by the fact that this case is not covered in common vulnerability databases. On the other hand, integrity loss is mapped to influence on the output of an active resource, and availability loss to the existence of an active resource.

## 4   Proof of Concept

We implemented a prototype to proof the applicability of automatic extraction from vulnerability databases. The prototype will use a designed data structure as an exchange format between components which extract information from various VDBs as well as components which output information for attack graph tools and related applications. As shown in Figure 6, the prototype is based on plugins: readers and writers. In the following, the extracting components will be referred to as readers, because they read information from a vulnerability database or some other source. Every reader is able to extract information from a specific data source. For example, an NVD reader is able to filter relevant attack information from the National Vulnerability Database (NVD) [11]. The counterpart of readers are writers, which output vulnerability information in different formats. Gathered data can be read by various source, e.g., attack graph tools or vulnerability analysis programs. Thus, it is reasonable to provide a writer for each target application.

The suggested design is based on plugins of readers and writers. Readers are able to read vulnerability information from a specific source, such as the NVD. Writers on the other hand can store vulnerability information, for example in a format interpretable by an attack graph tool. All plugins provide a simple interface and communicate based on a common data structure. This allows to link existing plugins into a tool chain and therefore convert vulnerability information from any source format for which a reader exists to any target format for which a writer exists. Finally, vulnerability information was transformed from vulnerability databases to serve as input for the MulVAL attack graph tool. Based on this, it was shown that sufficient data could be automatically transformed to identify an attack path in a company computer network, using the proposed

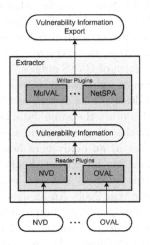

**Fig. 6.** Design of the Vulnerability Database Extractor

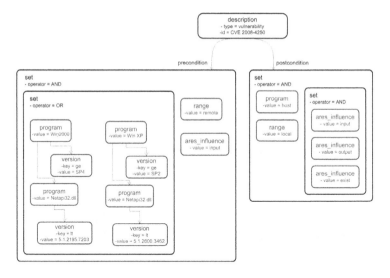

**Fig. 7.** Logical Data Model Example

data structure as an intermediary format. The internal data structure is based on sets and properties while sets can contain properties of further subsets, as shown in Figure 7.

Readers such as the NVD Reader or the OVAL Reader transform information from one XML representation into another XML representation, but the transformed information remains the same. The major benefit of this type of readers is the increased amount of available vulnerability information provided by a common vulnerability database which is based on the data structure used in the implementation. The CVE Reader on the other hand extracts information from textual descriptions of vulnerabilities. To be able to evaluate how much of the encoded information can be retrieved, it is useful to have a closer look at the extracted information. For this, the retrieved data will be compared to the data which is available in the form of CVSS entries. Those CVSS entries provide range and impact information of vulnerabilities in a standardized format. The NVD contains both, textual descriptions as well as CVSS values for all entries. Both information sets should contain the same data, therefore the comparison is based on these two sets. Note that this evaluation aims not at the evaluation of vulnerabilities itself, but rather at an analysis of how much of the information encoded in textual descriptions can be extracted correctly.

The data set on which this analysis is based consists of NVD entries for vulnerabilities identified in the year 2008. An analysis of the NVD vulnerability entries of the years 2006 and 2007 produces comparable and partially even better results. Figure 8 depicts the number of correctly identified attributes encoded in textual descriptions of vulnerabilities. The analyzed attributes are the range from which an attack can take place as well as which of the three security goals confidentiality, integrity, and availability can be violated by exploiting a

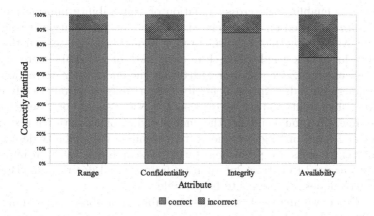

**Fig. 8.** Correctly Identified Attributes of Textual Description

vulnerability. The range information can be identified correctly in more than 90 percent of the cases, confidentiality violations in almost 82 percent of the cases, integrity violations in more than 85 percent, and availability violations in almost 75 percent of the analyzed descriptions.

When manually verifying the extracted information, a few interesting observations could be made, for example inconsistent CVSS classifications for same type of attacks. As this extraction process is based on a mapping of phrases such as "execute arbitrary code" to violations of security goals such as an integrity violation, not all phrase mappings matched the corresponding CVSS assessment. For example, vulnerabilities which lead to an arbitrary code execution are not necessarily categorized as integrity violations (e.g., CVE-2008-0387). Another example is the assessment of cross-site scripting vulnerabilities present in web applications. For the year 2008, 585 out of 602 cross-site scripting attacks are categorized as integrity violations, but only 72 percent of them as confidentiality loss and about 5 percent as availability loss. This assessment is at least disputable, because in the authors opinion being able to change the content and behavior of a website brings along a possible violation of both availability and confidentiality of the attacked service. Similar to these examples, a further analysis of how textual descriptions and CVSS entries correspond or contradict can be conducted, following up on the work done by Franqueira and van Keulen [15].

## 5    Conclusion

As MulVAL was the only available attack graph tool for this paper, we did not focus on additional writer plugins. In the future, writer plugins could be developed, providing further attack graph tools with the transformation capabilities of the proposed system. At the same time, reader plugins for additional vulnerability database could be developed, supplying a wider variety of attack information or enabling the cross-validation of vulnerability information from different sources. Based on the extended vulnerability information, which have been made available by the work presented, it should be evaluated whether new possibilities for attack graph research in particular, or vulnerability research in general have emerged. Because we realized that the information extraction from textual vulnerability description is feasible, algorithms to automatically create entries for vulnerability databases, such as the NVD, should be researched. At the same time, this would enhance the understanding of semantics of textual descriptions and could provide new means to a common vulnerability description. As the current results are promising, further research can focus on improving the algorithms used to extract information from textual descriptions. By using semantic techniques, this seems to be possible.

In this paper, we analyzed existing vulnerability databases concerning their usability in attack graph construction. The 10 most popular VDB providers were selected as the base for this evaluation. Most valuable attributes of vulnerability entries in this process include CVE identifiers, the impact of a vulnerability, the range from which an attack can be conducted, and the required or affected programs. The OVAL provides a framework to describe exploitable software configurations affected by a vulnerability. Similar to CVSS, OVAL is standardized and used by several organizations. For this paper, only vulnerability definitions are considered. Based on XML, such definitions consist of meta-data and criteria elements, whereas criteria elements are recursive and therefore allow configuration specifications at an arbitrary level of detail. Because important attributes, such as the attack range and the impact, are often described with a selection of English words, the interpretation of textual descriptions cannot be neglected. Not all information is available in CVSS format and OVAL definitions also rely on the use of English phrases. Nevertheless, it has been demonstrated that verbalization is often semi-formal and therefore easily parsable. Finally, a prototype implementation was analyzed in terms of correctness. The analyzed attributes are the range from which an attack can take place as well as which of the three security goals confidentiality, integrity, and availability can be violated by exploiting a vulnerability. The range information can be identified correctly in more than 90 percent of the cases, confidentiality violations in almost 82 percent of the cases, integrity violations in more than 85 percent, and availability violations in almost 75 percent of the analyzed descriptions. The extracted vulnerability information was used to construct attack graphs by means of the MulVAL tool.

# References

1. Schneier, B.: Attack Trees: Modeling Security Threats. Journal Dr. Dobb's Journal (December 1999), http://www.ddj.com/architect/184411129
2. Sheyner, O., Haines, J., Jha, S., Lippmann, R., Wing, J.M.: Automated Generation and Analysis of Attack Graphs. In: Proceedings of the 2002 IEEE Symposium on Security and Privacy (S&P 2002), pp. 273–284. IEEE Press, Washington (2002)

3. Noel, S., Jajodia, S.: Managing attack graph complexity through visual hierarchical aggregation. In: Proceedings of Workshop on Visualization and Data Mining for Computer Security (VizSEC/DMSEC 2004), pp. 109–118. ACM, Washington (2004)

4. Ou, X., Govindavajhala, S., Appel, A.: MulVAL: A Logic-based Network Security Analyzer. In: Proceedings of 14th USENIX Security Symposium, p. 8. USENIX Association, Baltimore (2005)

5. Secunia Advisories, http://secunia.com/advisories/ (accessed March 3, 2009)

6. OSV Database, Open source vulnerability database. OSVDB, http://osvdb.org/ (accessed March 2009)

7. Mitre Corporation, Common vulnerabilities and exposures. CVE, http://cve.mitre.org/ (accessed March 2009)

8. Mitre Corporation, Open Vulnerability and Assessment Language, OVAL, http://oval.mitre.org/ (accessed March 3, 2009)

9. SecurityFocus, Security Focus Bugtraq, http://www.securityfocus.com/ (accessed March 2009)

10. ISS, X-force, http://xforce.iss.net/ (accessed March 2009)

11. NIST, National Vulnerability Database, NVD, http://nvd.nist.gov/ (accessed March 3, 2009)

12. US CERT, US-CERT vulnerability notes database, http://www.kb.cert.org/vuls/ (accessed March 2009)

13. Mell, P., Scarfone, K., Romanosky, S.: A complete guide to the common vulnerability scoring system version 2.0 (2007), http://www.first.org/cvss/ (accessed March 3, 2009)

14. Mitre Corporation, CVE-2008-4250 (accessed March 2009)

15. Franqueira, V.N.L., van Keulen, M.: Analysis of the nist database towards the composition of vulnerabilities in attack scenarios. Technical Report TR-CTIT-08-08, University of Twente, Enschede (February 2008)

# Traitor Tracing without A Priori Bound on the Coalition Size

Hongxia Jin[1] and Serdar Pehlivanoglu[2,*]

[1] IBM Almaden Research Center
jin@us.ibm.com
[2] Computer Science and Engineering, University of Connecticut
Storrs, CT, USA
sep05009@cse.uconn.edu

**Abstract.** Traitor tracing is an essential mechanism for discouraging the piracy in digital content distribution. An adversarial model is identified as rebroadcasting the content encrypting keys or the content in the clear form. It is possible to fight against these piracy models by employing a fingerprinting code that gives a way to differentiate the encryption capability of each individual. We point three important characteristics of a fingerprinting code that affects its deployment in traitor tracing scheme against pirate rebroadcasting: (i) A robust fingerprinting code tolerates an adversary that chooses not to rebroadcast some messages. (ii) A tracing algorithm for fingerprinting code that does not require a priori upper-bound on coalition size to be successful in detecting a traitor. (iii) Extending the length of the fingerprinting code which refers to traitor-identification procedure of the code that doesn't depend on the length of the code or the distribution of the markings over the code.

We presented the first traitor tracing scheme with formal analysis of its success in traitor-identification that doesn't assume a priori bound on a traitor-coalition size while at the same time it is possible to extend the code without degrading the success of traitor identification due to non-extended part. This construction also supports the robustness without requiring a high pirate rebroadcasting threshold.

## 1 Introduction

Traitor tracing has been an active research area since its inception in [1]. It generally refers to a mechanism that detects the guilty users who have participated in a pirate attack when pirate evidences become available. Different types of traitor tracing schemes have been designed for different types of pirate attacks. One particular attack is called re-broadcasting attack (or anonymous attack) where the attackers re-distribute the content encrypting keys or the decrypted plain content to stay anonymous. When pirated copies of content or keys are recovered, traitor tracing scheme for re-broadcasting attack aims to detect the users who participated in the re-distribution of the pirated copies.

---

* Research is done as an Intern at Almaden Research Center.

P. Samarati et al. (Eds.): ISC 2009, LNCS 5735, pp. 234–241, 2009.

To defend against re-broadcasting attack, one naturally wants to build many versions of the content (typically using watermarking and different encryptions) and assign different users different copies of the content. The recovered pirated copies (as feedback) will link back to the original users who were assigned those versions. Typically a content is divided into multiple segments, each segment can have different versions created from different watermarking and different encryptions. There is a separate marking allocation table that specifies what marking version of each segment that a user should be assigned. While dynamic traitor tracing [2] allocates the versions in-real time, sequential traitor tracing of [3,4] suggests a pre-determined mark allocation table that is based on fingerprinting codes [5,6,7,8,9].

**Preliminaries**

*Fingerprinting Codes.* A codeword over an alphabet $Q$ is an $\ell$-tuple $x = (x_1, \ldots, x_\ell)$ where $x_i \in Q$ for $1 \le i \le \ell$. If a code $\mathcal{W} = \{a^1, \ldots, a^n\} \subseteq Q^\ell$ consists of $n$ codewords, then we will call it an $(\ell, n, q)$-code where $|Q| = q$.

For any set of indices $T \subseteq [n]$, we define the set of pirate codewords due to $T$, denoted $\mathbf{desc}(\mathcal{C}_T)$ by

$$\mathbf{desc}(\mathcal{C}_T) = \{x \in Q^\ell : x_i \in \{a_i : a \in \mathcal{C}_T\}, 1 \le i \le \ell\}$$

where $\mathcal{C}_T = \{a^j \mid j \in T\}$ and $x_i, a_i$ are the $i$-th symbol of the codewords.

In other words, the descendent set $\mathbf{desc}(\mathcal{C}_T)$ is the set of pirate codewords that could be produced by the coalition of the set $T$. We are also interested in partial codewords which are missing some positions in the pirate codeword. We will denote the missing part by $\perp$. Formally, we define $\mathbf{desc}^\perp(\mathcal{C}_T)$ accordingly:

$$\mathbf{desc}^\perp(\mathcal{C}_T) = \{x \in \{\{\perp\} \cup Q\}^\ell : x_i \in \{\perp\} \cup \{a_i : a \in \mathcal{C}_T\}, 1 \le i \le \ell\}$$

For the purpose of simplifying the notation, later in the paper, a pirate codeword due to the set $T$ of a length $i < \ell$ will refer to codeword in $\mathbf{desc}^\perp(\mathcal{C}_T)$ whose last $\ell - i$ positions are filled with $\perp$.

A fingerprinting code is a pair of algorithms (CodeGen, Tracing) that generates a code for which it is possible to detect piracy:

- CodeGen is an algorithm, given input $(n, \nu, q)$ with a security parameter $\nu = \log(\frac{1}{\epsilon})$ for some small $\epsilon$, that samples a pair $(\mathcal{C}, tk) \leftarrow \mathsf{CodeGen}(n, \nu, q)$ where $\mathcal{C}$ is an $(\ell, n, q)$-code defined over an alphabet $Q$, and tracing key $tk$ is some auxiliary information to be used for tracing that may be empty.
- Tracing is a deterministic algorithm that takes input the tracing key $tk$ as well as a pair $(c, T)$, where $c \in \{\{\perp\} \cup Q\}^\ell$ and $T \subseteq [n]$ , and it outputs a codeword-index $t \in [n] \setminus T$ or fails. The fingerprinting code is called open if $tk$ is empty.

*Sequential Traitor Tracing Scheme:* A sequential traitor tracing scheme against pirate rebroadcasting is based on an underlying fingerprinting code (CodeGen, Identify) as a marking table for matching the content-versions to the receivers. A sequential traitor tracing scheme consists of the following algorithms:

- KeyDist: Given $(n, \nu, q)$ it first produces a $(\ell, n, q)$-code $\mathcal{W} = \{w^1, \ldots, w^n\}$ where $(\mathcal{W}, tk) \leftarrow \mathsf{Codegen}(n, \nu, q)$ over an alphabet $Q$. It, then, produces a collection of keys $ek = \{k_j^i\}_{j \in [q]}^{i \in [\ell]} \subseteq K$. The user key $sk_u$ is set to $\langle k_{w_1^u}^1, k_{w_2^u}^2, \ldots, k_{w_\ell^u}^\ell \rangle$ where $w^u = \langle w_1^u, w_2^u, \ldots, w_\ell^u \rangle \in \mathcal{W}$.
- Transmit: Given a message $m$ and transmission state $i \in \{1, \ldots, \ell\}$; it transmits the encryption of the message $M$ with $ek$ by using a symmetric encryption scheme $(\mathsf{E}, \mathsf{D})$:

$$\langle i, \mathsf{E}_{k_1^i}(\mathsf{Emb}(1, m)), \mathsf{E}_{k_2^i}(\mathsf{Emb}(2, m)), \ldots, \mathsf{E}_{k_q^i}(\mathsf{Emb}(q, m)) \rangle$$

where $\mathsf{Emb} : \{1, \ldots, q\} \times \mathsf{M} \to \mathsf{M}$ is a watermarking embedding algorithm for which there exists a reading algorithm $\mathsf{Read}$ such that $\mathsf{Read}(\mathsf{Emb}(j, m)) = j$ for all $m \in \mathsf{M}$ and for $j = 1, \ldots, q$.
- Receive: Given the key-material $sk_u = \langle k_{w_1^u}^1, k_{w_2^u}^2, \ldots, k_{w_\ell^u}^\ell \rangle$ for some $u \in [n]$ and a transmission of the form:

$$\langle i, c_1, c_2, \ldots, c_q \rangle$$

it returns $\mathsf{e} = \mathsf{D}_{k_{w_i^u}^i}(c_{w_i^u})$; observe that $\mathsf{Read}(\mathsf{e}) = w_i^u$ holds.

**The Model**

*Tracing Pirate Rebroadcast.* We will consider a sequence of content transmission, specifically denote the number of messages in the sequence by $\ell$ which amounts to the length of the code generated by the KeyDist algorithm. Each message has $q$ different versions created by a $q$-ary robust watermarking system. The content provider sends each different version of all messages and observes the feedback coming from pirate rebroadcast. Figure 1 is the sequential traitor tracing scheme against pirate rebroadcasting. The scheme employs a fingerprinting code (CodeGen, Tracing). Here, the figure illustrates a pirate rebroadcasting of the clear content, it should be noted that the rebroadcasting of content key can also be considered in this same model.

Denoting the set of traitors detected on the $i$-th message by $\mathsf{T}_i$, the feedbacks observed in the $(i + 1)$-th message are consistent with the versions assigned to the set of traitors $\mathsf{T} \setminus (\bigcup_{k=0}^i \mathsf{T}_k)$. In this model, after detecting a traitor, the feedbacks due to that traitor are removed from the feedback sequence, i.e. we denote this partial feedback sequence by $\mathsf{p}^{(i)}$. A feedback sequence is possible if it can be generated by a set of traitors despite the fact that some of its traitors are detected in early phases. We define the feedback sequence as $\mathsf{T}$-possible, if that sequence is a possible outcome of the set of traitors in this model.

**Definition 1.** *Denoting the set of traitors by $\mathsf{T} \subseteq [n]$, a pirate rebroadcast $\mathsf{p} = \langle \mathsf{p}_1, \ldots, \mathsf{p}_\ell \rangle \in \mathbf{desc}(\mathcal{C}_\mathsf{T})$ against a sequential traitor tracing based on a (CodeGen, Tracing) fingerprinting code is called $\mathsf{T}$-possible, if there exists a collection $\cup_{k=1}^\ell \mathsf{T}_k \subseteq \mathsf{T}$ so that*

$$\mathsf{p}_i \in \{w_i^j \mid j \in \mathsf{T} \setminus (\bigcup_{k=1}^{i-1} \mathsf{T}_k)\} \qquad \mathsf{T}_i = \mathsf{Tracing}(\mathsf{p}^{(i)}, \cup_{k=1}^{i-1} \mathsf{T}_k)$$

Tracing Pirate Rebroadcast(A sequence of $\ell$ messages)
1.    $(\mathcal{W}, tk, ek, sk_1, \ldots, sk_n) \leftarrow \mathsf{KeyDist}(n, \nu, q)$
2.    Set $\mathsf{p}$ empty and suppose $\mathcal{W} = \{\mathsf{w}^1, \ldots, \mathsf{w}^n\}$.
3.    for $ctr = 1$ to $\ell$ do
4.        Broadcast $c_{ctr} = \langle ctr, c_1, c_2, \ldots, c_q \rangle \leftarrow \mathsf{Transmit}(ek, m_{ctr})$
5.        Observe a pirate rebroadcast $\mathsf{Emb}(a, m)$ for some $a \in \{1, \ldots, q\}$
6.        Set $\mathsf{p}_{ctr} = a$
7.        Append $\mathsf{p}_{ctr}$ to the pirate rebroadcast $\mathsf{p}$
8.        Let $\mathsf{T}' = \mathsf{Tracing}(\mathsf{p}, \mathsf{T})$
9.        Set $\mathsf{T} = \mathsf{T} \cup \mathsf{T}'$
10.       for $j = 1$ to $ctr$ do
11.           Set $\mathsf{p}_j = \bot$ if there is a traitor-index $t \in \mathsf{T}'$ for which $\mathsf{p}_j = \mathsf{w}_j^t$ holds.

**Fig. 1.** Traitor tracing in pirate rebroadcasting

holds for $0 \leq i \leq \ell$. Here, $\mathsf{T}_i$ refers to the set of traitors identified right after observing the $i$-th pirate rebroadcast. $\mathsf{p}^{(i)} \in \mathbf{desc}^{\perp}(\mathcal{C}_{\mathsf{T}^{(i)}})$ where $\mathsf{T}^{(i)} = \mathsf{T} \setminus (\bigcup_{k=0}^{i-1} \mathsf{T}_k)$ is the set of undetected traitors. $\mathsf{p}^{(i)}$ is defined as follows:

$$\mathsf{p}_k^{(i)} = \begin{cases} \mathsf{p}_i & \text{if } k = i \\ \bot & k < i \text{ and } \mathsf{p}_k^{(i-1)} \in \{\mathsf{w}_k^j \mid j \in \mathsf{T}_{i-1}\} \\ \mathsf{p}_k^{(i-1)} & \text{otherwise} \end{cases}$$

We define $\mathsf{Tracing}(\mathsf{p}) = \cup_{k=1}^{\ell} \mathsf{T}_k$ as the set of traitors identified.

Finally, we say a sequential traitor tracing scheme based on a fingerprinting code (CodeGen, Tracing) is $w$-traceability scheme if for any $|\mathsf{T}| \leq w$ it holds that $\mathsf{Tracing}(\mathsf{p}) = \mathsf{T}$ holds for all $\mathsf{T}$-possible pirate rebroadcast $\mathsf{p} \in \mathbf{desc}(\mathcal{C}_{\mathsf{T}})$.

## Remarks on Traceability:

1. REVOKING EARLY DETECTED TRAITORS: If a user-index $j \in [n]$ is detected as a traitor index after observing a feedback on the $i$-th message, then from message $i + 1$, his reception of version $\mathsf{w}_{i'}^j$, for $i' > i$, will be blocked. This is the same model that have been disposed by [2,10,3,4]. Furthermore, [11] illustrates the feasibility of revoking a user while tracing.

2. PIRATE REBROADCASTING THRESHOLD: The above model requires the existence of feedback for each message. It is desirable to tolerate a stronger adversary that omits rebroadcasting some messages. This is achievable by employing a robust fingerprinting code (e.g. [12]). A fingerprinting code that doesn't support such robustness will either fail in being successful or requires the adversary to rebroadcast with a high rebroadcasting threshold, i.e. at least $1 - \frac{\ell-1}{\ell} + \alpha$ probability for some nonnegligible $\alpha$.

3. TRAITOR COALITION SIZE: In general a fingerprinting code is said to be successful in identifying a traitor under a condition which bounds the coalition

size. For instance, a traitor tracing scheme based on a $w$-TA fingerprinting codes require the coalition size to be not bigger than $w$. It is desirable to have a scheme that doesn't depend on such a priori bound on coalition size.

4. EXTENDABLE TRACING: Once a fingerprinting code of length $\ell$ is sampled by a KeyDist algorithm, the tracing will terminate after $\ell$ messages by outputting a subset of traitor coalition. To continue identifying other undetected traitors, the tracing will restart with a new fingerprinting code. We say a fingerprinting code supports extendable tracing if the next stage in tracing can take advantage of the information available from previous tracing steps.

**Our Results**

In this work, we gave an improvement on the conversion of traceability codes based on an error-correcting code into the sequential tracing schemes. More specifically, [4] states that any $c$-TA error correcting code implies a sequential $c$-tracing scheme that is capable of tracing back all traitors in a coalition of size $c$. We show that it is possible to construct a sequential $c\sqrt{2}$-tracing scheme by employing a $c$-TA code.

We, also, presented a traitor tracing scheme based on a fingerprinting code without a priori bound on the traitor-coalition size. It is also possible to extend this code without harming the traitor-identifications made so far. This construction is also immune to any adversarial action that drops/omits some rebroadcasts and doesn't require any threshold.

To the best of our knowledge, the only traitor tracing scheme without a priori bound on coalition size is employed by AACS standard [13] and presented in [10,14]. But their traitor detection scheme is very different from ours. While our scheme follows the sequential one by one traitor detection paradigm as illustrated in Figure 1, their scheme aims to detect the entire guilty coalition together.

## 2    Improving the Construction of Sequential Tracing Schemes Out of Error Correcting Codes

[4] formally models the sequential traitor tracing against pirate rebroadcast. The authors also state a general construction that is capable of converting a $c$-TA code based on a linear error-correcting code into a sequential $c$-tracing scheme. In this subsection, we will improve the analysis of this conversion which would yield a stronger tracing scheme that is capable of tracing a factor of $\sqrt{2}$ more than the former conversion.

**Theorem 1.** *Let* $\mathcal{W} \leftarrow \mathsf{CodeGen}_E(n, \nu, q)$ *be an* $(\ell, n, q)_d$-*error correcting code that satisfies*

$$d \geq (1 - \frac{2}{c(c+1)})\ell + \frac{2}{c+1}$$

*The sequential traitor tracing scheme based on* $(\mathsf{CodeGen}_E, \mathsf{Tracing}_c)$ *is* $c$-*traceability scheme where* $\mathsf{Tracing}_c$ *is defined as in Figure 2.*

Tracing$_c$(Pirate Rebroadcast p, detected traitors T)
1.    Let $s_j = |\{i \mid \mathsf{p}_i = \mathsf{w}_i^j\}|$
2.    Reorder the users so that $s_i > s_{i+1}$ for $i \geq 1$ holds
3.    set $k = 1$ and $t = |\mathsf{T}|$
4.    repeat
5.      If $s_k < (c - t - k + 1) \cdot (\ell - d) + 1$
6.        then break
7.      k=k+1
8.    until break
9.    if $k \geq 2$ then output $\mathsf{T}' = \{u_1, \ldots, u_{k-1}\}$ where $u_i$ has score $s_i$

**Fig. 2.** An improved Sequential Traitor detection algorithm

The proof of the theorem relies on the following observation: the $i$-th detected user is an actual traitor with at most $(c + 1 - i)(\ell - d) + 1$ pirate rebroadcasting for $i = 1, \ldots, c$. This can be proven by induction; assuming that $i$-th detected user is not a traitor, then there would be a traitor among the undetected $c - i + 1$ traitors who has at least $\ell - d + 1$ overlaps which contradicts with the fact that the hamming distance of the code is given as $d$.Talking with the notation given in [4], it is possible to construct sequential $c\sqrt{2}$-traceability schemes by employing $c$-TA codes. The proof of the corollary is straightforward. A $c$-TA code satisfies the equation $d \geq (1 - \frac{1}{c^2})\ell + \frac{1}{c}$

**Corollary 1.** *Given a $c$-TA error correcting code, the* Tracing *given in Figure 2 is capable of tracing up to $c\sqrt{2} - 1$ traitors for the same error correcting code.*

## 3   Extendable Fingerprinting Code without a Priori Bound on Coalition Size

We will present, now, an extendable secret fingerprinting code without a priori bound on coalition size. The CodeGen$_R(n, \nu, q)$ algorithm creates a code $\mathcal{C}$ whose codewords are sampled from the codeword-space $Q^\ell$ randomly, i.e. for all $y \in Q$ it holds that $\mathbf{Pr}[c_i = y] = \frac{1}{q}$ for any $c \in \mathcal{C}$ and $i = 1, \ldots, \ell$. We will denote the code constructed randomly by $\mathcal{C}_R$. The corresponding tracing algorithm is given in Figure 3. After a feedback observed in the system, we score each user with the number of overlaps between the pirate rebroadcast and its unique codeword. The tracing algorithm checks the highest score: if it exceeds a threshold then the user is accused as a traitor. This threshold depends on the length of the pirate rebroadcast (the rebroadcast due to the undetected traitors), size of code and the size of the marking alphabet. If the user with highest score is accused, then the next user with highest score is checked if it exceeds the new threshold. This procedure continues until a user in the order found to be below the updated threshold. We discuss the correctness of the Tracing$_R$ algorithm in Theorem 2.

---

Tracing$_R$(Pirate Rebroadcast p, disabled set T)
1.    Let $l$ be the actual size(without $\perp$) of the rebroadcast p
2.    Let $s_j = |\{i \mid p_i = w_i^j\}|$
3.    Reorder the users so that $s_i > s_{i+1}$ for $i \geq 1$ holds
4.    set $k = 1$ and $s_0 = 0$
5.    repeat
6.        set $threshold = \log_{\frac{q \cdot c_k}{(l - \sum_{i=0}^{k-1} c_i) \cdot e}} \frac{n}{\epsilon}$
7.        If $s_k < threshold$
8.            then break
9.            k=k+1
10.    until break
11.    if $k \geq 2$ then output $\{u_1, \ldots, u_{k-1}\}$ where $u_i$ has score $s_i$

---

Fig. 3. Traitor detection algorithm without a priori bound on the coalition size

**Theorem 2.** *Consider a secret $(\ell, N, q)$-code $\mathcal{C}_R$ that is constructed randomly, sampled uniformly from the codeword space $Q^\ell$. Its corresponding Tracing$_R$ algorithm is given in Figure 3. If the Tracing$_R$ algorithm outputs a set of size $t$ upon detecting a pirate rebroadcast p, then the output set is a subset of traitors with probability at least $1 - t \cdot \epsilon$.*

The proof of the theorem uses a Chernoff bound very similar to the computations in [1] where the probability of a particular user having a score of *threshold* out of $l$ positions is computed. The analysis will show that if the score exceeds the threshold given in Figure 3, then the accused user is among the traitor coalition with high probability. Here, it should be noted that the tracing key *tk* is set as the code $\mathcal{C}_R$, i.e. the code is secret code.

Observe that the tracing algorithm given in Figure 3 does not require any a priori bound on the coalition size. Overall, the tracing algorithm will be able to disable all traitors after sufficient number of feedbacks. Of course this number depends on the coalition size, the below theorem relates the required number of feedbacks to detect all traitors. Note that the relation doesn't affect the deployment the tracing algorithm, meaningly the algorithm succeeds without any assumption on the coalition size. The relation does only states how successful can a traitor coalition be in generating pirate rebroadcasts.

**Theorem 3.** *Consider a secret $(\ell, N, q)$-code $\mathcal{C}_R$ that is constructed randomly, sampled uniformly from the codeword space $Q^\ell$. Its corresponding Tracing$_R$ algorithm is given in Figure 3. There is no T-possible pirate rebroadcast p with $|T| = t$ and error rate less than $t\epsilon$ for $\ell \geq t \log_{\frac{q}{e \cdot t}} \frac{n}{\epsilon}$.*

The proof of the theorem can be observed as follows: At any time the $i$-th detected user would have a score (number of overlaps with the pirate rebroadcast)

of $s_i = \log_{\frac{q\alpha_i}{e}} n/\epsilon$ exceeding the threshold where $\alpha_i$ corresponds to ratio of the score of that particular user to the partial pirate rebroadcast due to the undetected traitors. Since, this user has the highest score among the other undetected traitors, it satisfies that $\alpha_i \geq 1/t$. Hence, it holds that $s_i < \log_{\frac{q}{e \cdot t}} n/\epsilon$.

In general; the length of the pirate rebroadcast due to the traitor coalition T would be bounded by the below formula:$\ell \leq \sum_{i=1}^{t} s_i \leq t \log_{\frac{q}{e \cdot t}} \frac{n}{\epsilon}$

*Properties of the Code.* Not only, the code doesn't depend on a priori bound for the coalition size but also the tracing algorithm given in Figure 3 does not depend on the length of the actual code. Moreover, each position of the code can be considered as an independent distribution from other positions in the code $\mathcal{C}_R$. This gives us an opportunity to extend the length of the fingerprinting code without affecting the success of tracing algorithm, and even tracing would continue smoothly. Moreover, the code supports a low pirate rebroadcasting threshold.

# References

1. Chor, B., Fiat, A., Naor, M.: Tracing traitors. In: Desmedt, Y.G. (ed.) CRYPTO 1994. LNCS, vol. 839, pp. 257–270. Springer, Heidelberg (1994)
2. Fiat, A., Tassa, T.: Dynamic traitor tracing. In: Wiener, M. (ed.) CRYPTO 1999. LNCS, vol. 1666, pp. 354–371. Springer, Heidelberg (1999)
3. Safavi-Naini, R., Wang, Y.: Sequential traitor tracing. In: Bellare, M. (ed.) CRYPTO 2000. LNCS, vol. 1880, pp. 316–332. Springer, Heidelberg (2000)
4. Safavi-Naini, R., Wang, Y.: Sequential traitor tracing. IEEE Transactions on Information Theory 49(5), 1319–1326 (2003)
5. Boneh, D., Shaw, J.: Collusion-secure fingerprinting for digital data (extended abstract). In: Coppersmith, D. (ed.) CRYPTO 1995. LNCS, vol. 963, pp. 452–465. Springer, Heidelberg (1995)
6. Le, T.V., Burmester, M., Hu, J.: Short c-secure fingerprinting codes. In: Boyd, C., Mao, W. (eds.) ISC 2003. LNCS, vol. 2851, pp. 422–427. Springer, Heidelberg (2003)
7. Staddon, J., Stinson, D.R., Wei, R.: Combinatorial properties of frameproof and traceability codes. IEEE Transactions on Information Theory 47(3), 1042–1049 (2001)
8. Stinson, D.R., Wei, R.: Combinatorial properties and constructions of traceability schemes and frameproof codes. SIAM J. Discrete Math. 11(1), 41–53 (1998)
9. Tardos, G.: Optimal probabilistic fingerprint codes. In: STOC, pp. 116–125. ACM, New York (2003)
10. Jin, H., Lotspiech, J.: Renewable traitor tracing: A trace-revoke-trace system for anonymous attack. In: Biskup, J., López, J. (eds.) ESORICS 2007. LNCS, vol. 4734, pp. 563–577. Springer, Heidelberg (2007)
11. Kiayias, A., Pehlivanoglu, S.: Tracing and revoking pirate rebroadcasts. In: ACNS, pp. 253–271 (2009)
12. Boneh, D., Naor, M.: Traitor tracing with constant size ciphertext. In: ACM Conference on Computer and Communications Security, pp. 501–510 (2008)
13. AACS Specifications (2006), http://www.aacsla.com/specifications/
14. Jin, H., Lotspiech, J., Megiddo, N.: Efficient coalition detection in traitor tracing. In: SEC, pp. 365–380 (2008)

# SISR – A New Model for Epidemic Spreading of Electronic Threats

Boris Rozenberg[1,2], Ehud Gudes[1,2], and Yuval Elovici[1,3]

[1] Deutche Telekom Laboratories at BGU
[2] Department of Computer Science
[3] Department of Information System Engineering,
Ben Gurion University, Beer Sheva 84105, Israel

**Abstract.** Epidemic spreading in complex networks has received much attention in recent years. Previous research identified a propagation scenario of electronic threats which has not been described by any of the existing analytical models. In this scenario an infected node instead of being removed contributes to the infection spreading upon the reinfection attempt (for example, Sober, Sobig, and Mydoom Worms). In this paper we formally define and describe analytically a new model, Susceptible-Infected-Suspended-Reinfected (SISR), which complies with this scenario of epidemic spreading in both homogeneous and complex networks. We then evaluate the model by comparing it to the SIR model and by comparing its estimations with simulation results.

## 1   Introduction

Modeling worm spreading is a major element of research into worm detection. It's done by using various simulation tools or analytical models. The primary strength of the analytical models is computational efficiency - they can be applied on networks of millions of hosts. A practical use of such models was also shown recently in [7], where authors describe a general multi-agent architecture which is able to detect the existence of worms by computing their propagation gradient and comparing it to the analytical model predictions, thus showing the usefulness of such a model. In this paper we focus on the worm propagation over email social networks. This kind of worm spreads via infected email messages [3]. The worm may be in the form of an attachment or the email may contain a link to an infected website. However, in both cases email is the vehicle. In the first case the worm will be activated when the user clicks on the attachment. In the second case the worm will be activated when the user clicks on the link leading to the infected site. Once activated, the worm infects the victim machine (install a backdoor for example), harvests email addresses from it and sends itself to all obtained addresses (machine's neighbors). In recent years a lot of new email worms have been discovered. The list includes Storm worm, Stration worm, Nugache Worm, Warezov worm and others [3]. This shows that the problem is still current and a serious one. In [6], Zou et al presented two main strategies of email worm propagation: reinfection strategy and nonreinfection strategy. Reinfection means

P. Samarati et al. (Eds.): ISC 2009, LNCS 5735, pp. 242–249, 2009.

that if some already infected user receives the worm instance again and open it, the worm will send itself again to all user's neighbors. Nonreinfection means that the infected user sends the worm to his neighbors only once. Some email worms belong to the reinfection type such as Sober, Sobig, Mydoom [3,6], while others exhibit the nonreinfection behavior (Melissa, Netsky, Swen [3,6]). The second strategy (nonreinfection) is described by the SIR analytical model, while the first one is currently not described analytically by any model. In this paper we present a new analytical model, Susceptible-Infected-Suspended-Reinfected (SISR), that describes the reinfection strategy of worm propagation. The rest of the paper is structured as follows: Section 2 reviews the SIR model of epidemic propagation in homogeneous and complex networks, Section 3 presents our new models, Section 4 describes the evaluation of the new models and Section 5 concludes the paper.

## 2   Background

### 2.1   The SIR Propagation Model

In the SIR epidemic propagation model [1], each individual can be in Susceptible, Infected or Removed state. Susceptible individuals become infected with probability if at least one of the neighbors is infected. Infected individuals become Removed with probability one (other probabilities can be considered). This model describes the spreading of those kinds of infections for which exists a permanent immunity or infection kills the infected individuals. In both cases, the individual in the Removed state can not be infected again and cannot infect others. This analytical model can be used to analyze the propagation of email worms that employ the nonreinfection propagation strategy only.

**The SIR model for homogeneous networks.** For homogeneous networks, the SIR model can be described by the following four equations [4]:

$$\rho(t) + S(t) + R(t) = 1 \tag{1}$$

$$\frac{d\rho(t)}{dt} = -\rho(t) + \lambda k \rho(t) S(t) \tag{2}$$

$$\frac{dS(t)}{dt} = -\rho(t) - \lambda k \rho(t) S(t) \tag{3}$$

$$\frac{dR(t)}{dt} = \rho(t) \tag{4}$$

where $\rho(t)$, $S(t)$ and $R(t)$ are the densities of infected, susceptible, and removed individuals at time $t$, respectively.

**The SIR model for complex networks.** Moreno et al [4] have presented the Susceptible-Infected-Removed (SIR) model that describes the dynamics of epidemic spreading in the complex networks. The model is represented by the following equations:

$$\rho_k(t) + S_k(t) + R_k(t) = 1 \tag{5}$$

$$\frac{d\rho_k(t)}{dt} = -\rho_k(t) + \lambda k S_k(t)\Theta(t) \tag{6}$$

$$\frac{dS_k(t)}{dt} = -\lambda_k S_k(t)\Theta(t) \tag{7}$$

$$\frac{dR_k(t)}{dt} = \rho_k(t) \tag{8}$$

$$\Theta(t) = \frac{\sum_k (k-1)P(k)\rho_k(t)}{\sum_k kP(k)} \tag{9}$$

where $\rho_k(t)$, $S_k(t)$ and $R_k(t)$ are the densities of infected, susceptible, and removed nodes of degree $k$ at time $t$, respectively, $P(k)$ is the fraction of nodes with degree $k$ and $\lambda$ is the probability that a susceptible node is infected by one infected neighbor. The factor $\Theta(t)$ gives a probability that any given link leads to an infected individual [5,2].

## 3   The SISR Propagation Model

In this section we describe a new model of epidemic propagation. In this model (see Fig. 1), there are four possible states for each individual in the network: Susceptible, Infected, Suspended and Reinfected (SISR). As in the SIR model, susceptible individuals become infected with probability $\lambda$ if at least one of their neighbors is infected. Infected individuals try to infect their neighbors. In contrast to the SIR model, infected individuals do not move to the Removed state, but to the Suspended state with probability one (other probabilities can be considered). Individuals in the Suspended state are infected and can contribute to the propagation upon reinfection. Thus, upon reinfection attempt, individuals in the Suspended state move to the Reinfected state with probability $\lambda$, try to infect their neighbors and move back to the Suspended state with probability one (other probabilities can be considered). Note that countermeasures are not considered in this study (e. g., application of antivirus software, etc.). When we say that some node is in the Suspended state, we mean that this node became suspended due to the nature of the worm and not due to the countermeasures application (from the propagation point of view, infected node passes to the

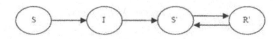

**Fig. 1.** SISR propagation model

Suspended state immediately after the sending of infectious message to all its neighbors).

The SISR model describes closely the reinfection propagation strategy of email worms, which till now were modeled by the SIR model. Following subsections give a detailed description of the SISR model for homogeneous and complex networks.

### 3.1 SISR Model for Homogeneous Networks

For homogeneous networks, the SISR model can be described by the following five equations:

$$\rho(t) + S(t) + S'(t) + R'(t) = 1 \tag{10}$$

$$\frac{d\rho(t)}{dt} = -\rho(t) + \lambda k(\rho(t) + R'(t))S(t) \tag{11}$$

$$\frac{dS(t)}{dt} = -\lambda k(\rho(t) + R'(t))S(t) \tag{12}$$

$$\frac{dS'(t)}{dt} = \rho(t) \tag{13}$$

$$\frac{dR'(t)}{dt} = -R'(t) + \lambda k(\rho(t) + R'(t))S'(t) \tag{14}$$

where $\rho(t)$, $S(t)$, $S'(t)$ and $R'(t)$ are the densities of infected, susceptible, suspended and reinfected individuals at time $t$, respectively. These equations can be explained as follows: *susceptible* individuals become *infected* with probability proportional to the density of *infectious* and *susceptible* individuals. The density of currently *infectious* nodes is given by the term $(\rho(t) + R'(t))$. *Infected* individuals become *suspended* with probability one (other probabilities can be considered). Equation (14) states that *suspended* individuals become *reinfected* with probability proportional to the density of currently *infectious* and *suspended* individuals.

### 3.2 SISR Model for Complex Networks

For complex networks, the SISR model is described by the following equations:

$$\rho_k(t) + S_k(t) + S'_k(t) + R'_k(t) = 1 \tag{15}$$

$$\frac{d\rho_k(t)}{dt} = -\rho_k(t) + \lambda k S_k(t)\Theta(t) \qquad (16)$$

$$\frac{dS_k(t)}{dt} = -\lambda_k S_k(t)\Theta(t) \qquad (17)$$

$$\frac{dS'_k(t)}{dt} = \rho_k(t) \qquad (18)$$

$$\frac{dR'_k(t)}{dt} = -R'_k(t) + \lambda k S'_k(t)\Theta(t) \qquad (19)$$

$$\Theta(t) = \frac{\sum_k (k-1)P(k)(\rho_k(t) + R'_k(t))}{\sum_k kP(k)} \qquad (20)$$

where $\rho_k(t)$, $S_k(t)$, $S'_k(t)$ and $R'_k(t)$ are the densities of infected, susceptible, suspended and reinfected nodes of degree $k$ at time $t$, respectively, $P(k)$ is the fraction of nodes with degree $k$ and $\lambda$ is the probability that a susceptible node is infected by one infected neighbor. The probability that a new individual with $k$ neighbors will be infected is proportional to the infection rate $\lambda$, the density of susceptible individuals of degree $k$ ($S_k(t)$) and the degree $k$. As in the homogeneous network case, we add one more density $R'_k(t)$ - is the density of reinfected nodes that should be taken into account during the $\Theta(t)$ computation. In our model $\Theta(t)$ is the probability that any given link points to currently infectious node. This probability depends not only on the fraction of infected nodes of degree $k$, but also on the fraction of reinfected nodes of degree $k$.

## 4   Evaluation

We evaluate the SISR model presented in this paper in two steps. First, we compare the SISR model to the SIR model. Secondly, we compare results produced by the SISR model with results of simulations that simulate real worms spreading on real networks.

### 4.1   SISR vs. SIR

We use the discrete-time method to calculate the numerical solutions of both models for various $< k >$ and $\lambda$ values. Figure 2 and Figure 3 show the obtained results. From Fig. 2 we can understand the impact of various $\lambda$ values on propagation dynamics. It is not surprising that for larger $\lambda$ values, the infection spreads faster and infects larger fraction of the population in both models. Figure 3 analyzes the impact of $< k >$ (average degree) values. It shows that in networks with higher connectivity, the infection spreads faster and infects a larger fraction of the population in both models. It is clear from Fig. 2 and Fig. 3 that in the SISR model the number of infected nodes at any time t is greater or equal to the number of infected nodes in the SIR model in the same time for the same parameters values. This is consistent with behavior reported for reinfection email worms (see [6]).

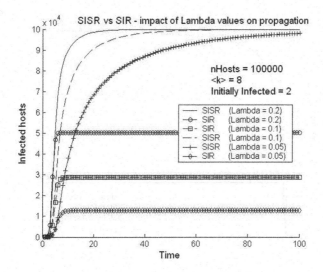

**Fig. 2.** The impact of $\lambda$ values on propagation dynamics

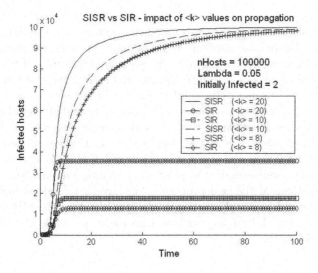

**Fig. 3.** The impact of $< k >$ (average degree) values on propagation dynamics

## 4.2 SISR vs. Simulation

Now we'll show that SISR model presented in this paper provides a reasonable estimation of the real epidemic spreading process. In order to see this, we compare the SIR and SISR analytical models with results of the simulations that simulate reinfection and noreinfection propagation strategies of real worms on real networks. We start from generating the power-law network. Using the

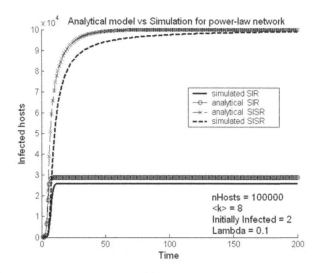

**Fig. 4.** Comparison of simulation results to SIR and SISR analytical models

obtained network we compute the numerical solutions for the SIR and SISR models. In parallel we have implemented a malware propagation simulation tool and run it on the same network. Finally, we compare the results of the analytical models with the simulation's results. We run the simulation 100 times for each one of the strategies separately (reinfection and noreinfection) and calculated the average number of infected hosts as function of time. On the other hand we average 100 executions of the analytical models, where degrees of the initially infected hosts were chosen randomly. Figure 4 presents the obtained results for a network with 100000 nodes, average degree 8 and $\lambda$=0.1. We can see that both analytical models overestimate the simulation results, but provide the reasonable approximations of the real processes. We have obtained similar dynamics for various values of $< k >$, $I_0$ and $\lambda$. We can see also that the SIR model can't be used to describe the reinfecting worm propagation, while the SISR model describes it very closely.

## 5    Conclusions

In this paper we have presented SISR - a new model that describes analytically the propagation of epidemics that use the reinfection strategy, the strategy where an infected node instead of being removed, contributes to the infection spreading upon the reinfection attempt (for example, Sober, Sobig, and Mydoom Worms). We have studied the impact of various parameters on the propagation dynamics. We compared the SISR model to the SIR model and the results match our expectations: in the SISR model the number of infected nodes at any time $t$, is greater or equal to the number of infected nodes in the SIR model, in the same time, for the same parameters values. This observation determines that, similar to the SIR model, there is no epidemic threshold in the SISR model too. Moreover, we compare results produced by the SISR model with results of simulations that simulate real worms spreading on real networks. We show that the model produces reasonable results relative to the simulation results.

# References

1. Anderson, R.M., May, R.M.: Infectious diseases in humans. Oxford University Press, Oxford (1992)
2. Boguna, M., Pastor-Satorras, R., Vespignani, A.: Epidemic Spreading in Complex Networks with Degree Correlations. Lecture Notes in Physics: Statistical Mechanics of Complex Networks (2003)
3. http://www.viruslist.com/
4. Moreno, Y., Pastor-Satorras, R., Vespignani, A.: Epidemic outbreaks in complex heterogeneous networks. Eur. Phys. J. B 26, 521–529 (2002)
5. Pastor-Satorras, R., Vespignani, A.: Epidemic spreading in scale-free networks. Phys. Rev. Lett. 86 (2001)
6. Zou, C.C., Towsley, D., Gong, W.: Modeling and Simulation Study of the Propagation and Defense of Internet E-mail Worms. Proceedings of IEEE Transactions on dependable and secure computing 4(2) (2007)
7. Rozenberg, B., Gudes, E., Elovici, Y.: A Distributed Framework for the Detection of new worm-related Malware. In: Ortiz-Arroyo, D., Larsen, H.L., Zeng, D.D., Hicks, D., Wagner, G. (eds.) EuroISI 2008. LNCS, vol. 5376, pp. 179–190. Springer, Heidelberg (2008)

# An Efficient Distance Bounding RFID Authentication Protocol: Balancing False-Acceptance Rate and Memory Requirement

Gildas Avoine[1] and Aslan Tchamkerten[2]

[1] Université catholique de Louvain, Louvain-la-Neuve, Belgium
[2] Telecom ParisTech, Paris, France

**Abstract.** The Mafia fraud consists in an adversary transparently relaying the physical layer signal during an authentication process between a verifier and a remote legitimate prover. This attack is a major concern for certain RFID systems, especially for payment related applications.

Previously proposed protocols that thwart the Mafia fraud treat relaying and non-relaying types of attacks equally: whether or not signal relaying is performed, the same probability of false-acceptance is achieved. Naturally, one would expect that non-relay type of attacks achieve a lower probability of false-acceptance.

We propose a low complexity authentication protocol that achieves a probability of false-acceptance essentially equal to the best possible false-acceptance probability in the presence of Mafia frauds. This performance is achieved without degrading the performance of the protocol in the non-relay setting. As an additional feature, the verifier can make a rational decision to accept or to reject a proof of identity even if the protocol gets unexpectedly interrupted.

**Keywords:** Authentication, false-acceptance rate, proximity check, mafia fraud, memory, relay attack, RFID.

## 1 Introduction

Radio Frequency Identification (RFID) allows to identify and authenticate objects or subjects wirelessly, using transponders — micro-circuits with an antenna — queried by readers through a radio frequency channel. This technology is one of the most promising of this decade and is already widely used in practice (e.g., access cards, public transportation passes, payment cards, passports). This success is partly due to the steadily decrease in both size and cost of passive transponders called *tags*. The characteristics of this technology — ubiquity, low-resource, wireless — open a security breach that is seriously considered by the US National Institute of Standards and Technology, which recently published guidelines on how to securely develop RFID systems [1].

P. Samarati et al. (Eds.): ISC 2009, LNCS 5735, pp. 250–261, 2009.

In 1987 Desmedt *et al.* [2] introduced the *Mafia fraud*[1] that defeated any authentication protocol. In this attack, the adversary successfully passes the authentication by relaying the messages between the verifier and a remote legitimate prover. When it was introduced, the Mafia fraud appeared somewhat unrealistic since the prover is supposed unaware of the manoeuvre.

Nowadays, the Mafia fraud is a major issue of concern for RFID systems. We illustrate this in the following example. Consider an RFID-based ticket machine in a theater. To buy a ticket, the customer needs to be close enough to the machine (RFID reader) such that his pass (RFID tag) is in the field of the machine. The pass can be kept in the customer's pocket during the transaction. A ticket is delivered by the machine if the pass is able to prove its authenticity. Assume there is a line of customers waiting for a ticket, including Alice the victim. Bob and Charlie are the adversaries: Bob is far in the queue close to Alice, while Charlie faces the machine. When the machine initiates the transaction with Charlie's card, Charlie forwards the received signal to Bob who transmits it to Alice. The victim's tag automatically answers since a passive RFID tag — commonly used for such applications — responds without requiring the agreement of its holder. The answer is then transmitted back from Alice to the machine through Bob and Charlie who act as relays. The whole communication is transparently relayed and the attack eventually succeeds: Alice pays Charlie's ticket. Note that Bob must be close to the victim in order to query her tag. In such an application, the communication distance is either a few centimeters (when the tag is ISO 14443-compliant [3]) or a few decimeters (when the tag is ISO 15693-compliant [4]). This is more than enough to enable an adversary to illegitimately query the tag of a passerby. In 2005, Hancke [5] successfully performed a Mafia fraud against an RFID system where the two colluders where 50 meters apart and connected through a radio-channel.

In 2007, Halváč and Rosa [6] noticed that the standard ISO 14443 [3] for proximity cards and widely deployed in secure applications, can easily be abused by a Mafia fraud due to the untight timeouts in the communication. Indeed, ISO 14443 specifies a *frame waiting time* (FWT) such that the reader is allowed to retransmit or give up the communication if the queried tag remains unresponsive while the FWT is over. The FWT is equal to $FWT = (256 \times 16/fc) \times 2^{FWI}$, where fc is the frequency carrier (13.56 MHz in almost all secure RFID applications), and where FWI is the Frame Waiting time Integer, a value chosen between 0 and 14. By default FWI = 4, which means that FWT = 4.8 ms. However, when the tag needs more time to process the information it receives, it can impose the reader to increase the FWI up to 14, which corresponds to FWT = 4949 ms. (This feature is used for example by electronic passports that implement active authentication [7]. Passports are not able to compute an RSA or ECC signature on the fly within 4.8 ms and so require a larger FWT.) During a Mafia fraud the adversary can request the reader to increase its timeout up to 4949 ms, which gives her enough time to perform the attack over a long distance using for instance Internet.

---

[1] Sometimes referred to as 'relay attack.'

## 2   State of the Art and Contributions

In 1990 Brands and Chaum [8] proposed a protocol that thwarts the Mafia fraud and which is based on the idea of a *proximity check* introduced in [9]. The protocol, depicted in Figure 1, consists of a *fast phase* followed by a *slow phase*. During the fast phase, the verifier and the prover exchange random one-bit messages and the verifier measures the round trip time (RTT) of the exchanges. After $n$ rounds, where $n$ is a security parameter, the slow phase is engaged. The verifier asks the prover to sign the received and sent bits, and, upon reception of the signature, and given the measured RTT, the verifier decides whether or not to accept the proof of identity. The probability that a Mafia fraud succeeds is then $(1/2)^n$.

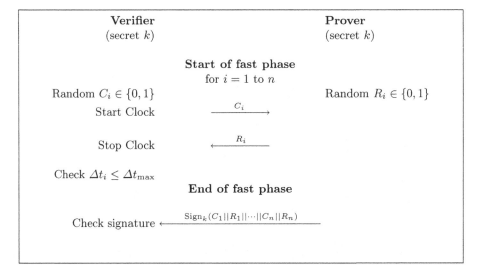

**Fig. 1.** Brands and Chaum's protocol

It is only in 2005, after Hancke put into practice a mafia fraud [5] that proximity check protocols[2] came back under the spotlights. The same year, Hancke and Kuhn [10] published a new distance bounding protocol that is today a key reference. Given in Figure 2, their protocol consists of a slow phase followed by a fast phase. In the slow phase, the verifier and the prover first exchange random nonces, then, based on the nonces and the secret key, they compute two secret registers in the form of $n$-bit strings $V$ and $W$. The fast phase consists of $n$ rounds. During the $i$th round, the verifier sends a random bit and the prover answers the $i$th bit $V_i$ of $V$ if the challenge is 0, and the $i$th bit $W_i$ of $W$ if the challenge is 1.

---

[2] In the literature often referred to as 'distance bounding protocols.'

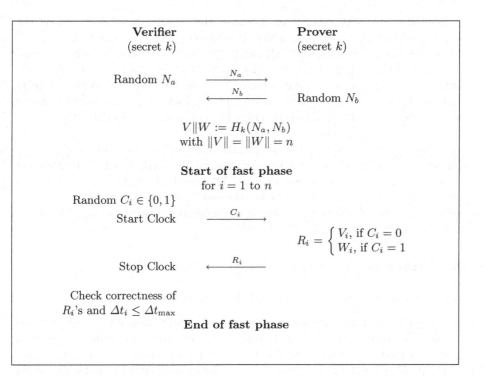

**Fig. 2.** Hancke and Kuhn's protocol

As explained in [10], the false-acceptance rate (FAR) is $(3/4)^n$ instead of $(1/2)^n$, as in Brands and Chaum's protocol, because an adversary can query the prover between the slow phase and the fast phase in order to obtain one full register. However, the protocol has interesting properties such as the absence of a signature at the final stage which allows the verifier to make a 'rational' decision on whether to accept or to reject a proof of identity even in cases where the protocol gets unexpectedly interrupted. In practice, one could imagine the situation where the verifier accepts a proof of identity provided that a minimal number of correct fast phase replies are given, so that to allow some flexibility in the event of an interrupted authentication. In contrast, with the Brands and Chaum protocol, if the protocol does not end properly, i.e., if the final signature is not received by the verifier, it is difficult for the verifier to infer about the validity of the proof of identity.

Since 2005, several protocols have been proposed. Either they are based on the approach of Brands and Chaum (BC), require a final signature, and target $(1/2)^n$ as FAR ( [11,12,8,13,14,15,16,17,18,19,20]), or, they follow the Hancke and Kuhn (HK) approach, have no final signature, and target $(3/4)^n$ as FAR ([10,21]).[3]

---

[3] A comparison of most of these protocols is given in [18].

Note that for both families of protocols the security is solely based on the number of fast phase rounds, which in practice cannot be made very large.[4] Moreover, for both families a FAR of $(1/2)^n$ (or $(3/4)^n$) can be achieved even without carrying a mafia fraud. In other words, these protocols do not distinguish between an attacker that relay signals from an attacker that does not relay signals. As a consequence, because $n$ can't be made large, these protocols are not suitable for applications where a high level of security is demanded, yet mafia frauds are hard to perform.

Below we provide a new low complexity distance bounding protocol that, in particular, combines the advantages of the BC and HK families. It does not require a final signature, it achieves a FAR essentially equal to $(1/2)^n$ in the presence of Mafia frauds, and achieves the same level of security with respect to non-Mafia type of attacks as common challenge-response authentication protocols (e.g., compliant with ISO 9798 [22]).

# 3  Protocol

## 3.1  Protocol Requirements and Assumptions

In the presence of a legitimate prover, the authentication protocol must guarantee that the verifier always accepts his proof of identity. The protocol must also prevent an adversary of being falsely identified, assuming she can participate either passively or actively in protocol executions with either or both the prover and the verifier. This means that the adversary can both eavesdrop protocol executions between the legitimate prover and the verifier (passive attack), and be involved in protocol executions with the verifier and the legitimate prover separately or simultaneously (active attack). We assume that neither the prover nor the verifier colludes with the adversary, i.e., the only information the adversary can obtain from the prover or the verifier is through protocol executions.

## 3.2  Protocol Description and Initialization

The protocol we describe in this section may, for certain RFID applications, require too much memory. Nevertheless, to simplify the exposition, we present and analyze this version of the protocol and later (Section 5) provide a twist that allows to drastically reduce the memory requirement while not affecting the security of the protocol.

The protocol consists of a 'slow' authentication phase followed by a 'fast' proximity check phase. Both phases have their own security parameters: $m$ (credential size) for the authentication and $n$ (number of rounds) for the proximity check.

**Initialization.** Prior to the protocol execution, the legitimate prover and the verifier agree on the security parameters $m$ and $n$ and a common secret key $k$.

---

[4] To the best of our knowledge, distance bounding protocol haven't been implemented yet.

**Authentication.** The verifier first sends a random nonce $N_a$ to the prover, in the form of a bit string. The prover then generates a random nonce $N_b$ and, based on $N_a$ and $N_b$, computes a keyed-hash value $H_k(N_a, N_b)$ whose output size is at least $m + 2^{n+1} - 2$ bits. The prover sends to the verifier both $N_b$ and $[H_k(N_a, N_b)]_1^m$, which denotes the first $m$ bits of $H_k(N_a, N_b)$. (The length of the bit strings $N_a$ and $N_b$ is discussed in Section 4.)

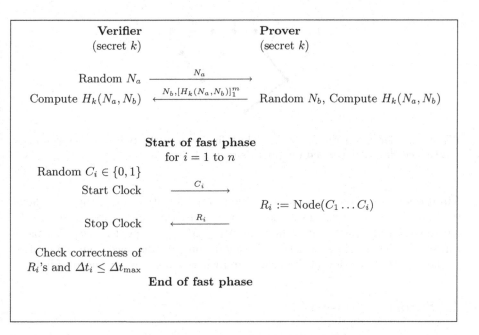

**Fig. 3.** Tree-based RFID distance bounding protocol

**Proximity check.** Using the subsequent $2^{n+1} - 2$ bits of the hash value $H_k(N_a, N_b)$, denoted $[H_k(N_a, N_b)]_{m+1}^{m+2^{n+1}-2}$, the prover and the verifier label a full binary tree of of depth $n$ as follows (see Figure 4 for an example). The left and the right edges are labeled 0 and 1, respectively, and each node (except the root) is associated with the value of a particular bit in $[H_k(N_a, N_b)]_{m+1}^{m+2^{n+1}-2}$ in a one-to-one fashion.[5] An $n$-round fast bit exchange between the verifier and the prover proceeds using the tree: the edge and the node values represent the verifier's challenges and the prover's replies, respectively. At each step $i \in \{1, 2, \ldots, n\}$ the verifier generates a challenge in the form of a random bit $C_i$ and sends it to the prover. The prover replies $R_i = \text{Node}(C_1 \ldots C_i)$, the value of the node in the tree whose edge path from the root is $C_1, C_2, \ldots, C_i$.

---

[5] To do this one can sequentially assign the bit values of $[H_k(N_a, N_b)]_{m+1}^{m+2^{n+1}-2}$ to all the nodes of the tree by starting with the lowest level nodes, moving left to right, and moving up in the tree after assigning all the nodes of the current level.

In the example illustrated by Figure 4, the verifier always replies 0 in the second round unless the first and the second challenges are equal to 1 in which case the verifier replies 1, i.e., Node(00) = Node(01) = Node(10) = 0 and Node(11) = 1.

Finally, for all $i \in \{1, 2, \ldots, n\}$, the verifier measures the time interval $\Delta t_i$ between the instant $C_i$ is sent until $R_i$ is received.

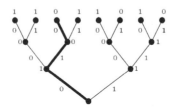

**Fig. 4.** Decision tree with $n = 3$. The thick line path in the tree corresponds to the verifier's challenges $0, 1, 0$ and the prover's replies $1, 0, 0$.

**Final decision.** The verifier accepts the prover's identity only if the $m$ authentication bits are correct and if the $n$ replies of the fast phase are correct while meeting the time constraint of the form $\Delta t_i \leq \Delta t_{\max}$, $i \in \{1, 2, \ldots, n\}$. A typical threshold value for $\Delta t_{\max}$ is $2d/c$, where $d$ denotes the distance from the verifier to the expected position of the prover and where $c$ denotes the speed of light.

## 4    Security Analysis

Protocols belonging to the HK family do not distinguish authentication from proximity check, which means that the security level of the proximity check is as high as the authentication one, in other words the credential parameter $m$ is equal to the number of fast phase rounds $n$. While $m = 64$ is a realistic assumption,[6] $n = 64$ seems to be unpracticable due to the limited transaction time and because a proximity check over many bits seems already a practical challenge. In our protocol, authentication and proximity check are distinct. We can keep $m = 64$ while choosing a smaller $n$. A conservative value for the nonces' lengths is $|N_a| = |N_b| = m = 64$ bits.

We analyze our protocol by considering two cases, depending on whether or not the legitimate prover is reachable during the attack.

### 4.1    Attack in the Absence of a Legitimate Prover

The case where the legitimate prover is unreachable right during the attack is similar to the classical cryptographic model. To succeed the adversary must pass both the authentication and the proximity check, without knowing the

---

[6] Note that attacks cannot be performed off-line.

secret key. Since the hash function $H_k$ is supposed to be cryptographically secure, we can consider that $[H_k(N_a, N_b)]_1^m$ provides no information about $[H_k(N_a, N_b)]_{m+1}^{m+2^{n+1}-2}$, i.e., the authentication reveals nothing about the proximity check and vice versa. The protocol thus achieves the same security level as any challenge-response protocol whose credential size is $m + n$ bits.

## 4.2 Attack in the Presence of a Legitimate Prover

When the legitimate prover is reachable during the attack, the adversary can execute a Mafia fraud in order to successfully pass the authentication step. The FAR is then computed as follows.

Due to the time constraint, the adversary cannot usefully relay information between the verifier and the prover during the fast phase without being detected; the adversary's reply at time $i$ must be independent of the verifier's challenge at time $i$, for any $i \in \{1, 2, \ldots, n\}$. However, there is no time measure before the fast phase, which allows the adversary to query the legitimate prover with one sequence of challenges $\tilde{C}^n \triangleq \tilde{C}_1 \ldots \tilde{C}_n$, hoping these will correspond to the challenges $C^n \triangleq C_1 \ldots C_n$ provided by the verifier during the fast phase.

Since the probability of false acceptance is the same given any $\tilde{C}^n$, without loss of generality we assume that the adversary queries the prover with the all-zero sequence, i.e., $\tilde{C}^n = 0^n$. The adversary is then successful only if $\tilde{R}_i = R_i$ for all $i \in \{1, 2, \ldots, n\}$, where $\tilde{R}_i$ denotes the adversary's reply at time $i$.

Letting $t$ be the first time $i \geq 1$ when $C_i = 1$, we have that $\tilde{R}_i = R_i$ for $i \in \{1, 2, \ldots, t-1\}$, and $\tilde{R}_i = R_i$ with probability $1/2$ for $i \in \{t, t+1, \ldots, n\}$, because the adversary can still try her chance by sending random replies once $C_i = 1$ is observed. Therefore, letting $R^n = R_1, R_2, \ldots, R_n$, the probability of a successful attack over one particular protocol execution can be computed as

$$\Pr(\tilde{R}^n = R^n) = \sum_{i=1}^{n} \Pr(\tilde{R}^n = R^n | t = i) \Pr(t = i)$$
$$+ \Pr(\tilde{R}^n = R^n | C^n = 0^n) \Pr(C^n = 0^n)$$
$$= \sum_{i=1}^{n} 2^{-(n-i+1)} 2^{-i} + 2^{-n}$$
$$= 2^{-n}(n/2 + 1) .$$

# 5 Multiple Trees: Balancing FAR and Memory Requirement

The second phase of the protocol is memory consuming; for $n$ fast phase rounds we need to store

$$2^{n+1} - 2$$

bits. We now provide a means to drastically reduce this memory requirement by means of multiple trees.

Consider a fast phase based on $\alpha$ small trees of depth $k$, rather than based on a single large tree of depth $n = \alpha k$. The fast phase proceeds in the same way than described in Section 3.2 except that now the verifier accepts a proof of identity only if the $k$ replies of each of the $\alpha$ trees are correct. Using multiple trees requires to store

$$\alpha(2^{k+1} - 2) \tag{1}$$

bits for the fast phase and the FAR guaranteed by the proximity check equals to

$$\left(2^{-k} (k/2 + 1)\right)^{\alpha}. \tag{2}$$

It is easily seen that the use of multiple trees in place of a single tree reduces the storage requirements at the expense of the false-acceptance rate. In general, among all pairs $(\alpha, k)$ that achieve a targeted probability of false-authentication in the presence of active attacks, one may want to pick the pair for which $\alpha$ is maximal so that to reduce the storage requirement. When $\alpha = 1$ and $k = n$ (single tree case), the storage requirement is maximal and the probability of false-acceptance is minimal. At the other extreme, when $\alpha = n$ and $k = 1$, the fast phase of our protocol corresponds to the Hancke and Kuhn protocol [10]. The storage requirement is minimal, equal to $2n$, and the probability of false-acceptance is maximal, i.e., $(3/4)^n$. Finally note that, in order for the FAR of the proximity check to decay as $(1/2)^n$ instead of $(3/4)^n$, it is necessary and sufficient that $k$ is a growing function of $n$.[7] Letting, for instance, $k = \log_2 n$ and $\alpha = n/\log_2 n$, the storage requirement becomes

$$\frac{2n^2}{\log_2 n}(1 - 1/n)$$

which is already a huge improvement compared to the single tree case $(2^{n+1} - 2)$ for $n \geq 2$.

The key in reducing the FAR from $(3/4)^n$, given by the Hancke and Kuhn protocol, to $(1/2)^n$ lies in the dependencies of the answers provided by the prover. In the Hancke and Kuhn protocol, the reply at time $i$ is only a function of the $i$th challenge. When using trees, the $i$th reply potentially also depends on challenges that are posterior to the $i$th challenge, making it less likely for an adversary to succeed. Interestingly, the past dependency for each reply need only be 'mild': to achieve $2^{-n}$ it is sufficient to consider many trees each of small depth $\log_2(n)$, i.e., each reply depends at most on the last $\log_2(n)$ challenges. As a consequence, the storage requirement can be maintained low; the storage requirement grows

---

[7] More precisely, when $k$ is a growing function of $n$, the exponential rate at which the FAR decreases with respect to $n$ approaches one as $n$ grows, i.e.,

$$-\frac{1}{n} \log_2(\text{FAR})$$

tends to 1 as $n$ tends to infinity.

quadratically with $n$ instead of exponentially as in the single tree case. The bottom line is that the use of multiple trees allows to drastically reduce the storage requirement without penalizing the false-acceptance rate.

As a numerical example, to achieve a FAR of 0.01% in the presence of Mafia frauds, the Hancke and Kuhn protocol requires 32 rounds, the Brands and Chaum 14 rounds, and ours 17 rounds (single tree). With these parameters, our protocol allows to reduce the FAR down to $0.01\% \cdot 2^{-m}$ ($m$ is typically equal to 64 or 128.) with respect to non-Mafia types of attacks, in contrast with the Hancke and Kuhn and the Brands and Chaum protocols.

For our protocol, the use of a single tree of depth 17 necessitates 32 Kbytes of memory, but a FAR of 0.01% in the presence of Mafia frauds can also be obtained by using two trees each of depth 9 (yielding 18 fast phase rounds). This decreases the needed memory down to 256 bytes (0.25 Kbytes). For comparison, a typical chip for ePassports contains roughly 40Kbytes of EEPROM and 6Kbytes of RAM.

## 6    Computation

Note that only one step of the protocol involves computation, the hash value. In particular, the labeling of the nodes involves no computation, and selectors can efficiently be implemented in wired logic to directly access these values.

Tags that include a microprocessor usually embed a hash function — this is for example mandatory for tags compliant with DOC 9303 [7] which imposes SHA-1. Note that some tags, e.g., Oberthur ID-One EPass 64 [23], implement even the SHA-256 hash function.

Tags without microprocessor usually do not implement a standardized hash function. Instead, a symmetric cipher is available, which can be either a stream cipher or a block cipher. The cipher can then be the building block of a hash function [24]. We note that in 2004, Feldhofer, Dominikus, and Wolkerstorfer [25] proposed a lightweight implementation of AES in less than 4000 logic gates, enabling its implementation with wired logic only. We are not aware of commercial products using this implementation, though.

## 7    Concluding Remarks

The contribution of this paper consists in a low complexity tree-based RFID distance bounding protocol that combines the advantages of the protocols belonging to the Brands and Chaum's family with the advantages of the protocols belonging to the Hancke and Kuhn's family. In particular, it essentially achieves the optimal false-acceptance probability in the presence of Mafia frauds and it allows the verifier to make a rational decision even if the protocol does not end properly. In contrast with previously proposed distance bounding protocols, the security of the present protocol when the adversary can perform relay attacks does not come at the expense of the security of the protocol when the adversary cannot perform relay attacks. Our protocol achieves the same level of security with respect to non-Mafia type of attacks as common challenge-response authentication protocols.

Our protocol is suited, in terms of memory and computation, to current RFID tags designed for secure applications. It is so a solid candidate for environments where on-the-fly authentication is needed while dealing with Mafia type of frauds, e.g., in e-payment and public transportation.

# References

1. Karygiannis, T., Eydt, B., Barber, G., Bunn, L., Phillips, T.: Guidelines for securing radio frequency identification (RFID) systems – special publication 800-98. Recommandations of the National Institute of Standards and Technology (April 2007)
2. Desmedt, Y., Goutier, C., Bengio, S.: Special uses and abuses of the fiat-shamir passport protocol. In: Pomerance, C. (ed.) CRYPTO 1987. LNCS, vol. 293, pp. 21–39. Springer, Heidelberg (1988)
3. ISO/IEC 14443: Identification cards – contactless integrated circuit(s) cards – proximity cards
4. ISO/IEC 15693: Identification cards – contactless integrated circuit(s) cards – vicinity integrated circuit(s) card
5. Hancke, G.: A practical relay attack on ISO 14443 proximity cards (February 2005) (manuscript)
6. Halváč, M., Rosa, T.: A Note on the Relay Attacks on e-Passports: The Case of Czech e-Passports. Cryptology ePrint Archive, Report 2007/244 (2007)
7. ICAO DOC–9303: Machine readable travel documents, part 1, vol. 2 (November 2004)
8. Brands, S., Chaum, D.: Distance-bounding protocols. In: Helleseth, T. (ed.) EUROCRYPT 1993. LNCS, vol. 765, pp. 344–359. Springer, Heidelberg (1994)
9. Beth, T., Desmedt, Y.: Identification tokens – or: Solving the chess grandmaster problem. In: Menezes, A., Vanstone, S.A. (eds.) CRYPTO 1990. LNCS, vol. 537, pp. 169–176. Springer, Heidelberg (1991)
10. Hancke, G., Kuhn, M.: An RFID distance bounding protocol. In: Conference on Security and Privacy for Emerging Areas in Communication Networks – SecureComm 2005, Athens, Greece. IEEE, Los Alamitos (2005)
11. Bussard, L., Roudier, Y.: Embedding distance-bounding protocols within intuitive interactions. In: Hutter, D., Müller, G., Stephan, W., Ullmann, M. (eds.) Security in Pervasive Computing. LNCS, vol. 2802, pp. 119–142. Springer, Heidelberg (2004)
12. Bussard, L., Bagga, W.: Distance-bounding proof of knowledge to avoid real-time attacks. In: Ryoichi, S., Sihan, Q., Eiji, O. (eds.) Security and Privacy in the Age of Ubiquitous Computing, Chiba, Japan. IFIP International Federation for Information Processing, vol. 181, pp. 223–238. Springer, Heidelberg (2005)
13. Munilla, J., Ortiz, A., Peinado, A.: Distance Bounding Protocols with Void-Challenges for RFID. Printed handout of Workshop on RFID Security – RFIDSec 2006 (July 2006)
14. Singelée, D., Preneel, B.: Distance bounding in noisy environments. In: Stajano, F., Meadows, C., Capkun, S., Moore, T. (eds.) ESAS 2007. LNCS, vol. 4572, pp. 101–115. Springer, Heidelberg (2007)
15. Munilla, J., Peinado, A.: Attacks on Singelee and Preneel's protocol. Cryptology ePrint Archive, Report 2008/283 (June 2008)

16. Nikov, V., Vauclair, M.: Yet Another Secure Distance-Bounding Protocol. Cryptology ePrint Archive, Report 2008/319 (2008), http://eprint.iacr.org/
17. Capkun, S., Buttyan, L., Hubaux, J.P.: SECTOR: secure tracking of node encounters in multi-hop wireless networks. In: 1st ACM Workshop on Security of Ad Hoc and Sensor Networks – SASN 2003, pp. 21–32 (2003)
18. Kim, C.H., Avoine, G., Koeune, F., Standaert, F.X., Pereira, O.: The Swiss-Knife RFID Distance Bounding Protocol. In: International Conference on Information Security and Cryptology – ICISC, Seoul, Korea. LNCS. Springer, Heidelberg (2008)
19. Tu, Y.J., Piramuthu, S.: RFID Distance Bounding Protocols. In: First International EURASIP Workshop on RFID Technology, Vienna, Austria (September 2007)
20. Meadows, C., Poovendran, R., Pavlovic, D., Chang, L., Syverson, P.: 2. In: Distance Bounding Protocols: Authentication Logic Analysis and Collusion Attacks. Advances in Information Security series, Secure Localization and Time Synchronization for Wireless Sensor and Ad Hoc Networks, vol. 30, pp. 279–298. Springer, Heidelberg (2007)
21. Reid, J., Gonzalez Neito, J., Tang, T., Senadji, B.: Detecting relay attacks with timing based protocols. In: Bao, F., Miller, S. (eds.) ACM symposium on Information, computer and communications security – ASIACCS, Singapore, pp. 204–213. ACM, New York (2007)
22. ISO/IEC 9798: Information technology – security techniques – entity authentication
23. Oberthur Card Systems: Id-one epass
24. ISO/IEC 10118-2: Information technology – security techniques – hash-functions – part 2: Hash-functions using an n-bit block cipher
25. Feldhofer, M., Dominikus, S., Wolkerstorfer, J.: Strong Authentication for RFID Systems using the AES Algorithm. In: Joye, M., Quisquater, J.-J. (eds.) CHES 2004. LNCS, vol. 3156, pp. 357–370. Springer, Heidelberg (2004)

# Robust Authentication Using Physically Unclonable Functions*

Keith B. Frikken[1], Marina Blanton[2], and Mikhail J. Atallah[3]

[1] Computer Science and Software Engineering, Miami University
frikkekb@muohio.edu
[2] Department of Computer Science and Engineering, University of Notre Dame
mblanton@cse.nd.edu
[3] Department of Computer Science, Purdue University
mja@cs.purdue.edu

**Abstract.** In this work we utilize a physically unclonable function (PUF) to improve resilience of authentication protocols to various types of compromise. As an example application, we consider users who authenticate at an ATM using their bank-issued PUF and a password. We present a scheme that is provably secure and achieves strong security properties. In particular, we ensure that (i) the user is unable to authenticate without her device; (ii) the device cannot be used by someone else to successfully authenticate as the user; (iii) the device cannot be duplicated (e.g., by a bank employee); (iv) an adversary with full access to the bank's personal and authentication records is unable to impersonate the user even if he obtains access to the device before and/or after the setup; (v) the device does not need to store any information. We also give an extension that endows the solution with emergency capabilities: if a user is coerced into opening her secrets and giving the coercer full access to the device, she gives the coercer alternative secrets whose use notifies the bank of the coercion in such a way that the coercer is unable to distinguish between emergency and normal operation of the protocol.

## 1 Introduction

Recent work has demonstrated the existence and practicality of physically unclonable functions (PUFs), but many of their security implications remain to be explored. PUFs have both advantages and limitations compared to more traditional security devices. E.g., compared to a smartcard, a PUF has the advantage that it cannot be cracked and its secrets revealed, or replicated by an insider who has the blueprint. But unlike a smartcard, one can no longer assume the convenient existence of multiple copies that all contain the same key, nor can one assume any storage capacity within a device other than the PUF functionality.

The focus of this work is authentication, where a physically unclonable function (PUF) is used to provide superior resilience against various forms of compromise. A PUF is a function that is tied to a device and cannot be reproduced on another device, even another device from the same manufacturing batch. That is, a PUF is computed

---

* Portions of this work were supported by grants NSF-CNS-0627488 and AFOSR-FA9550-09-1-0223, and by sponsors of CERIAS.

P. Samarati et al. (Eds.): ISC 2009, LNCS 5735, pp. 262–277, 2009.

using unique physical characteristics of the device, and any attempts to tamper with the device change the behavior of the device and therefore destroy the PUF. This function is often assumed to be evaluated on a challenge $c$ which is sent to the device. Upon receiving $c$, the response is computed as $r = PUF(c)$ and is assumed to be unpredictable to anyone without access to the device. Schemes exist for using in different contexts (e.g., for protection of intellectual property and authentication), where the inability to clone the function improves the properties of a solution.

Here we use PUFs for authentication in contexts such as bank ATMs, through the use of a device with a built-in PUF. The ATM communicates with the bank to establish authenticity of the user before any transaction. We are able to achieve strong security properties which are not simultaneously achieved by previous protocols. In particular, our protocol provably has the following properties (in the random oracle model):

- a user is unable to successfully authenticate without her device;
- a stolen device cannot be used to authenticate as the user;
- the device functionality cannot be duplicated (e.g., by an employee of the bank even if that employee has access to the card);
- an adversary with full access to the bank's data with user information and authentication records is unable to impersonate the user even if she obtains access to the device before and/or after the account is setup.

Furthermore, our design requirements are to avoid placing any sensitive information on the device, to eliminate any possibility of data compromise (i.e., the PUF, which measures a physical characteristic of the device, will be destroyed in the event of tampering with the device, while the data stored on the device might not be erased). In fact, our protocols do not require the device to store *any* information not related to the PUF functionality, which introduces a challenge in the protocol design.

**Our Contributions**

1. We provide a protocol for one-factor authentication with PUFs (Section 4.1). It provides only a weak form of security in that to authenticate the adversary needs to have had physical access to the PUF at some point in time. One limitation of this protocol (and any one-factor "what you have" authentication mechanism) is that in order to impersonate a user, the adversary only needs physical access to the device.
2. We provide a stronger protocol for two-factor authentication that combines PUFs with passwords (Section 4.2). The adversary must have had access to the PUF and to the user's password in order to impersonate the user, even if the adversary has compromised the bank's servers. A unique feature of this protocol is that the password is not stored in either the PUF or the bank, but is integrated into the PUF challenge, and thus in order to perform a dictionary attack one must have physical access to the PUF.
3. One limitation of the previous schemes is that an adversary can clone the PUF in software by having physical access to the PUF. That is, the adversary can obtain the PUFs response to a challenge, and then build a piece of software that impersonates the user. To mitigate this software cloning attack, we introduce a protocol which requires the authenticator to currently have physical access to the PUF in order to authenticate (Section 4.3). This protocol requires a stronger assumption than those

required by the previous schemes: We assume an integrated PUF (or computational PUF) where the device performs some computation with the PUF.

4. We give an extension which additionally improves robustness of the protocol when a user is coerced into giving her device and secret data (e.g., her password), which permits an adversary to authenticate on behalf of the user. We provide a mechanism for a user to give a false secret to the coercer that will lead to successful authentication, but will trigger an alarm at the bank. Solutions of this type are common in physical security systems, but do not appear in cryptographic protocols[1].

## 2    Related Work

Existing literature on PUF-based authentication is not extensive and can be divided into three categories: (i) implementation-based publications that consider the feasibility of reliably computing a PUF response to a challenge; (ii) PUF-based authentication for IP (intellectual property) protection; and (iii) enhancing properties of lightweight authentication solutions using PUF.

Publications from the first category include [2,3] and others and are complementary to our work. They also provide support for using public-key cryptography with PUF-based authentication. Publications from the second category (e.g., [4,5,6]) are also largely implementation-based, often implementing existing authentication protocols for reconfigurable FPGA and are not suitable for our purposes. The last category covers PUF-based protocols for RFID (Radio-frequency identification) systems [7,8,9] and human protocols HB [10,11]. The RFID publications are implementation-based realizing simple authentication constructions. Recent results [10,11] strengthen the HB protocol by using PUFs and are not suitable in our context (i.e., do not achieve the properties we seek).

Authentication protocols based on smart-cards can also be viewed as related to our framework. However, the nature of PUF-based authentication places unique requirements: For a smartcard protocol to fit our model, the smartcard must implement a PUF and have no other information stored, yet satisfy our security requirements – there are no such previous smartcard protocols.

Multi-factor authentication protocols, which often use a password and a mobile device, have been explored in prior literature (see, e.g., [12,13,14,15] among others – some have insufficient security analysis). Resilience to user impersonation in the event of database compromise (the "insider threat"), however, is not considered and not achieved in previous work. In our case both factors (i.e., the user password and the token) are inaccessible to the server in their plain form, so that an insider with full access to the server is unable to recover either of them.

Boyen [16] uses biometrics and fuzzy extractors (i.e., biometric-based key derivation) to provide zero-storage authentication that achieves insider security. Our solution then can be viewed as an authentication mechanism with similar security properties, but which is based on a different technique and type of device (instead of using a device that captures biometrics) and additionally includes passwords as the second security

---

[1] The only publication on panic passwords in computer systems we are aware of is [1] that treats the general framework of panic passwords and is discussed later in this section.

factor. Note that we desire the same level of security even when the PUF is misused (by either a bank employee who temporarily gets access to the PUF or the user herself). This means that, to ensure that the device is present during each authentication session, we would like to make the raw information output of a PUF inaccessible to the user and use computational capabilities of a PUF. This problem is not a threat in case of biometric-based authentication, when the user is interested in erasing her personal biometric data output by the device and used in the protocol.

Recent work of Clark and Hengartner [1] defines the framework for panic passwords, where any user has a regular password and another, panic, password which can be used when the user is coerced into giving her password to the adversary. They define the adversarial model in terms of the response the user receives from the authenticator upon using a panic password, and goals/capabilities of the adversary. Our solution was designed independently of this recent model, but in section 5 we briefly discuss how it fits the Clark-Hengartner framework.

# 3   Security Model

## 3.1   Problem Description

There are three principal entities: server $S$ (or another entity authenticating the user on behalf of the server), user $\mathcal{U}$, and device $\mathcal{D}$. Before authentication can take place, the user obtains a device with a PUF built into it and participates in the registration or enrollment protocol with the server. Once the registration is complete, the user will be able to authenticate with the help of the device. Thus, we specify two procedures:

Enroll: is a protocol between $S$ and $\mathcal{U}$, where the user $\mathcal{U}$ registers with the server with the aid of $\mathcal{D}$. If enrollment is successful, the server obtains and stores a token $\mathrm{cred}_{\mathcal{U}}$ that can be used in subsequent authentications.

Auth: is a protocol between $S$ and $\mathcal{U}$, where $\mathcal{U}$ uses $\mathcal{D}$ and $S$ uses its stored credentials $\mathrm{cred}_{\mathcal{U}}$ to make its decision to either accept or reject the user.

## 3.2   Modeling PUFs

Prior literature does not contain a lot of cryptographic constructions where PUFs are used in a provably secure scheme. We are aware of the following uses of such functions. In what follows, we will generically refer to the entity trying to authenticate (i.e., user, device, tag, etc.) as a client and to the entity verifying authentication as a server.

1. *Straightforward authentication.* This is the most common form found in the PUF literature, where the server sends a challenge $c$ and the client responds with $r = PUF(c)$. At the enrollment phase, the server stores $n$ challenges $c_1, \ldots, c_n$ and their corresponding responses $r_1, \ldots, r_n$ for each client. During authentication, the client is challenged on one of the $c_i$'s at random and that $(c_i, r_i)$ is removed from the database. If the server runs out of challenge-response pairs (CRPs), there are protocols for updating the server's database with new CRPs [17].

2. *PUF as a random oracle.* Modeling a PUF as a random oracle (as in [8]) might be unnecessary if the full features of the random oracle model are not used.

3. *PUF as a computable function.* Hammouri and Sunar [10] define a delay-based PUF that can be represented using a linear inequality. This means that the server does not need to store CRPs, but instead can compute the expected responses. While it might be possible to model the specific PUF used in the above paper, for general functions it is commonly assumed that the function cannot be modeled and its behavior cannot be predicted by any entity without physical access to it.

4. *PUF in previously published identification protocols.* Some papers gave implementations where a PUF response is used as a part of known identification protocols. E.g., Tuys and Batina [7] use PUFs in Schnorr's identification protocol, where the user's secret key is set to be PUF's response to a challenge. Similarly, Batina et al. [9] use Okamoto identification protocol with PUF-derived secrets. We, however, aim to design a PUF-based protocol specific to our security goals.

As in the previous PUF literature, we make the standard assumption that, without having the physical device, the behavior of a PUF is impossible to predict. Let PUF be a function $PUF : \{0,1\}^{\kappa_1} \to \{0,1\}^{\kappa_2}$ that on input of length $\kappa_1$ produces a string of length $\kappa_2$. Before giving the definition, let us first define the following *PUF response game*:

**Phase 1:** Adversary $\mathcal{A}$ requests and gets the PUF response $r_i$ for any $c_i$ of its choice.

**Challenge:** $\mathcal{A}$ chooses a challenge $c$ that it has not queried thus far.

**Phase 2:** $\mathcal{A}$ is allowed to query the PUF for challenges other than $c$.

**Response:** Eventually, $\mathcal{A}$ outputs its guess for $r'$ for PUF's response to $r = PUF(c)$.

$\mathcal{A}$ wins if $r = r'$. Let $Adv_{\mathcal{A}}^{puf}(\kappa_2) = \Pr[r = r']$ denote the probability of $\mathcal{A}$ winning.

Under different conditions and in different environments, PUF responses to the same challenge can contain noise resulting in non-perfect match. We measure such noise in terms of hamming distance between two binary strings $x_1$ and $x_2$ of equal length $\kappa$, i.e., $\text{dist}(x_1, x_2)$ is the number of positions such that $i$th bit of $x_1$ is different from the $i$th bit of $x_2$. In what follows, let $U_\kappa$ denote the set of strings chosen uniformly at random from $\{0,1\}^\kappa$. Now we are ready to define a PUF:

**Definition 1.** *A physically unclonable function* $PUF_D : \{0,1\}^{\kappa_1} \to \{0,1\}^{\kappa_2}$ *bound to a device D is a function with the following properties:*

1. Efficient: *$PUF_D$ is easy to evaluate;*
2. Hard to characterize: *for any probabilistic polynomial time (PPT) adversary $\mathcal{A}$, $Adv_{\mathcal{A}}^{puf}(\kappa_2)$ is negligible in $\kappa_2$;*
3. Bounded noise: *in a wide variety of environments, the distance between two responses from $PUF_D$ on the same challenge is at most t, e.g., $\Pr[\text{dist}(y,z) > t \mid x \leftarrow U_{\kappa_1}, y \leftarrow PUF_D(x), z \leftarrow PUF_D(x)] \leq \varepsilon_1$ for a negligibly small $\varepsilon_1$;*
4. Unique: *the $PUF_D$ is unique for each D (even those from the same manufacturing batch), e.g., for any other function $PUF_{D'}$, $\Pr[\text{dist}(y,z) \leq t \mid x \leftarrow U_{\kappa_1}, y \leftarrow PUF_D(x), z \leftarrow PUF_{D'}(x)] \leq \varepsilon_2$ for sufficiently small $\varepsilon_2$.*

We call such a function a $(t, \varepsilon_1, \varepsilon_2)$ PUF (i.e., $\varepsilon_1$ and $\varepsilon_2$ are false rejection rate and false acceptance rate, respectively). Some of our constructions furthermore assume that a PUF is inseparatable from the device to which it is bound, i.e., our latter schemes make the strong assumption that the device (circuit) can do computation based on PUF responses. More specifically:

**Definition 2.** *An integrated PUF (I-PUF) has the following additional properties:*

1. *It is bound to the chip – any attempt to remove it changes its behavior.*
2. *Its communication with the chip cannot be accessed from outside the chip.*
3. *The output of the PUF cannot be accessed.*

I-PUFs have been used in prior literature (see, e.g., [7]), and the best known examples of them are silicon PUFs [17] and coating PUFs [18].

Because the output of a PUF is noisy, PUF-based authentication must either tolerate a certain threshold of errors at the protocol level or implement a mechanism for correcting the errors prior to using the response of the PUF. We choose the second option. Prior literature [7,6] already contains examples of using functions such as *fuzzy extractors* to remove the noise and extract responses that are close to uniform. Fuzzy extractors [19] can be defined for different metric spaces, and throughout this work we will assume we are dealing only with Hamming distance as the distance metric. The definition furthermore assumes a sufficient amount of uncertainty of the noisy string from which a random string is being extracted, defined in terms of min-entropy $m$ (see [19] for more precise definitions). The construction generates a public helper string $P$ that permits correction of errors and reconstruction of the extracted string and ensures that, even after releasing $P$, the statistical distance between the extracted string and a uniformly chosen string of the same length is less than a (negligibly small) threshold $\varepsilon$ (likewise, we refer the reader to [19] for precise definitions).

**Definition 3 ([19]).** *An $(m,\ell,t,\varepsilon)$ fuzzy extractor is given by procedures* Gen *and* Rep:

Gen: *is a probabilistic algorithm that on input $W$ outputs a string $R \in \{0,1\}^{\ell}$ and a helper string $P$, such that for any distribution of $W$ with min-entropy $m$, if $(R,P) \leftarrow$ Gen$(W)$, then the statistical difference between $(R,P)$ and $(U_{\ell},R)$ is at most $\varepsilon$.*

Rep: *is a deterministic algorithm that, given $P$ and $W'$ such that* dist$(W,W') \leq t$, *allows to exactly reproduce $R$: if $(R,P) \leftarrow$ Gen$(W)$, then* Rep$(W',P) = R$.

Our discussion does not necessitate the details of fuzzy extractors, as we refer to them only at a high level. They make possible the construction of an exact I-PUF having $(t,\varepsilon_1,\varepsilon_2)$ $PUF : \{0,1\}^{\kappa_1} \rightarrow \{0,1\}^{\kappa_2}$ such that:

1. An I-PUF bound to device $D$ is associated with a $(m,\ell,t,\varepsilon_3)$ fuzzy extractor (Gen, Rep), where Gen, Rep, and $PUF_D$ are efficient procedures.
2. During the enrollment phase, given a challenge $c$, I-PUF computes $(R,P) \leftarrow$ Gen$(r)$, where $r \leftarrow PUF_D(c)$ and outputs $P$.
3. In a wide variety of environments, given a pair $(c,P)$ where $c \leftarrow U_{\kappa_1}$ and $P$ was produced by Gen$(PUF_D(c))$, the exact extracted string can be recovered: $\Pr[x \neq y \mid x \leftarrow$ Rep$(PUF_D(c),P), y \leftarrow$ Rep$(PUF_D(c),P)] \leq \varepsilon_1$.
4. Any PPT adversary $\mathcal{A}$ cannot distinguish I-PUF's output from a random value with more than a negligible probability, i.e., $Adv_{\mathcal{A}}^{puf-ind}(\ell) \leq \varepsilon_3$ as defined below.

The last property can be viewed as a decisional version of the PUF response game, which we call *PUF response indistinguishability game* and it is defined as follows:

**Enroll:** $\mathcal{A}$ executes the enrollment phase on any values $c_i$ of its choice receiving the corresponding $P_i$ values from the PUF. Let $CP$ be the set of these $(c_i, P_i)$ pairs.

**Phase 1:** $\mathcal{A}$ requests and receives PUF response $R_i$ for any $(c_i, P_i) \in CP$ of its choice.

**Challenge:** $\mathcal{A}$ chooses a challenge $c$ that it queried in **Enroll** phase but not in **Phase 1**. A random bit $b$ is chosen. If $b = 0$, $\mathcal{A}$ receives $R = \text{Rep}(PUF_D(c), P)$ where $(c, P) \in CP$, otherwise it receives a string uniformly chosen from from $\{0, 1\}^{\ell}$.

**Phase 2:** $\mathcal{A}$ is allowed to query the PUF for challenges in $CP$ other than $(c, P)$.

**Response:** Eventually, $\mathcal{A}$ outputs a bit $b'$.

$\mathcal{A}$ wins if $b = b'$. Let $Adv_{\mathcal{A}}^{puf\text{-}ind}(\ell) = \Pr[b = b']$ denote the probability of $\mathcal{A}$ winning the game. We assume that $Adv_{\mathcal{A}}^{puf\text{-}ind}(\ell) - 1/2$ is negligible [2].

In addition to the above properties, as before, we have that two I-PUFs will produce the same output on a $(c, P)$ pair with probability at most $\varepsilon_2$. Finally, our protocols rest on the difficulty of the discrete logarithm in certain groups. That is, we assume that any PPT adversary $\mathcal{A}$, when given group $G$ of large order $q$, group generator $g$, and element $g^x$ for some $x \in \mathbb{Z}_q$, has a negligible change in outputting $x$.

### 3.3  Security Requirements

We place strict security requirements on the authentication process to achieve a solution robust to various types of misuse. In particular, we target to achieve the following properties beyond the traditional infeasibility to impersonate a user:

- Authentication by an adversary is not successful even with the possession of the device. Here we assume a powerful adversary who has access to all stored information at the server's side, including all information stored by the server during the enrollment phase such as recorded $(c, P)$ pairs for the device and other user's information cred$_{\mathcal{U}}$, as well as information belonging to other users. This strong notion of security is necessary in realistic scenarios, when, for example, the device originally resides with a bank, is consequently issued to a user, and a bank employee might later temporarily get access to the device and attempt to impersonate the user.
- Authentication by an honest user without the device is not successful with more than a negligible probability. This is important because, if this property holds, it is equivalent to a strong form of unclonability, i.e., even if an adversary knows all bank and user information, it cannot create a clone of the device in question. This adds resilience to a "what you have" form of authentication, because it guarantees that one must have the device during a successful login.

The use of I-PUFs ensures that the device cannot be duplicated or cloned, and tampering with the PUF effectively makes it unusable. Thus, the above requirements guarantee that both the original device and the user must be present at the time of authentication for authentication to succeed (except with negligible probability).

Furthermore, to ensure that tampering with the device does not reveal any sensitive information, our design stores no such information on the device. In fact we assume that the device does not store any information at all, and can be used for authentication with a number of servers. Thus, all necessary information is provided as input to the device, which makes the design of the protocols particularly challenging in presence of adversaries who can query the device's response on various inputs.

---

[2] We assume the PUF is built with a security parameter that allows tuning this probability.

# 4 Schemes

## 4.1 Preliminary Scheme

In this section we introduce a scheme that does not satisfy the security criteria, but that is a warm-up to the later schemes that do. In this and subsequent solutions we assume that the server $S$ sets up and announces a group $\mathbb{G}_q$ of prime order $q$, in which the discrete logarithm problem is hard, and its generator $g$. That is, $\mathbb{G}_q$ could be a subgroup of the multiplicative group $\mathbb{Z}_p^*$ for a prime $p$. We assume either that the PUF is constructed to use $\mathbb{G}_q$ or the user submits the group to the PUF whenever it queries the PUF[3]

The authentication protocol given below uses a zero-knowledge proof of knowledge (ZKPK) of discrete logarithm. In a nutshell, a ZKPK of a discrete logarithm $y$ to the base $g$ allows the prover to convince the verifier that she knows $x$ such that $y = g^x$ without revealing any information about $x$. Because standard and well-known solutions for several discrete logarithm based ZKPKs exist, we do not list their details in this work and refer the reader to, e.g., [20].

Enroll :
1. Server $S$ sends challenge $c$ to user $\mathcal{U}$.
2. $\mathcal{U}$ sends $c$ to device $\mathcal{D}$ for Gen protocol.
3. $\mathcal{D}$ sends to $\mathcal{U}$ $(r, P)$.
4. $\mathcal{U}$ sends $(g^r, P)$ to $S$ who stores the information along with $c$.

Auth :
1. $S$ sends challenge $(c, P)$ to the user $\mathcal{U}$.
2. $\mathcal{U}$ sends $(c, P)$ to device $\mathcal{D}$ for Rep protocol.
3. $\mathcal{D}$ sends $r$ to $\mathcal{U}$.
4. $\mathcal{U}$ and $S$ engage in a ZKPK of discrete logarithm $g^r$ to the base $g$.

Clearly, the above scheme does not satisfy either authentication goal. That is, if the adversary has access to the device and knows the server's challenge, then it can obtain the response and can impersonate the user without the device.

## 4.2 Preliminary Scheme Revisited

The problem with the previous scheme is that having the PUF (at any point in time) allows an adversary to impersonate the user. In this section we modify the previous scheme by adding a user password (and thus the adversary must have the device and guess the user's password). The password is integrated into the Enroll and Auth protocols so that the password is necessary to do authentication. Furthermore, the password is not stored anywhere. This prevents an adversary from being able to login in to the protocol even if it has the device. While this scheme still does not require that the user have the device, it does prevent a malicious outsider from impersonating the user (assuming that the user can choose a strong password). In what follows, $H : \{0,1\}^* \to \mathbb{Z}_q$ is a cryptographic hash function and $||$ denotes concatenation of strings.

---

[3] To prevent tampering with this value the PUF could take the hash of the description of this group with the challenge to form a modified challenge which it then responds to. So as not to clutter the exposition we have omitted this step from our scheme.

Note that in this scheme the password is bound to the PUF-challenge in order to get the PUF-response. Thus the password is not really stored anywhere, and thus in order to perform a dictionary attack the adversary must have physical access to the PUF.

Enroll :
1. Server $S$ sends challenge $c$ to user $\mathcal{U}$.
2. $\mathcal{U}$ sends $H(c\|pwd)$, where $pwd$ is the password, to device $\mathcal{D}$ for Gen protocol.
3. Device $\mathcal{D}$ sends $(r,P)$ to $\mathcal{U}$..
4. $\mathcal{U}$ sends $(g^r,P)$ to server $S$ who stores the information along with $c$.

Auth :
1. Server $S$ sends challenge $c$ and $P$ to the user $\mathcal{U}$.
2. User sends $(H(c\|pwd),P)$ to device $\mathcal{D}$ for Rep protocol.
3. $\mathcal{D}$ sends $r$ to $\mathcal{U}$.
4. User $\mathcal{U}$ and server $S$ engage is ZKPK of discrete logarithm for $g^r$ to the base $g$.

At a high level, this scheme requires that the adversary enter the user's password in order to the actual challenge sent to the PUF, thus this prevents an adversary with the PUF from being able to find the response $r$.

The proof of security of this approach has two parts. First, it is shown that if the response is generated independently from the PUF then breaking the above authentication scheme implies that the discrete log problem can be solved. Thus this implies (assuming discrete log problem is hard) that a computationally-bounded adversary has a negligible success probability (in the security parameter for the size of the prime $q$) in breaking the above scheme. The second part of the proof shows that if $\mathcal{A}$ can break the scheme with non-negligible probability when a real PUF is used, then this could be used to win the *PUF response indistinguishability game* with non-negligible probability. In other words, to determine if a specific response came from the PUF, the adversary uses $\mathcal{A}$ and if $\mathcal{A}$ succeeds then we assume that we are dealing with a real PUF (because if we are not the success probability is negligible).

**Lemma 1.** *If $H$ is a random oracle and there exists an adversary, $\mathcal{A}$, that successfully authenticates with the above authentication protocol with probability $1/p(|q|)$ when given a randomly generated challenge (that is independent from the PUF), then $\mathcal{A}$ contains a knowledge extractor that can solve the discrete log problem with non-negligible probability (in the length of $q$).*

**Proof.** Assume that such an adversary $\mathcal{A}$ exists, and that an adversary $\mathcal{B}$ is given a discrete log problem instance $g,q,g^r$ and is given access to a PUF. $\mathcal{B}$ sends the challenge $c_s, g^r, P$ for a randomly chosen $c_s$ and $P$ to $\mathcal{A}$. To simulate $H$, $\mathcal{B}$ creates a set of tuples $HSET$ and initializes it by choosing a random password $pwd$ and adding $(c_s\|pwd, h_s)$ to $HSET$ for a randomly chosen $h_s$. When $\mathcal{A}$ queries $H$ on a value $x$, $\mathcal{B}$ does the following: If there is a tuple $(x,y)$ already in $HSET$, it responds with $y$; otherwise, it chooses a random $r'$, adds $(x,r')$ to $HSET$, and responds with $r'$. When $\mathcal{A}$ queries $PUF$ with $(c_{\mathcal{A}},P_{\mathcal{A}})$, $\mathcal{B}$ does the following: If $c_{\mathcal{A}} = h_s$ and $P_{\mathcal{A}} = P$, $\mathcal{B}$ outputs FAIL. Otherwise $\mathcal{B}$ queries its PUF with $(c_{\mathcal{A}},P_{\mathcal{A}})$ and receives $r_A$. $\mathcal{B}$ then sends to $\mathcal{A}$ the value $r_A$.

It is straightforward to show that if $\mathcal{B}$ does not output FAIL, then the above view is the same as the view when engaging in the protocol. In the following we show that: (i)

$\mathcal{B}$ outputs FAIL with negligible probability, and (ii) if $\mathcal{B}$ does not output FAIL and $\mathcal{A}$ succeeds with non-negligible probability, then $\mathcal{B}$ can use $\mathcal{A}$ to obtain $r$.

$\mathcal{B}$ outputs FAIL only when $\mathcal{A}$ asks for the PUF response for a challenge $(d,P)$ and $d = h_s$. There are two situations: (i) $\mathcal{A}$ queries $H$ on $c\|pwd$ or (ii) $\mathcal{A}$ does not query $H$ on $c_s\|pwd$. The first case implies that $\mathcal{A}$ knows $pwd$ (which we assume is a negligible event), and the second case corresponds to $\mathcal{A}$ randomly guessing $h_s$ which is negligible. Thus, $\mathcal{B}$ outputs FAIL with negligible probability.

Now if $\mathcal{B}$ does not output FAIL and $\mathcal{A}$ can create a ZKPK of the discrete log of $g^r$, then by the properties of zero-knowledge, there must a be knowledge extractor for $\mathcal{A}$ that produces the secret $r$. $\mathcal{B}$ uses this knowledge extractor to solve the discrete log problem. Notice that if $\mathcal{A}$ succeeds then so does $\mathcal{B}$, and therefore assuming discrete log problem is hard, an adversary $\mathcal{A}$ does not exist.                                     □

We now utilize the above lemma to show that an adversary cannot break the above protocol if a real PUF is used (except with negligible probability). The lynchpin to this argument is that if such an adversary exists, then this adversary could be used to distinguish between a fake and real PUF, which violates the assumed security of the *PUF response indistinguishability game*.

**Theorem 1.** *Any polynomial-time adversary with access to the PUF (with security parameter $\ell$) and server information has a negligible probability of passing the authentication protocol for a previously generated enrollment, assuming that $H$ is a random oracle, the discrete log problem is hard, and the passwords are chosen from a large enough domain to make guessing the password succeed with negligible probability.*

**Proof.** Assume that such an adversary $\mathcal{A}$ exists, we then use this as a black-box to construct an adversary $\mathcal{B}$ for the *PUF response indistinguishability game* that succeeds with non-negligible probability. $\mathcal{B}$ proceeds as follows: it chooses a random challenge value $c_s$ and a random password $pwd$. It computes $c' = H(c_s\|pwd)$ and chooses $c'$ as its challenge. $\mathcal{B}$ then receives a pair $(r,P)$ where with probability $1/2$ the value $r$ is $PUF_D(c)$ and is otherwise a randomly chosen value. $\mathcal{B}$ constructs server information $c_s, g^r, P$ and invokes the adversary $\mathcal{A}$ on these values while $\mathcal{B}$ provides oracle access to the PUF and to the random oracle $H$ in the exact same manner as in Lemma 1. Eventually $\mathcal{A}$ will output a proof of knowledge. If this proof of knowledge is correct, then $\mathcal{B}$ outputs 0; otherwise, $\mathcal{B}$ chooses a random guess for $b'$ and outputs this value.

We now analyze the probability $\Pr[b = b']$. Let $F$ be the event the $\mathcal{B}$ outputs FAIL. Since $F$ was shown to be a negligible event in Lemma 1, we concentrate on $Pr[b = b'|\overline{F}]$. We condition it based on event $b = 0$ or $b = 1$, which gives us:

$$\Pr[b = b'|\overline{F}] = \frac{1}{2}\Pr[b = b'|\overline{F}, b = 0] + \frac{1}{2}\Pr[b = b'|\overline{F}, b = 1]$$

Let $G$ be the event that $\mathcal{A}$ outputs a correct proof of knowledge. We condition both of the above cases on $G$. In case of $b = 1$:

$$\Pr[b = b'|\overline{F}, b = 1] = \Pr[b = b'|\overline{F}, b = 1, G]\Pr[G|\overline{F}, b = 1] + \Pr[b = b'|\overline{F}, b = 1, \overline{G}]\Pr[\overline{G}|\overline{F}, b = 1].$$

Here, $\Pr[b = b'|\overline{F}, b = 1, G] = 0$, $\Pr[b = b'|\overline{F}, b = 1, \overline{G}] = \frac{1}{2}$, and $\Pr[G|\overline{F}, b = 1]$ is negligible (by Lemma 1). This gives us:

$$\Pr[b = b' | \overline{F}, b = 1] > \frac{1}{2} - neg_2(\ell)$$

for some negligible function $neg_2$. Next, let us consider $b = 0$, in which case we have:

$$\Pr[b = b' | \overline{F}, b = 0] = \Pr[b = b' | \overline{F}, b = 0, G] \Pr[G | \overline{F}, b = 0] + \Pr[b = b' | \overline{F}, b = 0, \overline{G}] \Pr[\overline{G} | \overline{F}, b = 0].$$

Here, $\Pr[b = b' | \overline{F}, b = 0, G] = 1$, $\Pr[b = b' | \overline{F}, b = 0, \overline{G}] = \frac{1}{2}$, and $\Pr[G | \overline{F}, b = 0] > \frac{1}{f(\ell)}$ for some polynomial $f$ (this follows from our assumption that $\mathcal{A}$ breaks the authentication with non-negligible probability). Putting all of this together, it is straightforward to show that $\Pr[b = b' | \overline{F}, b = 0] > \frac{1}{2} + \frac{1}{h(\ell)}$ for some polynomial $h$.

In summary, $\Pr[b = b' | \overline{F}] - \frac{1}{2}$ is non-negligible, hence so is $\Pr[b = b'] - \frac{1}{2}$. □

### 4.3   Final Scheme

Here we present the final scheme, where any user is required to currently possess the device in order to be able to successfully authenticate. The principal idea behind this approach is that the device does not reveal the response $r$ in any protocol. The device produces only zero-knowledge proofs that it possesses the secret $r$. And since the proofs are zero-knowledge an adversary cannot learn $r$ by observing the device. Unlike the previous two protocols, this protocol assumes that the PUF can also perform computation.

Enroll :
1. Server $\mathcal{S}$ sends challenge $c$ to user $\mathcal{U}$ along with description of the group $\mathbb{G}_q$, denoted by $\langle \mathbb{G}_q \rangle$ and which could consists of a pair $(p, q)$, and its generator $g$.
2. User $\mathcal{U}$ sends $H(c \| pwd), \langle \mathbb{G}_q \rangle, g$, where $pwd$ is a user password, to device $\mathcal{D}$ for a modified Gen protocol.
3. Device $\mathcal{D}$ calculates a challenge $d = H(H(c \| pwd), \langle \mathbb{G}_q \rangle, g)$ and runs Gen on this value to obtain response $r, P$. $\mathcal{D}$ then sends to the user $(g^r, P)$.
4. User forwards $(g^r, P)$ to server $\mathcal{S}$, which stores the information along with $c, g, \langle \mathbb{G}_q \rangle$.

Auth :
1. Server $\mathcal{S}$ sends challenge $c, \langle \mathbb{G}_q \rangle, g, P$, and a nonce $N$ to the user $\mathcal{U}$.
2. $\mathcal{U}$ sends $(H(c \| pwd), \langle \mathbb{G}_q \rangle, g, P, N)$ to device $\mathcal{D}$ for Rep protocol.
3. Device $\mathcal{D}$ calculates a challenge $d = H(H(c \| pwd), g, p)$ and runs Rep on this value to obtain response $r$. $\mathcal{D}$ chooses a random value $v \in \mathbb{Z}_q$ and calculates $t = g^v$. $\mathcal{D}$ then calculates $c' = H(g, g^r, t, N)$ and $w = v - c'r \bmod q$, and sends $c', w$ to the $\mathcal{U}$.
4. User $\mathcal{U}$ sends these values to the server $\mathcal{S}$. $\mathcal{S}$ calculates $t' = g^w g^{rc'}$ and accepts the authentication if $c' = H(g, g^r, t', N)$, and otherwise rejects the value.

What is implicit in this and previous schemes is the step where the user provides its account number or some other identifying information that permits the server to locate the user's record with the corresponding helper data $P$ and authentication verification information. What form this account information takes is not essential in our solution, and different mechanisms would be acceptable. For example, since we assume that the device does not store information permanently, the account number can be computed at the user side as a function of the user's identity and the bank name.

We first show security of a simpler (but very similar) system that uses an oracle. In this system, the oracle is initialized by obtaining the group setup $\langle \mathbb{G}_q \rangle$ and $g$, after which it chooses a random value $r$ and publishes $g^r$. This operation is performed once, and all consecutive interactions with the oracle will use the same value $r$. After the setup stage, a user can query the oracle with a nonce $N$. On each query, the oracle chooses a random value $v \in \mathbb{Z}_q$ and calculates $t = g^v$. It then computes $c' = H(g,g^r,t,N)$ and $w = v - c'r \bmod q$, and replies with $c',w$ to the querier. We denote this oracle by $O_{auth}$.

The difference between this system and the PUF-based system is that the value $r$ is randomly chosen (rather than produced by the PUF) and the system cannot be used for authentications on different challenges $c$. Let an adversary be given black box access to oracle $O_{auth}$ and a challenge nonce $N$. The adversary is allowed to query the oracle on all values except the challenge nonce. We now argue that, assuming that $H$ is a random oracle, the adversary cannot forge a proof for the nonce $N$ to the challenger.

The core of the computation performed by the above oracle (and the device in our Auth protocol) is basically a proof of knowledge of the exponent of $g^r$ to the base $g$, where the proof uses a priori specified value of nonce $N$. The basic form of this proof of knowledge was used in different authentication protocols, including the standard Schnorr identification protocol [21]. It is well known that in Schnorr's protocol, if an adversary can produce the proof with a non-negligible probability, then there is a knowledge extractor that can produce $r$ with non-negligible probability. That basic argument is that $t$ must be chosen before $c'$ and thus for a given value of $t$ there must be a non-negligible portion of $c'$ values for which an adversary can construct a proof. Furthermore, if the adversary can construct two proofs for the same $t$ value but different $c'$ values, then they can obtain $r$. We now argue that, if such a knowledge extractor exists, then assuming the random oracle model, there is a polynomial time adversary that can solve the discrete logarithm problem.

**Lemma 2.** *Any polynomial-time user has at most negligible probability of success authenticating in the above modified system with oracle $O_{auth}$.*

**Proof.** Let $A$ be a proof generator with oracle access to $O_{auth}$ that succeeds in answering the challenge for nonce $N$ with non-negligible probability. Assume that algorithm $B$ with access to $A$ is given a value $g^r$ and is asked to provide $r$. $B$ provides $A$'s access to random oracles $H$ and $O_{auth}$ and answers such queries as follows. Recall that in all queries to $O_{auth}$ the same $g,g^r$ are used.

1. $B$ creates a list of values $L_{OH}$ that will store queries to $O_{auth}$ and $H$. The list is initially empty.
2. When $A$ queries $O_{auth}$ on a nonce value $N$, $B$ does the following: it chooses a random response $w \in \mathbb{Z}_q$ and a random value $c \in \mathbb{Z}_q$. It sets $t = g^w g^{rc}$ and then adds the pair $(t,N,c)$ to $L_{OH}$; but if there is already a tuple $(t,N,\hat{c})$ in $L_H$, $\hat{c} \neq c$, then $B$ outputs FAIL. It returns $w,c$ to $A$.
3. When $A$ queries $H$ on $(g,g^r,t,N)$, $B$ searches $L_{OH}$ for a value of the form $(t,N,c)$ for some $c$. If it exists, it responds with $c$. If not, $B$ chooses a random value $c$, adds $(t,N,c)$ to $L_{OH}$, and responds with $c$.

We first argue that $B$ outputs FAIL with negligible probability. The only way it happens is if, when answering a query to $O_{auth}$, the chosen value $t$ is already in $L_{OH}$. However, $t$

will be a randomly chosen value in $\mathbb{G}_q$ and, since there are at most a polynomial number of tuples in $L_{OH}$, the probability of an overlap is negligible. Therefore, if $A$ succeeds with non-negligible probability, a knowledge extractor would exist that allows $B$ to obtain $r$. Thus, no such $A$ exists, assuming the discrete logarithm problem is hard.    □

Now consider a challenger that provides an adversary with oracle access to a PUF. The adversary queries a challenger with either Enroll or Auth queries. The challenger answers all queries with the PUF. Eventually the adversary asks for a challenge and is given $c, \langle \mathbb{G}_q \rangle, g$, and a nonce $N$. The adversary can then continue to ask Enroll and Auth queries (but cannot ask for Auth on the specific nonce $N$ and the specific challenge). The goal of the adversary is to be able to construct a response to the challenge that would pass the authentication verification at the server. The adversary wins if the authentication is successful.

**Theorem 2.** *Any polynomial-time adversary without the proper I-PUF device is unable to successfully authenticate with more than negligible probability in the* Auth *protocol.* Proof omitted due to page constraints.

## 5   Adding Emergency Capabilities

We would like to provide a user under duress with the possibility to lie about her secrets in such a way that a "silent alarm" is triggered at the server. The coercer should be unable to distinguish between an authentication protocol with real password and one with an emergency password; nor should it be detectable that the authentication protocol has provisions for using different secrets. More precisely, we consider an adversary who can record the user's communication with the server during successful authentication protocols, but does not have access to the communication between the user and the server at the enrollment stage. The adversary then forces to the user to reveal all information the user possesses in relation to authentication, including all secrets such as passwords, and also obtains physical access to the user's device. The adversary engages in an authentication protocol with the server on behalf of the user. We require that all information the adversary observes with full access to the user-provided data and the device does not allow it to distinguish its communication with the bank from the previously-recorded communication of the user with more than negligible probability. This means that all messages must follow exactly the same format and the distributions of data on different executions are not distinguishable.

We next present a scheme that has this capability. Often the above problem of coercion is addressed by letting the user choose two different passwords (or PINs), the first for normal operation and second for emergencies (i.e., it also sets off an alarm). This simple approach no longer works for PUFs because of the noisy nature of their responses. That is, the server will need to send the appropriate helper data $P$ prior to knowing what password is being used; sending two helpers would be a tipoff to the coercer. We solve this problem by splitting each password (real and false) in two parts: the first part is identical in both passwords and it used by PUF to compute its challenge and response. The second halves are different, but the PUF is not queried on their values.

Enroll :
1. Server $S$ sends challenge $c$ to user $\mathcal{U}$ along with $\langle \mathbb{G}_q \rangle$ and generator $g$ of $\mathbb{G}_q$.
2. User $\mathcal{U}$ sends $c, \langle \mathbb{G}_q \rangle, g, pwd_1, pwd_2, pwd_3$, where $pwd_i$'s are three user passwords, to device $\mathcal{D}$ for a modified Gen protocol.
3. Device $\mathcal{D}$ calculates a challenge $d = H(H(c \| pwd_1), \langle \mathbb{G}_q \rangle, g)$ and runs Gen on this value to obtain response $r, P$. $\mathcal{D}$ then sends to the user $(g^{H(r\|pwd_2)}, g^{H(r\|pwd_3)}, P)$.
4. User $\mathcal{U}$ forwards $(g^{H(r\|pwd_2)}, g^{H(r\|pwd_3)}, P)$ to server $S$, which stores the information along with $c, g, \langle \mathbb{G}_q \rangle$.

Auth :
1. Server $S$ sends challenge $c, \langle \mathbb{G}_q \rangle, g, P$, and a nonce $N$ to the user $\mathcal{U}$.
2. $\mathcal{U}$ sends $(c, \langle \mathbb{G}_q \rangle, g, P, N, pwd)$ to device $\mathcal{D}$ for Rep protocol, where $pwd = pwd_1 \| pwd_2$ in a normal login and $pwd = pwd_1 \| pwd_3$ in an emergency login.
3. Device $\mathcal{D}$ splits $pwd$ into two parts $pwd'$ and $pwd''$. $\mathcal{D}$ then calculates its challenge $d = H(H(c \| pwd'), g, p)$ and runs Rep on this value to obtain response $r$. $\mathcal{D}$ chooses a random value $v \in \mathbb{Z}_q$ and calculates $t = g^v$. $\mathcal{D}$ then calculates $c' = H(g, g^{H(r, pwd'')}, t, N)$ and $w = v - c'H(r\|pwd'') \bmod q$, and sends $c', w$ to the $\mathcal{U}$.
4. User $\mathcal{U}$ sends these values to the server $S$. $S$ calculates $t' = g^w g^{H(r\|pwd_2)c'}$ and accepts the authentication as normal if $c' = H(g, g^{H(r\|pwd_2)}, t, N)$. If this fails, then $S$ calculates $t' = g^w g^{H(r\|pwd_3)c'}$ and accepts the authentication as an emergency login if $c' = H(g, g^{H(r\|pwd_3)}, t, N)$.

We now argue the coercer's inability distinguish between normal and emergency executions. As stated earlier, the coercer can record $\mathcal{U}$'s normal authentications prior to coercion. The communication between $\mathcal{U}$ and $S$ during Auth consists of values $(c, \langle \mathbb{G}_q \rangle, g, P, N, c', w)$, where the first five are sent by the server as a challenge and the last two are the user's reply. Coercion gives the adversary the device $\mathcal{D}$ and user's password $pwd = pwd_1 \| pwd_3$, that he then uses in the protocol. We now formally state (the proof is omitted due to page constraints) that the adversary's view of the protocol after the coercion is indistinguishable from its view of previous invocations of the protocol by the user.

**Theorem 3.** *A polynomial-time coercer with access to private user data and I-PUF has negligible probability of distinguishing between normal and emergency executions.*

In the above solution, our goal was to provide an authentication mechanism where the communication during the protocol upon use of emergency password cannot be distinguished from normal communication, i.e., the observable response remains the same regardless of what password is used. The actions taken by the server, however, can be different depending on what password is used (e.g., in emergency, a silent alarm can sound at the bank and the ATM can issue marked bills). The work of Clark and Hengartner [1] ties the use of a panic password (or passwords) to the context in which this functionality is used, as well as the goals and capabilities of the adversary. It is assumed that the system design is open, in which case the adversary will be aware of the emergency capabilities of the system. The adversary is also capable of forcing the user to authenticate several times, possibly using different passwords. The adversary thus can force the user to open all (regular and panic) passwords he has. In our case, the goals of the adversary can be to avoid detection (i.e., not trigger the alarm at the bank) or escape with unmarked money (i.e., authenticate at least once with the regular password). We refer the reader to [1] for more information on how such goals can be achieved with one or more panic passwords. The goal of this section is to provide a protocol to support emergency capabilities that can be combined with any policy the system wants to employ in terms of how to use and respond to panic passwords.

One weakness of our protocol is that the adversary could force a user to reveal two passwords, and then choose one of the passwords at random. Once the user reveals multiple passwords, the adversary would then either have a 50% chance of either catching the user in a lie (if the user provided a bad password) or a 50% chance of using the non-emergency password (if the user did not provide a bad password). We leave the mitigation of this problem for future work.

# 6   Conclusions

In this work we describe authentication solutions based on a PUF device that provide stronger security guarantees to the user than what previously could be achieved. In particular, in our solution each user is issued a device that aids in authentication and cannot be copied or cloned. We ensure that: (i) the device alone is not sufficient for authenticating; (ii) the user must have the device in order to successfully authenticate; (iii) anyone with complete access to the authentication data at the server side and the device itself is still unable to impersonate the user (even if the access to the device is possible prior to account setup). These guarantees hold in the random oracle model.

As another contribution of this work, we add protective mechanisms to the protocol that allow institutions to quickly recognize attacks when a user is coerced into revealing her secrets. We allow the user to have an alternative secret that triggers an alarm at the corresponding institution, but allows for successful authentication in such a way that the adversary is unable to distinguish between protocol executions that use the regular and alternative secrets.

A future direction of research is to achieve similar results, but without the random oracle model. More broadly, there is a need for a systematic investigation of the implications of PUFs for security functionalities other than authentication, such as fighting piracy, policy enforcement, tamper-resistance, and anti-counterfeiting.

## Acknowledgments

The authors thank the anonymous reviewers for their comments and useful suggestions.

## References

1. Clark, J., Hengartner, U.: Panic passwords: Authenticating under duress. In: USENIX Workshop on Hot Topics in Security, HotSec 2008 (2008)
2. Suh, G., Devadas, S.: Physical unclonable functions for device authentication and secret key generation. In: DAC, pp. 9–14 (2007)

3. Ozturk, E., Hammouri, G., Sunar, B.: Towards robust low cost authentication for pervasive devices. In: IEEE International Conference on Pervasive Computing and Communications, pp. 170–178 (2008)
4. Simpson, E., Schaumont, P.: Offline hardware/software authentication for reconfigurable platforms. In: Goubin, L., Matsui, M. (eds.) CHES 2006. LNCS, vol. 4249, pp. 311–323. Springer, Heidelberg (2006)
5. Guajardo, J., Kumar, S., Schrijen, G.J., Tuyls, P.: Physical unclonable functions and public-key crypto for FPGA IP protection. In: International Conference on Field Programmable Logic and Applications, pp. 189–195 (2007)
6. Guajardo, J., Kumar, S.S., Schrijen, G.J., Tuyls, P.: FPGA intrinsic PUFs and their use for IP protection. In: Paillier, P., Verbauwhede, I. (eds.) CHES 2007. LNCS, vol. 4727, pp. 63–80. Springer, Heidelberg (2007)
7. Tuyls, P., Batina, L.: RFID-tags for anti-counterfeiting. In: Pointcheval, D. (ed.) CT-RSA 2006. LNCS, vol. 3860, pp. 115–131. Springer, Heidelberg (2006)
8. Bolotnyy, L., Robins, G.: Physically unclonable function-based security and privacy in RFID systems. In: IEEE International Conference on Pervasive Computing and Communications (PerCom 2007), pp. 211–220 (2007)
9. Batina, L., Guajardo, J., Kerins, T., Mentens, N., Tuyls, P., Verbauwhede, I.: Public-key cryptography for RFID-tags. In: Pervasive Computing and Communications Workshops, pp. 217–222 (2007)
10. Hammouri, G., Sunar, B.: PUF-HB: A tamper-resilient HB based authentication protocol. In: Bellovin, S.M., Gennaro, R., Keromytis, A.D., Yung, M. (eds.) ACNS 2008. LNCS, vol. 5037, pp. 346–365. Springer, Heidelberg (2008)
11. Hammouri, G., Ozturk, E., Birand, B., Sunar, B.: Unclonable lightweight authentication scheme. In: Chen, L., Ryan, M.D., Wang, G. (eds.) ICICS 2008. LNCS, vol. 5308, pp. 33–48. Springer, Heidelberg (2008)
12. Park, Y.M., Park, S.K.: Two factor authenticated key exchange (TAKE) protocol in public wireless LANs. IEICE Transactions on Communications E87-B(5), 1382–1385 (2004)
13. Pietro, R.D., Me, G., Strangio, M.: A two-factor mobile authentication scheme for secure financial transactions. In: International Conference on Mobile Business (ICMB 2005), pp. 28–34 (2005)
14. Bhargav-Spantzel, A., Sqicciarini, A., Modi, S., Young, M., Bertino, E., Elliott, S.: Privacy preserving multi-factor authentication with biometrics. Journal of Computer Security 15(5), 529–560 (2007)
15. Stebila, D., Udupi, P., Chang, S.: Multi-factor password-authenticated key exchange. Technical Report ePrint Cryptology Archive 2008/214 (2008)
16. Boyen, X.: Reusable cryptographic fuzzy extractors. In: ACM Conference on Computer and Communications Security (CCS 2004), pp. 82–91 (2004)
17. Gassend, B., Clarke, D., van Dijk, M., Devadas, S.: Silicon physical random functions. In: ACM Conference on Computer and Communications Security (CCS 2002), pp. 148–160 (2002)
18. Skoric, B., Tuyls, P.: Secret key generation from classical physics. Philips Research Book Series (2005)
19. Dodis, Y., Reyzin, L., Smith, A.: Fuzzy extractors: How to generate strong keys from biometrics and other noisy data. In: Cachin, C., Camenisch, J.L. (eds.) EUROCRYPT 2004. LNCS, vol. 3027, pp. 523–540. Springer, Heidelberg (2004)
20. Chaum, D., Evertse, J.H., van de Graaf, J.: An improved protocol for demonstrating possession of discrete logarithms and some generalizations. In: Price, W.L., Chaum, D. (eds.) EUROCRYPT 1987. LNCS, vol. 304, pp. 127–141. Springer, Heidelberg (1988)
21. Schnorr, C.: Efficient signature generation by smart cards. Journal of Cryptology 4(3), 161–174 (1991)

# Risks of the CardSpace Protocol

Sebastian Gajek[1], Jörg Schwenk[1], Michael Steiner[2], and Chen Xuan[1]

[1] Horst Görtz Institute for IT-Security
Ruhr-University Bochum, Germany
{sebastian.gajek,joerg.schwenk,chen.xuan}@rub.de
[2] IBM T.J. Watson Research Center, USA
msteiner@watson.ibm.com

**Abstract.** Microsoft has designed a user-centric identity metasystem encompassing a suite of various protocols for identity management. CardSpace is based on open standards, so that various applications can make use of the identity metasystem, including, for example, Microsoft Internet Explorer or Firefox (with some add-on). We therefore expect Microsoft's identity metasystem to become widely deployed on the Internet and a popular target to attack. We examine the security of CardSpace against today's Internet threats and identify risks and attacks. The browser-based CardSpace protocol does not prevent against replay of security tokens. Users can be impersonated and are potential victims of identity theft. We demonstrate the practicability of the flaw by presenting a proof of concept attack. Finally, we suggest several areas of improvement.

**Keywords:** CardSpace, identity management, analysis.

## 1 Introduction

Microsoft has introduced the CardSpace identity metasystem [1]. In essence, CardSpace follows the line of identity management (for short, IM) protocols (e.g. [2,3,4,5,6]). These protocols adapt the idea of Needham-Schroeder's third party model of authentication (for short, NS) to the Web, in which two players wish to negotiate a shared secret and ask a trusted third party for assistance. In the IM setting, the share is a security token containing the user's identity information (dubbed *claims*), such as email and shipping address. Whereas NS protocols are self-contained, IM protocols are composite protocols: They are restricted to standard Internet technologies and few browsing functionalities, such as message parsing and cross-domain access control. These design constraints fit into existing browser-server technologies. However, they also turned out to be a security challenge for identity management.

RELATED WORK. The literature of IM protocols is peppered with vulnerability results. Kormann and Rubin [7] analyze CardSpace's predecessor .NET Passport, and disclose several risks and attacks. The authors demonstrate that the adversary may steal the security token concealed in a ticket granting ticket cookie by mounting DNS attacks. In their attack description, they assume that the

P. Samarati et al. (Eds.): ISC 2009, LNCS 5735, pp. 278–293, 2009.

average user does not properly understand SSL and is probably not aware of SSL certificates. Groß [8] analyzes SAML, an alternative identity management protocol, and shows that the protocol is vulnerable to adaptive attacks. The adversary intercepts the authentication token contained in the URL. Groß makes use of the fact that browsers add the URL in the HTTP referrer tag when they are redirected. Hence, a man-in-the-middle adversary signaling the browser to redirect the request to a rogue server retrieves the authentication token from the referrer tag. The previously described deficiencies in the SAML protocol have led to a revised version of SAML. Groß and Pfitzmann analyze this version, again finding the need for improvements [9]. Similar vulnerabilities have been found in the analysis of the Liberty single sign on protocol [10]. Pfitzmann and Waidner point out some weaknesses in presence of man-in-the-middle attacks.

CONTRIBUTIONS. We analyze the security of CardSpace focusing our attention on the browser-based protocols. These protocols have in common that the user employs a commodity Web browser to participate in the protocol. This is the crux with CardSpace. Attacks that contaminate the security of commodity browsers carry over to CardSpace. We expose some security vulnerability in the way the protocol is interfaced to distinguished browser functionalities and show that the vulnerability is exploitable under reasonable assumptions. We describe an attack where the security token is extracted from the protocol execution. This is a crucial security problem. By replaying the token, the attacker may impersonate the user and gain access to her services. In order to demonstrate not only the feasibility, but also the attack's practicability, we present a proof of concept attack implementation. We discuss countermeasures in this work and demonstrate that minor modifications to CardSpace achieve the desired protection against the identified risk. Our recommendations include (a) refinement of the protocol by binding the CardSpace token to the underlying communication channels, and (b) a strengthening of the browser security model by shaping a cross-access policy to permit only authenticated players to retrieve the security token. Both provisions are efficiently realizable and guarantee that no feasible attacker replays the security token or injects malicious scripts into the hijacked communication. We remark that our technique is independent of the underlying protocol and therefore is valuable measure to protect any browser-based protocol against SOP contamination and replay of credentials.

ORGANIZATION. The remainder of this paper is organized as follows: In Section 2, we briefly introduce and discuss the browser-based CardSpace protocol. In Section 3, we describe risks and attacks of the protocol and present in Section 4 some countermeasures to fix the problem. In Section 5, we draw our conclusions.

## 2    Microsoft's Identity Metasystem CardSpace

### 2.1    Roles and Interfaces

The CardSpace identity management protocol for browser-based protocols involves the following participants:

- The *user* is a subject who can have a digital identity and need to prove her identity to a relying party in order to access some authorized services or resources.
- The *identity selector* plays an important role in the whole CardSpace concept. It is a user interface, which is used to manage a user's *InfoCards*[1] and retrieve an associated security token from an identity provider. Technically, the identity selector serves as browser plug-in and enhances the (cryptographic) mechanisms of commodity browsers.
- Another user interface and active participant of the CardSpace protocol is the *client application*. In this paper, it is a commodity Web browser that interoperates with the identity selector. The browser mediates the messages between the user, identity selector and the relying party.
- The *identity provider* provides a digital identity for a user and assures that the user really is who she claims to be. An identity provider is commonly a trusted third party who defines a user's identity in the form of an InfoCard and provides some security token services, which may issue the associated security token on the fly.
- The *relying party* is a Web site or an application that in some way accepts and relies on a digital identity from a user. A relying party provides some resources or services, consumes an identity presented as claims and contained in a security token to authenticate a user, and then makes an authorization decision, such as allowing this user to access some of its resources or services.

REMARK. The essential design difference between the Passport and related browser-based identity management protocols is that CardSpace takes advantages of an additional party, namely the identity selector. This selector considerably improves the functionalities of commodity browsers in order to relieve the user's burden to make a wrong security discussion (see Section 3 for more discussions). In essence, all the user has to do is to select an InfoCard. The protocol complexity is shifted to the identity selector who performs the remaining security tasks. Obviously, CardSpace's design is in the spirit of the average non-expert computer user. Nowadays, users are target to various attacks and it is widely accepted that users are the weakest link in the identity system. Since CardSpace can become an important player in the future Internet infrastructure it is important to provide a particularly *robust* solution against today's risks and threats, including attacks against the user. Apparently, CardSpace strives to achieve this goal. Otherwise, there is no reason to move away from the former 3-party model, which is realizable with arbitrary browser in arbitrary location.

---

[1] InfoCards are a concept introduced in CardSpace that is very similar to business cards issued from identity providers. Roughly speaking, an InfoCard contains information to locate an identity provider. An InfoCard does not contain any security critical data so that compromise of InfoCards or careless behavior leads to a security breach. The user pre-installs the InfoCard in a setup phase where she provides the identity provider with her claims.

## 2.2    How Does CardSpace Work?

We now take a closer look into the browser-based protocol details. See Fig. 1. The protocol proceeds in four stages:

**Retrieving RP's Policy (Step 1).** First of all, the user calls the relying party's login page in her browser. A relying party expresses its token requirement directly in the HTML text of the Web site. HTML extensions are used to signal to the browser when to invoke the identity selector. Essentially, there are two different syntaxes for calling into the identity selector: An HTML and an XHTML object tag. The HTML syntax is used by an ActiveX object, and the XHTML object called *CardSpaceToken* is used by a binary behavior object. Because existing sites more commonly adopt the HTML tag, we have built our attack firstly on this syntax. The other syntax will not be addressed any more in this paper. (For more information about the XHTML syntax, we refer to [11].) Further, the relying party can define its policy in this object. Most importantly, the relying party specifies in its policy the claims it requires from the user. Technically, the CardSpaceToken object is wrapped into a standard HTML form. Forms are commonly used in Web sites to collect information from a user for processing.

If the user decides to click on the login button on the relying party's login page, the browser retrieves the relying party's policy, calls the identity selector and forwards the relying party's policy to the identity selector. The identity selector displays first the identity of relying party's Web site to help the user make an informed decision. If the Web site is secured over HTTPS, the relying party's certificate is displayed to the user and the user is asked whether she wants to accept it. If the user does not trust the site, she can directly abort the protocol. Otherwise, the protocol continues and the identity selector displays all the InfoCards that satisfy the relying party's policy. The user is then asked to select an InfoCard from this set. At the end of this phase, the identity selector knows what claims the relying party is actually asking for and which InfoCards can be used for this authentication.

**Retrieving IP's Policy (Step 2).** After the user has selected an InfoCard, the identity selector extracts the identity provider's *security token service (STS)* information and retrieves the identity provider's policy. An STS is the service from which the actual security tokens are retrieved. Every identity provider runs one or more STSs and issues security tokens for users on the fly. The transport of the identity provider's policy is secured with HTTPS to prevent policy tampering attacks. At the end of this phase, the identity selector gets the identity provider's policy from its STS and knows the security requirement of the identity provider.

**Retrieving the Security Token (Step 3).** Depending on the policy of identity provider's STS service, the identity selector prompts the user to enter her authentication information as user credential. This authentication information will later be wrapped into an authentication token and handed to the identity

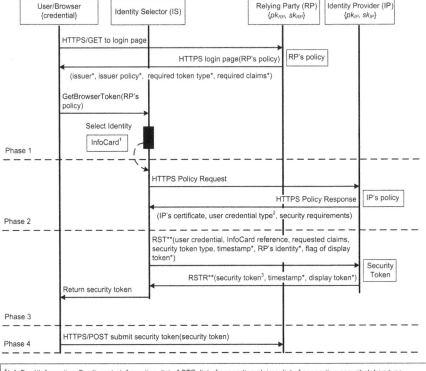

**Fig. 1.** Browser-based CardSpace Identity Management Protocol

provider within the *request security token (RST)* message. Four user credential types are supported in the current CardSpace implementation: They are username/password, KerberosV5 service ticket, X.509v3 certificate and self-issued token.[2] Apart from the user authentication information, the identity selector also includes the following information in the RST message: An InfoCard reference, required claims which are requested from the relying party, the type of the desired security token, and optionally a timestamp to ensure the freshness of the message, the relying party's identity and a flag indicating if a display token is required. A display token contains the claims to be displayed to the user, before sending them to the relying party (see below for more discussions). The RST message is secured in the way that defined in the identity provider's policy, i.e. over HTTPS. Note that the identity selector will always be able to authenticate the identity provider because it knows the identity selector's public key being

---

[2] Self-issued tokens are generated by the user herself. She acts as her own identity provider.

part of the InfoCard. This is an essential design decision to thwart impersonation attacks in the light of naive users.

After the identity provider has successfully checked the user's credential, it responses with a *request security token response (RSTR)* message, which is also secured with an identity provider preferred method. As its name suggests, this message should include a security token in the desired format. According to the user's InfoCard reference, the STS service finds the user's identity information in its database and adds the required claims into the security token. If the relying party's identity is disclosed to the identity provider, the relying party's URL should also be included in the security token. Then the security token should be signed by the identity provider with its private key and encrypted to the relying party's public key. Note that if the identity provider does not encrypt the security token, the identity selector will automatically encrypt it to the relying party's public key.

The RSTR message optionally includes a timestamp to ensure the freshness of the message and a display token, if the display token flag is set in the RST message. Since the identity selector is not able to read the content of the security token, a display token, which contains the security token's content in a textual format, is used to display to the user the content of the security token. The user may check the content of the security token before she submits it to the relying party. If the user is not satisfied with it, she may choose to abort the protocol. The purpose of the display token is to give the user a feeling of transparency. She shall be aware of the claims to be submitted to the relying party.

**Submitting the Security Token (Step 4).** The browser submits the token to the relying party with the HTTPS/POST method. After the relying party receives the security token, it decrypts the security token with its private key, verifies the identity provider's signature and checks the user's identity claims. If everything is correct, the relying party returns back the requested service or resource. At the end of the protocol, the user is successfully authenticated to the relying party and gains access to the relying party's Web site.

# 3   Risks of the CardSpace Protocol

We identify a risk in the design of the browser-based CardSpace protocol. Whenever the relying party receives a security token, the protocol does not ensure that the user has been in fact involved in the protocol execution. The token is encrypted to the relying party's public key and signed by the identity provider. Our first observation is that there is neither a cryptographic binding to the user nor to the underlying secure channel (where the user implicitly proved her identity). The token does not contain the user's identity; it contains the user's identity information in form of claims.

Our second observation is that this construction offers no protection against token replay. Consequently, any party in possession of the security token—be it the honest user or the adversary—may act as legitimate user and gain access to the relying party's service, thus contradicting the security of mutual authentication[3] between user and relying party. Note, however, that the adversary learns nothing about the claims from the security token (because they are encrypted to the relying party's public key.)

Our third observation is that the browser mediates the security token. When the identity selector forwards the token, it embeds the token value into the *CardSpaceToken* object. Like any HTML object, this object becomes part of the browser's document object model (DOM[4]). Browsers offer a standardized DOM API interface that permits scripting languages (e.g. JavaScript) to access parts of the Web document. The *same-origin policy (SOP)*, a security policy universally supported in browsers, enforces that the access of scripts is limited to objects originating from the same source. Access across different Web objects is evaluated based on the object's *protocol name, domain name* and *port*. (The port number is implicitly defined by the protocol in use.)

The risk of the browser-based CardSpace protocol is that an attacker thwarting the SOP check gets the privilege to access the CardSpace token. This is a general problem of browsers and carries over to the identity management protocol. An attacker subverting the SOP efficiently contaminates the security of CardSpace. Exactly this dependency is exploited in our attack (as we will see below).

## 3.1   The SOP Problem

The SOP is a legacy security policy. It is widely believed that the SOP does provide weak isolation of Web content [12]. Given the fact that today's browsers heavily decide on the basis of protocol and domain name whether to trust some source, we revise the problems with enforcing the same-origin policy.

**Problems with SOP's Domain Name Check.** The same-origin policy is unexceptionally based on DNS host names. But actually the network access is performed with the help of IP addresses. When the browser starts loading network content, the DNS system first resolves the host name and after this the request is sent with the help of the IP address, which is technically defining the destination. The origin of the content will be still determined as DNS host name and the SOP does only work properly in case there is no mismatch between DNS host name and technical IP address. Unfortunately the today's used DNS system is vulnerable to a number of known attacks that can be classified as server-sided and client-sided attacks.

Server-sided DNS spoofing attacks like *DNS name chaining* or *DNS cache poisoning*[5] stand for attacks that lead directly to a DNS domain name-IP address mismatch. Another type of server-sided attacks that also circumvent the SOP are

---

[3] Informally, a protocol is said to be secure in the sense of mutual authentication (matching conversation), if (a) the relying party receives in the same protocol session messages the user has sent, and (b) the user receives in the same protocol session messages the relying party has sent.

[4] http://www.w3c.org/dom

[5] DNS cache poisoning is well-known for a long time, but has recently gained much attention under the headline "DNS debacle". See also [13].

attacks like cross-site-scripting or cross-side-request-forgery [14,15,16], enabled by improper input validation.

Client-sided attacks are recently discussed because they do not require attacker's control regarding the server side. They gained much attention under the terms *Drive-By pharming* [17] and *DNS rebinding attacks* [18,19] and have in common that browser-based languages are applied to alter home router DNS configurations and connect to different IP addresses with the same host name using multiple DNS A records or time-varying DNS responses, respectively. A relaxed attack thereof is *Wi-Fi spoofing* [20], where the adversary offers free Internet access points at public places. When the user connects to the wireless network, the adversary can assign arbitrary DNS server to her.

These attacks reflect the today's view that the adversary controls the network, delays, alters and sends arbitrary network messages including *unauthenticated* DNS resolution responses.

**Problems with SOP's Protocol Check.** Secure socket layer (SSL) is a countermeasure facing MITM attacks in the field of Internet communication: It targets confidentiality, integrity and sender authentication between server and client at transport layer. Combined with HTTP the protocol is called HTTPS. Although the security of the TLS protocols is proved [21], there are well-known problems with trusting and verifying certificates: weak issuing policies and insufficient verification of certificate requesters by the CAs have shown that attackers receive valid certificates for their rogue servers. Very recently, it has been shown that chosen-prefix attacks are constructible with reasonably computational resources that allow creation of rogue CA root certificates [22]. With a rogue root certificate, the attacker may create arbitrarily many valid server certificates. This discloses an inflation of CAs' verification trustworthy—usual validation loses its trustworthiness more and more.

Another problem is the interaction between browser and user and her lack of awareness. If certificates are self-signed, out-dated, or do not match the host name, browsers can warn the user. Users tend to ignore these warnings [23,24] and may even declare that they do not want to see such a warning again. New browser versions make it harder to ignore such warnings. Despite improved user interfaces, it turned out that users still tend to ignore certificate warnings. This assumption is justified by several usability studies [23,24,25,26] and reflected by the increasing number of phishing attacks in practice.

A further unsolved problem appears in mashup web applications, where different SSL states cannot be related to different parts of the displayed web page. At most one SSL information can be valuable if the user can relate it to the whole page he can see in the browser window. For instance, an HTTPS-loaded HTML file can load a (plain) HTTP-loaded gadget that is then treated as loaded via HTTPS. This fact should be seen as a kind of side-channel attack vector against SOP and care must be taken upon composing gadgets. Jackson and Barth pointed out that an adversary may exploit this side-channel to inject malicious content [27]. Still, the user cherishes the illusion that a private and authenticated channel protects the communication.

## 3.2  Replaying the Security Token

Our attack borrows ideas from Karlof et al. by deploying dynamic pharming [19]. We remark that any attack vector (or set thereof) subverting the SOP is useful to mount the attack. The essence of dynamic pharming is to thwart the domain name resolution, and lure the user to a rogue site, in which the relying party's site is loaded within an inline frame and utilize some script in order to access the relying party's frame. Therefore, the adversary needs to change the DNS entry of his own site back to the relying party's one so that the original login page is referred into an inline frame of the malicious site. According to the SOP, since the attacker's site and the referenced relaying party's site appear to have the same "origin", the attacker's site has access to the legitimate site. With malicious JavaScript codes in the malicious site, the adversary can hijack the user's login session and steal the user's security token.

An attack illustration is depicted in Fig. 2. Our demonstrator has been successfully tested on Windows XP SP2, running Internet Explorer Version 7.0.5730.13 and Windows Vista, running Internet Explorer 7.0.6000.16643. In detail, our dynamic pharming attack proceeds as follows:

1. The adversary manipulates the user's DNS server and adds a round robin entry for the relying party's domain to accomplish updating of the DNS entry. Two IP addresses are associated to the domain. For instance, the relying party's IP address is 1.1.1.1, and the attacker's IP address is 2.2.2.2.
2. When the user requests the URL, such as https://goodsite.com, the DNS server returns first of all the adversary's IP address (2.2.2.2). In order to complete an SSL connection, the adversary must use a certificate for his server. Either, he is lucky and receives a certificate from a certificate authority, enforcing weak issuing policies, or he switches to self-signed certificates. In our implementation, we opted for the latter. Then, the browser displays a warning to the user. The certificate is invalid because a trusted certificate authority did not sign it. As mentioned before, users ignore such security warnings continuously [23,25,26]. Though IE7 uses a full page warning instead of an unimpressive window alert offering standard options (i.e., ignore and continue, or cancel connection), studies show that users will ignore a full page warning as well [24]. Accordingly, we consider the attacker to overcome the warnings.
3. The adversary's malicious Web site is loaded into the user's browser instead of the legitimate site. The malicious site contains three Frames: Frame0 contains codes to stop the adversary's Web server. Frame1 contains the malicious JavaScript codes for stealing the security token and displays a text area which gathers the security token prior to its submission. The text area shall simply mimic the fact that the adversary has intercepted the security token. Frame2 hosts the legitimate login page of the relying party.
4. The adversary's Web server is stopped by Frame0 after the malicious site is loaded into the browser. One big challenge for adversaries is the browser's use of DNS pinning. With DNS pinning, a Web browser caches the result of a DNS query for a fixed period of time, regardless of the DNS entry's specified

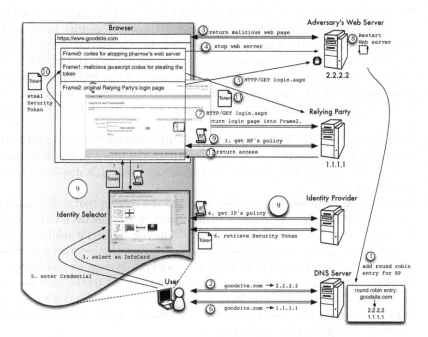

**Fig. 2.** Dynamic Pharming Attack against CardSpace

lifetime. Considering this in our attack, once the user has loaded the malicious site, the browser will keep using the adversary's IP address regardless of the update of entries in the DNS server. As a result, the legitimate Web site cannot be loaded correctly anymore. Fortunately for the adversaries, the browser will drop the current DNS pinning and refresh its DNS entry for a given domain, if the Web server on this domain is not reachable any more [18]. That means, after the adversary has delivered his malicious site to the user, he can now stop his Web server to reject the user's subsequent requests. Then, the user's browser first tries to load the content from the adversary's Web server because of the DNS pinning but fails of course, then it drops the DNS pinning, refreshes the DNS entry, receives the server's IP address and loads the legitimate site. The attack continues as before.

5. Because of the DNS pinning, `Frame2` first tries to request its content from the malicious server. The request is rejected and the browser drops the current DNS pinning.

6. Then the browser queries a new entry for the domain and this time the DNS server returns the relying party's original IP (1.1.1.1).

7. The legitimate login page is referenced into `Frame2`.

8. The malicious Web server is started again and waits for the next request.

9. After the user clicks on the *Sign in* button, the usual CardSpace authentication process begins as described in Section 2. The browser gets the relying party's policy, forwards it to the identity selector, the user selects an Info-

Card and retrieves an associated security token from an identity provider. At the end of this step, the identity selector returns the security token to the browser. Note that the CardSpace object hosting the relying party's policy in the legitimate site also does an internal check to enforce the SOP. The identity selector can only be displayed to users if the login page has the same "origin" as the root page. But because this check is also limited on the SOP, once the malicious site has passed the browser's check, it can pass CardSpace's check too.

10. With the malicious codes in `Frame1`, the security token is first recorded by the adversary. For example, the `Frame1` can submit the token to the adversary's Web server. In our implementation, the security token is displayed in the upper text area.

11. The adversary submits the security token to the relying party. We simulate this adversarial behavior in our proof of concept by pasting the security token from the text area into the clipboard. Next, one can choose an arbitrary other browser, navigate to our demonstration site, paste the security token into the text area, and press the 'login' button without ever using CardSpace.

12. Finally, the relying party returns the requested access to the user and the user successfully authenticates herself to the relying party. The adversary can reuse the security token anytime within the token's lifetime and authenticate to the relying party on behalf of the user.

### 3.3   Discussions

Clearly, one may argue that DNS spoofing can be used to hijack any TCP/IP connection and any protocol whose security depends on opening an SSL web page will be broken when a careless user does not verify the server certificate. However, under these two "phishing" assumptions identity theft proliferated to the fastest growing Internet crime. Since CardSpace aims at protecting average Internet users against identity theft, it is of prime importance to address this type of attacker. Otherwise, the identity system fails short like standard password authentication over SSL and does not improve the present situation. In the light of this attacker, the average Internet user does not take advantages from CardSpace.

Our attack model raises another challenging question. Is it possible for the attacker to inject malicious scripting code that is more powerful such that the adversary performs an even more sophisticated attack? For instance, is it possible to embed script code that generates an entire online transaction on behalf of the user? Then, it is needless to steal the token. The attacker could operate as man-in-the-middle, wait until the user redeems the security token, and then do some harm.

In the next section, we show that minor modifications to CardSpace achieve the desired robustness against this stronger attacker. We propose two improvements in this area. The first counteracts replay attacks by introducing a stronger cryptographic binding of the security token to its underlying secure channel. We show that breaking the security of our revised CardSpace protocol (in the sense of replaying the token) is as hard as breaking the security of the SSL handshake protocol. Our second improvement addresses the browser security model and prevents the attacker from embedding malicious scripts into the security context of the relying party.

# 4 Countermeasures

## 4.1 Protection against Replay Attacks

A weakness of the CardSpace protocol is that it countenances reuse of tokens. In possession of the security token, the attacker simply replays the token to gain access to the user's services. We wish to construct a protocol that foils the replay and makes the token undeniable. By undeniable we mean that (a) the identity provider can limit who can verify the validity of the security token and (b) user and most importantly relying party determine whether a security token is sincere in the sense that it truly origins from the user. The first goal is already realized in CardSpace by encrypting the security token to the relying party's public key. The latter, however, is neither addressed.

For the achievement of token undeniability, we recommend the binding of the security token with some high-entropy secret extracted from the channel between the user and the relying party. This provision makes sure that the relying party checks that the party requesting the login in the present session is in fact the party that has asked for the issuing of the security token. The secret is a *cryptographic channel identifier (cid)*, determining the active protocol session. The purpose of the identifier is to fingerprint the channel and make it aware of the session [28], while validating the fingerprint is only feasible by the session participants, i.e. browser and relying party. Ideally, both players compute *cid* from a shared secret. Otherwise, an interactive protocol would be necessary to synchronize the identifier. Technically, we deduce the identifier from the SSL handshake. In a nutshell, the SSL handshake provides a comprehensive framework of messages and cryptographic tools to negotiate a common secret called the *master secret k*. The master secret is then used to derive cryptographic keys for the instantiation of secure channels. A candidate approach would be to reuse the master key and compute the channel identifier in the spirit of the SSL keying material extraction specification [29]

$$cid \longleftarrow \mathrm{PRF}_k(\text{``channel identifier''})$$

where $\mathrm{PRF}()$ denotes the pseudo-random function as specified in SSL and 'channel identifier' is a string delimiter. A closer comparison with the native SSL specification [30] reveals that the pseudo-random function is additionally parameterized with the concatenation of nonces transmitted by the players in the initial handshake protocol rounds or a function thereof. We leave the nonces from the definition to assure that the *cid* is identically computed in the abbreviated SSL handshake, where the parties reuse the master secret to infer new session keys. The rationality behind our construction is that we wish to ensure a consistent channel identifier between the same instance of the full and abbreviated SSL handshake session. In some cases, an abbreviated handshake can be initiated to transmit the security token. Such a construction also appeals to single-sign on for multiple use of the security token. (The expiration date of the token is restricted to the time until either party erases the master key.) Assuming a secure pseudo-random function, it can be shown that *cid* is a fresh value and indistinguishable from a random number. The only potential mismatch to occur is that the adversary computes the master secret. However, it can be shown by reduction to the security of the SSL handshake itself that this event occurs with negligible probability in front of a probabilistic polynomial time-bounded adversary. See [21] for a proof. (Recall, we simply utilize the master key as seed for the pseudo-random function, but make no changes in the native handshake protocol.)

## 4.2   Protection against SOP Contamination

The crux with browsers is the enforcement of the legacy same-origin policy and the fact that the adversary gains access to the DOM, either bearing the security token or embedding arbitrary malicious script into the secure channel. We recommend to augment browser's same-origin policy with some authentication and isolation mechanisms by providing stronger object separation. The main idea is to prevent the attacker from accessing the relying party's DOM (including the token) via any scripting language. Nowadays, commodity Web browsers enforce some kind of "authentication" and "isolation" through the cryptographic functionalities provided by the SSL protocol. Web browsers offer the SSL protocol to securely exchange messages between two principals and prohibit that any feasible adversary eavesdropping the network alters messages. Evidently, SSL operates on transport layer and messages are secured while in transit on the network. Upon reception, the message plain text is forwarded to the application layer, meaning the browser's rendering engine. Unfortunately, objects are then processed according to the mature SOP and the attacker goes wild.

Karlof et al. propose a stronger browser policy using cryptographic identifiers to protect against dynamic pharming attacks [19]. Apart from checking the common "origin", the browser additionally validates the certificate chain corresponding to the SSL connection. Those adversaries who do not possess a valid certificate for the legitimate domain cannot apply their attacks anymore. The deficiency with the proposal is that in some cases the stronger SOP fails short. Examples discussed earlier include "side-channel" or rogue CA certificate attacks.

We recommend stronger isolation of HTML objects from scripting languages. To this end, we introduce the *Session-Correlated Cross-Communication Policy (SCPO)*. The SCPO ensures that objects from the same SSL session have the privilege to access each other. Therefore, the SOP check is augmented with the channel identifier *cid* introduced in the previous section. Informally, the SCPO policy enforces the following access rule:

*When the browser processes a DOM object for the very first time, it tags the object with protocol name, domain name, port number, and channel identifier. If and only if protocol name, domain name, port number, and channel identifier match, then the browser permits access to this object. Else, it ensures that no other object has access to this object.*

An ultimate requirement for the safe deployment of SCOP is to ensure that the channel identifier is not accessible by the DOM API. Then, the policy wraps script code into a secure compartment for the document objects triggered over different SSL sessions. It blinds the objects within the compartment. Consequently, a malicious script from a concurrent session will never perceive this object. Further, it is infeasible to address any object and descendant within this compartment, write into the objects, identify any wrapped objects, or intercept any events. Our proposal realizes a compartment ship for concealing security critical functionalities and data. The session-correlated cross-communication model simply ensures that objects have a privileged environment, where the execution of functionalities is prevented from any alternation. Applying the SCPO security policy to CardSpace means that we isolate the CardSpace object. Any access to the CardSpace element and its descendants is strictly prohibited by scripting languages (loaded from other SSL sessions). Reconciling our attack from Section 3.2, stealing the security token fails because `Frame1` containing the malicious script and `Frame2` including the relying party's site do not have matching channel identifiers.

The advantage of SCPO over the stronger policy proposed by Karloff et al. and the SOP is that a considerably stricter isolation mechanism is provided. The decision to grant an object access is based on the SSL session. It is more fine-grained. Another advantage is that a cryptographic channel identifier is used although we do not assume the presence of a CA. The channel identifier is an ephemeral secret. Thus, the SCPO is independent of the user's behavior or any PKI problem. That means, an attacker injecting malicious scripts from a concurrent session will be unable to gain access to the compartment—even if the user does not properly verify the SSL certificate or the attacker has a rogue CA certificate. Moreover, in some cases the SCOP protects against cross-site scripting and request forgery attacks. These are the cases, where the "cross site" is loaded from another SSL session. Clearly, the nitty-gritty idea of allowing scripts to dynamically alter the DOM is limited. However, we do not see the necessity for CardSpace to have that property. Recall, the only purpose of browsers is to mediate the security token from the identity selector to the relying party. Otherwise, the token's privacy cannot be guaranteed.

## 5  Conclusion

We have described and analyzed the browser-based CardSpace identity management protocol and identified some risks. We have built a proof-of-concept implementation of a dynamic pharming attack which observes a user's authentication process, steals the user's security token and impersonates the user to the legitimate relying party. Our proof-of-concept attack builds on identical adversarial assumptions Kormann and Rubin made in their analysis of CardSpace's predecessor .NET Passport [7]. In fact, the potential difference between the passport and CardSpace lies in the browser's handling of security tokens. Passport employs cookies, whereas CardSpace utilizes a new HTML object. From a security point of view, there are slight changes because browsers treat both mechanisms in the same way, i.e. scripting functionalities may gain the privilege to access the tokens. Despite the fact that CardSpace does an internal check to control the access by enforcing the same-origin policy, attacks which defeat the browser's SOP can also defeat the CardSpace's policy. No countermeasures have been made to protect the security token from interception and replay.

Microsoft has introduced with CardSpace a beautiful identity management system that considerably improves user authentication and protects against identity theft on the Internet. In order to make this identity metasystem robust against more sophisticated attacks of identity theft, we have deduced several countermeasures for the improvement of CardSpace design. First of all, the security token should be linked to the user's SSL connection to the relying party. This measure prevents adversaries from reusing the security token in another session. Second of all, an advanced SOP check should be deployed. We discussed a strengthening of the proposal to guard against SOP contamination, namely to completely isolate the CardSpace object and prohibit that any scripting functionality can access the relying party's DOM objects including the security token.

# References

1. Nanda, A.: A technical reference for the information card profile v1.0 (2006)
2. Liberty Alliance Project: Liberty Phase 2 final specifications (2003)
3. Kaler, C. (ed.): A.N.: Web Services Federation Language (WS-Federation), Version 1.0, BEA and IBM and Microsoft and RSA Security and VeriSign (2003)
4. OASIS Standard: Security assertion markup language, SAML (2002), http://www.oasis-open.org/committees/security/docs/
5. Cantor, S., Erdos, M.: Shibboleth-architecture draft v05 (2002)
6. Microsoft Corporation: .NET Passport documentation, in particular Technical Overview, and SDK 2.1 Documentation (started 1999) (2001)
7. Kormann, D., Rubin, A.: Risks of the passport single signon protocol. Computer Networks 33(1-6), 51–58 (2000)
8. Groß, T.: Security analysis of the SAML single sign-on browser/artifact profile. In: ACSAC 2003. IEEE Computer Society, Los Alamitos (2003)
9. Groß, T., Pfitzmann, B.: SAML artifact information flow revisited. In: Workshop on Web Services Security. IEEE Computer Society, Los Alamitos (2006)
10. Pfitzmann, B., Waidner, M.: Analysis of liberty single-sign-on with enabled clients. IEEE Internet Computing 7(6), 38–44 (2003)
11. Bertocci, V., Garrett Serack, C.B.: Understanding windows cardspace, pp. 224–247. Addison-Wesley, Reading (2007)
12. Personal communication with participants of dagstuhl seminar 09141 on web application security (March 2009)

13. Kaminsky, D.: It's the end of the cache as we know it (2008),
    http://www.doxpara.com/DMK_BO2K8.ppt
14. Zuchlinski, G.: The anatomy of cross site scripting (2003)
15. Jovanovic, N., Kirda, E., Kruegel, C.: Preventing cross site request forgery attacks.
    In: Securecomm and Workshops, pp. 1–10 (2006)
16. Kirda, E., Krügel, C., Vigna, G., Jovanovic, N.: Noxes: a client-side solution for
    mitigating cross-site scripting attacks, pp. 330–337. ACM, New York (2006)
17. Stamm, S., Ramzan, Z., Jakobsson, M.: Drive-by pharming. In: Qing, S., Imai, H.,
    Wang, G. (eds.) ICICS 2007. LNCS, vol. 4861, pp. 495–506. Springer, Heidelberg
    (2007)
18. Jackson, C., Barth, A., Bortz, A., Shao, W., Boneh, D.: Protecting browsers from
    dns rebinding attacks. In: CCS 2007, pp. 421–431. ACM, New York (2007)
19. Karlof, C., Shankar, U., Tygar, J.D., Wagner, D.: Dynamic pharming attacks and
    locked same-origin policies for web browsers. In: CCS 2007, pp. 58–71. ACM, New
    York (2007)
20. Akritidis, P., Chin, W.Y., Lam, V.T., Sidiroglou, S., Anagnostakis, K.G.: Proximity
    breeds danger: emerging threats in metro-area wireless networks. In: SS 2007, pp.
    1–16. USENIX Association (2007)
21. Gajek, S., Manulis, M., Pereira, O., Sadeghi, A.R., Schwenk, J.: Universally com-
    posable security analysis of TLS. In: Baek, J., Bao, F., Chen, K., Lai, X. (eds.)
    ProvSec 2008. LNCS, vol. 5324, pp. 313–327. Springer, Heidelberg (2008)
22. Stevens, M., Sotirov, A., Appelbaum, J., Lenstra, A., Molnar, D., Osvik, D.A., de
    Weger, B.: Short chosen-prefix collisions for md5 and the creation of a rogue ca
    certificate. In: Crypto 2009. Springer, Heidelberg (to appear, 2009)
23. Dhamija, R., Tygar, J.D., Hearst, M.A.: Why phishing works. In: CHI, pp. 581–590.
    ACM, New York (2006)
24. Schechter, S., Dhamija, R., Ozment, A., Fischer, I.: The emperor's new security
    indicators. In: Symposium on Security and Privacy, pp. 51–65. IEEE Computer
    Society, Los Alamitos (2007)
25. Herzberg, A.: Why Johnny can't surf (safely)? attacks and defenses for web users.
    Elsevier Computers & Security 28(1-2), 63–71 (2009)
26. Jackson, C., Simon, D.R., Tan, D.S., Barth, A.: An evaluation of extended valida-
    tion and picture-in-picture phishing attacks. In: Dietrich, S., Dhamija, R. (eds.) FC
    2007 and USEC 2007. LNCS, vol. 4886, pp. 281–293. Springer, Heidelberg (2007)
27. Jackson, C., Barth, A.: Beware of finer-grained origins. In: W2SP 2008 (2008)
28. Oppliger, R., Hauser, R., Basin, D.: Ssl/tls session-aware user authentication. Com-
    puter 41(3), 59–65 (2008)
29. Rescorla, E.: Keying material extractors for transport layer security (tls). IEFT
    Internet-Draft (2008)
30. Dierks, T., Allen, C.: RFC2246, The tls protocol version 1.0 (1999)

# Fair E-Cash: Be Compact, Spend Faster[*]

Sébastien Canard[1], Cécile Delerablée[2], Aline Gouget[3], Emeline Hufschmitt[4],
Fabien Laguillaumie[5], Hervé Sibert[6], Jacques Traoré[1], and Damien Vergnaud[7]

[1] Orange Labs R&D, 42 rue des Coutures, BP6243, F-14066 Caen Cedex, France
[2] UVSQ, 45 Avenue des Etats-Unis, 78035 Versailles Cedex, France
[3] Gemalto, 6 rue de la Verrerie, 92190 Meudon, France
[4] Thalès Communications, 160 boulevard de Valmy, 92704 Colombes, France
[5] GREYC - Université de Caen-Basse Normandie, France
[6] ST-Ericsson, 9-11 rue Pierre-Felix Delarue, 72100 Le Mans Cedex 9, France
[7] École normale supérieure – C.N.R.S. – I.N.R.I.A., France

**Abstract.** We present the first *fair e-cash system* with a compact wallet that enables users to spend efficiently $k$ coins while only sending to the merchant $\mathcal{O}(\lambda \log k)$ bits, where $\lambda$ is a security parameter. The best previously known schemes require to transmit data of size at least linear in the number of spent coins. This result is achieved thanks to a new way to use the Batch RSA technique and a tree-based representation of the wallet. Moreover, we give a variant of our scheme with a less compact wallet but where the computational complexity of the spend operation does not depend on the number of spent coins, instead of being linear at best in existing systems.

**Keywords:** Fair e-cash, privacy-preserving, batch RSA, blind signature.

## 1 Introduction

Electronic cash systems allow users to withdraw electronic coins from a bank, and then to pay merchants using these coins preferably in an off-line manner, i.e. with no need to communicate with the bank or a trusted party during the payment. Finally, the merchant deposits the coins he has received to the bank.

An e-cash system should provide user anonymity against both the bank and the merchant during a purchase in order to emulate the perceived anonymity of regular cash. However, it seems that the necessity to fight against money laundering encourages the design of fair e-cash systems where a trusted party can, at any time when it's needed, revoke the anonymity of users. We thus focus on the design of fair e-cash systems. In order to reach the privacy target while being reasonably practical, it is necessary to focus on the efficiency of the most repeated protocol, namely the spending one between the user and the merchant. It should also be possible to withdraw or spend several coins more efficiently than repeating a single withdrawal or spending protocol. At last, we must pay attention to the compactness of the data that are exchanged in all protocols.

---

[*] This work has been financially supported by the French Agence Nationale de la Recherche and the TES Cluster under the PACE project while 2nd author was working at Orange Labs and 4th author at ENS.

P. Samarati et al. (Eds.): ISC 2009, LNCS 5735, pp. 294–309, 2009.

*Related Works.* The compact e-cash system [1] has recently aroused a new interest in e-cash by proposing the first e-cash system permitting a user to efficiently withdraw a wallet with $2^L$ coins such that the space required to store these coins, and the complexity of the withdrawal protocol, are proportional to $L$ rather than to $2^L$. Another possibility of efficient withdrawal is also given in [2]. These schemes fulfill all security properties usually required in the non-fair setting but do not consider the efficiency of the spending phase. One solution to improve it is to manage a wallet that contains coins with several monetary values [3]. The main drawback of this solution is that the user must choose during the withdrawal protocol how many coins he wants for each monetary value. In [4], the initial compact e-cash scheme is modified to improve the spending phase; however, the overall cost is still linear in the number of spent coins and, again, the paper only consider non-fair e-cash. Consequently, there exists no privacy-preserving fair e-cash system allowing the user to both (i) withdraw compact wallets and (ii) spend several coins while the transmitted data size is less than linear in the number of spent coins.

*Our Contributions.* This paper presents a fair e-cash system with a compact wallet that allows users to spend efficiently $k$ coins while sending to the merchant only $\mathcal{O}(\lambda \log k)$ bits, with $\lambda$ a security parameter, while preserving the privacy of the users. Our proposal makes use of two main cryptographic building blocks: *blind signatures* [5] and *batch cryptography* [6]. The concept of blind signature is the essence of many e-cash systems [7,8,9]. However, many of these suffer from a lack of efficiency since they usually use the cut-and-choose method in order to identify double-spenders [7]. The Batch RSA method makes it possible to efficiently obtain multiple RSA signatures of multiple messages. Batch cryptography has been used to build several e-cash systems, in order to get additional properties [10,11], to decrease the amount of processing done by the merchant [12], or to improve the efficiency of the withdrawal process at the cost of the linkability of coins withdrawn together [13].

To the best of our knowledge, our proposal is the most efficient (fair) e-cash system in terms of wallet storage size, computational complexity of spending and spending transfer size, which is strongly unforgeable. Note that the level of anonymity achieved by our scheme is strong but it is not perfect. Indeed it is strong because it is impossible to link (i) a withdrawal protocol with a user identity, (ii) a spending protocol to a withdrawal protocol, and (iii) two spending protocols but only under specific constraints. The anonymity property achieved by our scheme cannot be perfect since some information related to the coin number (with respect to the wallet) leaks during the spending phase.

## 2   Security Model

### 2.1   Algorithms

A fair e-cash system involves four kinds of players: a user $\mathcal{U}$, a bank $\mathcal{B}$, a merchant $\mathcal{M}$ and a judge $\mathcal{J}$. Each user is able to withdraw a wallet with $\ell$ coins. Such wallet consists of an identifier and a proof of validity. A fair e-cash scheme is defined by the following algorithms, where $\lambda$ is a security parameter.

- ParamGen($1^\lambda$) is a probabilistic algorithm that outputs the parameters of the system *params*. In the sequel, all algorithms take as input $1^\lambda$ and *params*.
- JKeyGen(), BKeyGen() and UKeyGen() are key generation algorithms for $\mathcal{J}$, $\mathcal{B}$ and $\mathcal{U}$, respectively. The key pairs are denoted by $(sk_\mathcal{J}, pk_\mathcal{J})$, $(sk_\mathcal{B}, pk_\mathcal{B})$, and $(sk_\mathcal{U}, pk_\mathcal{U})$. Note that UKeyGen() also provides the keys of merchants that can be seen as users in e-cash systems.
- Register($\mathcal{J}(sk_J, pk_\mathcal{U}), \mathcal{U}(sk_\mathcal{U}, pk_\mathcal{J})$) is an interactive protocol whose outcome is a notification decision of $\mathcal{J}$ together with a certificate of validity of $\mathcal{U}$'s public key which guarantee that $\mathcal{U}$ knows his secret key.
- Withdraw($\mathcal{U}(pk_\mathcal{B}, sk_\mathcal{U}, \ell), \mathcal{B}(pk_\mathcal{U}, sk_\mathcal{B})$) is an interactive protocol that allows $\mathcal{U}$ to withdraw a wallet $W$ of $\ell$ coins. The output of $\mathcal{U}$ is a wallet $W$, i.e. an identifier $I$ and a proof of validity $\Pi$, or an error message $\perp$. The output of $\mathcal{B}$ is its view $\mathcal{V}_\mathcal{B}^{\text{Withdraw}}$ of the protocol.
- Spend($\mathcal{U}(W, pk_\mathcal{M}, pk_\mathcal{B}, k), \mathcal{M}(sk_\mathcal{M}, pk_\mathcal{B})$) is an interactive protocol enabling $\mathcal{U}$ to spend $k$ coins. $\mathcal{M}$ outputs the serial numbers $S_0, \cdots, S_{k-1}$ and a proof of validity $\pi$. $\mathcal{U}$'s output is an updated wallet $W'$ or an error message $\perp$.
- Deposit($\mathcal{M}(sk_\mathcal{M}, (S_0, \ldots, S_{k-1}), \pi, pk_\mathcal{B}), \mathcal{B}(pk_\mathcal{M}, sk_\mathcal{B})$) is an interactive protocol allowing $\mathcal{M}$ to deposit the coins, i.e. $S_0, \ldots, S_{k-1}$ and $\pi$. $\mathcal{B}$ adds the coins to the list of spent coins or outputs an error message $\perp$.
- Identify($S, \pi_1, \pi_2, sk_\mathcal{J}$) is an algorithm executed by $\mathcal{J}$ which outputs a proof $\Pi_G$ and either a registered public key $pk_\mathcal{U}$ or $\perp$.
- VerifyGuilt($S, pk_\mathcal{U}, \Pi_G, pk_\mathcal{J}$) is an algorithm allowing to publicly verify the proof $\Pi_G$ that the Identify has been done correctly.

## 2.2 Security Properties

We informally describe the security statements of a fair e-cash scheme.

**Unforgeability.** From the bank point of view, what matters is that no coalition of users can ever spend more coins than they have withdrawn:
- let $\mathcal{A}$ be an adversary that has access to the public key $pk_\mathcal{B}$ of the system;
- $\mathcal{A}$, playing a user, executes in a concurrent manner Withdraw and Deposit protocols with the bank. $\mathcal{A}$ can legitimately withdraw $f$ wallets; we denote by $w_f$ the number of coins withdrawn during these executions.
- the adversary $\mathcal{A}$ wins the game if, at any time, the honest bank accepts more than $w_f$ coins (without detecting a double-spending).

We require that no PPT adversary succeeds in this game with non-negligible probability.

**Anonymity.** From the user privacy point of view, the bank, even when cooperating with malicious users and merchants, should not learn anything about a user's spending other than from the environment. We capture a weaker notion of anonymity by assuming that the targeted users withdraw and spend the same number of coins (see discussion in Section 5.2):

- let $\mathcal{A}$ be an adversary that has access to the secret key $sk_{\mathcal{B}}$ of the bank;
- $\mathcal{A}$ executes Withdraw (as the bank) and Spend (as the merchant) protocols any number of times. $\mathcal{A}$ can also corrupt players;
- at any time of the game, $\mathcal{A}$ chooses two honest users $\mathcal{U}_0$ and $\mathcal{U}_1$ such that both $\mathcal{U}_0$ and $\mathcal{U}_1$ has withdrawn and spent the *same* number of coins. Then, a bit $b \in \{0,1\}$ is chosen and a Spend protocol is played between $\mathcal{U}_b$ and $\mathcal{A}$. At the same time, we assume that $\mathcal{U}_{\bar{b}}$ also plays a Spend protocol that is not observed by $\mathcal{A}$. Next, $\mathcal{A}$ can again executes Withdraw (as the bank) and Spend (as the merchant) protocols;
- the adversary $\mathcal{A}$ finally outputs a bit $b'$.

We require that for any PPT adversary, the probability that $b' = b$ differs significantly from $1/2$ is negligible.

**Identification of double-spenders.** From the bank's point of view, no collection of users should be able to double-spend a coin without revealing one of their identities:

- let $\mathcal{A}$ be a an adversary that has access to $pk_{\mathcal{B}}$;
- $\mathcal{A}$ executes, as a user, Withdraw and Spend protocols as many time as it wishes;
- $\mathcal{A}$ wins the game if, at any time, the bank outputs $\perp$ while the merchant executes the Deposit protocol and Identify outputs $\perp$.

We require that no PPT adversary succeeds with non-negligible probability.

**Exculpability.** The bank, even cooperating with malicious users, cannot falsely accuse honest users from having double-spent a coin, and only users who double-spent a coin can be convicted:

- let $\mathcal{A}$ be an adversary that has access to both the secret key $sk_{\mathcal{B}}$ of the bank and the one $sk_{\mathcal{J}}$ of the judge;
- the adversary $\mathcal{A}$ can create as many users as he wants and corrupt some of them. All along the game, $\mathcal{A}$ plays the bank side of the Withdraw and Deposit protocols, $\mathcal{A}$ can play either the role of the user (as a corrupted user) or the role of the merchant during Spend protocols;
- the adversary $\mathcal{A}$ wins the game if, at any time, the Identify algorithm outputs the public key of an honest user together with a valid proof $\Pi_G$.

We require that no PPT adversary succeeds with non-negligible probability.

# 3    Useful Tools, Notations and Conventions

In the sequel, $\lambda$ is the general security parameter. In a withdrawal protocol, the user withdraws $\ell \leq K = 2^L$ coins from the bank, and every coin is labeled with a serial number $S_j, 0 \leq j < \ell$. In a spending protocol, the number of remaining coins in the wallet before spending and the number of coins to be spent is denoted by $K'$ and $k$, respectively.

## 3.1    Batch RSA Method

The Batch RSA method [6] makes it possible, for a given RSA modulus, to efficiently obtain multiple RSA signatures whose public exponents are coprime pairwise.

Let $n$ be an RSA modulus for which the factorization is only known by the signer. Let $e_0, \ldots, e_{\ell-1}$ be $\ell$ exponents, coprime both pairwise and with $\phi(n)$, with $\ell \leq K = 2^L$. As the efficiency of the Batch RSA depends on the size of these exponents, a generic suitable choice is the $\ell$ first odd prime numbers. Let $E = \prod_{i=0}^{\ell-1} e_i$. Given messages $S_0, S_1, \ldots, S_{\ell-1}$, it is possible to generate the $\ell$ roots $S_0^{1/e_0} \pmod{n}, \ldots, S_{\ell-1}^{1/e_{\ell-1}} \pmod{n}$ in $\mathcal{O}(\log K \log E + \log n)$ modular multiplications and $\mathcal{O}(K)$ divisions. We sketch the steps of the Batch RSA description and complexity proof described in [6]:

- (B1) compute the product $M = \prod_{i=0}^{\ell-1} S_i^{E/e_i}$ along a binary tree as shown in Figure 1 for the case $\ell = 5$. Every complete binary tree with $\ell$ leaves is suitable. However, for efficiency purpose, we suppose the height of the tree is $\mathcal{O}(\log K) = \mathcal{O}(L)$. Each node in the tree contains a value $M_{[i_1 \ldots i_2]} = \prod_{i=i_1}^{i_2} S_i^{E_{[i_1 \ldots i_2]}/e_i}$ with $E_{[i_1 \ldots i_2]} = \prod_{i=i_1}^{i_2} e_i$. In order to compute this tree, the number of operations is $\mathcal{O}(\log K \log E + \log n)$ multiplications;
- (B2) compute the batch signature $M^{1/E} = \prod_{i=0}^{\ell-1} S_i^{1/e_i}$, as a usual RSA signature with public exponent $E$;
- (B3) decompose $M^{1/E}$ in order to obtain the values $S_i^{1/e_i}$. In this step, the binary tree built at the first step is parsed down, and at each node of the tree the value $M_{[i_1 \ldots i_2]}^{1/E_{[i_1 \ldots i_2]}} = \prod_{i=i_1}^{i_2} S_i^{1/e_i}$ is computed and broken into two factors (one for each son) by using the Chinese remainder theorem and the values computed in (B1). The cost of this last step is $\mathcal{O}(K)$ modular divisions and $\mathcal{O}(\log E \log K)$ operations.

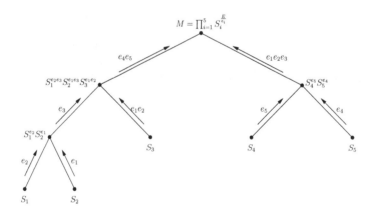

**Fig. 1.** Withdrawal binary tree for the computation of $M$

**Use of Batch RSA in our proposal.** The messages signed using Batch RSA are the serial numbers of coins. For efficiency purpose, the Batch RSA exponents $e_i$ are the $K$ first prime numbers. Therefore, we have $\log E = \mathcal{V}(e_{K-1})$, where $\mathcal{V}$ is the Chebyshev function[1]. This yields $\log E \sim K \ln K$.

---

[1] We recall that the Chebyshev function is $\mathcal{V}(x) = \sum_{p \leq x \ \text{prime}} \log(p)$.

During the withdrawal, the user has to perform steps (B1) and (B2) (see Section 3.2) in order to receive an aggregated signature on all the serial numbers that he has chosen. The aggregated value $M^{1/E}$ represents his wallet.

One novel aspect of our scheme is that it is never necessary to fully decompose the aggregated signature into all the signatures of spent coins during the spending phase. Indeed, at each spending, the current aggregated signature is split into two parts following a single node operation from step (B3), the first part being the aggregated signature of the coins to be spent, and the second part being the new wallet signature representing the remaining coins. Suppose that a user still owns an aggregated signature $M_F^{1/E'} = \prod_{i \in F} S_i^{1/e_i}$, with $F \subset \{0, \dots, \ell - 1\}$ and $E' = \prod_{i \in F} e_i$. This user wants to spend a subset $F_1$ of the coins in $F$. Let $F_2 = F \backslash F_1$. In order to compute the aggregated signature $M_{F_1}^{1/E_1'} = \prod_{i \in F_1} S_i^{1/e_i}$, the user creates two binary trees, corresponding to the subsets $F_1$ and $F_2$, respectively, and connects them at the root of a new binary tree. Then, the user computes the resulting tree as in step (B1) above in order to obtain the two factors $M_{F_1}$ and $M_{F_2}$. The cost is $\mathcal{O}(\log \#F \log E' + \log n)$. Using the values computed for the roots of each subset $F_i$, the user can now retrieve the aggregated signature to be spent and the remainder as another aggregated signature. The cost of this operation is 2 modular divisions and $\mathcal{O}(\log E')$ multiplications. An example is shown in Figure 2.

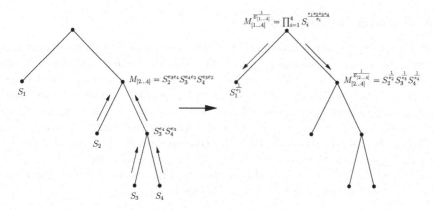

**Fig. 2.** Binary tree built to spend coins $2, 3, 4$ from a wallet with 4 remaining coins

This technique allows a user to carry a very small amount of data and to transfer reduced signature data. Indeed, in this case, only the non-spent interval and the remaining aggregated signature must be stored in the wallet, while a single aggregated signature is sent to the merchant. There are several trade-offs related to how we use the Batch RSA signatures. We detail them in Section 6.

## 3.2 RSA Blind Signature Scheme

A blind signature [5] is a protocol between a user and a signer where the user gets a signature from the signer in a way that the signer does not know the content of the message he is signing. Furthermore, the signer cannot link afterward his views of the protocol to the resulting signatures.

A common blind signature is the RSA blind signature scheme from Chaum [5,14]. This three-move blind signature scheme is defined by a set of five algorithms $BS=(\mathsf{KeyGen}, \mathsf{Blind}, \mathsf{Sign}, \mathsf{UnBlind}, \mathsf{Verif})$, where $\mathsf{Blind}$ corresponds to the computation of $\tilde{M} = r^e.\mathcal{H}(M) \pmod{n}$ where $r$ is a secret random value, $M$ is the message to be blindly signed and $\mathcal{H}$ is a one-way collision-resistant hash function, while $\mathsf{UnBlind}$ consists in computing $\sigma = \tilde{\sigma}/r \pmod{n}$, where $\tilde{\sigma}$ is a classical RSA signature on the message $\tilde{M}$. Thus, it is obvious that $\sigma$ is also a classical RSA signature of the message $M$.

**Use of the RSA blind signature scheme in our proposal.** Our scheme relies on blind RSA signatures using the Batch RSA technique, for which we choose a modulus $n$, where $\log n$ is polynomial in $\lambda$. The messages signed using the RSA blind signature are serial numbers of coins. During step (B2), the batch signature is replaced by a blind signature process. Thus, for $M = \prod_{i=0}^{\ell-1} \mathcal{H}(S_i)^{E/e_i}$, instead of simply computing the message $M^{1/E} = \prod_{i=0}^{\ell-1} \mathcal{H}(S_i)^{1/e_i}$, the signer obtains from the user $\tilde{M} = r^E M \pmod{n}$ and computes $\tilde{\sigma} = \tilde{M}^{1/E} = r \prod_{i=0}^{\ell-1} \mathcal{H}(S_i)^{1/e_i}$ $\pmod{n}$. The user finally computes, as for the traditional RSA blind signature scheme, $\sigma = \tilde{\sigma}/r \pmod{n}$, which corresponds to $\prod_{i=0}^{\ell-1} \mathcal{H}(S_i)^{1/e_i}$, as desired.

### 3.3   Signature of Knowledge

Zero-knowledge proofs of knowledge (ZKPK) are interactive protocols between a verifier and a prover allowing a prover to assure the verifier his knowledge of a secret, without any leakage of it. In the following, we use proofs of knowledge of a discrete logarithm [15,16], of a representation, proof of equality of two known representations in the same or in different groups [17]. In the following, we denote by $PK(\alpha_1, \ldots, \alpha_q : \mathsf{R}(\alpha_1, \ldots, \alpha_q))$ a proof of knowledge of the secrets $\alpha_1, \ldots, \alpha_q$ verifying the relation $\mathsf{R}$. Note that the combination of these proofs and the underlying security have been studied in [18,19] and refined in [20].

These interactive proofs can also be used non interactively (a.k.a. *signatures of knowledge*) by using the Fiat-Shamir heuristic [21].

### 3.4   Camenisch-Lysyanskaya Type Signature Schemes

Camenisch and Lysyanskaya have proposed in [22] various signature schemes which include new features. These signatures, called CL signatures for short, are based on Pedersen's commitment scheme which allows a user to commit some values without revealing them. CL signatures should satisfy the unforgeability property and have the following protocols.

- KeyGen: a key generation algorithm which outputs a key pair $(sk, pk)$.
- Sign: an efficient protocol between a user and a signer that permits the user to obtain from the signer a signature $\Sigma$ of some commitment $C = \mathsf{Commit}(x_1, \ldots, x_k)$ such that $(x_1, \ldots, x_k)$ are unknown from the signer. The

latter uses the CLSign algorithm on input $C$ and the user obtains a signature $\Sigma$ on the messages $(x_1, \ldots, x_k)$, such that $\mathsf{Verif}(\Sigma, (x_1, \ldots, x_k)) = 1$.

- ZKPK: an efficient ZKPK of a signature of some values that are moreover (may be independently) committed.
- Verif: a procedure verifying the signature $\Sigma$ on the messages $(x_1, \ldots, x_k)$.

One possible choice is to take the construction from [22], which is secure under the flexible RSA assumption (a.k.a. strong RSA assumption), and where the signature on values $(x_0, \ldots, x_k)$ is $(A, e, s)$ such that $A^e = a_0 a_1^{x_1} \cdots a_k^{x_k} b^s$, where the $a_i$'s and $b$ are public.

## 4   Compact Spending

In this section, we first give a high level description of our proposal before describing the procedure and protocols of our scheme.

### 4.1   Overview of Our Scheme

In e-cash systems, a withdrawal protocol allows a user to get from the bank, a wallet of coins that can be represented by a set of *serial numbers* and a signature of the bank that will allow him to prove the validity of the coins. The spending protocol of a fair e-cash system usually includes the generation of $\ell$ valid serial numbers $S_0, \ldots, S_{\ell-1}$ (to allow the detection of double-spending by the bank during the deposit protocol), a verifiable encryption of the spender public key, and a proof of validity of the $S_i$'s and of the encryption of the user public key without revealing any information about his identity.

*Serial numbers.* As we have seen, the Batch RSA technique can be used to obtain compact spendings by aggregating signatures. However, the transmission of the serial numbers also has to get more compact in order to decrease the overall spending complexity. In order to compact data related to serial numbers, we use a tree with a derivation mechanism from the root to the leaves which represent the serial numbers of the coins. In our scheme, the maximal number of coins that can be withdrawn during a protocol is a fixed parameter of the system $K = 2^L$. Each wallet of monetary value $\ell \leq K = 2^L$ withdrawn from the bank is mapped to a binary tree of $L + 1$ levels[2]. The tree root is assigned a *compact serial number* $S_{0,0}$. For every level $i$, $0 \leq i < L$, the $2^i$ nodes are assigned each a *compact* serial number denoted by $S_{i,j}$ with $0 \leq j < 2^i$. The values $S_{L,j}$ with $0 \leq j < 2^L$ related to the leaves of the tree are called the *serial numbers* of the purse and denoted $S_j$.

The derivation is illustrated by Figure 3 and it works as follows: the descendants from a node $S_{i,j}$ are given by a public function $\mathcal{F}(\cdot, \cdot)$ that, on input a

---

[2] The user may withdraw less than $2^L$ coins, but still has to work with a tree of depth $L + 1$, because the number of derivations to get the serial number of a coin must be the same for all users in order to prevent linking.

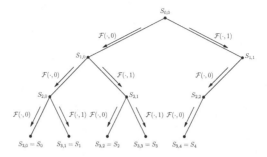

**Fig. 3.** Serial number binary tree for $\ell = 5$ and $K = 2^3$

compact serial number $S_{i,j}$ and a bit $b \in \{0,1\}$ to indicate *left* or *right*, outputs the (compact) serial number $S_{i+1,2j+b}$ of the left or right descendant of $S_{i,j}$ in the tree. Thus, from the tree root $S_{0,0}$, it is possible to compute all the serial numbers $S_{i,j}$, $0 \leq i \leq L$, $0 \leq j < 2^i$. The idea used to obtain compact spendings with serial numbers is that it is possible to send the serial number of a node $S_{i,j}$ instead of the serial numbers of all the leaves that come from him. Conversely, once a node $S_{i,j}$ is revealed, none of its descendants or ascendants can be spent, and no node can be spent more than once. This rule is necessary to protect against over-spending. It must also be impossible to compute a serial number without the knowledge of one of its ascendants. Finally, for security reasons, function $\mathcal{F}$ must be collision-free.

*Withdrawal.* During the withdrawal protocol, the user chooses a number $\ell \leq K = 2^L$ of coins to withdraw. For every $j$, $0 \leq j \leq \ell - 1$, the serial number $S_j$ is the message related to the exponent $e_j$ (see Section 3.1). The user computes the $\ell$ serial numbers $S_0, \ldots, S_{\ell-1}$ from a compact serial number $S_{0,0} = s$, where $s$ is a random value known only by the user but computed jointly by the bank and the user, so as to prevent an attack where two users use the same compact serial number. The user at last obtains from the bank both a blind Batch RSA signature on the serial numbers $S_0, \ldots, S_{\ell-1}$ with exponents $e_0, \ldots, e_{\ell-1}$ and a CL signature on $s$ and her identity $u$.

*Spending.* When a user wants to spend $k$ coins, she does not need to send $k$ serial numbers and $k$ proofs of validity but only one batch signature (see Section 3.1) and $\mathcal{O}(\lambda \log(k))$ nits corresponding to "compact serial numbers", assuming that the user spends the coins by increasing (or decreasing) exponents. As the size of the remaining values transmitted during spending is at most $\mathcal{O}(\lambda \log k)$ bits, this is also the overall size of the data transmitted during the spending protocol.

Finally, the merchant can verify the correctness of the serial numbers (w.r.t. the bank) using a ZKPK of the CL signature on the values s and u done by the user, following a technique given in [9] which permits us not to prove that the spent serial numbers are indeed generated from the value $s$ signed by the bank.

## 4.2   Setup Procedure

The ParamGen procedure first sets $2^L = K$ as the maximum number of coins in a wallet and $e_0, \ldots, e_{K-1}$ as $K$ distinct small prime numbers. For all $i \in [1, K]$, $E_i = \prod_{j=0}^{i-1} e_j$. Next $\mathsf{Enc}_{\mathcal{J}}(\cdot)$ is an encryption function of the judge's IND-CPA public key cryptosystem (e.g. the El Gamal encryption scheme), $\mathcal{H}(\cdot)$ and $\mathcal{F}(\cdot, \cdot)$ are two one-way collision resistant (hash) functions, $g$ is a generator of a cyclic group $G$ of prime or unknown order (the structure of the group depends on the chosen CL signature scheme). Next, the bank $\mathcal{B}$ (resp. the judge $\mathcal{J}$) executes the BKeyGen (resp. JKeyGen) procedure by executing the KeyGen algorithms of the CL and blind signature schemes (resp. of the encryption scheme).

During the UKeyGen procedure, each user $\mathcal{U}$ is finally associated to a long-term private key $sk_{\mathcal{U}} = u$ and a corresponding public key $pk_{\mathcal{U}} = g^u$, where $g$ is a public parameter.

## 4.3   Withdrawal Protocol

Let $\mathcal{U}$ be a user who wants to withdraw $\ell$ (with $0 < \ell \le K$) coins to the bank $\mathcal{B}$. The protocol between $\mathcal{U}$ and $\mathcal{B}$ is described in Figure 4. Note that $\mathcal{B}$ can compute the commitment $C$ on $u$, $s = s' + s''$ and $w$ using only $C'$ and $s''$ and without needing to know $s'$ and thus $s$. Next, the computation of $E_\ell$ and the serial numbers $S_0, \ldots, S_{\ell-1}$ is done using the tree structure we described above with $\mathcal{F}$ as function and $S_{0,0} = s$ as the tree root (see Sections 3.1 and 4.1 for details). The user $\mathcal{U}$ now possesses a wallet determined by the set $(s, u, w, \Sigma, \sigma)$.

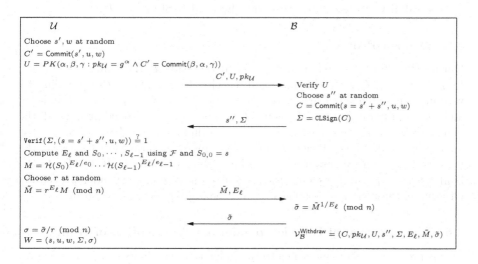

**Fig. 4.** Withdrawal Protocol

## 4.4    The Spend Protocol

Assume that a user $\mathcal{U}$ owns a wallet $(s, u, w, \Sigma, \sigma)$ and wants to spend $k$ coins to a merchant $\mathcal{M}$. The spend protocol works as follows:

1. $\mathcal{M}$ sends some public information *info* concerning the transaction (typically the time and date of the ongoing transaction);
2. $\mathcal{U}$ knows the smallest $i$ such that $S_i, \cdots, S_{i+k-1}$ are unspent serial numbers;
3. $\mathcal{U}$ does not need to compute the values of the serial numbers $S_i, \cdots, S_{i+k-1}$. Indeed, she only needs to compute the smallest number of master serial numbers necessary to allow the computation by the merchant of $S_i, \cdots, S_{i+k-1}$. In the worst case, we need $2\lfloor \log k \rfloor$ values $S_{i_1,j_1}, \ldots, S_{i_n,j_n}, 0 \le i_1, \ldots, i_n$ and $0 \le j_1 \le 2^{i_1} - 1, \ldots, 0 \le j_n \le 2^{i_n} - 1$. $\mathcal{U}$ sends to the merchant $S_{i_1,j_1}, \ldots, S_{i_n,j_n}$ and the index value $i$;
4. using the batch RSA signature described in Section 3.1, $\mathcal{U}$ computes the batch signature $\sigma_{[i,i+k-1]}$ on $S_i, \cdots, S_{i+k-1}$ (further denoted $\sigma_k$);
5. $\mathcal{U}$ computes $R = \mathcal{H}(info \| pk_{\mathcal{M}} \| \sigma_k)$ which is used as a freshness indicator;
6. next $\mathcal{U}$ computes two values $C_1 = \mathsf{Enc}_{\mathcal{J}}(pk_{\mathcal{U}})$ and $C_2 = \mathsf{Enc}_{\mathcal{J}}(s)$;
7. $\mathcal{U}$ produces a signature of knowledge $\Pi$ which proves that:
   - $C_1$ and $C_2$ are well-formed, that is $C_1$ is an encryption of $pk_{\mathcal{U}} = g^u$ and $C_2$ is an encryption of $s$ under the judge's public key encryption scheme, without revealing $pk_{\mathcal{U}}$ nor $s$;
   - $\mathcal{U}$ knows a CL bank's signature $\Sigma$ on $u$, $s$ and $w$ without revealing $u$, $s$, $w$ nor $\sigma$.

   She uses $c = \mathcal{H}(S_{i_1,j_1} \| \ldots \| S_{i_n,j_n} \| \sigma_k \| R \| C_1 \| C_2)$ as a challenge;
8. at the end, the user has sent $(i, S_{i_1,j_1}, \ldots, S_{i_n,j_n}, \sigma_k, C_1, C_2, \Pi, R)$;
9. the merchant $\mathcal{M}$ computes $S_i, \cdots, S_{i+k-1}$ from $S_{i_1,j_1}, \ldots, S_{i_n,j_n}$ and checks the validity of the coin by verifying the validity of $\sigma_k$ and $\Pi$;

## 4.5    Deposit Protocol

During this step, a merchant $\mathcal{M}$ sends to the bank $\mathcal{B}$ the values $(i, S_i, \ldots, S_{i+k-1}, \sigma_k, C_1, C_2, \Pi, R)$. The bank checks the validity of the spending by verifying the batch signature $\sigma_k$ on the values $S_i, \ldots, S_{i+k-1}$ using the index $i$, and the validity of the proof $\Pi$ using $R$, $C_1$ and $C_2$. If the spending is valid, the bank checks whether at least one of the serial numbers $S \in \{S_i, \ldots, S_{i+k-1}\}$ is already in its database. If not, $\mathcal{B}$ adds them into the database. Otherwise, the bank verifies the freshness of the spending using the value $R$. If it is fresh, the bank asks the judge to execute the identification of double spender procedure. Otherwise, the merchant is a cheater and the bank rejects the deposit.

## 4.6    Identification of Double Spender and Verification of Guilt

In this procedure, the bank sends to the judge two spendings $(i, S_i, \ldots, S_{i+k-1}, \sigma_k, C_1, C_2, \Pi, R)$ and $(i', S'_{i'}, \ldots, S'_{i'+k'-1}, \sigma'_{k'}, C'_1, C'_2, \Pi', R')$ such that there exists $i_0$ and $i'_0$ with $i \le i_0 \le i + k - 1$ and $i' \le i'_0 \le i' + k' - 1$ with $S_{i_0} = S'_{i'_0} = S$. This latter verifies the validity of both spendings, decrypts $C_2$ and $C'_2$ to retrieve $s$ and $s'$, and next decrypts $C_1$ and/or $C'_1$ if necessary.

- If $S$ cannot be computed from $s$ (resp. $s'$), then the judge decrypts $C_1$ (resp. $C_1'$) and concludes that $pk_{\mathcal{U}}$ (resp. $pk_{\mathcal{U}'}$) is guilty.
- Else, with high probability $s = s'$ (since $\mathcal{H}$ and $\mathcal{F}$ are collision-free) and $pk_{\mathcal{U}} = pk_{\mathcal{U}'}$ (since it is unlikely that two different users obtain the same wallet secret $s$ in the withdrawal phase and since $\mathcal{F}$ is collision-free). Thus, the judge concludes that $pk_{\mathcal{U}} = pk_{\mathcal{U}'}$ is guilty. Note that if the case $s = s'$ and $pk_{\mathcal{U}} \neq pk_{\mathcal{U}'}$ happens, that means that user $\mathcal{U}$ has proven the knowledge of a bank's signature on the values $(s, u)$ and user $\mathcal{U}'$ has proven the knowledge of a bank's signature on the values $(s, u')$. In this case, the two spendings are valid and the judge sends back a false alarm message since there is no double-spending.
- At the end, the judge produces a proof $\Pi_G$ that the public key of the guilty user has been correctly decrypted. The proof consists of the values ($s$ and $pk_{\mathcal{U}}$) related to the cheater and of a ZKPK that the secret key $sk_{\mathcal{J}}$ embedded in $pk_{\mathcal{J}}$ has correctly been used to decrypt $s$ and $pk_{\mathcal{U}}$.

The verification of guilt consists in verifying the judge's proof $\Pi_G$ on $pk_{\mathcal{U}}$ and $s$.

# 5  Security Analysis

In this section, we give the security arguments for our construction. We first detailed the security assumptions we use and next give the security theorem; security proofs are not included in the paper due to space restrictions.

## 5.1  Security Assumptions

**One-More Unforgeability.** In 2001, Bellare et al. [23] introduced the notion of *one-more one-way function*, and showed how it leads to a proof of security of Chaum's RSA-based blind signature scheme [14] in the random oracle model. We now introduce a variant of the one-more RSA problem in order to prove the security of the Batch variant of Chaum's blind signatures. The one-more flexible (or strong) RSA-problem is defined by the following game for an algorithm $\mathcal{A}$.

- the adversary $\mathcal{A}$ gets an RSA modulus $n$ and a public exponent $E$ made of the product of $\ell$ prime numbers $E = e_0 \ldots e_{\ell-1}$;
- it is given access to an *inversion* oracle that given $y \in \mathbb{Z}_n^*$ returns $x \in \mathbb{Z}_n^*$ such that $x^E = y \mod N$;
- it is given access to a *challenge* oracle that returns $\ell$ random challenges point from $\mathbb{Z}_n^*$;
- eventually, $\mathcal{A}$ wins the game if it succeeds in inverting $q \cdot \ell + 1$ points output by the challenge oracle using less than $q$ queries to the inversion oracle[3].

The *strong one-more RSA assumption* states that no probabilistic polynomial-time algorithm $\mathcal{A}$ may win the previous game with non-negligible probability.

---

[3] Using $q$ times the inversion oracle and the batch RSA technique given in Section 3.1, the adversary can easily invert $q \cdot \ell$ points.

Following, Bellare *et al.*'s technique from [23], it is readily seen that in the random oracle model, the Batch-RSA blind signature scheme is one-more unforgeable under the *strong one-more RSA assumption*:

**Lemma 1.** *If the one-more flexible RSA problem is hard, then the Batch-RSA blind signature scheme is polynomially-secure against one-more forgery in the random oracle model.*

*Proof.* It is almost identical to the one of [23, Theorem 16].          □

**Strong Blindness Property.** In the security proof of our e-cash system, we need a *Strong Blindness* property for this Batch-RSA blind signature scheme. More precisely, we have the following experiment:

- let $\mathcal{A}$ be a PPT Turing Machine having access to the signer's key pair and being able to participate to the blind process from the signer's point of view, obtain resulting message/signature $(M, \sigma)$ and obtain chosen partial pairs message/signature, that is all $S_i \in F$ and the signature $\prod_{i \in F} \mathcal{H}(S_i)^{1/e_i}$ for any $F \subset \{0, \cdots, \ell - 1\}$ of the adversary's choice (see Section 3.1 for details);
- at any time of the game, the adversary outputs two transcripts $I_0$ and $I_1$ of a blind signature process (from the signer's point of view) and a challenge $\tilde{F} \subset \{0, \cdots, \ell - 1\}$. The challenger next chooses at random a bit $b \in \{0, 1\}$ and outputs the messages and the signature corresponding to the transcript $I_b$ and the set $\tilde{F}$;
- the adversary finally outputs a bit $b'$.

The *Strong Blindness* property says that the probability that $b' = b$ differs significantly from $1/2$ is negligible.

**Lemma 2.** *The Batch-RSA Blind signature scheme unconditionally verifies the Strong Blindness property.*

*Proof.* Straightforward as the proof is similar to the security proof of the initial RSA blind signature scheme, which is unconditionally blind.          □

**Unforgeability of signature of knowledge.** In our construction, we use the Fiat-Shamir heuristic to make non-interactive traditional interactive zero-knowledge proofs of knowledge. In [24], Pointcheval and Stern prove that this transformation is secure in the random oracle model.

**Camenisch-Lysyanskaya type signature schemes.** We need the CL type signature scheme to be unforgeable, saying that even if an adversary has oracle access to the signing algorithm which provides signatures on messages of the adversary's choice, the adversary cannot create a valid signature on a message not explicitly queried. If we choose the CL signature scheme in [22], we need to assume that the flexible RSA problem is hard.

**The One-more discrete logarithm assumption.** The one-more discrete logarithm problem [23] is the following one. Given $l + 1$ values and having access to a discrete logarithm oracle at most $l$ times, find the discrete logarithm of all these values.

## 5.2   Security Statement

**Theorem 1.** *Our e-cash system is a secure fair e-cash system:*

- *unforgeability under the one-more unforgeability of the Batch-RSA blind signature scheme and the non-malleability of the signature of knowledge, in the random oracle model;*
- *anonymity under the strong blindness of the Batch-RSA blind signature scheme and the indistinguishability of the encryption scheme, in the random oracle model;*
- *identification of double-spenders under the unforgeability of the CL signature scheme, in the random oracle model;*
- *exculpability under the one-more discrete logarithm assumption, in the random oracle model.*

Note that our construction does not provide a perfect anonymity property since it is possible to know which leaves in the serial number binary tree are used during the spending. For example, if two spendings are from the same part of the tree, everyone can conclude that the spendings are from different wallets.

# 6   Efficiency Considerations

In order to simplify the complexity statements, we consider $\ell = K$, so that the exponents used for a wallet are the first $K = 2^L$ prime numbers; we have $\log E \sim K \ln K$. The coins are spent following the decreasing order of exponents. We denote by $E'$ the product of exponents corresponding to the number $K'$ of coins remaining in the wallet. As seen in Section 4, the data transfer size is always at least $\mathcal{O}(\lambda \log k)$.

Using Batch RSA as described in Section 3.1 as our default variant ($V0$) for the scheme yields the following efficiency trade-off: only the highest remaining exponent and one aggregated signature have to be stored in the wallet, with storage size $\mathcal{O}(\log n)$. During the spending phase, a binary tree has to be rebuilt, requiring $\mathcal{O}(\log K' \log E') = \mathcal{O}(K' \log^2 K' + \log n)$ multiplications, and the current signature has to be broken up in two pieces, which costs $\mathcal{O}(1)$ modular divisions plus $\mathcal{O}(\log E') = \mathcal{O}(K' \log K')$ modular multiplications. At last, a single aggregated signature is sent to the merchant, together with the number of coins and the biggest exponent, thus requiring transfer of $\mathcal{O}(\log n)$ bits. As this variant is targeted at reduced storage, it is relevant to store also the root serial number only and compute the needed serial numbers at each spending, thus minimizing the storage cost.

Instead of reducing the storage cost, we can also manage the Batch RSA tree similarly to the tree of serial numbers. This yields variant ($V1$): we store the initial withdrawal binary tree so that, during the spending, the user sends the aggregated signatures corresponding to the nodes of the tree closest to the root and such that all the corresponding leaves are in the spending set. The whole binary tree is stored, hence the initial storage size is $\mathcal{O}(K \log n)$. During the

**Table 1.** Efficiency trade-offs

|  | Default variant (V0) | Variant (V1) |
|---|---|---|
| Wallet storage size | $\mathcal{O}(\lambda + \log n)$ | $\mathcal{O}(K(\lambda + \log n))$ |
| Computational complexity of spending | $\mathcal{O}(K' \log^2 K' + \log n)M$ $+\mathcal{O}(1)D + \mathcal{O}(\log k)F$ | $\mathcal{O}(1)$ |
| Spending transfer size | $\mathcal{O}(\lambda \log k + \log n)$ | $\mathcal{O}((\lambda + \log n) \log k)$ |

spending phase, the user needs to send at most $2\lfloor \log_2(k+1) \rfloor$ aggregated signatures corresponding to tree nodes to the merchant, hence a data transfer of size $\mathcal{O}(\log n \log k)$. The computational cost for the user is the cost of retrieving the aggregated signatures corresponding to the nodes spent and to their remaining counterparts. At most, this requires $\mathcal{O}(\log K)$ signature break-ups (in case single coins must be retrieved), each of which costs $\mathcal{O}(1)$ modular divisions plus at most (for nodes closest to the tree root) $\mathcal{O}(\log E') = \mathcal{O}(K' \log K')$ modular multiplications. However, these values can be pre-computed off-line after the withdrawal of the wallet, and stored in the tree, thus achieving a $\mathcal{O}(1)$ on-line computational cost. This variant aims at reducing computations during spending, so it is relevant to store also the whole serial number tree in order to retrieve the needed serial numbers at each spending in $\mathcal{O}(1)$.

The relative storage, spending computational complexity and data transfer size of our schemes are summed up in Table 1; $M$ and $D$ are the respective costs of exponentiation, multiplication and division modulo $n$, $F$ is the cost of derivation with function $\mathcal{F}$, $\lambda$ is a security parameter, $K$ is the number of withdrawn coins, $k$ the number of spent coins and $K'$ the number of remaining coins in the wallet after spending. They take into account the complexities related to the serial numbers mentioned in Section 4, which provides the overall picture as the proof $\Pi$ and the remaining data only have a constant complexity.

# References

1. Camenisch, J., Hohenberger, S., Lysyanskaya, A.: Compact E-Cash. In: Cramer, R. (ed.) EUROCRYPT 2005. LNCS, vol. 3494, pp. 302–321. Springer, Heidelberg (2005)
2. Au, M.H., Wu, Q., Susilo, W., Mu, Y.: Compact E-Cash from Bounded Accumulator. In: Abe, M. (ed.) CT-RSA 2007. LNCS, vol. 4377, pp. 178–195. Springer, Heidelberg (2006)
3. Canard, S., Gouget, A., Hufschmitt, E.: A Handy Multi-coupon System. In: Zhou, J., Yung, M., Bao, F. (eds.) ACNS 2006. LNCS, vol. 3989, pp. 66–81. Springer, Heidelberg (2006)
4. Au, M.H., Susilo, W., Mu, Y.: Practical Compact E-Cash. In: Pieprzyk, J., Ghodosi, H., Dawson, E. (eds.) ACISP 2007. LNCS, vol. 4586, pp. 431–445. Springer, Heidelberg (2007)

5. Chaum, D.: Blind Signatures for Untraceable Payments. In: CRYPTO 1982, pp. 199–203 (1982)
6. Fiat, A.: Batch RSA. J. Crypt. 10(2), 75–88 (1997)
7. Chaum, D., Fiat, A., Naor, M.: Untraceable Electronic Cash. In: Goldwasser, S. (ed.) CRYPTO 1988. LNCS, vol. 403, pp. 319–327. Springer, Heidelberg (1990)
8. Brands, S.: Untraceable Off-line Cash in Wallets with Observers (Extended Abstract). In: Stinson, D.R. (ed.) CRYPTO 1993. LNCS, vol. 773, pp. 302–318. Springer, Heidelberg (1994)
9. Traoré, J.: Group Signatures and Their Relevance to Privacy-Protecting Off-Line Electronic Cash Systems. In: Pieprzyk, J.P., Safavi-Naini, R., Seberry, J. (eds.) ACISP 1999. LNCS, vol. 1587, pp. 228–243. Springer, Heidelberg (1999)
10. Ferguson, N.: Single term off-line coins. In: Helleseth, T. (ed.) EUROCRYPT 1993. LNCS, vol. 765, pp. 318–328. Springer, Heidelberg (1994)
11. Brands, S., Chaum, D.: Distance-Bounding Protocols (Extended Abstract). In: Helleseth, T. (ed.) EUROCRYPT 1993. LNCS, vol. 765, pp. 344–359. Springer, Heidelberg (1994)
12. Pavlovski, C., Boyd, C., Foo, E.: Detachable Electronic Coins. In: Varadharajan, V., Mu, Y. (eds.) ICICS 1999. LNCS, vol. 1726, pp. 54–70. Springer, Heidelberg (1999)
13. Boyd, C., Foo, E., Pavlovski, C.: Efficient Electronic Cash Using Batch Signatures. In: Pieprzyk, J.P., Safavi-Naini, R., Seberry, J. (eds.) ACISP 1999. LNCS, vol. 1587, pp. 244–257. Springer, Heidelberg (1999)
14. Chaum, D.: Blind Signature System. In: CRYPTO 1983, pp. 153 (1983)
15. Schnorr, C.: Efficient identification and signatures for smart cards. In: Brassard, G. (ed.) CRYPTO 1989. LNCS, vol. 435, pp. 239–252. Springer, Heidelberg (1990)
16. Girault, M., Poupard, G., Stern, J.: On the Fly Authentication and Signature Schemes Based on Groups of Unknown Order. J. Crypt. 19(4), 463–487 (2006)
17. Boudot, F., Traoré, J.: Efficient Publicly Verifiable Secret Sharing Schemes with Fast or Delayed Recovery. In: Varadharajan, V., Mu, Y. (eds.) ICICS 1999. LNCS, vol. 1726, pp. 87–102. Springer, Heidelberg (1999)
18. Kiayias, A., Tsiounis, Y., Yung, M.: Traceable Signatures. In: Cachin, C., Camenisch, J.L. (eds.) EUROCRYPT 2004. LNCS, vol. 3027, pp. 571–589. Springer, Heidelberg (2004)
19. Canard, S., Coisel, I., Traoré, J.: Complex zero-knowledge proofs of knowledge are easy to use. In: Susilo, W., Liu, J.K., Mu, Y. (eds.) ProvSec 2007. LNCS, vol. 4784, pp. 122–137. Springer, Heidelberg (2007)
20. Camenisch, J., Kiayias, A., Yung, M.: On the portability of generalized schnorr proofs. In: Joux, A. (ed.) EUROCRYPT 2009. LNCS, vol. 5479, pp. 425–442. Springer, Heidelberg (2009)
21. Fiat, A., Shamir, A.: How to Prove Yourself: Practical Solutions to Identification and Signature Problems. In: Odlyzko, A.M. (ed.) CRYPTO 1986. LNCS, vol. 263, pp. 186–194. Springer, Heidelberg (1987)
22. Camenisch, J., Lysyanskaya, A.: A Signature Scheme with Efficient Protocols. In: Cimato, S., Galdi, C., Persiano, G. (eds.) SCN 2002. LNCS, vol. 2576, pp. 268–289. Springer, Heidelberg (2003)
23. Bellare, M., Namprempre, C., Pointcheval, D., Semanko, M.: The One-More-RSA-Inversion Problems and the Security of Chaum's Blind Signature Scheme. J. Crypt. 16(3), 185–215 (2003)
24. Pointcheval, D., Stern, J.: Security Arguments for Digital Signatures and Blind Signatures. J. Crypt. 13(3), 361–396 (2000)

# On the Security of Identity Based Ring Signcryption Schemes

S. Sharmila Deva Selvi, S. Sree Vivek\*, and C. Pandu Rangan\*\*

Indian Institute of Technology Madras,
Theoretical Computer Science Laboratory,
Department of Computer Science and Engineering,
Chennai, India
{sharmila,svivek,prangan}@cse.iitm.ac.in

**Abstract.** Signcryption is a cryptographic primitive which offers authentication and confidentiality simultaneously with a cost lower than signing and encrypting the message independently. Ring signcryption enables a user to signcrypt a message along with the identities of a set of potential senders (that includes him) without revealing which user in the set has actually produced the signcryption. Thus a ring signcrypted message has anonymity in addition to authentication and confidentiality. Ring signcryption schemes have no group managers, no setup procedures, no revocation procedures and no coordination: any user can choose any set of users (ring), that includes himself and signcrypt any message by using his private and public key as well as other users (in the ring) public keys, without getting any approval or assistance from them. Ring Signcryption is useful for leaking trustworthy secrets in an anonymous, authenticated and confidential way.

To the best of our knowledge, seven identity based ring signcryption schemes are reported in the literature. Two of them were already proved to be insecure in [1] and [2]. In this paper, we show that four among the remaining five schemes do not provide confidentiality, to be specific, two schemes are not secure against chosen plaintext attack and other two schemes do not provide adaptive chosen ciphertext security. We then propose a new scheme and formally prove the security of the new scheme in the random oracle model. A comparison of our scheme with the only existing correct scheme by Huang et al. shows that our scheme is much more efficient than the scheme by Huang et al.

**Keywords:** Ring Signcryption, Cryptanalysis, Provable Security, Confidentiality, Chosen Plaintext Attack, Adaptive Chosen Ciphertext Attack, Bilinear Pairing, Random Oracle Model.

---

\* Work supported by Project No. CSE/05-06/076/DITX/CPAN on Protocols for Secure Communication and Computation, sponsored by Department of Information Technology, Government of India.

\*\* Work supported by Project No. CSE/05-06/075/MICO/CPAN on Research in Cryptography, sponsored by Microsoft Research India.

P. Samarati et al. (Eds.): ISC 2009, LNCS 5735, pp. 310–325, 2009.

# 1   Introduction

Let us consider a scenario, where a member of a cabinet wants to leak a very important and juicy information, regarding the president of the nation to the press. He has to leak the secret in an anonymous way, else he will be black spotted in the cabinet. The press will not accept the information unless it is authenticated by one of the members of the cabinet. Here, if the information is so sensitive and should not be leaked until the authorities in the press receives it, we should have confidential transmission of information. Thus, we require anonymity to safeguard the cabinet member who sends the information, authentication for the authorities in the press to believe the information and confidentiality until the information reaches the hands of the right person in the press. All the three properties are together achieved by a single primitive called "Ring Signcryption". The first identity based ring signcryption scheme was proposed by Huang et al. [3].

**Related Work and Our Contribution.** Huang et al.'s scheme [3] was inefficient because the sender has to compute $n + 2$ pairing for signcrypting a message and three pairing operations for unsigncrypting a ring signcryption. Subsequently, identity based ring signcryption schemes were reported in [4, 5, 1, 6, 7, 8] and these papers were attempts to design schemes more efficient than Huang et al.'s [3] scheme.

Among these seven schemes, the security weakness of [1] was shown in [7] and the weakness of [6] was shown in [2]. In this paper, we show that the schemes in [4], [8], [7] and [5] are insecure. Specifically, we show that [7] and [5] does not withstand adaptive chosen ciphertext attack, [4] and [8] are not secure against chosen plain text attack. This leaves the scheme by Huang et al. [3] as the only correct existing scheme. Then, we propose a new scheme and prove its security formally in a stronger security model. Moreover, our scheme is much more efficient than Huang et al.'s [3] scheme.

# 2   Preliminaries

## 2.1   Bilinear Pairing

Let $\mathbb{G}_1$ be an additive cyclic group generated by $P$, with prime order $q$, and $\mathbb{G}_2$ be a multiplicative cyclic group of the same order $q$. A bilinear pairing is a map $\hat{e} : \mathbb{G}_1 \times \mathbb{G}_1 \to \mathbb{G}_2$ with the following properties.

- **Bilinearity.** For all $P, Q, R \in \mathbb{G}_1$,
  - $\hat{e}(P + Q, R) = \hat{e}(P, R)\hat{e}(Q, R)$
  - $\hat{e}(P, Q + R) = \hat{e}(P, Q)\hat{e}(P, R)$
  - $\hat{e}(aP, bQ) = \hat{e}(P, Q)^{ab}$
- **Non-Degeneracy.** There exist $P, Q \in \mathbb{G}_1$ such that $\hat{e}(P, Q) \neq I_{\mathbb{G}_2}$, where $I_{\mathbb{G}_2}$ is the identity in $\mathbb{G}_2$.
- **Computability.** There exists an efficient algorithm to compute $\hat{e}(P, Q)$ for all $P, Q \in \mathbb{G}_1$.

## 2.2    Computational Bilinear Diffie-Hellman Problem (CBDHP)

**Definition 1.** *Given* $(P, aP, bP, cP) \in \mathbb{G}_1^4$ *for unknown* $a, b, c \in \mathbb{Z}_q^*$ *, the CBDH problem in* $\mathbb{G}_1$ *is to compute* $\hat{e}(P, P)^{abc} \in \mathbb{G}_2$.

*The advantage of any probabilistic polynomial time algorithm* $\mathcal{A}$ *in solving the CBDH problem in* $\mathbb{G}_1$ *is defined as*

$$Adv_{\mathcal{A}}^{CBDH} = Pr \left[ \mathcal{A}(P, aP, bP, cP) = \hat{e}(P, P)^{abc} | a, b, c \in \mathbb{Z}_q^* \right]$$

*The CBDH Assumption is that, for any probabilistic polynomial time algorithm* $\mathcal{A}$, *the advantage* $Adv_{\mathcal{A}}^{CBDH}$ *is negligibly small.*

## 2.3    Computation Diffie-Hellman Problem (CDHP)

**Definition 2.** *Given* $(P, aP, bP) \in \mathbb{G}_1^3$ *for unknown* $a, b \in \mathbb{Z}_q^*$, *the CDH problem in* $\mathbb{G}_1$ *is to compute* $abP$.

*The advantage of any probabilistic polynomial time algorithm* $\mathcal{A}$ *in solving the CDH problem in* $\mathbb{G}_1$ *is defined as*

$$Adv_{\mathcal{A}}^{CDH} = Pr \left[ \mathcal{A}(P, aP, bP) = abP \mid a, b \in \mathbb{Z}_q^* \right]$$

*The CDH Assumption is that, for any probabilistic polynomial time algorithm* $\mathcal{A}$, *the advantage* $Adv_{\mathcal{A}}^{CDH}$ *is negligibly small.*

## 2.4    Notations Used in This Paper

To have a better understanding and to enhance the readability and clarity, we use the following notations throughout the paper.

$\mathcal{U}_i$ - User with identity $ID_i$.

$\mathcal{U} = \{\mathcal{U}_i\}_{(i=1\,to\,n)}$ - Group of users in the ring (including the actual sender).
$\mathcal{M}$ - Message space.
$m$ - Message.
$l$ - Number of bits used to represent $m$.
$Q_i$ - Public key corresponding to $ID_i$.
$D_i$ - Private key corresponding to $ID_i$.
$ID_{\mathbb{S}}$ - Identity of the sender.
$ID_{\mathbb{R}}$ - Identity of the receiver.
$Q_{\mathbb{S}}$ - Public key of the sender.
$Q_{\mathbb{R}}$ - Public key of the receiver.
$D_{\mathbb{S}}$ - Private key of the sender.
$D_{\mathbb{R}}$ - Private key of the receiver.

# 3    Formal Security Model for Identity Based Ring Signcryption

## 3.1    Generic Scheme

A generic identity based ring signcryption scheme consists of the following four algorithms.

- **Setup($\kappa$):** Given a security parameter $\kappa$, the private key generator (PKG) generates the systems public parameters *params* and the corresponding master private key *msk* that is kept secret.
- **Extract($ID_i$):** Given a user identity $ID_i$, the PKG computes the corresponding private key $D_i$ and sends $D_i$ to $ID_i$ via a secure channel.
- **Signcrypt($m, \mathcal{U}, D_\mathbb{S}, ID_\mathbb{R}$):** This algorithm takes a message $m \in \mathcal{M}$, a receiver with identity $ID_\mathbb{R}$, the senders private key $D_\mathbb{S}$ and an ad-hoc group of ring members $\mathcal{U}$ with identities $\{ID_1, \ldots, ID_n\}$ as input and outputs a ring signcryption $C$. This algorithm is executed by a sender with identity $ID_\mathbb{S} \in \mathcal{U}$. $ID_\mathbb{R}$ may or may not be in $\mathcal{U}$.
- **Unsigncrypt($C, \mathcal{U}, D_\mathbb{R}$):** This algorithm takes the ring signcryption $C$, the ring members(say $\mathcal{U} = \{\mathcal{U}_i\}_{(i=1\,to\,n)}$) and the private key $D_\mathbb{R}$ of the receiver $ID_\mathbb{R}$ as input and produces the plaintext $m$, if $C$ is a valid ring signcryption of $m$ from the ring $\mathcal{U}$ to $ID_\mathbb{R}$ or "*Invalid*", if $C$ is an invalid ring signcryption. This algorithm is executed by a receiver $ID_\mathbb{R}$.

## 3.2    Security Notion

The formal security definition of signcryption was given by Baek et al. in [9]. The security requirements for identity based ring signcryption were defined by Huang et al. [3]. We extend the security model given in [3] by incorporating security against insider attacks. The security model is defined as follows.

**Definition 3.** *An identity based ring signcryption (IRSC) is indistinguishable against adaptive chosen ciphertext attacks (IND-IRSC-CCA2) if there exists no polynomially bounded adversary having non-negligible advantage in the following game:*

1. **Setup Phase:** *The challenger $\mathcal{C}$ runs the Setup algorithm with a security parameter $\kappa$ and sends the system parameters params to the adversary $\mathcal{A}$ and keeps the master private key msk secret.*
2. **First Phase:** *$\mathcal{A}$ performs polynomially bounded number of queries to the oracles provided to $\mathcal{A}$ by $\mathcal{C}$. The description of the queries in the first phase are listed below:*

    - **Key Extraction query:** *$\mathcal{A}$ produces an identity $ID_i$ corresponding to $U_i$ and receives the private key $D_i$ corresponding to $ID_i$.*
    - **Signcryption query:** *$\mathcal{A}$ produces a set of users $\mathcal{U}$, a receiver identity $ID_\mathbb{R}$ and a plaintext $m \in_R \mathcal{M}$ to the challenger $\mathcal{C}$. $\mathcal{A}$ also specifies the sender $\mathcal{U}_\mathbb{S} \in \mathcal{U}$ whose identity is $ID_\mathbb{S}$. Then $\mathcal{C}$ signcrypts $m$ from $ID_\mathbb{S}$ to $ID_\mathbb{R}$ with $D_\mathbb{S}$ and sends the result to $\mathcal{A}$.*

- **Unsigncryption query:** $\mathcal{A}$ produces a set of users $\mathcal{U}$, a receiver identity $ID_{\mathbb{R}}$, and a ring signcryption $C$. $C$ generates the private key $D_{\mathbb{R}}$ by querying the Key Extraction oracle. $C$ unsigncrypts $C$ using $D_{\mathbb{R}}$ and returns $m$ if $C$ is a valid ring signcryption from $\mathcal{U}$ to $ID_{\mathbb{R}}$ else outputs "Invalid".

   $\mathcal{A}$ queries the various oracles adaptively, i.e. the current oracle requests may depend on the response to the previous oracle queries.

3. **Challenge:** $\mathcal{A}$ chooses two plaintexts $\{m_0, m_1\} \in \mathcal{M}$ of equal length, a set of $n$ users $\mathcal{U}$ and a receiver identity $ID_{\mathbb{R}}$ and sends them to $C$. $\mathcal{A}$ should not have queried the private key corresponding to $ID_{\mathbb{R}}$ in the first phase. $C$ now chooses a bit $\delta \in_R \{0,1\}$ and computes the challenge ring signcryption $C^*$ of $m_\delta$, and sends $C^*$ to $\mathcal{A}$.

4. **Second Phase:** $\mathcal{A}$ performs polynomially bounded number of requests just like the first phase, with the restrictions that $\mathcal{A}$ cannot make Key Extraction query on $ID_{\mathbb{R}}$ and should not query for unsigncryption query on $C^*$. It should be noted that $ID_{\mathbb{R}}$ can be included as a ring member in $\mathcal{U}$, but $\mathcal{A}$ cannot query the private key of $ID_{\mathbb{R}}$.

5. **Guess:** Finally, $\mathcal{A}$ produces a bit $\delta'$ and wins the game if $\delta' = \delta$. The success probability is defined by:

$$Succ_{\mathcal{A}}^{IND-IRSC-CCA2}(\kappa) = \frac{1}{2} + \epsilon$$

Here, $\epsilon$ is called the advantage for the adversary in the attack.

**Note:** The difference between the security model for confidentiality in [3] and our model is, we allow the adversary to access the private key of the ring members (selected by the adversary during the challenge phase) and restrict access to the private key of the receiver of the challenge ring signcryption. But in [3], the adversary is not allowed to access the private keys of the ring members and the receiver (of the challenge ring signcryption).

**Definition 4.** An identity based ring signcryption scheme (IRSC) is said to be existentially unforgeable against adaptive chosen messages attacks (EUF-IRSC-CMA) if no polynomially bounded adversary has a non-negligible advantage in the following game:

1. **Setup Phase:** The challenger runs the Setup algorithm with a security parameter $\kappa$ and gives the system parameters to the adversary $\mathcal{A}$.

2. **Training Phase:** $\mathcal{A}$ performs polynomially bounded number of queries as described in First Phase of definition 3. The queries may be adaptive, i.e. the current query may depend on the previous query responses.

3. **Existential Forgery:** Finally, $\mathcal{A}$ produces a new triple $(\mathcal{U}, ID_{\mathbb{R}}, C)$ (i.e. a triple that was not produced by the signcryption oracle), where the private keys of the users in the group $\mathcal{U}$ were not queried in the training phase. $\mathcal{A}$ wins the game if the result of the Unsigncryption $(\mathcal{U}, ID_{\mathbb{R}}, C)$ is not "Invalid", in other words $C$ is a valid signcrypt of some message $m \in \mathcal{M}$. It should be noted that $ID_{\mathbb{R}}$ can also be member of the ring $\mathcal{U}$ in that case the private key of $ID_{\mathbb{R}}$ should not be queried by $\mathcal{A}$.

**Note:** The difference between the security model for unforgeability in [3] and our model is, we do not allow the adversary to access the private key of the ring members (selected by the adversary during the generation of the forgery) and the adversary is given access to the private key of the receiver of the forged ring signcryption. But in [3], the adversary is not allowed to access the private keys of the ring members as well as the receiver (of the forged ring signcryption).

## 4    Attacks on Various Ring Signcryption Schemes

This section gives an overview of four identity based ring signcryption schemes and the attacks corresponding to them. First we consider Yu et al.'s [4] anonymous signcryption scheme, followed by Fagen Li et al.'s [7] authenticatable anonymous signcryption scheme, next we take up Lijun et al.'s [8] ring signcryption scheme and conclude this section with the review and attack on Zhu et al.'s [5] scheme.

### 4.1    Overview of Anonymous Signcryption (ASC) Scheme of Yu et al.

Yu et al.'s ASC scheme [4] consists of four algorithms namely: *Setup, KeyGen, Signcryption* and *Unsigncryption*, which we describe below.

1. **$Setup(\kappa, l)$:** Here, $\kappa$ and $l$ are the security parameters.
   (a) The PKG selects $\mathbb{G}_1$, $\mathbb{G}_2$ of same order $q$ and a random generator $P$ of $\mathbb{G}_1$ .
   (b) Selects the master private key $s \in_R \mathbb{Z}_q^*$.
   (c) The master public key is computed as $P_{pub} = sP$.
   (d) Selects three strong public one-way hash functions: $H_1 : \{0,1\}^* \to \mathbb{G}_1^*$, $H_2 : \mathbb{G}_2 \to \{0,1\}^l$, $H_3 : \{0,1\}^* \to \mathbb{Z}_q^*$.
   (e) Selects an admissible pairing $\hat{e} : \mathbb{G}_1 \times \mathbb{G}_1 \to \mathbb{G}_2$.
   (f) The public parameters of the scheme are given by $params$=($\mathbb{G}_1$, $\mathbb{G}_2$, $\hat{e}$, $P$, $P_{pub}$, $H_1$, $H_2$, $H_3$,$q$).
2. **$KeyGen(ID_i)$:** Here, $ID_i$ is the identity of the user $\mathcal{U}_i$. The PKG performs the following.
   (a) The user public key is computed as $Q_i = H_1(ID_i)$
   (b) The corresponding private key $D_i = sQ_i$.
   (c) The PKG sends $D_i$ to the user $\mathcal{U}_i$ via a secure channel.
3. **$Signcryption(\mathcal{U}, m, ID_\mathbb{R}, ID_\mathbb{S}, D_\mathbb{S})$:** In order to signcrypt a message $m$, the sender does the following:
   (a) Chooses $r \in_R \mathbb{Z}_q^*$ and, computes $R = rP$, $R' = \hat{e}(P_{pub}, Q_\mathbb{R})^r$, $t = H_2(R')$ and $c = m \oplus t$.
   (b) For all $i = 1$ to $n$ and $i \neq \mathbb{S}$, chooses $U_i \in_R \mathbb{G}_1$ and computes $h_i = H_3(m, t, \mathcal{U}, U_i)$.
   (c) For $i = \mathbb{S}$ chooses $r'_\mathbb{S} \in_R \mathbb{Z}_q^*$ and, computes $U_\mathbb{S} = r'_\mathbb{S} Q_\mathbb{S} - \Sigma^n_{i=1, i \neq \mathbb{S}}(U_i + h_i Q_i)$, $h_\mathbb{S} = H_3(m, t, \mathcal{U}, U_\mathbb{S})$ and $V = (h_\mathbb{S} + r'_\mathbb{S})D_\mathbb{S}$.

Finally, the sender outputs the ring signcryption $C = (\mathcal{U}, c, R, h_1, \ldots, h_n, U_1, \ldots, U_n, V)$.

4. **Unsigncrypt**$(C = (\mathcal{U}, c, R, h_1, \ldots, h_n, U_1, \ldots, U_n, V), D_{\mathbb{R}})$**:** In order to unsigncrypt a ring signcryption $C$, the receiver does the following:

(a) Computes $t' = H_2(\hat{e}(R, D_{\mathbb{R}}))$ and $m' = c \oplus t'$.

(b) For $i = 1$ to $n$, checks whether $h'_i \overset{?}{=} H_3(m', t', \mathcal{U}, U_i)$.

(c) Checks whether $\hat{e}(P_{pub}, \Sigma_{i=1}^{n}(U_i + h'_i Q_i)) \overset{?}{=} \hat{e}(P, V)$.

If all the $n$ checks in (b) and the check in (c) are true, then output $m'$ as the message, else output "*Invalid*".

**Attack on ASC Scheme of Yu et al.:** During the challenge phase of the confidentiality game, the challenger $\mathcal{C}$ receives two messages $m_0$ and $m_1$ from the adversary $\mathcal{A}$. The challenger chooses $\delta \in_R \{0, 1\}$ and produces the challenge ring signcryption $C^*$ using the message $m_\delta$ and delivers $C^*$ to $\mathcal{A}$. Upon receipt of $C^* = (\mathcal{U}, c^*, R^*, h_1^*, \ldots, h_n^*, U_1^*, \ldots, U_n^*, V^*)$, $\mathcal{A}$ does the following to check whether $C^*$ is a signcryption of $m_0$ or $m_1$. (Since $\mathcal{A}$ knows both messages $m_0$ and $m_1$, $\mathcal{A}$ can perform the following computations.)

- Computes $t^* = c^* \oplus m_0$ and checks whether $h_i \overset{?}{=} H_3(m_0, t^*, \mathcal{U}, U_i^*)$, for $i = 1$ to $n$. If all the $n$ checks hold, then $C^*$ is the ring signcryption corresponding to $m_0$.
- If the above check does not hold, $\mathcal{A}$ computes $t^* = c^* \oplus m_1$, checks whether $h_i \overset{?}{=} H_3(m_1, t^*, \mathcal{U}, U_i^*)$, for $i = 1$ to $n$. If all the $n$ checks hold then $C^*$ is a valid ring signcryption for message $m_1$.
- At least one of the checks should hold *true*, else $C^*$ is an invalid ring signcryption.

Thus, $\mathcal{A}$ distinguishes the ring signcryption with out solving any hard problem. Here $\mathcal{A}$ does not interact with the challenger $\mathcal{C}$ after receiving the challenge ring signcryption $C^*$. Thus, our attack is indeed against the CPA security of the ASC scheme by Yu et al. reported in [4].

**Remark:** Informally, $\mathcal{A}$ is able to distinguish the ring signcryption because, the key component required to evaluate the hash value $h_i$ is $t'$ and it is available in $c = m_b \oplus t'$. $\mathcal{A}$ knows that $m_\delta$ is either $m_0$ or $m_1$ because $m_0$ and $m_1$ were chosen by $\mathcal{A}$ and submitted to $\mathcal{C}$ during the challenge phase by $\mathcal{A}$. Hence, $\mathcal{A}$ can find $t'$ without having access to the private key of the receiver and this led to the break in confidentiality (CPA).

## 4.2 Overview of Authenticatable Anonymous Signcryption Scheme (AASC) of Fagen Li et al.

The AASC scheme of Fagen Li et al. [7] consists of the five algorithms. A secure symmetric key encryption scheme $(E, D)$ is employed in this scheme where, $E$ and $D$ are the secure symmetric key encryption and decryption algorithms respectively.

1. **Setup($\kappa$):** Here, $\kappa$ is the security parameter.
   (a) The PKG selects $\mathbb{G}_1$, $\mathbb{G}_2$ of same order $q$ and a random generator $P$ of $\mathbb{G}_1$ .
   (b) Selects the master private key $s \in_R \mathbb{Z}_q^*$.
   (c) Computes the master public key $P_{pub} = sP$.
   (d) Selects three strong public one-way hash functions $H_1 : \{0,1\}^* \to \mathbb{G}_1$, $H_2 : \mathbb{G}_2 \to \{0,1\}^l$, $H_3 : \{0,1\}^* \to \mathbb{Z}_q^*$.
   (e) Selects an admissible pairing $\hat{e} : \mathbb{G}_1 \times \mathbb{G}_1 \to \mathbb{G}_2$ and a secure symmetric key encryption system $(E, D)$.
   (f) The public parameters of the scheme are set to be $params=(\mathbb{G}_1, \mathbb{G}_2, \hat{e}, P, P_{pub}, H_1, H_2, H_3, E, D)$.

2. **Extract($ID_i$):** Similar to the **Extract($ID_i$)** algorithm in 4.1.

3. **Signcrypt($\mathcal{U}, m, ID_\mathbb{R}, ID_\mathbb{S}, D_\mathbb{S}$):** In order to signcrypt a message $m$, the sender does the following:
   (a) Chooses $r \in_R \mathbb{Z}_q^*$, and computes $R = rP$, $k = H_2(\hat{e}(P_{pub}, Q_\mathbb{R})^r)$ and $c = E_k(m)$.
   (b) For $i = 1$ to $n$, $i \neq \mathbb{S}$, chooses $a_i \in_R \mathbb{Z}_q^*$, computes $U_i = a_i P$ and $h_i = H_3(c, \mathcal{U}, U_i)$.
   (c) For $i = \mathbb{S}$, chooses $a_\mathbb{S} \in_R \mathbb{Z}_q^*$, computes $U_\mathbb{S} = a_\mathbb{S} Q_\mathbb{S} - \Sigma_{i=1 i \neq \mathbb{S}}^n (U_i + h_i Q_i)$.
   (d) Computes $h_\mathbb{S} = H_3(c, \mathcal{U}, U_\mathbb{S})$ and $\sigma = (h_\mathbb{S} + a_\mathbb{S}) D_\mathbb{S}$.
   Finally, the sender outputs the ring signcryption as $C = (\mathcal{U}, c, R, U_1, \ldots, U_n, \sigma)$.

4. **Unsigncrypt($C = (\mathcal{U}, c, R, U_1, \ldots, U_n, \sigma), D_\mathbb{R}$):** To unsigncrypt $C$, the receiver does the following.
   (a) Computes $k' = H_2(\hat{e}(R, D_\mathbb{R}))$ and recover $m' = D_{k'}(c)$.
   (b) For $i = 1$ to $n$, computes $h'_i = H_3(c, \mathcal{U}, U_i)$.
   (c) Accepts $C$ and the message $m'$ if and only if $\hat{e}(P_{pub}, \Sigma_{i=1}^n (U_i + h'_i Q_i)) \overset{?}{=} \hat{e}(P, \sigma)$, else output "*Invalid*".

5. **Authenticate(C):** The actual sender $ID_\mathbb{S}$ can prove that the message $m$ was indeed signcrypted by him by running this protocol.
   (a) The sender chooses $x \in_R \mathbb{Z}_q^*$, computes $\mu = \hat{e}(P, \sigma)^x$ and sends $\mu$ to the verifier.
   (b) The verifier chooses $y \in_R \mathbb{Z}_q^*$ and sends it to the sender.
   (c) The sender computes $v = (x + y)(h_\mathbb{S} + a_\mathbb{S})$ and returns $v$ to the verifier.
   (d) The verifier checks whether $\hat{e}(P_{pub}, Q_\mathbb{S})^v \overset{?}{=} \mu.\hat{e}(P, \sigma)^y$ and accepts if the check holds.

**Attack on AASC Scheme of Fagen Li et al.:** The attack on AASC scheme is quite tricky one and it shows that the model considered by the authors did not address explicitly the scenario of the attack we propose. On receiving the challenge ring signcryption $C^* = (\mathcal{U}^*, c^*, R^*, U_1^*, \ldots, U_n^*, \sigma^*)$, in the challenge phase of the confidentiality game, $\mathcal{A}$ can find the message used for generating $C^*$. $\mathcal{A}$ knows the private keys of all the users except the receiver $ID_\mathbb{R}$ and the members of $\mathcal{U}^*$ (here, $\mathcal{U}^*$ is the group of ad-hoc members in the challenge ring signcryption $C^*$). Now, $\mathcal{A}$ chooses $\mathcal{U}'_E \notin \mathcal{U}^*$ with identity string $ID_E$ for which

$\mathcal{A}$ knows the private key $D_E$. $\mathcal{A}$ performs the following steps to distinguish $C^*$ as, whether it is a signcryption of $m_0$ or $m_1$, during the second phase of oracle queries by performing the following.

- $\mathcal{A}$ forms a new group with $\eta$ users who are totally different from $\mathcal{U}^*$. Let the new group be $\mathcal{U}' = \{\mathcal{U}'_1, \ldots, \mathcal{U}'_\eta\}$, where $\mathcal{U}'_E \in \mathcal{U}'$ and $\mathcal{U}' \neq \mathcal{U}^*$.
- For $i = 1$ to $\eta$, $i \neq E$, $\mathcal{A}$ chooses $a_i \in_R \mathbb{Z}_q^*$, computes $U'_i = a_i P$ and $h'_i = H_3(c^*, \mathcal{U}', U'_i)$.
- For $i = E$, $\mathcal{A}$ chooses $a_E \in_R \mathbb{Z}_q^*$, computes $U'_E = a_E Q_E - \Sigma^\eta_{i=1, i \neq E}(U'_i + h'_i Q_i)$.
- $\mathcal{A}$ computes $h'_E = H_3(c^*, \mathcal{U}', U'_E)$ and $\sigma' = (h'_E + a_E)D_E$.
- Now, $C' = (\mathcal{U}', c^*, R^*, U'_1, \ldots, U'_\eta, \sigma')$ is also a valid ring signcryption on the same message $m_\delta$, which was used by $\mathcal{C}$ to generate $C^*$ and $C'$ is entirely different from $C^*$, since $\mathcal{U}' \neq \mathcal{U}^*$. Thus, $\mathcal{A}$ can legally query the unsigncryption of $C'$ during the second phase of the confidentiality game.
- $\mathcal{A}$ gets the unsigncryption to $C'$ from $\mathcal{C}$ as the message $m_b$ and from this $\mathcal{A}$ concludes correctly whether $C^*$ is the signcryption of $m_0$ or $m_1$.

Distinguishing the ring signcryption after the start of the second phase of interaction and a decryption query leads to a break in CCA2 security of the system. Thus, we claim that the AASC scheme by Fagen Li et al. [7] is not adaptive chosen ciphertext secure.

**Remark:** In this scheme, ring signcryption is achieved by using the *Encrypt-then-Sign* paradigm, where the signature part is a ring signature algorithm. This scheme lacks the binding between the encryption and signature; any adversary can alter the signature component of any ring signcryption and with the same receiver, i.e., the output of the encryption is alone used as input to for signature generation. This facilitates the adversary to generate a new valid signature an use it with the remaining components of the challenge ring signcryption, which forms a totally different valid ring signcryption. Now, the adversary can make use of the unsigncryption oracle to unsigncrypt the newly formed ring signcryption. Note that since the encryption part is same as the challenge ring signcryption and the signature part is varied, the newly formed ring signcryption yields the same message as in the challenge ring signcryption and this query is legal with respect to the security model..

### 4.3 Overview of Identity Based Ring Signcryption (IRSC) Scheme of Lijun et al.

The IRSC scheme of Lijun et al. [8] consists of the following four algorithms.

1. **Setup($\kappa$):** Here, $\kappa$ is the security parameters.
   (a) The PKG selects $\mathbb{G}_1, \mathbb{G}_2$ of same prime order - $q$ and a random generator $P$ of $\mathbb{G}_1$ .

(b) Selects the master private key $s \in_R \mathbb{Z}_q^*$.

(c) The master public key is set to be $P_{pub} = sP$.

(d) Selects three cryptographic hash functions $H_1 : \{0,1\}^* \to \mathbb{G}_1$, $H_2 : \{0,1\}^* \to \mathbb{Z}_q^*$, $H_3 : \{0,1\}^* \to \mathbb{Z}_q^*$.

(e) Selects an admissible pairing $\hat{e} : \mathbb{G}_1 \times \mathbb{G}_1 \to \mathbb{G}_2$.

(f) The public parameters of the scheme are set to be $params = (\mathbb{G}_1, \mathbb{G}_2, \hat{e}, P, P_{pub}, H_1, H_2, H_3, q)$.

2. **KeyGen** $(ID_i)$: Similar to the **Extract**$(ID_i)$ algorithm in 4.1.

3. **Signcrypt**$(\mathcal{U}, m, ID_\mathbb{R}, ID_\mathbb{S}, D_\mathbb{S})$: In order to signcrypt the message $m$ the sender does the following:

(a) Chooses $r_0 \in_R \mathbb{Z}_q^*$ and computes $R_0 = r_0 P$, $W = r_0 P_{pub}$.

(b) For $i = 1$ to $n$, $i \neq \mathbb{S}$, chooses $r_i \in_R \mathbb{Z}_q^*$, computes $R_i = r_i P$ $h_i = H_2(m\|\mathcal{U}\|R_i\|R_0)$.

(c) For $i = \mathbb{S}$, chooses $r_\mathbb{S} \in_R \mathbb{Z}_q^*$, computes $R_\mathbb{S} = r_\mathbb{S} P - \Sigma_{i=1, i \neq \mathbb{S}}^n (h_i Q_i)$, $h_\mathbb{S} = H_2(m\|\mathcal{U}\|R_\mathbb{S}\|R_0)$ and $V = h_\mathbb{S} D_\mathbb{S} + \Sigma_{i=1}^n r_i P_{pub}$.

(d) Computes $y = \hat{e}(W, Q_\mathbb{R})$, $t = H_3(y)$, $c = m \oplus t$.

Finally the sender outputs the ciphertext as $C = (\mathcal{U}, c, V, R_0, R_1, \ldots, R_n)$.

4. **Unsigncrypt**$(C = (\mathcal{U}, c, V, R_0, R_1, \ldots, R_n), D_\mathbb{R})$: In-order to unsigncrypt $C$, the receiver does the following.

(a) Computes $t' = H_3(\hat{e}(D_\mathbb{R}, R_0))$ and recovers $m' = c \oplus t'$.

(b) For $i = 1$ to $n$, computes $h_i' = H_2(m\|\mathcal{U}\|R_i\|R_0)$.

(c) Checks whether $\hat{e}(P_{pub}, \Sigma_{i=1}^n (R_i + h_i' Q_i)) \overset{?}{=} \hat{e}(P, V)$.

If all the $n$ checks in (b) and the check in (c) are true, then output $m'$ as the message, else output "*Invalid*".

**Attack on IRSC Scheme of Lijun et al.:** During the challenge phase of the confidentiality game, the challenger $\mathcal{C}$ receives two messages $m_0$ and $m_1$ from the adversary $\mathcal{A}$. The challenger chooses $\delta \in_R \{0,1\}$ and generates the challenge ring signcryption $C^*$ using the message $m_\delta$ and delivers $C^*$ to $\mathcal{A}$. Upon receipt of $C^* = (\mathcal{U}, c^*, V^*, R_0^*, R_1^*, \ldots, R_n^*)$, $\mathcal{A}$ does the following to distinguish $C^*$ as, whether $C^*$ is the signcryption of $m_0$ or $m_1$. Since $\mathcal{A}$ knows both messages $m_0$ and $m_1$, $\mathcal{A}$ can perform the following computations.

- $\mathcal{A}$ can compute $h_i = H_2(m_0\|\mathcal{U}\|R_i^*\|R_0^*)$ for $i = 1$ to $n$. (since $R_i^*$, $R_0^*$ are known from the ring signcryption $C^*$).

- Checks whether $\hat{e}(P_{pub}, \Sigma_{i=1}^n (R_i^* + h_i Q_i)) \overset{?}{=} \hat{e}(P, V^*)$. If this check holds, then $C^*$ is a valid ring signcryption of $m_0$.

- If the above check does not hold, perform the previous two steps with $m_0$ replaced by $m_1$. If the ring signcryption was formed with one of the two messages $m_0$ or $m_1$, any one of the above checks will hold, else the ring signcryption $C^*$ is an invalid one.

Thus, $\mathcal{A}$ can distinguish the challenge signcryption without knowing the key of the receiver in the challenge ring signcryption $C^*$.

**Remark:** The intuition behind the attack is, in the ring signcryption proposed by Lijun et al. [8] the ring signcryption can be verified if the message and the corresponding ring signcryption is known. During the confidentiality game the adversary $\mathcal{A}$ knows the message, which is either $m_0$ or $m_1$, with these information $\mathcal{A}$ concludes whether $C^*$ is a ring signcryption of $m_0$ or $m_1$.

### 4.4   Overview of IRSC Scheme of Zhu et al.

The IRSC scheme of Zhu et al. [5] consists of the following four algorithms.

1. **Setup**$(\kappa, l)$: Here, $\kappa$ and $l$ are the security parameters.
   (a) The PKG selects $\mathbb{G}_1$, $\mathbb{G}_2$ of same order $q$ and a random generator $P$ of $\mathbb{G}_1$ .
   (b) Selects the master private key $s \in_R \mathbb{Z}_q^*$ and computes the master public key to be $P_{pub} = sP$.
   (c) Selects four cryptographic hash functions $H_1 : \{0,1\}^* \rightarrow \mathbb{G}_1^*$, $H_2 : \mathbb{G}_1^* \rightarrow \{0,1\}^l$, $H_3 : \{0,1\}^l \times \mathbb{G}_1 \rightarrow \{0,1\}^l$, $H_4 : \{0,1\}^* \rightarrow \mathbb{Z}_q^*$.
   (d) Selects an admissible pairing $\hat{e} : \mathbb{G}_1 \times \mathbb{G}_1 \rightarrow \mathbb{G}_2$.
   (e) The public parameters of the scheme are set to be $params$=$(\mathbb{G}_1, \mathbb{G}_2, \hat{e}$, $P, P_{pub}, H_1, H_2, H_3, H_4, q)$.
2. **KeyGen** $(ID_i)$: Similar to the **Extract**$(ID_i)$ algorithm in 4.1.
3. **Signcrypt**$(\mathcal{U}, m, ID_\mathbb{R}, ID_\mathbb{S}, D_\mathbb{S})$: In order to signcrypt the message $m$, the sender does the following:
   (a) Chooses $r \in_R \mathbb{Z}_q^*$, $\hat{m} \in_R \mathcal{M}$ and, computes $R_0 = rP$, $R^{'} = \hat{e}(rP_{pub}, Q_\mathbb{R})$, $k = H_2(R^{'})$, $c_1 = \hat{m} \oplus k$ and $c_2 = m \oplus H_3(\hat{m}\|R_0)$.
   (b) For $i = 1$ to $n$, $i \neq \mathbb{S}$, chooses $U_i \in_R \mathbb{G}_1^*$ and computes $h_i = H_4(c_2\|U_i)$.
   (c) For $i = \mathbb{S}$, chooses $r' \in_R \mathbb{Z}_q^*$, computes $U_\mathbb{S} = r'Q_\mathbb{S} - \Sigma_{i=1,i\neq\mathbb{S}}^n(U_i + h_iQ_i)$, $h_\mathbb{S} = H_4(c_2\|U_\mathbb{S})$ and $V = (h_\mathbb{S} + r')D_\mathbb{S}$.

   Finally, outputs the ring signcryption as $C = (\mathcal{U}, R_0, c_1, c_2, U_1, \ldots, U_n, V)$.
4. **Unsigncrypt**$(C = (\mathcal{U}, R_0, c_1, c_2, U_1, \ldots, U_n, V)$ , $D_\mathbb{R})$: To unsigncrypt a ring signcryption $C$, the receiver does the following.
   (a) For $i = 1$ to $n$, computes $h_i' = H_4(c_2\|U_i)$.
   (b) Checks whether $\hat{e}(P_{pub}, \Sigma_{i=1}^n(U_i + h_i'Q_i)) \stackrel{?}{=} \hat{e}(P, V)$, if so, computes $k' = H_2(\hat{e}(R_0, D_\mathbb{R}))$, and recovers $\hat{m}' = c_1 \oplus k'$ and $m' = c_2 \oplus H_3(\hat{m}'\|R_0)$. Accept $m'$ as the valid message.

**Note:** The actual scheme in [10] had typos in setup, keygen as well as signcryption algorithms. The definition of the hash function $H_3$ was inconsistent. Instead, of $H_2$, it was written $H_1$, instead of $H_1$, it was written $H_0$ and instead of $U_\mathbb{S} = r'Q_\mathbb{S} - \Sigma_{i=1,i\neq\mathbb{S}}^n(U_i + h_iQ_i)$, it was written $U_\mathbb{S} = r'Q_\mathbb{S} - \Sigma_{i=1,i\neq\mathbb{S}}^n(U_i + h_iQ_\mathbb{S})$. We have corrected all of them in our review, in order to maintain the consistency of the scheme.

**Attack on IRSC Scheme of Zhu et al.:** On receiving the challenge ring signcryption $C^* = (\mathcal{U}^*, R_0^*, c_1^*, c_2^*, U_1^*, \ldots, U_n^*, V^*)$, in the challenge phase of the confidentiality game, $\mathcal{A}$ can find the message used for generating $C^*$. $\mathcal{A}$ knows the private keys of all the users except the receiver $ID_{\mathbb{R}}$ and the members of $\mathcal{U}^*$ (here, $\mathcal{U}^*$ is the group of ad-hoc members in the challenge ring signcryption $C^*$). Now, $\mathcal{A}$ chooses $\mathcal{U}_E' \notin \mathcal{U}^*$ with identity string $ID_E$ for which $\mathcal{A}$ knows the private key $D_E$. $\mathcal{A}$ performs the following steps to distinguish $C^*$ as, whether it is a signcryption of $m_0$ or $m_1$, during the second phase of oracle queries by performing the following.

- $\mathcal{A}$ forms a new group $\mathcal{U}'$ with $\eta$ members who are totally different from the users in $\mathcal{U}^*$ present in the challenge ring signcryption. Consider $\mathcal{U}' = \{\mathcal{U}_1', \ldots, \mathcal{U}_\eta'\}$ and $\mathcal{U}_E' \in \mathcal{U}'$ (The private key of $\mathcal{U}_E'$ is known to $\mathcal{A}$).
- Chooses a message $m'$ and computes $c_2' = c_2^* \oplus m'$.
- For all $i = 1$ to $\eta$ and $i \neq E$, chooses $U_i' \in_R \mathbb{G}_1^*$ and computes $h_i' = H_4(c_2' || U_i')$.
- For $i = E$, chooses $r' \in_R \mathbb{Z}_q^*$ and computes $U_E' = r' Q_A - \Sigma_{i=1}^\eta (U_i' + h_i' Q_i)$.
- Computes $h_E' = H_4(c_2' || U_E')$ and $V' = (r' + h_E') D_E$
- Now, $C' = (\mathcal{U}', R_0^*, c_1^*, c_2', U_1', \ldots, U_n', V')$ is a valid ring signcryption on message $m_b \oplus m'$.

Now, during the second phase of training, $\mathcal{A}$ requests the unsigncryption of $C'$ to $\mathcal{C}$. Note that it is legal for $\mathcal{A}$ to ask for unsigncryption of $C'$ because it is derived from $C^*$ and not exactly the challenge ring signcryption $C^*$. $\mathcal{C}$ responds with $M = m_b \oplus m'$ as the output for the query. $\mathcal{A}$ now obtains $m_b = M \oplus m'$ and thus identifies the message in the challenge ring signcryption $C^*$.

**Remark:** This attack is possible due to the same reason as described in the remark for the attack stated in section 4.2.

# 5 New Ring Signcryption Scheme (New-IBRSC)

In this section, we present a new improved identity based ring signcryption scheme (New-IBRSC), taking into account the attacks carried out in the previous section. New-IBRSC consists of the following four algorithms:

1. **Setup($\kappa$):** This algorithm is executed by the PKG to initialize the system by taking a security parameter $\kappa$ as input.
   - Selects $\mathbb{G}_1$ an additive group and $\mathbb{G}_2$ a multiplicative group, both cyclic with same prime order - $q$ and a random generator $P$ of the group $\mathbb{G}_1$.
   - Selects $s \in_R \mathbb{Z}_q^*$ as the master private key and computes the master public key $P_{pub} = sP$.
   - Selects four cryptographic hash functions $H_1 : \{0,1\}^* \to \mathbb{G}_1$, $H_2 : \mathbb{G}_2 \to \{0,1\}^{|\mathcal{M}|} \times \mathbb{Z}_q^* \times \mathbb{G}_1$, $H_3 : \{0,1\}^* \to \mathbb{Z}_q^*$ and $H_4 : \{0,1\}^{|\mathcal{M}|} \times \mathbb{Z}_q^* \to \mathbb{Z}_q^*$.
   - Picks a bilinear pairing $\hat{e} : \mathbb{G}_1 \times \mathbb{G}_1 \to \mathbb{G}_2$ with the appropriate properties specified in section 2.

- The public parameter of the scheme is $params = (\mathbb{G}_1, \mathbb{G}_2, \hat{e}, P, P_{pub}, H_1, H_2, H_3, H_4, q)$.

2. **Keygen**$(ID_i)$: This algorithm takes $ID_i$, the identity of a user $\mathcal{U}_i$ as input. The PKG who executes this algorithm computes the private key and public key for the user with identity $ID_i$ as follows:
   - The public key is computed as $Q_i = H_1(ID_i)$
   - The corresponding private key $D_i = sQ_i$.
   - PKG sends $D_i$ to user $\mathcal{U}_i$ via a secure channel.

3. **Signcrypt**$(\mathcal{U}, m, ID_\mathbb{R}, Q_\mathbb{R}, ID_\mathbb{S}, D_\mathbb{S})$: For signcrypting a message $m$ to the receiver $\mathcal{U}_\mathbb{R}$ with public key $Q_\mathbb{R}$ the sender with private key $D_\mathbb{S}$ and public key $Q_\mathbb{S}$ performs the following:
   - Selects $n$ potential senders and forms an ad-hoc group $\mathcal{U}$, including its own identity $ID_\mathbb{S}$.
   - Chooses $w \in_R \mathbb{Z}_q^*$, computes $r = H_4(m, w)$, $U = rP$ and $\alpha = \hat{e}(P_{pub}, Q_\mathbb{R})^r$.
   - For $i = 1$ to $n$, $i \neq \mathbb{S}$, chooses $U_i \in_R \mathbb{G}_1$ and computes $h_i = H_3(m, U_i, \alpha, \mathcal{U}, Q_\mathbb{R})$.
   - For $i = \mathbb{S}$, chooses $r_\mathbb{S} \in_R \mathbb{Z}_q^*$ and, computes $U_\mathbb{S} = r_\mathbb{S}Q_\mathbb{S} - \Sigma_{i=1, i\neq\mathbb{S}}^n (U_i + h_iQ_i)$, $h_\mathbb{S} = H_3(m, U_\mathbb{S}, \alpha, \mathcal{U}, Q_\mathbb{R})$ and $V = (h_\mathbb{S} + r_\mathbb{S})D_\mathbb{S}$.
   - Computes $y = (m\|w\|V) \oplus H_2(\alpha)$.
   Finally, the sender outputs the ring signcryption $C = (y, \mathcal{U}, U, U_1, \ldots, U_n)$.

4. **Unsigncrypt**$(C = (y, \mathcal{U}, U, U_1, \ldots, U_n), D_\mathbb{R})$: The receiver $\mathcal{U}_\mathbb{R}$ with identity $ID_\mathbb{R}$ does the following to unsigncrypt the ring signcryption $C$:
   - Computes $\alpha' = \hat{e}(U, D_\mathbb{R})$, retrieves $m'$, $w'$ and $V'$ as $(m'\|w'\|V') = y \oplus H_2(\alpha')$, and checks whether $U \stackrel{?}{=} H_4(m', w')P$.
   - For $i = 1$ to $n$, computes $h_i' = H_3(m', U_i, \alpha', \mathcal{U}, Q_\mathbb{R})$ and checks whether $\hat{e}(P_{pub}, \Sigma_{i=1}^n(U_i + h_i'Q_i)) \stackrel{?}{=} \hat{e}(P, V')$.
   If both the above checks hold, then the receiver $\mathcal{U}_\mathbb{R}$ accepts $C$ as the valid ring signcryption and the message $m'$ as the valid message.

**Correctness:** We show the correctness of the unsigncryption algorithm here:

$$\text{LHS} = \hat{e}(P_{pub}, \Sigma_{i=1}^n(U_i + h_i'Q_i))$$
$$= \hat{e}(sP, \Sigma_{i=1}^n(U_i + h_i'Q_i))$$
$$= \hat{e}(sP, \Sigma_{i=1, i\neq\mathbb{S}}^n(U_i + h_i'Q_i)).\hat{e}(sP, (U_\mathbb{S} + h_\mathbb{S}'Q_\mathbb{S}))$$
$$= \hat{e}(sP, \Sigma_{i=1, i\neq\mathbb{S}}^n(U_i + h_i'Q_i)).\hat{e}(sP, r_\mathbb{S}Q_\mathbb{S} - \Sigma_{i=1, i\neq\mathbb{S}}^n(U_i + h_iQ_i) + h_\mathbb{S}'Q_\mathbb{S})$$
$$= \hat{e}(sP, \Sigma_{i=1, i\neq\mathbb{S}}^n(U_i + h_i'Q_i) - \Sigma_{i=1, i\neq\mathbb{S}}^n(U_i + h_iQ_i)).\hat{e}(sP, r_\mathbb{S}Q_\mathbb{S} + h_\mathbb{S}'Q_\mathbb{S})$$
$$= \hat{e}(sP, r_\mathbb{S}Q_\mathbb{S} + h_\mathbb{S}'Q_\mathbb{S})$$
$$= \hat{e}(P, V') = \text{RHS}$$

Note that the above correctness holds only if $h_i = h_i'$ for $(i = 1$ to $n)$.

# 6   Security Results for New-IBRSC

The anonymity proof for the new identity based ring signcryption scheme (New-IBRSC) follows from the underlying identity based ring signature [11]. The composition of encryption and ring signature scheme to form the ring signcryption scheme (New-IBRSC) does not induce a weakness in the anonymity property. Encryption components used in the ring signature generation do not reveal any information about the ring members. The binding between the encryption and the ring signature is obtained with the help of the session key that is used for encrypting the message. Even though, the session key is an input to the message hash in the ring signature it does not contain any information that reveal the identity of the sender and hence forth we concentrate only on the security against adaptive chosen ciphertext attack (CCA2) and security against chosen message attack (CMA). We formally prove the security of the new identity based ring signcryption scheme (New-IBRSC), indistinguishable under chosen ciphertext attack (IND-IBRSC-CCA2) and existentially unforgeable under chosen message and identity attack (EUF-IBRSC-CMA) in the random oracle model. We consider the security model given in section 3 to prove the security of the New-IBRSC.

### 6.1   Confidentiality Proof of New-IBRSC (IND-IBRSC-CCA2)

**Theorem 1.** *If an IND-IBRSC-CCA2 adversary $\mathcal{A}$ exists against New-IBRSC scheme, asking $q_{H_i}$ ($i = 1$, $2$, $3,4$) hash queries to random oracles $H_i$ ($i = 1$, $2$, $3$, $4$), $q_e$ extract queries, $q_{sc}$ signcryption queries and $q_{us}$ unsigncryption queries, then there exist an algorithm $\mathcal{C}$ that solves CBDHP.*

Proof of this theorem is omitted due to page restriction and will be given in the full version of this paper.

### 6.2   Unforgeability Proof of New-IBRSC (EUF-IBRSC-CMA)

**Theorem 2.** *If an EUF-IBSC-CMA forger $\mathcal{A}$ exists against New-IBRSC scheme, asking $q_{H_i}$ ($i = 1$, $2$, $3$, $4$) hash queries to random oracles $H_i$ ($i = 1$, $2$, $3$, $4$), $q_e$ extract secret key queries, $q_{sc}$ signcryption queries and $q_{us}$ unsigncryption queries, then there exist an algorithm $\mathcal{C}$ that solves CDHP.*

Proof of this theorem is omitted due to page restriction and will be given in the full version of this paper.

## 7   Conclusion

As a concluding remark we summarize the work in this paper. Ring signcryption is a primitive which enables a user to transmit authenticated messages anonymously and confidentially. To the best of our knowledge there were seven ring signcryption schemes in the identity based setting. Already it was shown in [2] that [6] was not CCA2 secure and in [7] it was shown by Fagen Li et al. that, [1] was not CCA2 secure. So, five out of seven identity based ring signcryption schemes were believed to be secure till date. We have shown that [4] and [8] does not even provide security against chosen plaintext attack (CPA); [7] and [5] does not provide security against adaptive chosen ciphertext attack (CCA2), by demonstrating attacks on confidentiality of these schemes. This leaves Huang et al.'s [3] scheme as the only secure identity based ring signcryption scheme. We have proposed a new identity based ring signcryption scheme for which we proved the security against chosen ciphertext attack and existential unforgeability in the random oracle model. Also we have compared our scheme with Huang et al.'s scheme below. In the comparison table, $n$ represents the number of members in the ring.

**Table 1.** Efficiency Comparison with [3]

| Scheme | Signcryption | | | | | Unsigncryption | | | | |
|---|---|---|---|---|---|---|---|---|---|---|
| | $SPM$ | $BP$ | $EXP$ | $\mathbb{G}_2 M$ | $PA$ | $SPM$ | $BP$ | $EXP$ | $\mathbb{G}_2 M$ | $PA$ |
| New-IBRSC | $n+2$ | 1 | 1 | – | $2n-2$ | $n$ | 3 | - | – | $2n-1$ |
| Scheme in [3] | $2n+2$ | $n+2$ | – | 1 | $2n$ | $n$ | 3 | - | $n+1$ | $n$ |

$SPM$ - Scalar Point Multiplication, $BP$ - Bilinear Pairing, $EXP$ - Exponentiation in $\mathbb{G}_2$, $\mathbb{G}_2 M$ - Multiplication of two $\mathbb{G}_2$ elements and $PA$ - Point Addition.

**Table 2.** Ciphertext Size Comparison with [3]

| Scheme | Ciphertext Size |
|---|---|
| New-IBRSC | $2|\mathcal{M}| + (n+2)|\mathbb{G}_1|$ |
| Scheme in [3] | $2|\mathcal{M}| + (n+1)|\mathbb{G}_1| + n|\mathbb{Z}_q^*|$ |

Thus, our new identity based ring signcryption scheme (New-IBRSC) is a significant improvement over the scheme proposed by Huang et al. [3]

# References

1. Zhang, M., Yang, B., Zhu, S., Zhang, W.: Efficient secret authenticatable anonymous signcryption scheme with identity privacy. In: Yang, C.C., Chen, H., Chau, M., Chang, K., Lang, S.-D., Chen, P.S., Hsieh, R., Zeng, D., Wang, F.-Y., Carley, K.M., Mao, W., Zhan, J. (eds.) ISI Workshops 2008. LNCS, vol. 5075, pp. 126–137. Springer, Heidelberg (2008)
2. Sree Vivek, S., Sharmila Deva Selvi, S., Rangan, P.: On the security of two ring signcryption schemes. Cryptology ePrint Archive, Report 2009/052. To appear in the proceedings of SECRYPT 2009 (2009)
3. Huang, X., Susilo, W., Mu, Y., Zhang, F.: Identity-based ring signcryption schemes: Cryptographic primitives for preserving privacy and authenticity in the ubiquitous world. In: AINA 2005: Proceedings of the 19th International Conference on Advanced Information Networking and Applications, pp. 649–654. IEEE Computer Society, Los Alamitos (2005)

4. Yu, Y., Li, F., Xu, C., Sun, Y.: An efficient identity-based anonymous signcryption scheme. Wuhan University Journal of Natural Sciences 13(6), 670–674 (2008)
5. Zhu, Z., Zhang, Y., Wang, F.: An efficient and provable secure identity based ring signcryption scheme. Computer Standards & Interfaces, 649–654 (2008), http://dx.doi.org/10.1016/j.csi.2008.09.023
6. Li, F., Xiong, H., Yu, Y.: An efficient id-based ring signcryption scheme. In: International Conference on Communications, Circuits and Systems, ICCCAS 2008, May 2008, pp. 483–487 (2008)
7. Li, F., Shirase, M., Takagi, T.: Analysis and improvement of authenticatable ring signcryption scheme. In: International Conference ProvSec 2008, Paper appears in Journal of Shanghai Jiaotong University (Science), December 2008, vol. 13-6, pp. 679–683 (2008)
8. Zhun, L., Zhang, F.: Efficient identity based ring signature and ring signcryption schemes. In: International Conference on Computational Intelligence and Security, CIS 2008, December 2008, vol. 2, pp. 303–307 (2008)
9. Baek, J., Steinfeld, R., Zheng, Y.: Formal proofs for the security of signcryption. In: Naccache, D., Paillier, P. (eds.) PKC 2002. LNCS, vol. 2274, pp. 80–98. Springer, Heidelberg (2002)
10. Chung, Y.F., Wu, Z.Y., Chen, T.S.: Ring signature scheme for ecc-based anonymous signcryption. Computer Standards & Interfaces Journal (2008)
11. Chow, S.S.M., Yiu, S.M., Hui, L.C.K.: Efficient identity based ring signature. In: Ioannidis, J., Keromytis, A.D., Yung, M. (eds.) ACNS 2005. LNCS, vol. 3531, pp. 499–512. Springer, Heidelberg (2005)

# A Storage Efficient Redactable Signature in the Standard Model

Ryo Nojima[1], Jin Tamura[1], Youki Kadobayashi[2], and Hiroaki Kikuchi[3]

[1] National Institute of Information and Communications Technology (NICT), Japan
[2] Nara Institute of Science of Technology (NAIST), Japan
[3] Tokai University, Japan

**Abstract.** In this paper, we propose a simple redactable signature scheme for super-sets whose message-signature size is $O(|M| + \tau)$, where $\tau$ is a security parameter and $M$ is a message to be signed. The scheme proposed by Johnson et al. in CT-RSA 2003 has the similar performance but this scheme was proven secure based on the RSA assumption in the random oracle model. In this paper, we show that such a scheme can be constructed based on the RSA assumption without the random oracles.

## 1 Introduction

In this paper, we consider a *redactable* signature scheme that comprises a third party called a *redactor* in addition to the signer and the verifier. This signature scheme attracts much attention since it allows the redactor to hide the partial information of the original message without disabling the verifier from verifying the integrity of the resultant message. More concretely, it allows the redactor to replace arbitrary bit positions in the message with a special symbol, say "#", which means it has a *hiding* property as well as a certain *weak unforgeable* property. These special properties are useful in many scenarios, e.g., publishing medical data which should be done in an anonymized and yet authentic form.

Extensive studies have been carried out on this signature scheme (e.g., [1,2,3]). The main drawback of the previously proposed schemes lies in their overhead. In the "standard" signature schemes, the message-signature size is $O(\tau + |M|)$, where $\tau$ and $M$ are the sizes of the security parameter (signature) and the message $M$, respectively. The overhead of the redactable signatures proposed in [2,3] is high. For example, the overhead of the scheme proposed in [2] is $O(\tau \cdot |M|)$. To the best of our knowledge, the best solution in terms of the size is [1], where the message-signature size is $O(\tau + |M|)$. However, from the security point of view, it was based on the hardness of the RSA problem in the *random oracle* model.

In this paper, we propose a scheme whose security can be proven by assuming the hardness of the RSA problem *without* the random oracle while keeping the message-signature size $O(\tau + |M|)$. Our scheme is based on the redactable signature proposed by Miyazaki, Hanaoka, and Imai [4]. In that paper, the authors showed a very interesting scheme based on a pseudorandom generator.[1]

---

[1] The security proof was done later by Yang et al. in [5].

P. Samarati et al. (Eds.): ISC 2009, LNCS 5735, pp. 326–337, 2009.
© Springer-Verlag Berlin Heidelberg 2009

However, the signature scheme only has restricted redaction ability. The scheme only allows the redactor to hide one string from the original message. In this paper, we remove such constraint by employing the technique used in the verifiable random function [6].

**Related Works:** Many redactable (or sanitizable) signature schemes have been proposed so far (e.g., [7,8,9,10]). In most of the schemes, the authors concentrate on the redactable signature schemes which have some additional useful properties. In this paper, we have not provided such additional properties in our scheme so as to keep it simple as possible.

Recently, in CT-RSA 2009, the authors proposed the redactable signature for strings [11]. Their scheme was aimed at hiding the length of the redacted sub-strings. Hence they constructed a redactable signature with some additional properties. In fact, this scheme used the redactable signature scheme proposed in [1,2] as a building block. More precisely, the authors regard [1,2] as a redactable signature for *super-sets* and propose the signature scheme by using these as a black-box. The scheme proposed by us can be considered as a redactable signature for super-sets; thus, by combining our scheme with that in [11], we can obtain a redactable signature scheme for strings as well.

**Organization:** We provide the definitions and the notations used in this paper in the next section. The proposed scheme is shown in Section 3. In the section, we utilize the verifiable random function proposed in [6] to construct the redactable signature scheme. In Section 4, we prove the security of the proposed scheme and then conclude this paper in Section 5.

## 2   Preliminaries

### 2.1   Notations

We denote a security parameter by $\tau \in \mathbb{N}$. A function $\mu : \mathbb{N} \to [0,1]$ is said to be *negligible*, if for every $c$ there exists $\tau_c > 0$ such that for all $\tau > \tau_c$, $\mu(\tau) < \frac{1}{\tau^c}$. A function $\mu : \mathbb{N} \to [0,1]$ is said to be *non-negligible* if it is not a negligible function. We denote "probabilistic polynomial-time" by PPT . All of the algorithms run in PPT in the security parameter $\tau$. We often do not write $1^\tau$ explicitly.

If $S$ is a set then we denote the experiment of selecting $x \in S$ according to the uniform distribution by $x \xleftarrow{\$} S$. If $\mathsf{A}(x)$ is a randomized algorithm with some input $x$, then we denote the experiment of $\mathsf{A}(x)$ producing $y$ over the internal coins by $y \xleftarrow{\$} \mathsf{A}(x)$. On the other hand, if $\mathsf{A}$ is a deterministic algorithm then we denote it simply by $y \leftarrow \mathsf{A}(x)$. Further, we denote the assignment of $b$ into $a$ by $a \leftarrow b$.

Let $s_1, s_2, \ldots, s_n$ be strings of finite length. We denote the concatenation of $s_1, s_2, \ldots, s_n$ by $\langle s_1, s_2, \ldots, s_n \rangle$. In this paper, we denote the message to be signed by $M = \langle M_1, M_2, \ldots, M_L \rangle$ with $M_i \in \{0,1\}$ for every $1 \le i \le L$. We denote $[L] = \{1, \ldots, L\}$ and if $S \subset [L]$ then $\overline{S}$ is $[L] \setminus S$.

## 2.2 RSA Assumption

Let $P, Q$ be primes of the same length $\tau$ and let $e$ be an integer such that $\gcd(e, \phi(N)) = 1$, where $\phi$ is the Euler function and $N = PQ$. The RSA permutation $f_{e,N} : \mathbb{Z}_N^* \to \mathbb{Z}_N^*$ is then defined by $f_{e,N} : x \mapsto x^e \bmod N$. Given randomly chosen $e, P, Q$, and $y \in \mathbb{Z}_N^*$, it is believed that producing $x$ such that $x^e = y \bmod N$ is hard for all PPT algorithms. In this paper, we assume that $e$ is a prime of length $|N| + 1 = 2\tau + 1$. We denote a set of $N$s which is composed of the products of two primes $P, Q$ of length $\tau$ by $\mathrm{MOD}_\tau$. That is, $|N| = 2\tau$ for every $N \in \mathrm{MOD}_\tau$. Further, we denote a set of primes of length $\tau$ by $\mathrm{Primes}_\tau$. Then, we can formally describe the assumption as follows: For every PPT algorithm $\mathsf{B}$

$$\Pr\left[N \xleftarrow{\$} \mathrm{MOD}_\tau, y \xleftarrow{\$} \mathbb{Z}_N^*, e \xleftarrow{\$} \mathrm{Primes}_{|N|+1}, x \xleftarrow{\$} \mathsf{B}(e, N, y) \middle| x^e = y \bmod N\right] \tag{1}$$

is negligible.

## 2.3 Definition of the Digital Signature Scheme

A "standard" digital signature scheme $\mathsf{SS} = (\mathsf{Gen}_S, \mathsf{Sig}_S, \mathsf{Ver}_S)$ consists of three algorithms. Given a security parameter $\tau$, the key-generation algorithm $\mathsf{Gen}_S$ outputs signing/verification keys $(sk_S, vk_S)$. Given a signing key $sk_S$ and a message $M$, the signing algorithm $\mathsf{Sig}_S$ outputs a signature $\sigma_S$. Given a verification key $vk_S$, a message $M$, and a signature $\sigma_S$, the verification algorithm $\mathsf{Ver}_S$ outputs a signature validity decision 1 or 0. Here, 1 and 0 mean acceptance and rejection, respectively. We require the signature scheme to satisfy the existential unforgeability with respect to chosen-message attacks (EU-CMA), which is defined as follows:

**Definition 1 ([12]).** *Let* $\mathsf{SS} = (\mathsf{Gen}_S, \mathsf{Sig}_S, \mathsf{Ver}_S)$ *be a digital signature scheme. We say that* $\mathsf{SS}$ *satisfies EU-CMA if, for every PPT* $\mathsf{A}$,

$$\Pr\left[\begin{array}{c} (vk_S, sk_S) \xleftarrow{\$} \mathsf{Gen}_S, \\ (M^*, \sigma_S^*) \xleftarrow{\$} \mathsf{A}^{\mathsf{Sig}_S(sk_S, \cdot)}(vk_S) \end{array} \middle| \begin{array}{c} \mathsf{Ver}_S(vk_S, M^*, \sigma_S^*) = 1 \\ \wedge M^* \notin Q \end{array}\right]$$

*is negligible, where* $Q$ *is the set of queries to the oracle* $\mathsf{Sig}_S(sk_S, \cdot)$ *produced by* $\mathsf{A}$.

Any signature scheme that satisfies EU-CMA can be employed in our scheme. Hence, we can employ the RSA based schemes such as [13,14,15].

## 2.4 Definitions of a Redactable Signature for Super Sets

In this section, we present the definition of the redactable signature for super sets. The definition is based on [11].

Let $\Sigma = \{0, 1, \#\}$. We define a partial order $\preceq$ on $\Sigma$ such that $\# \preceq 0, \# \preceq 1$, and $a \preceq a$ for each $a \in \Sigma$. This induces a partial order $\preceq$ on $\Sigma^*$ by pointwise comparison, namely, $\langle w_1 \cdots w_n \rangle \preceq \langle x_1 \cdots x_n \rangle$ holds if $w_i \preceq x_i$ for every $i$. In this paper, whenever $M \preceq M'$, we assume $|M| = |M'|$.

The redactable signature is composed of a quadruple of algorithms $\mathsf{RS} = (\mathsf{Gen}_R, \mathsf{Sig}_R, \mathsf{Red}_R, \mathsf{Ver}_R)$:

- The PPT algorithm $\mathsf{Gen}_R$ outputs $(vk_R, sk_R)$. We call $sk_R$ and $vk_R$ a signing-key and a verification-key, respectively.
- On input $(sk_R, M)$, the PPT algorithm $\mathsf{Sig}_R$ outputs a signature $\sigma_R$.
- On input $(vk_R, \sigma_R, M, M')$, the PPT algorithm $\mathsf{Red}_R$ outputs $\sigma'_R$, where we always assume $M' \preceq M$.
- On input $(vk_R, \sigma'_R, M')$, the PPT algorithm $\mathsf{Ver}_R$ outputs 1 or 0 meaning acceptance and rejection, respectively.

**Definition 2.** *A redactable signature scheme $\mathsf{RS}$ with respect to binary relation $\preceq$ is a tuple of PPT algorithms $(\mathsf{Gen}_R, \mathsf{Sig}_R, \mathsf{Red}_R, \mathsf{Ver}_R)$ such that*

- *for any message $M$,*

$$\sigma_R \xleftarrow{\$} \mathsf{Sig}_R(sk_R, M); \mathsf{Ver}_R(vk_R, \sigma_R, M) = 1$$

*for every $(vk_R, sk_R)$ produced by $\mathsf{Gen}_R$; and*
- *for any messages $M$ and $M'$ such that $M' \preceq M$,*

$$\sigma_R \xleftarrow{\$} \mathsf{Sig}_R(sk_R, M); \sigma'_R \leftarrow \mathsf{Red}_R(vk_R, \sigma_R, M, M'); \mathsf{Ver}_R(vk_R, \sigma'_R, M') = 1,$$

*for every $(vk_R, sk_R)$ produced by $\mathsf{Gen}_R$.*

*The above mentioned conditions are required for the correctness of the scheme. For the security, this signature scheme must satisfy the following conditions.*

- **Unforgeability:** *Let $Q$ be a set of messages and let $\mathrm{span}(Q) = \{M' \mid \exists M \in Q$ such that $M' \preceq M\}$. Then, for every PPT adversary $\mathsf{A}$*

$$\Pr\left[ \begin{array}{c} (vk_R, sk_R) \xleftarrow{\$} \mathsf{Gen}_R \\ (M', \sigma'_R) \xleftarrow{\$} \mathsf{A}^{\mathsf{Sig}_R(sk_R, \cdot)}(vk) \end{array} \middle| \begin{array}{c} \mathsf{Ver}_R(vk_R, \sigma'_R, M') = 1 \\ M' \notin \mathrm{span}(Q) \end{array} \right]$$

*is negligible, where $Q$ is a set of queries sent by $\mathsf{A}$ to the oracle $\mathsf{Sig}_R(sk_R, \cdot)$.*
- **Hiding:** *For every two-staged PPT adversary $\mathsf{A} = (\mathsf{A}_1, \mathsf{A}_2)$,*

$$\left| \Pr\left[ \begin{array}{c} (vk_R, sk_R) \xleftarrow{\$} \mathsf{Gen}_R, b \xleftarrow{\$} \{0,1\} \\ (M_0, M_1, M', \mathsf{st}) \xleftarrow{\$} \mathsf{A}_1(vk_R), \sigma_R \xleftarrow{\$} \mathsf{Sig}_R(sk_R, M_b) \\ \sigma'_R \xleftarrow{\$} \mathsf{Red}_R(vk_R, \sigma_R, M_b, M'), b' \xleftarrow{\$} \mathsf{A}_2(\mathsf{st}, \sigma'_R) \end{array} \middle| b = b' \right] - \frac{1}{2} \right|$$

*is negligible, where $M_0, M_1, M'$ satisfy $M' \preceq M_0$ and $M' \preceq M_1$.*

# 3   The Proposed Scheme

## 3.1   Prime Sequence Generator and Hard-Core Predicate

In the proposed scheme, we utilize a *prime sequence generator*

$$H : \{1, \ldots, L\} \to \text{Primes}_{|N|+1}$$

which on input $i$ outputs a prime $p_i$. This was first proposed in [16] as a building block for the private information retrieval and later utilized in [6] for the
verifiable random function. The notable property of this generator is that if $H$
is randomly chosen from an appropriate function family, denoted by $\mathcal{H}$, then
$H(1), H(2), \ldots, H(L)$ are distinct primes of length $|N| + 1$ with probability at
least $1 - 2^{-\Omega(\tau)}$. It is important to note that the random oracles are not used in
this generator. Instead, the $2k^2$-wise independent function is employed, where
$k = 2\tau$. (Refer to [6] and [16] for detailed information and also see Appendix for
its construction.)

We denote $p_i := H(i)$ for every $1 \le i \le L$ and $\widehat{p} = p_1 p_2 \cdots p_L$. We also define
$(\widehat{p}/p_i) = p_1 \cdots p_{i-1} p_{i+1} \cdots p_L \bmod \phi(N)$, where $N$ is the RSA modulus that will
be used in the proposed scheme.

We also use the Goldreich-Levin hard-core predicate [17] to redact (or hide)
the partial message. Let $\mathsf{hc}_\alpha : \mathbb{Z}_N^* \to \{0,1\}$ be a hash function defined by
$\mathsf{hc}_\alpha(y) := \alpha \cdot y$, where $\cdot$ is an inner product modulo 2 by viewing $y \in \mathbb{Z}_N$ as
a bit string of length $2\tau = |N|$.

## 3.2   Construction

In our scheme, to sign a message $M = \langle M_1, \ldots, M_L \rangle$, it is firstly exclusive-ORed
with a binary string $\langle \mathsf{hc}_\alpha(g^{(\widehat{p}/p_1)}), \ldots, \mathsf{hc}_\alpha(g^{(\widehat{p}/p_L)}) \rangle$ of length $L$, where $g \in \mathbb{Z}_N^*$
is randomly chosen. Then the encrypted message

$$C = \langle M_1, \ldots, M_L \rangle \oplus \langle \mathsf{hc}_\alpha(g^{(\widehat{p}/p_1)}), \ldots, \mathsf{hc}_\alpha(g^{(\widehat{p}/p_L)}) \rangle$$

and $g^{\widehat{p}} \bmod N$ are concatenated and signed by the standard signature scheme
SS. Here, the message is encrypted because $\langle \mathsf{hc}_\alpha(g^{(\widehat{p}/p_1)}), \ldots, \mathsf{hc}_\alpha(g^{(\widehat{p}/p_L)}) \rangle$ is
(pseudo)random by the RSA assumption [6].

The detail of a proposed scheme is as follows:

**Key Generation:** First, we describe the algorithm $\mathsf{Gen}_R$. Given $\tau$, it works as
follows:

- $N \xleftarrow{\$} \mathrm{MOD}_\tau$, where $N = PQ$ with $P, Q$ primes
- $(sk_S, vk_S) \xleftarrow{\$} \mathsf{Gen}_S(1^\tau)$ (Recall that this is a pair of keys for the standard
  digital signature scheme)
- $\alpha \xleftarrow{\$} \{0,1\}^{2\tau}$ (Note that $|N| = 2\tau$.)
- $H \xleftarrow{\$} \mathcal{H}$
- $sk_R \leftarrow (sk_S, P, Q, \alpha, H)$  and  $vk_R \leftarrow (vk_S, N, \alpha, H)$.
- Output $(vk_R, sk_R)$

**Sign:** To sign a message $M = \langle M_1, \ldots, M_L \rangle$ (with $M_i \in \{0,1\}$), the signing algorithm $\mathsf{Sig}_R(sk_R, M)$ works as follows:

- Parse $sk_R$ as $(sk_S, P, Q, \alpha, H)$
- $g \xleftarrow{\$} \mathbb{Z}_N^*$
- For every $1 \leq i \leq L$, compute

$$k_i \leftarrow g^{(\widehat{p}/p_i)} \bmod N, k_i' \leftarrow \mathsf{hc}_\alpha(k_i) \text{ and } C_i \leftarrow k_i' \oplus M_i.$$

- $\sigma_S \xleftarrow{\$} \mathsf{Sig}_S(sk_S, \langle C_1, \ldots, C_L, g^{\widehat{p}} \bmod N \rangle)$
- $\sigma_R \leftarrow (\sigma_S, g^{\widehat{p}} \bmod N, g \bmod N, \langle M_1, \ldots, M_L \rangle)$
- Output $\sigma_R$

**Redact:** Given $(vk_R, \sigma_R, M, M')$ such that $M' \preceq M$, the redaction algorithm $\mathsf{Red}_R(vk_R, \sigma_R, M, M')$ works as follows:

- Parse $\sigma_R$ as $(\sigma_S, g^{\widehat{p}} \bmod N, g \bmod N, \langle M_1, \ldots, M_L \rangle)$
- Parse $M'$ as $(M_1', \ldots, M_L')$
- $S \leftarrow \{1 \leq i \leq L \mid M_i' = \#\}$
- $\widehat{p}_S \leftarrow \prod_{i \in S} p_i$
- For each $i \in S$,

$$k_i \leftarrow g^{(\widehat{p}/p_i)} \bmod N, k_i' \leftarrow \mathsf{hc}_\alpha(k_i) \text{ and } C_i \leftarrow k_i' \oplus M_i.$$

- For each $1 \leq i \leq L$,

$$C_i' \leftarrow \begin{cases} C_i & \text{if } i \in S \\ M_i & \text{otherwise} \end{cases}$$

- $\sigma_R' \leftarrow (\sigma_S, g^{\widehat{p}} \bmod N, g^{\widehat{p}_S} \bmod N, \langle C_1', \ldots, C_L' \rangle)$
- Output $\sigma_R'$

**Verify:** Given $vk_R, \sigma_R', M'$, the verification algorithm $\mathsf{Ver}_R(vk_R, \sigma_R', M')$ works as follows:

- Parse $vk_R$ as $(vk_S, N, \alpha, H)$.
- Parse $\sigma_R'$ as $(\sigma_S, y_1 \bmod N, y_2 \bmod N, \langle C_1', \ldots, C_L' \rangle)$
- $S \leftarrow \{1 \leq i \leq L \mid M_i' = \#\}$
- Verify $y_1 \stackrel{?}{=} (y_2)^{\prod_{i \in \overline{S}} p_i} \bmod N$. If not then output 0 and terminate. Here $\overline{S} = \{1, \ldots, L\} \setminus S$.
- For each $i \in \overline{S}$,

$$k_i \leftarrow y_2^{\prod_{j \in \{1, \ldots, L\} \setminus (S \cup \{i\})} p_j} \bmod N, k_i' \leftarrow \mathsf{hc}_\alpha(k_i) \text{ and } C_i \leftarrow k_i' \oplus M_i'.$$

- For each $i \in S$, $C_i \leftarrow C_i'$ (this partial message is already encrypted.)
- Output $\mathsf{Ver}_S(vk_S, \langle C_1, C_2, \ldots, C_L, y_1 \rangle, \sigma_S)$.

It is clear that the size of the message-signature pair is $O(\tau + |M|)$. More precisely, $2\tau + c\tau + 2|M|$, where $c\tau$ is assumed to be the size of the signature of $\mathsf{SS}$ for some constant $c$.

## 4  Security Evaluation

We now show that the proposed scheme satisfies the unforgeable and the hiding properties.

**Theorem 1 (Unforgeability).** *The proposed scheme satisfies the unforgeable property if the underlying standard signature scheme* $\mathsf{SS} = (\mathsf{Gen}_S, \mathsf{Sig}_S, \mathsf{Ver}_S)$ *satisfies the EU-CMA property.*

*Proof.* We prove that if there exists a forger $\mathsf{A}$ against $\mathsf{RS}$ with non-negligible success probability, then we can use it to construct a forger $\mathsf{B}$ against $\mathsf{SS}$. Recall that $\mathsf{B}$ can access the oracle $\mathsf{Sig}_S(sk_S, \cdot)$. The algorithm $\mathsf{B}(vk_S)$ proceeds as follows:

- Generate primes $P, Q$ of length $\tau$ and set $N \leftarrow PQ$
- Generate $\alpha \xleftarrow{\$} \{0,1\}^{|N|}$ and $H \xleftarrow{\$} \mathcal{H}$
- Run $\mathsf{A}$ on input $vk_R \leftarrow (vk_S, N, \alpha, H)$
- When $\mathsf{A}$ queries $M$, generate $g \xleftarrow{\$} \mathbb{Z}_N$, compute $g^{\widehat{p}} \bmod N$, and, for every $i \in \{1, \ldots, L\}$, compute

$$k_i \leftarrow g^{(\widehat{p}/p_i)} \bmod N, k'_i \leftarrow \mathsf{hc}_\alpha(k_i) \text{ and } C_i \leftarrow k'_i \oplus M_i.$$

Obtain the signature $\sigma_S$ by accessing the oracle with the message

$$\langle C_1, \ldots, C_L, g^{\widehat{p}} \bmod N \rangle.$$

Return

$$\sigma_R \leftarrow (\sigma_S, g^{\widehat{p}} \bmod N, g \bmod N, \langle M_1, \ldots, M_L \rangle)$$

to $\mathsf{A}$
- When $\mathsf{A}$ outputs $(\sigma'_R, M')$, parse $\sigma'_R$ to obtain

$$(\sigma_S, y_1 \bmod N, y_2 \bmod N, C')$$

- If $0 = \mathsf{Ver}_R(vk, \sigma'_R, M')$ then terminate.
- $S \leftarrow \{1 \le i \le L \mid M'_i = \#\}$
- For every $i$, compute $C_i$ as follows:
  - $C_i \leftarrow C'_i$ if $i \in S$, and
  - $C_i \leftarrow M'_i \oplus \mathsf{hc}_\alpha(g_2^{\prod_{j \in [L] \setminus (S \cup \{i\})} p_j} \bmod N)$ otherwise.
- Output $\sigma_S$ and a message $\langle C_1, \ldots, C_L, y_1 \rangle$.

If $\mathsf{A}$ succeeds in forging, then $\mathsf{Ver}_S(vk_S, \langle C_1, \ldots, C_L, y_1 \rangle, \sigma_S) = 1$. We can then consider two cases:

- **(Case 1):** The message $\langle C_1, \ldots, C_L, y_1 \rangle$ has not been queried by $\mathsf{B}$ to the oracle; and
- **(Case 2):** The message $\langle C_1, \ldots, C_L, y_1 \rangle$ has been queried by $\mathsf{B}$ to the oracle.

**Case 1** implies that B successfully forges a new message-signature pair. Therefore, we concentrate on **Case 2**. In this case, the success of A in forging implies that there exist two distinct messages that generate the same $\langle C_1, \ldots, C_L, y_1 \rangle$ for some common $g$. However, this does not happen whenever $p_i \neq p_j$ for every $i \neq j$. Therefore, the probability of A succeeds must be bounded by $2^{-\Omega(\tau)}$ which comes from the property of the prime sequence generator.

Taking these in mind,

$$\Pr[\text{A forges}] = \Pr[\textbf{Case 1} \vee \textbf{Case 2}] \leq \Pr[\textbf{Case 1}] \vee \Pr[\textbf{Case 2}]$$

$$\leq \Pr[\text{B forges}] + \frac{1}{2^{\Omega(\tau)}},$$

which results in $\Pr[\text{A forges}]$ being negligible. It is clear that if A is PPT, then so is B. This concludes the proof. □

To prove the hiding property, we use the following lemma which has been implicitly used in [6].

**Lemma 1.** *For every* PPT $A = (A_1, A_2)$,

$$\left| \Pr\left[ \begin{array}{c} H \xleftarrow{\$} \mathcal{H}, N \xleftarrow{\$} \text{MOD}_\tau, \\ g \xleftarrow{\$} \mathbb{Z}_N^*, \alpha \xleftarrow{\$} \{0,1\}^{2\tau} \\ (s_T, \mathtt{st}) \xleftarrow{\$} A_1(H, N, \alpha, L, g^{\widehat{p}}) \end{array} \middle| \begin{array}{c} A_2(\mathtt{st}, g^{p_{s_T}}) = \mathsf{hc}_\alpha(g^{\widehat{p}/p_{s_T}})) \\ \wedge 1 \leq s_T \leq L \end{array} \right] - \frac{1}{2} \right| \quad (2)$$

*is negligible if the probability (1) is negligible for every* PPT B *and* L *is bounded by some polynomial* poly *in the security parameter* $\tau$, *i.e.,* $L \leq \texttt{poly}(\tau)$.

*Proof (Sketch).* For every $1 \leq s_T \leq L$, there is no algorithm which outputs $g^{\widehat{p}/p_{s_T}}$ given $g^{p_{s_T}}$, $H, N, \alpha, g^{\widehat{p}}$ with non-negligible probability. This is because if there exists such an algorithm, then we can obtain $g$ from $g^{p_{s_T}}$ and $g^{\widehat{p}/p_{s_T}}$ by the extended Euclid algorithm, and this contradicts to the RSA assumption. From this fact, we can observe that $g^{\widehat{p}/p_{s_T}}$ becomes pseudorandom by applying the Goldreich-Levin hard-core predicate $\mathsf{hc}_\alpha$[17]. □

We also we use the following useful lemma.

**Lemma 2 (Lemma B.13 in [18]).** *Let* $R, S, B$ *be jointly distributed random variables with values in* $\{0, 1\}$. *Assume that* $B$ *and* $S$ *are independent and that* $B$ *is uniformly distributed. Then*

$$\Pr[R = S] = \frac{1}{2} + \Pr[R = B \mid S = B] - \Pr[R = B].$$

By using these lemmas, we prove the hiding property of our proposed scheme:

**Theorem 2 (Hiding).** *If the RSA problem is hard for every* PPT *algorithm (in the sense of (1)) then the proposed scheme satisfies the hiding property.*

*Proof.* We first consider the following. For any $S \subseteq \{1, \ldots, L\}$, given

$$\{g^{\prod_{i \in S} p_i} \bmod N \mid i \in S\}, N, H, \alpha$$

the sequence

$$\{\mathsf{hc}_\alpha(g^{\widehat{p}/p_i}) \mid i \in S\}$$

is pseudorandom when $g, H$, and $\alpha$ are randomly chosen from an appropriate domain. More precisely, we can describe this as follows:

*Claim.* Assume that the RSA problem is hard in the sense of Lemma 1. Then, for every PPT algorithm $\mathsf{A} = (\mathsf{A}_1, \mathsf{A}_2)$,

$$\left| \Pr \left[ \begin{array}{c} N \xleftarrow{\$} \mathrm{MOD}_\tau, g \xleftarrow{\$} \mathbb{Z}_N^*, \alpha \xleftarrow{\$} \{0,1\}^{2\tau}, H \xleftarrow{\$} \mathcal{H}, \\ (S = \{s_1, \ldots, s_\omega\}, \mathsf{st}) \xleftarrow{\$} \mathsf{A}_1(N, H, L, \alpha, g^{\widehat{p}}), \\ R_0 \leftarrow (\mathsf{hc}_\alpha(g^{\widehat{p}/p_{s_1}}), \ldots, \mathsf{hc}_\alpha(g^{\widehat{p}/p_{s_\omega}})), \\ R_1 \xleftarrow{\$} \{0,1\}^\omega, b \xleftarrow{\$} \{0,1\}, \\ b' \xleftarrow{\$} \mathsf{A}_2(\mathsf{st}, g^{\prod_{i \in S} p_i}, R_b) \end{array} \middle| \begin{array}{c} b = b' \\ \wedge S \subseteq [L] \end{array} \right] - \frac{1}{2} \right| \quad (3)$$

is negligible.

*Proof.* We first assume that there is an adversary $\mathsf{A}$ that makes (3) non-negligible. We then construct the adversary $\mathsf{B} = (\mathsf{B}_1, \mathsf{B}_2)$ that makes (2) non-negligible. We prove this by using a hybrid-argument. The construction of $\mathsf{B}$ is as follows.

- Given $(N, H, L, \alpha, g^{\widehat{p}})$, feed $(N, H, L, \alpha, g^{\widehat{p}})$ into $\mathsf{A}$ to obtain $S = \{s_1, \ldots, s_\omega\}$ and $\mathsf{st}$.
- Randomly choose $s_T \xleftarrow{\$} \{s_1, \ldots, s_\omega\}$ and output $(s_T, \mathsf{st})$ to obtain $g^{p_{s_T}} \bmod N$
- For every $i \in \{s_{T+1}, \ldots, s_\omega\}$, compute $k_i \leftarrow \mathsf{hc}_\alpha(g^{(\widehat{p}/p_i)})$
- For every $i \in \{s_1, \ldots, s_{T-1}\}$, compute $k_i \xleftarrow{\$} \{0,1\}^\kappa$
- Set $k_T \xleftarrow{\$} \{0,1\}$
- Feed $k_1, \ldots, k_\omega$ and $g^{p_{s_1} \cdots p_{s_\omega}} \bmod N$ to $\mathsf{A}$. Then output $k_T$ if the output of $\mathsf{A}$ equals 1, and output $1 - k_T$ otherwise

We analyze the probability of $\mathsf{A}$ outputting 1. For every $0 \le i \le \omega$, let $\mathcal{E}_i$ be an event such that $\mathcal{E}_i$ is true if and only if $\mathsf{A}$ outputs 1 when the input to $\mathsf{A}$ is

$$(k_1, \ldots, k_{i-1}, k_i, \ldots, k_\omega) = (k_1, \ldots, k_{i-1}, \mathsf{hc}_\alpha(g^{(\widehat{p}/p_i)}), \ldots, \mathsf{hc}_\alpha(g^{(\widehat{p}/p_\omega)})). \quad (4)$$

That is, $k_1, \ldots, k_{i-1} \in \{0,1\}$ are randomly chosen values and the rest are real values generated by $g$. We can then rewrite the assumption as

$$\epsilon \le |\Pr[\mathcal{E}_0] - \Pr[\mathcal{E}_\omega]| \quad (5)$$

for some non-negligible function $\epsilon$. By modifying (5), we obtain

$$\begin{aligned} \epsilon &\le |\Pr[\mathcal{E}_0] - \Pr[\mathcal{E}_\omega]| \\ &= |\Pr[\mathcal{E}_0] - \Pr[\mathcal{E}_1] + \cdots + \Pr[\mathcal{E}_i] - \Pr[\mathcal{E}_{i+1}] + \cdots + \Pr[\mathcal{E}_{\omega-1}] - \Pr[\mathcal{E}_\omega]| \\ &= \left| \sum_{i=0}^{\omega-1} \Pr[\mathcal{E}_i] - \Pr[\mathcal{E}_{i+1}] \right|. \end{aligned}$$

By using Lemma 2,

$$\Pr\left[\mathsf{B} \text{ outputs } \mathsf{hc}_\alpha(g^{\widehat{p}/p_{s_T}})\right]$$

$$= \frac{1}{2} + \Pr\left[\mathsf{B} \text{ outputs } k_T \mid \mathsf{hc}_\alpha(g^{\widehat{p}/p_{s_T}}) = k_T\right] - \Pr\left[\mathsf{B} \text{ outputs } k_T\right]$$

$$= \frac{1}{2} + \sum_{i=0}^{\omega-1} \Pr\left[T = i\right] \left(\Pr\left[\mathsf{B} \text{ outputs } k_T \mid \mathsf{hc}_\alpha(g^{\widehat{p}/p_{s_T}}) = k_T, T = i\right]\right.$$

$$\left. - \Pr\left[\mathsf{B} \text{ outputs } k_T \mid T = i\right]\right)$$

$$= \frac{1}{2} + \frac{1}{\omega} \sum_{i=0}^{\omega-1} (\Pr[\mathcal{E}_i] - \Pr[\mathcal{E}_{i+1}]).$$

Putting these together, we obtain

$$\left| \Pr\left[\mathsf{B} \text{ outputs } \mathsf{hc}_\alpha(g^{\widehat{p}/p_{s_T}})\right] - \frac{1}{2} \right| \geq \frac{\epsilon}{\omega}.$$

This concludes the proof of the claim. ☐

We now return back to the proof of the theorem. To prove this theorem, we assume the existence of $\mathsf{A} = (\mathsf{A}_1, \mathsf{A}_2)$, which breaks the hiding property of the proposed scheme. Then, we show the adversary $\mathsf{B} = (\mathsf{B}_1, \mathsf{B}_2)$ which makes (3) non-negligible by using $\mathsf{A}$. The construction of $\mathsf{B}$ is as follows.

- Given $(N, H, L, \alpha, g^{\widehat{p}})$, run $\mathsf{Gen}_S$ to obtain $(vk_S, sk_S)$ and set $vk_R \leftarrow (vk_S, N, \alpha, H)$
- Feed $vk_R$ into $\mathsf{A}_1$ and obtain $M$, $M^{(0)}, M^{(1)}$ and $\mathsf{st}_A$ such that $M \preceq M^{(0)}, M^{(1)}$. Then obtain redacted positions $S = \{s_1, \ldots, s_\omega\}$
- Output $S = \{s_1, \ldots, s_\omega\}$ to obtain $g^{p_{s_1} \cdots p_{s_\omega}} \bmod N$ and $r_1, \ldots, r_\omega$ (with state information)
- For every $i \in \overline{S} = [L] \setminus S$, compute

$$k_i \leftarrow g^{(\widehat{p}/p_i)} \bmod N, k_i' \leftarrow \mathsf{hc}_\alpha(k_i) \text{ and } C_i \leftarrow k_i' \oplus M_i.$$

- Choose $d \in \{0, 1\}$ at random and set $C_{s_i} \leftarrow r_i \oplus M_{s_i}^{(d)}$ for every $1 \leq i \leq \omega$
- Compute

$$g^{\widehat{p}} \leftarrow (g^{p_{s_1} \cdots p_{s_\omega}})^{\Pi_{i \in \overline{S}} p_i} \bmod N,$$

and set $g^{\widehat{p}_S} = g^{p_{s_1} \cdots p_{s_\omega}} \bmod N$
- Obtain $\sigma_S \overset{\$}{\leftarrow} \mathsf{Sig}_S(sk, \langle C_1, \ldots, C_L, g^{\widehat{p}} \bmod N \rangle)$
- Set $C_i' \leftarrow M_i$ for every $i \in \overline{S}$ and $C_{s_i}' \leftarrow C_{s_i}$ for every $1 \leq i \leq \omega$
- Feed $(\mathsf{st}_A, \sigma_S, g^{\widehat{p}} \bmod N, g^{\widehat{p}_S} \bmod N, \langle C_1', \ldots, C_L' \rangle)$ into $\mathsf{A}_2$, and obtain $d'$
- If $d' = d$ then output $b' \leftarrow 1$, and output $b' \leftarrow 0$ otherwise

If $b = 0$ then $C_{s_i} \leftarrow r_i \oplus M_{s_i}^{(d)}$ is completely random for every $i \in S$. Therefore, $\Pr[d = d' \mid b = 0] = 1/2$ in this case. On the other hand, if $b = 1$, then B perfectly computes the (redacted) signature of the proposed scheme. From the assumption that $|\Pr[d = d' \mid b = 1] - 1/2|$ is non-negligible,

$$|\Pr[d = d' \mid b = 1] - \Pr[d = d' \mid b = 0]| = |\Pr[d = d' \mid b = 1] - 1/2|$$

becomes non-negligible. This contradicts to the result of the claim. The remaining part of the proof deals with showing that the running time of B is PPT. However, it is easy to see that if A is PPT, B is also PPT. This concludes the proof.    □

# 5    Concluding Remarks

In this paper, we showed the redactable signature scheme whose message-signature size is $O(\tau + |M|)$. The construction of our scheme was inspired from the verifiable random function of [6] and the redactable signature of [4]. Compared to the one proposed in [1], the signature size in our scheme is bigger. However, we showed that the random oracles are not needed in our proposal.

The main problem of our proposal is the inefficiency of the primes sequence generator [6]. Hence, in our future research, we are intend to replace the prime sequence generator with a division intractable hash function [19]. However, since there are problems in instantiating such functions [20], we are interested in constructing a *weaker* version of the division intractable hash function and proving the security by assuming the hardness of the strong RSA problem in the standard model.

We are also interested in the generic construction of the redactable signature schemes from the verifiable random functions.

# References

1. Johnson, R., Molnar, D., Song, D.X., Wagner, D.: Homomorphic signature schemes. In: Preneel, B. (ed.) CT-RSA 2002. LNCS, vol. 2271, pp. 244–262. Springer, Heidelberg (2002)
2. Steinfeld, R., Bull, L., Zheng, Y.: Content extraction signatures. In: Kim, K.-c. (ed.) ICISC 2001. LNCS, vol. 2288, pp. 285–304. Springer, Heidelberg (2002)
3. Miyazaki, K., Iwamura, M., Matsumoto, T., Sasaki, R., Yoshiura, H., Tezuka, S., Imai, H.: Digitally signed document sanitizing scheme with disclosure condition control. IEICE Transactions 88-A(1), 239–246 (2005)
4. Miyazaki, K., Hanaoka, G., Imai, H.: Bit-by-bit sequence sanitizable digitally signed document sanitizing scheme. In: SCIS (2006)
5. Yang, P., Hanaoka, G., Matsuura, K., Imai, H.: Security notions and proof of a bitwise sanitizable signature scheme from any one-way permutation. In: SCIS (2008)
6. Micali, S., Rabin, M.O., Vadhan, S.P.: Verifiable random functions. In: FOCS, pp. 120–130 (1999)

7. Izu, T., Kanaya, N., Takenaka, M., Yoshioka, T.: Piats: A partially sanitizable signature scheme. In: Qing, S., Mao, W., López, J., Wang, G. (eds.) ICICS 2005. LNCS, vol. 3783, pp. 72–83. Springer, Heidelberg (2005)
8. Miyazaki, K., Hanaoka, G., Imai, H.: Digitally signed document sanitizing scheme based on bilinear maps. In: ASIACCS, pp. 343–354 (2006)
9. Haber, S., Hatano, Y., Honda, Y., Horne, W., Miyazaki, K., Sander, T., Tezoku, S., Yao, D.: Efficient signature schemes supporting redaction, pseudonymization, and data deidentification. In: ASIACCS, pp. 353–362 (2008)
10. Brzuska, C., Fischlin, M., Freudenreich, T., Lehmann, A., Page, M., Schelbert, J., Schröder, D., Volk, F.: Security of sanitizable signatures revisited. In: Public Key Cryptography, pp. 317–336 (2009)
11. Chang, E.C., Lim, C.L., Xu, J.: Short redactable signatures using random trees. In: CT-RSA, pp. 133–147 (2009)
12. Goldwasser, S., Micali, S., Rivest, R.L.: A digital signature scheme secure against adaptive chosen-message attacks. SIAM J. Comput. 17(2), 281–308 (1988)
13. Cramer, R., Damgård, I.: New generation of secure and practical RSA-based signatures. In: Koblitz, N. (ed.) CRYPTO 1996. LNCS, vol. 1109, pp. 173–185. Springer, Heidelberg (1996)
14. Hohenberger, S., Waters, B.: Realizing hash-and-sign signatures under standard assumptions. In: EUROCRYPT, pp. 333–350 (2009)
15. Hohenberger, S., Waters, B.: Short and stateless signatures from the rsa assumption. To appear in CRYPTO (2009)
16. Cachin, C., Micali, S., Stadler, M.: Computationally private information retrieval with polylogarithmic communication. In: Stern, J. (ed.) EUROCRYPT 1999. LNCS, vol. 1592, pp. 402–414. Springer, Heidelberg (1999)
17. Goldreich, O., Levin, L.A.: A hard-core predicate for all one-way functions. In: STOC, pp. 25–32 (1989)
18. Delfs, H., Knebl, H.: Introduction to Cryptography. Springer, Heidelberg (2002)
19. Gennaro, R., Halevi, S., Rabin, T.: Secure hash-and-sign signatures without the random oracle. In: Stern, J. (ed.) EUROCRYPT 1999. LNCS, vol. 1592, pp. 123–139. Springer, Heidelberg (1999)
20. Coron, J.S., Naccache, D.: Security analysis of the gennaro-halevi-rabin signature scheme. In: Preneel, B. (ed.) EUROCRYPT 2000. LNCS, vol. 1807, pp. 91–101. Springer, Heidelberg (2000)

# A    The Construction of a Prime Sequence Generator [16]

We introduce the prime sequence generator $P(\cdot, \cdot, \cdot)$ used in [6].

- Input: An $\log_2 L$ bit string $x$, a polynomial $Q$ of degree at most $2k^2 - 1$ over $GF(2^k)$, and an $l$-bit string coin.
- Output: a $(k + 1)$-bit integer $p_x$ (a prime with overwhelming probability)
- Code:
  - For $j = 1, \ldots, 2k^2$, let $y_j$ be the $(k + 1)$-bit string $\langle 1, Q(\langle x, \bar{j} \rangle) \rangle$, where $\bar{j}$ denotes the $j$'s string in $\{0, 1\}^{k - \log_2 L}$ under the lexicographic order.
  - Use the primality testing algorithm with random coins coin to test each $y_j$ for primality, and let $p_x$ be the first prime in the sequence $y_1, \ldots, y_{2k^2}$. Output $p_x$. (If there is no prime then the output is $y_{2k^2}$.)

In the proposed scheme, $k = 2\tau$ and the generator is defined by $H(x) = P(x, Q, \text{coin})$. Also $H \xleftarrow{\$} \mathcal{H}$ means choosing $Q$ and coin at random.

# Generic Construction of Stateful Identity Based Encryption

Peng Yang[1], Rui Zhang[2], Kanta Matsuura[1], and Hideki Imai[2]

[1] The University of Tokyo
{pengyang,kanta}@iis.u-tokyo.ac.jp
[2] Research Center for Information Security,
Advanced Industrial Science and Technology, Japan
{r-zhang,h-imai}@aist.go.jp

**Abstract.** The concept of stateful encryption was introduced by Bellare et al. in 2006. Compared with conventional public key encryption scheme, stateful encryption can achieve much better encryption performance. In this paper, we introduce a related primitive called stateful identity based key encapsulation mechanism (SIBKEM). SIBKEM is a simpler primitive, however, together with multi-time use IND-CCA secure symmetric encryption, it implies secure stateful identity based encryption. We then demonstrate there is a generic construction of SIBKEM from a wide class of identity based non-interactive key exchange schemes.

## 1 Introduction

*Public key encryption* (PKE) is an important tool for securing digital communicabilities [1,2]. PKE schemes are often much slower than *symmetric encryption* (SE) schemes. In resource-constrained environments like mobile communication and sensor networks, this disadvantage of PKE will be quite undesirable, since system performance will drop greatly due to the high computational cost from frequent discrete modular exponentiations.

To improve the performance of PKE, Bellare, Kohno and Shoup [3] introduced the concept of *stateful PKE* (SPKE) in ACM-CCS'06. In such a scheme, a sender maintains some state information. Without loss of generality, the state information can be viewed as two parts: the secret part and the public part. Then the encryption algorithm takes as input not only a message and the public key of receiver, but also his current secret state to produce a ciphertext. As a result, the sender's computational cost for encryption is dramatically reduced. Decryption performance remains unchanged from stateless scheme, and the receivers need not even necessarily notice whether the sender is stateful if the public state is included in the ciphertext.

Recently, Baek, Zhou and Bao [4] proposed a "generic" construction, and demonstrated many efficient instantiations. We remark that the "generic" construction of [4] requires additionally that underlying *key encapsulation mechanism* (KEM) [5] meets two non-standard properties: "partitioned" and "reproducibility". Thus their approach is not necessarily a real simplification for scheme designing.

P. Samarati et al. (Eds.): ISC 2009, LNCS 5735, pp. 338–346, 2009.
© Springer-Verlag Berlin Heidelberg 2009

On the other hand, an *identity based encryption* (IBE) scheme is a special public key encryption scheme, where public keys can be arbitrary strings, advocated by Shamir [6] to simplify public key certificate management. The model of *Stateful IBE* (SIBE) was first formalized by Phong, Matsuoka and Ogata [7], as the stateful counterpart of IBE. Yang, Zhang and Matsuura proposed variants of SPKE and SIBE schemes [8], trading assumptions/generality with computation costs. Currently there is not any generic construction of SIBE.

OUR CONTRIBUTIONS. The main contribution of this paper could be considered as the explanation of the essence of such SIBE schemes. This research was motivated to remove the symmetric encryption part from the security model and proof of stateful schemes. We hope that this work could help to understand the original idea of [3]. To achieve this goal, we introduce a simpler primitive called *stateful identity based KEM* (SIBKEM), which can be used to achieve SIBE, together with multi-time use IND-CCA secure symmetric encryption. We formally state a composition theorem for such an approach.

Our secondary contribution is a construction of SIBKEM based from a well-studied cryptographic primitive, so-called *identity based non-interactive key exchange* (IBNIKE). As its name suggests, an IBNIKE scheme is a non-interactive key exchange scheme that two players set up their shared key. Our construction is totally black-box: given a class of IBNIKE scheme, we can construct an SIBKEM scheme without essential modifications of the algorithms nor resorting to random oracles. Recently, Paterson and Srinivasan [9,10] proposed a transforma from IBNIKE to a CPA secure IBE. Their work could be adapted to produce a stateless IBKEM, then the standard composition result by Bentahar et al. [11] would then allow the construction of IND-CCA secure IBE.

## 2 Preliminaries

In this section, we review the security models of identity based non-interactive key exchange, stateful identity based encryption, and symmetric encryption.

### 2.1 Conventions

NOTATIONS. Let $y \leftarrow A(x_1, ..., x_n)$ denote the experiment of assigning the result of $A$ to $y$. If $S$ is a finite set then let $x \leftarrow S$ denote the operation of picking an element at random and uniformly from $S$. If $\alpha$ is neither an algorithm nor a set then let $x \leftarrow \alpha$ denote a simple assignment statement.

NEGLIGIBLE FUNCTION. We say a function $\epsilon : \mathbb{N} \to \mathbb{R}$ is negligible if for every constant $c \geq 0$ there exits an integer $k_c$ such that $\epsilon(k) < k^{-c}$ for all $k > k_c$.

### 2.2 Identity Based Non-interactive Key Exchange

IBNIKE is not a new concept, since it is only a natural extension of its PKI counterpart. The first IBNIKE was proposed by Sakai, Ohgishi and Kasahara [12]. We first review the model of IBNIKE, and then define the security notion.

ALGORITHMS. An identity based non-interactive key exchange scheme is specified by three algorithms. $\mathcal{IBNIKE} = \{\text{Setup, Ext, Shr}\}$, where

Setup: The randomized setup algorithm takes as input security parameter $1^\lambda$ where $\lambda \in \mathbb{N}$. It outputs the system parameters $sp$ and the master key $mk$. It also specifies the shared key space $\mathcal{SHK}$ by $sp$. ($\mathcal{SHK}$ may be included in $sp$.) We write $(sp, mk) \leftarrow \text{Setup}(1^\lambda)$.

Ext: The (possibly randomized) extract algorithm takes as input $sp$, $mk$ and an identity $id \in \{0, 1\}^n$. It outputs a secret key $sk_{id}$ corresponding to $id$. We write $sk_{id} \leftarrow \text{Ext}(sp, mk, id)$.

Shr: The deterministic sharing algorithm takes as inputs $sp$, a private key $sk_{id_A}$ and a user's identities $id_B$, where $id_A \neq id_B$. It outputs the shared key $K_{A,B} \in \mathcal{SHK}$ between $A$ and $B$. This algorithm has symmetry. We write $K_{A,B} \leftarrow \text{Shr}(sp, sk_{id_A}, id_B) = \text{Shr}(sp, sk_{id_B}, id_A)$.

SECURITY NOTION. We establish the IND (i.e., indistinguishability against adaptive chosen identity attack and adaptively reveal attack) game for IBNIKE between an adversary $\mathcal{A}$ and a challenger $\mathcal{C}$.

**Setup:** $\mathcal{C}$ takes the security parameter $\lambda$ and runs Setup alogrithm. It passes the resulting system parameter $sp$ to $\mathcal{A}$ and keeps the master key $mk$.

**Phase 1:** $\mathcal{A}$ issues two types of oracle queries $q_1, \cdots, q_i$ where a query is one of

⋄ extraction queries on an identity $id$. $\mathcal{C}$ responds with a corresponding secrete key $sk_{id}$.
⋄ reveal queries on a pair of identities $(id_1, id_2)$, $\mathcal{C}$ responds with the key $K_{1,2}$ shared between these two identities.

These queries may be asked adaptively, that is, each query $q_i$ may depends on the replies to $q_1, \cdots, q_{i-1}$.

**Challenge:** Once $\mathcal{A}$ decides phase 1 is over, he outputs two target identities $id_A, id_B$, with restriction that pair $(id_A, id_B)$ has not appeared in previous reveal queries, and neither $id_A$ nor $id_B$ has appeared in previous extraction queries. Then $\mathcal{C}$ flips a coin $b \in \{0, 1\}$. If $b = 0$, $\mathcal{C}$ returns $\mathcal{A}$ a random value from key space $\mathcal{SHK}$; otherwise $\mathcal{C}$ returns the real key $K_{A,B}$.

**Phase 2:** $\mathcal{A}$ issues more queries $q_{i+1}, \cdots, q_j$ where a query is one of

⋄ Extraction queries on an identity $id \notin \{id_A, id_B\}$. $\mathcal{C}$ responds as in phase 1.
⋄ Reveal queries on a pair of identities $(id_1, id_2) \neq (id_A, id_B)$, $\mathcal{C}$ responds as in phase 1.

**Guess:** Finally, $\mathcal{A}$ outputs a bit $b' \in \{0, 1\}$.

We refer such an adversary $\mathcal{A}$ as an IND adversary. $\mathcal{A}$'s advantage in this IND security game is defined to be $\mathbf{Adv}(\mathcal{A}) = |\Pr[b' = b] - 1/2|$. We say that an IBNIKE scheme is secure in the sense of IND if the advantage is negligible for any *probabilistic polynomial-time* (PPT) algorithm $\mathcal{A}$.

## 2.3   Stateful Identity Based Encryption

Due to the space limitation, algorithms and security notion of SIBE are omitted. For further particulars, please refer to [7,8].

## 2.4   Symmetric Encryption

Here, we briefly review the model and the security requirements of symmetric encryption (SE).

An SE scheme consists of three algorithms, $\mathcal{SE} = (\mathsf{K}, \mathsf{E}, \mathsf{D})$. The randomized key generation algorithm $\mathsf{K}$ takes as input the security parameter $\lambda$ and outputs a session key $dk$. We write $dk \leftarrow \mathsf{K}(\lambda)$. The (possibly randomized) encryption algorithm $\mathsf{E}$ takes as input a session key $dk$ and a plaintext $m$ and computes a ciphertext $C$. We write $C \leftarrow \mathsf{E}(dk, m)$. The decryption algorithm $\mathsf{D}$ takes as input a session key $dk$ and a ciphertext $C$ and outputs a plaintext $m$ (or "$\perp$" for invalid). We write $m/\perp \leftarrow \mathsf{D}(dk, C)$. The standard consistency constraint is that $\forall dk : m \leftarrow \mathsf{D}(dk, \mathsf{E}(dk, m))$.

In this paper, we only consider symmetric encryption which guarantees indistinguishability against chosen ciphertext attack. Due to the space limitation, the description of *IND-CCA* security notion of SE is omitted. Interested reader are referred to [13]. In this paper, we require SE to be multiple time secure, and such SE schemes can be generically built from standard block ciphers and message authentication codes (MAC) [13].

# 3   Stateful Identity Based Key Encapsulation Mechanism

In this section, we introduce the model and the security notions of SIBKEM. Roughly speaking, SIBKEM is the "stateful version" of *conventional identity based key encapsulation mechanism* (IBKEM). In particular, in SIBKEM, the sender maintains state information. For a specified identity, the session key encapsulated by the sender remains the same unless the state is updated. Since it is deterministic, SIBKEM is weaker than IBKEM, i.e., the adversary can issue neither encapsulation query nor decapsulation query on the target identity.

## 3.1   Algorithms

An SIBKEM scheme is specified by five algorithms. $\mathcal{SIBKEM} = \{\mathsf{Setup}, \mathsf{Ext}, \mathsf{NwSt}, \mathsf{Enc}, \mathsf{Dec}\}$.

Setup: The randomized setup algorithm takes as input security parameter $1^\lambda$ where $\lambda \in \mathbb{N}$. It outputs the system parameters $sp$ and the master key $mk$. It also specifies the key space $\mathcal{SHK}$ by $sp$. ($\mathcal{SHK}$ may be included in $sp$.) We write $(sp, mk) \leftarrow \mathsf{Setup}(1^\lambda)$.

Ext: The (possibly randomized) key extraction algorithms takes as input $sp$, $mk$ and a user's identity $id$. It outputs a secret key $sk_{id}$ corresponding to $id$. We write $sk_{id} \leftarrow \mathsf{Ext}(sp, mk, id)$.

NwSt: The randomized new state algorithm takes as input $sp$. It outputs a new state $st$ of a sender. We write $st \leftarrow \mathsf{NwSt}(sp)$.

Enc: The deterministic encapsulation algorithm takes as input $sp$, $id$ and $st$, where $id$ is the receiver's identity. It outputs the corresponding ciphertext $c$ of a session key $dk$. We write $(c, dk) \leftarrow \mathsf{Enc}(sp, st, id)$.

Dec: The deterministic decapsulation algorithm takes as $sp$, $sk_{id}$ and a ciphertext $c$. It outputs the session key $dk$. We write $dk \leftarrow \mathsf{Dec}(sp, sk_{id}, c)$.

## 3.2   IND-ID-CCA Security

We establish the IND-ID-CCA (indistinguishability against adaptive chosen identity attack and adaptive chosen ciphertext attack) game for SIBKEM between an adversary $\mathcal{A}$ and a challenger $\mathcal{C}$. The game is described as follows.

**Setup:** $\mathcal{C}$ takes the security parameter $\lambda$ and runs Setup of SIBE. It passes the the resulting system parameters $sp$ to $\mathcal{A}$ and keeps the masker key $mk$ to himself. The state $st$ is decided a-priori by $\mathcal{C}$ .

**Phase 1:** $\mathcal{A}$ issues three types of queries $q_1, \cdots, q_i$ where a query is one of
  ◇ extraction queries on an identity $id$. $\mathcal{C}$ responds with a corresponding secret private key $sk_{id}$ of $id$.
  ◇ encapsulation queries on an identity $id$. $\mathcal{C}$ responds with ciphertext $c$ and a decryption key $dk$ under $id$ and the current state $st$.
  ◇ decapsulation queries on an identity and a ciphertext $(id, c)$. $\mathcal{C}$ responds with the corresponding decryption key $dk$ of $c$.

These queries may be asked adaptively, that is, each query $q_i$ may depends on the replies to $q_1, \cdots, q_{i-1}$.

**Challenge:** Once $\mathcal{A}$ decides that phase 1 is over, he outputs an $id^*$ on which he wishes to be challenged. The only restriction is that $id^*$ must not appear in any query in phase 1. Then $\mathcal{C}$ computes a valid key-ciphertext pair $(c^*, dk_1^*)$ and flips a coin $b \in \{0, 1\}$. If $b = 0$, then $\mathcal{C}$ chooses a random key $dk_0^*$ from the key space and returns $(c^*, dk_0^*)$ to $\mathcal{A}$; otherwise $\mathcal{C}$ returns $(c^*, dk_1^*)$.

**Phase 2:** $\mathcal{A}$ issues more queries $q_{i+1}, \cdots, q_j$ where a query is one of
  ◇ extraction queries on an identity $id \neq id^*$. $\mathcal{C}$ responds as in phase 1.
  ◇ encapsulation queries on an identity $id \neq id^*$. $\mathcal{C}$ responds as in phase 1.
  ◇ decapsulation queries on an identity and a ciphertext $(id, c) \neq (id^*, c^*)$. $\mathcal{C}$ responds as in phase 1. Note that since the decapsulation algorithm is deterministic on fixed $id$ and $st$, the restriction is actually $id \neq id^*$.

**Guess:** Finally, $\mathcal{A}$ outputs a bit $b' \in \{0, 1\}$.

$\mathcal{A}$'s advantage in this IND-ID-CCA game is defined to be $\mathbf{Adv}(\mathcal{A}) = |\Pr[b = b'] - 1/2|$. We say that an SIBKEM scheme is secure in the sense of IND-ID-CCA if the advantage is negligible for any PPT algorithm $\mathcal{A}$.

## 3.3 The Composition Theorem

By combining an IND-ID-CCA secure $\mathcal{SIBKEM} = \{$SIBKEM.Setup, SIBKEM.Ext, SIBKEM.NwSt, SIBKEM.Enc, SIBKEM.Dec$\}$ and a multi-time use IND-CCA secure $\mathcal{SE} = \{$SE.K, SE.E, SE.D$\}$, we can obtain an IND-ID-CCA secure $\mathcal{SIBE} = \{$Setup, Ext, NwSt, Enc, Dec$\}$. We omit composition details since it is straightforward. At a high level, the SIBE sender uses SE.E to encrypt a message by using the key $dk$ encapsulated by SIBKEM.Enc, and the SIBE receiver runs SE.D to decrypt with $dk$ recovered by SIBKEM.Dec.

**Theorem 1.** *Suppose $\mathcal{SIBKEM}$ is IND-ID-CCA secure, and $\mathcal{SE}$ is IND-CCA secure. Then the hybrid encryption scheme $\mathcal{SIBE}$ is IND-ID-CCA secure.*

*Due to the space limitation, the detailed proof is omitted.*

# 4 A Generic Construction

In this section, we propose a generic construction of stateful identity based key encapsulation mechanism. Our building block is identity based non-interactive key exchange (with mild requirements). As previous work, similar requirements to convert an IBNIKE scheme to an IND-ID-CPA secure IBE scheme were discussed [9,10]. By applying our generic construction to various IBNIKE schemes, we can obtain SIBKEM schemes which provide various functionalities.

## 4.1 Preparation

As described in Section 2, an IBNIKE scheme is specified by three basic algorithms, Setup, Ext, and Shr. To show the generic construction, in addition to these three basic algorithms, we require three additional algorithms which can be derived from the basic algorithms.

Sample: The randomized sample algorithm takes input as $sp$ and output a temporary key pair $(pk, sk) \in \{\mathcal{PK}\} \times \{\mathcal{SK}\}$, where $sk$ is the corresponding secret key to the public key $pk$. And the identifier of $pk$ cannot be revealed. One can imagine that $pk$ is the image of a virtual identifier $id$, and $id$ must not be in collision with other realistic identities in the identity space.

Shr': If a party $B$ has neither an identity nor an secret key, and $B$ wants to exchange a key to a target party $A$ with identity $id_A$, then Shr' takes as input $(sp, sk_B, id_A)$, where $sk_B$ is $B$'s temporary secret key generated in Sample. It outputs a key $K_{A,B}$. Shr' is a deterministic algorithms.

Shr'': If a party $A$ with identity $id_A$ and secret key $sk_{id_A}$ wants to exchange a key with a party $B$ who does not have an identity but a temporary public key $pk_B$, then Shr'' takes as input $(sp, sk_{id_A}, pk_B)$, where $pk_B$ is generated in Sample. It outputs a key $K_{A,B}$. Shr'' is a deterministic algorithms.

We require the consistency of Shr' and Shr'' algorithms, i.e., if $sk_{id_A}$ is secret key of $id_A$, and $sk_B$ is secret key of $pk_B$, then Shr'$(sp, sk_B, id_A) = $ Shr''$(sp, sk_{id_A}, pk_B)$, where $(pk_B, sk_B) \leftarrow$ Sample$(sp)$ and $sk_{id_A} \leftarrow$ Ext$(sp, mk, id_A)$.

At a first glance, these algorithms seem to require special properties to IB-NIKE schemes, but to the best of our knowledge, it is easy to construct such algorithms for almost all currently known IBNIKE schemes.

## 4.2 From IBNIKE to SIBKEM

Let $\mathcal{IBNIKE}$ = {Setup, Ext, Shr, Sample, Shr', Shr"} be an IBNIKE scheme. By employing $\mathcal{IBNIKE}$ as buiding block, we show a generic construction of an SIBKEM scheme $\mathcal{SIBKEM}$ = {K.Setup, K.Ext, K.NwSt,K.Enc,K.Dec} as follows:

K.Setup: It takes as input $1^\lambda$, and runs Setup of $\mathcal{IBNIKE}$ to obtain $sp, mk$, where $sp$ contains a description of the shared key space $\mathcal{SHK}$. The output is $(sp, mk)$.

K.Ext: It takes as input $(sp, mk, id)$, and runs Ext of $\mathcal{IBNIKE}$ on $(sp, mk, id)$ to obtain $sk_{id}$ of an identity. The output is $sk_{id}$.

K.NwSt: It takes as input $sp$, and runs Sample of $\mathcal{IBNIKE}$ to obtain a temporary key pair $(\hat{pk}, \hat{sk})$. It sets $st \leftarrow (\hat{pk}, \hat{sk})$ and outputs $st$.

K.Enc: It takes as input $(sp, id, st)$, parses $st$ as $(\hat{pk}, \hat{sk})$, and then runs Shr' of $\mathcal{IBNIKE}$ on input $(sp, \hat{sk}, id)$ to obtain a key $K$. It sets the ciphertext $c \leftarrow \hat{pk}$, $dk \leftarrow K$, and outputs $(c, dk)$.

K.Dec: It takes as input $sp, sk_{id}, c$, and runs Shr" on input $(sp, sk_{id}, c)$ to obtain the key $K$. It sets $dk \leftarrow K$, and outputs $dk$. According to the consistency of Shr' and Shr", $dk$ is the valid key outputed by K.Enc.

## 4.3 Security Analysis

Here, we analyze the security of our generic construction. For convenience, we use the simulation-based proof technique. As described below, our proof has perfect simulation.

**Theorem 2.** *Suppose $\mathcal{IBNIKE}$ is IND secure. Then $\mathcal{SIBKEM}$ is IND-ID-CCA secure.*

*Main idea of the proof.* Our strategy is as follows. Towards contradiction, we prove that if a scheme $\mathcal{SIBKEM}$ we constructed is *not* secure in the IND-ID-CCA sense, then the underlying scheme $\mathcal{IBNIKE}$ is *not* secure in the IND. So we first assume there exists an IND-ID-CCA adversary $\mathcal{A}$ who can successfully break IND-ID-CCA with an advantage which is not negligible, then we show that we can construct an IND adversary $\mathcal{B}$ who can successfully break IND with an advantage which is not negligible.

*Due to the space limitation, the detailed proof is omitted.*

# 5    Conclusions

In this paper, we firstly proposed a cryptographic primitive called stateful identity based key encapsulation mechanism (SIBKEM). We defined the security notion, and showed that by combining secure SIBKEM and secure symmetric encryption, we can obtain secure stateful identity based encryption.

Secondly, we showed how to generically construct such SIBKEM scheme from a well-studied cryptographic primitive named identity based non-interactive key exchange (IBNIKE). Although our discussion was only in identity based settings, but we note that part of our results could be applied to conventional public key settings.

## Acknowledgements

We thank the ISC 2009 committees and anonymous reviewers for helpful comments. We especially thank Kenneth Paterson for constructive suggestions.

## References

1. Goldwasser, S., Micali, S.: Probabilistic encryption. J. Comput. Syst. Sci. 28(2), 270–299 (1984)
2. Naor, M., Yung, M.: Public-key cryptosystems provably secure against chosen ciphertext attacks. In: STOC, pp. 427–437. ACM, New York (1990)
3. Bellare, M., Kohno, T., Shoup, V.: Stateful public-key cryptosystems: how to encrypt with one 160-bit exponentiation. In: ACM CCS 2006, pp. 380–389. ACM, New York (2006)
4. Baek, J., Zhou, J., Bao, F.: Generic constructions of stateful public key encryption and their applications. In: Bellovin, S.M., et al. (eds.) ACNS 2008. LNCS, vol. 5037, pp. 75–93. Springer, Heidelberg (2008)
5. Shoup, V.: A standard for public-key encryption. ISO 18033-2 (2006)
6. Shamir, A.: Identity-based cryptosystems and signature schemes. In: Blakely, G.R., Chaum, D. (eds.) CRYPTO 1984. LNCS, vol. 196, pp. 47–53. Springer, Heidelberg (1985)
7. Phong, L.T., Matsuoka, H., Ogata, W.: Stateful identity-based encryption scheme: Faster encryption and decryption. In: ASIACCS 2008, pp. 381–388. ACM, New York (2008)
8. Yang, P., Zhang, R., Matsuura, K.: Stateful public key encryption: How to remove gap assumptions and maintaining tight reductions. In: ISITA 2008. IEEE, Los Alamitos (2008)
9. Paterson, K.G., Srinivasan, S.: On the relations between non-interactive key distribution, identity-based encryption and trapdoor discrete log groups. Journal version at DCC [10]
10. Paterson, K.G., Srinivasan, S.: On the relations between non-interactive key distribution, identity-based encryption and trapdoor discrete log groups. Designs, Codes and Cryptography 52, 219–241 (2009); Preliminary versions at Cryptology ePrint Archive: Report 2007/453 [9]
11. Bentahar, K., Farshim, P., Malone-Lee, J., Smart, N.P.: Generic constructions of identity-based and certificateless kems. J. Cryptology 21(2), 178–199 (2008)

12. Sakai, R., Ohgishi, K., Kasahara, M.: Cryptosystems based on pairing. In: SCIS 2000, pp. 26–28 (2000)
13. Bellare, M., Namprempre, C.: Authenticated encryption: Relations among notions and analysis of the generic composition paradigm. In: Okamoto, T. (ed.) ASIACRYPT 2000. LNCS, vol. 1976, pp. 531–545. Springer, Heidelberg (2000); Full version appeared in [14]
14. Bellare, M., Namprempre, C.: Authenticated encryption: Relations among notions and analysis of the generic composition paradigm. J. Cryptology 21(4), 469–491 (2008); Preliminary version appeared in [13]

# Privacy-Aware Attribute-Based Encryption with User Accountability

Jin Li[1], Kui Ren[1], Bo Zhu[2], and Zhiguo Wan[3]

[1] Department of ECE, Illinois Institute of Technology, USA
{jinli,kren}@ece.iit.edu
[2] Canada Concordia University, Canada
zhubo@ciise.concordia.ca
[3] Tsinghua University, China
wanzhiguo@tsinghua.edu.cn

**Abstract.** As a new public key primitive, attribute-based encryption (ABE) is envisioned to be a promising tool for implementing fine-grained access control. To further address the concern of user access privacy, privacy-aware ABE schemes are being developed to achieve hidden access policy recently. For the purpose of secure access control, there is, however, still one critical functionality missing in the existing ABE schemes, which is user accountability. Currently, no ABE scheme can completely prevent the problem of illegal key sharing among users. In this paper, we tackle this problem by firstly proposing the notion of accountable, anonymous, and ciphertext-policy ABE (CP-A$^3$BE, in short) and then giving out a concrete construction. We start by improving the state-of-the-art of anonymous CP-ABE to obtain shorter public parameters and ciphertext length. In the proposed CP-A$^3$BE construction, user accountability can be achieved in black-box model by embedding additional user-specific information into the attribute private key issued to that user, while still maintaining hidden access policy. The proposed constructions are provably secure.

**Keywords:** Access control, Anonymity, Attribute-based, Ciphertext-policy, Accountability.

## 1 Introduction

Today's computing and electronic technology innovations have unprecedentedly enabled ubiquitous information generation, processing, and distribution in both volume and speed. Vast amounts of information resources are made available and readily accessible to individuals and organizations through various computer systems and the Internet. This trend, however, also poses new challenges in designing suitable secure access control mechanisms. Generally, among the various requirements, today's access control schemes should at least meet the following ones: 1) fine-grained access policy, 2) protection of user privacy, and 3) assurance of user accountability.

P. Samarati et al. (Eds.): ISC 2009, LNCS 5735, pp. 347–362, 2009.

Recently, the notion of ABE, which was proposed by Sahai and Waters [1], has attracted much attention in the research community to design flexible and scalable access control systems. For the first time, ABE enables public key based one-to-many encryption. Therefore, it is envisioned as a highly promising public key primitive for realizing scalable and fine-grained access control systems, where differential yet flexible access rights can be assigned to individual users. To address complex and general access policy, two kinds of ABE have been proposed : key-policy ABE (KP-ABE) and ciphertext-policy ABE (CP-ABE). In KP-ABE, access policy is assigned in attribute private key, whereas, in CP-ABE, the access policy is specified in the ciphertext.

Besides fine-grained access policy, there is an increasing need to protect user privacy in today's access control systems. To address this problem, anonymous ABE was introduced in [2,3] and further improved by [4]. Anonymous ABE has a wide range of applications. For example, in some military circumstances, the access policy itself could be sensitive information. Therefore, to share resources with users possessing certain attribute-policy, anonymous ABE scheme can be applied to encrypt the resources while keeping the access policy specified in the ciphertext hidden.

Although the anonymous ABE can provide secure anonymous access control, before its widely deployment, another important security aspect, user accountability, has to be formally addressed. In particular, the problem of key abuse, $i.e.$, illegal key sharing among users, should be prevented. This problem is extremely important as in an ABE-based access control system, the attribute private keys directly imply users' privileges to the protected resources. The dishonest users may share their attribute private keys with other users, who do not have these privileges. They can just directly give away part of their original or transformed keys such that nobody can tell who has distributed these keys. Consequently, it renders the system useless. To the best of our knowledge, the issue of user accountability in access control system based on ABE is quite new in the literature and has not been solved yet. Such key abuse problems exist in all current access control schemes constructed from ABE as the attribute private keys assigned to users are never designed to be linked to any user specific information except the commonly shared user attributes. This is the reason why attribute private key can be abused by users without being detected.

To construct privacy-aware fine-grained ABE with user accountability, in this paper, the notion of accountable and anonymous CP-ABE (CP-A$^3$BE) is proposed. This is achieved by binding user identity in the attribute private key. CP-A$^3$BE can be applied to prevent the key sharing among users based on the following observation. If the user shares his attribute private key, the user's identity will be detected from the pirate device embedded with the shared private key. In normal encryption of CP-A$^3$BE, the message is encrypted with respect to some ciphertext-policy, in which the identity part is for all users. Any users can decrypt the ciphertext as long as their attribute private keys satisfy this policy. In tracing encryption, a message is encrypted to users with some ciphertext-policy, in which the identity part is for the suspicious users. In this algorithm, only the suspicious users with attribute private keys that satisfy this ciphertext-policy can decrypt the ciphertext. Due to the anonymity of CP-A$^3$BE, the tracing encryption algorithm and normal encryption algorithm are indistinguishable from the viewpoint of any user. Specifically, given a pirate device and the detected attributes embedded, the attribute center, who is in charge of the attribute private key issuing, can find the suspicious identity list of users possessing these attributes. To pinpoint the identity of the user sharing the attribute private key in the pirate device, the attribute center applies the tracing algorithm to encrypt a message with respect to the attributes and identities in the suspicious list. Computing in this way, the identity could be found if the ciphertext for some specific identity can be decrypted by this pirate device.

## 1.1 Related Work

Since the introduction of ABE in implementing fine-grained access control systems, a lot of works have been proposed to design flexible ABE schemes. There are two methods to realize the fine-grained access control based on ABE: KP-ABE and CP-ABE. They were both mentioned in [5] by Goyal et al. In KP-ABE, each attribute private key is associated with an access structure that specifies which type of ciphertexts the key is able to decrypt, and ciphertext is labeled with sets of attributes. In a CP-ABE system, a user's key is associated with a set of attributes and an encrypted ciphertext will specify an access policy over attributes. CP-ABE is different from KP-ABE in the sense that, in CP-ABE, it is the encryptor who assigns certain access policy for the ciphertext. When a message is being encrypted, it will be associated with an access structure over a predefined set of attributes. In CP-ABE, user will only be able to decrypt a given ciphertext if its attributes pass through the corresponding access structure specified in the ciphertext. The first KP-ABE construction [5] realized the monotonic access structures for key policies. To enable more flexible access policy, Ostrovsky et al. [6] presented the first KP-ABE system that supports the expression of non-monotone formulas in key policies. However, KP-ABE is less flexible than CP-ABE because the policy is determined once the user's attribute private key is issued. Later, Bethencourt et al. [7] proposed the first CP-ABE construction. However, the construction [7] is only proved secure under the generic group model. To overcome this weakness, Cheung and Newport [8] presented another construction that is proved to be secure under the standard model. The construction supports the types of access structures that are represented by AND of different attributes. Later, in [9], the authors gave another construction for more advanced access structures based on number theoretic assumption. To further achieve receiver-anonymity, Boneh and Waters [10] proposed a predicate encryption scheme based on the primitive called Hidden Vector Encryption. The scheme in [10] can also realize the anonymous CP-ABE by using the opposite semantics of subset predicates. Katz, Sahai, and Waters [11] proposed a novel predicate encryption scheme supporting inner product predicates. Their scheme is very general and can achieve both KP-ABE and hidden CP-ABE schemes. However, the constructions of [10,11] are very inefficient compared to [4]. Recently, several attempts [12,13,14] have been made to address the accountability problem in ABE-based access control. In [14], they considered how to defend the key-abuse problem in KP-ABE schemes while only achieving privacy for part of the attributes. In [13], another trusted party was introduced in the protocol and each decryption operation should get assistance from the trusted party. As a result, the third party has to handle a huge amount of load, which greatly limits their application in the real world. The work [12] does not rely on the existence of trusted party. Instead, they used the technique of identity-based wildcard encryption [15] to achieve the accountability for the user. However, a strong assumption of well-formedness decryption key is required in the pirate device. Therefore, the result in [12] is still not practical enough. In our work, these two drawbacks: the introduction of trusted party and strong assumption of white-box, can be avoided. In addition to the accountability, the user privacy, is also considered in our constructions, which cannot be realized in [12,13].

ORGANIZATION. Some preliminaries are given in Section 2, including the syntax and basic mathematic tools used in the paper. In Section 3, we propose two improved constructions of privacy-aware CP-ABE. In Section 4, the CP-A$^3$BE construction is proposed to realize the fine-grained access control system with user privacy and accountability. This paper ends with concluding remarks.

## 2    Preliminaries

### 2.1    Syntax

**System Model.** Before introducing CP-A$^3$BE, we first give the system model for anonymous CP-ABE. In the anonymous CP-ABE architecture, there are two entities: attribute center (AC) and user. AC is in charge of the issue of attribute private key to users requesting them. The user, who wants to access data, should get the attribute private key from AC in advance. The encryptor can specify the ciphertext-policy such that only users whose attribute private keys satisfy the policy are able to decrypt the ciphertext. In addition, the ciphertext-policy is kept hidden. The users with an attribute private key are able to check whether his attributes satisfy the ciphertext-policy or not. In our system model, a binary relation $R$ is defined as part of public parameter according to the concrete requirements of anonymous CP-ABE. We denote it by $R(L, W) = 1$ if the attribute list $L$ satisfies ciphertext-policy $W$.

**Definition 1.** *An anonymous CP-ABE system consists of four algorithms, namely, Setup, KeyGen, Encryption, and Decryption, which are defined as follows:*

*Setup($1^\lambda$). The setup algorithm, on input security parameter $1^\lambda$, outputs a master secret key sk and public key pk.*

*KeyGen(L, sk). The key generation algorithm, on input attribute list L and master key sk, outputs $sk_L$ as the attribute private key for L.*

*Enc(M, W, pk). The encryption algorithm, on input a message M together with ciphertext-policy W, outputs $\mathcal{C}$, as the encryption on M with respect to W.*

*Dec(C, sk$_L$). The decryption algorithm, on input the ciphertext C and the attribute private key sk$_L$, outputs M if R(L, W) = 1. Otherwise, it returns ⊥.*

**Adversary Model.** The goal of adversary in anonymous CP-ABE system can be either one of the following 1) Extracting information of plaintext from the ciphertext. Here, the adversary is allowed to control some users and access their attribute private keys that do not match the ciphertext-policy; 2) Distinguishing underlying access-policy in the ciphertext.

The two goals of adversary can be integrated in the indistinguishability against ciphertext-policy and chosen ciphertext attacks (CP-IND-CCA). In this work, a weaker notion, called indistinguishability against selective ciphertext-policy and chosen message attack (sCP-IND-CPA) [7,8,5], will be used. The definition is the same with CP-IND-CCA, except in sCP-IND-CPA, the adversary has to submit its challenge attributes before the setup phase. Furthermore, the decryption oracle is not available to the adversary. The formal definition is given based on the following sCP-IND-CPA game involving an adversary $\mathcal{A}$:

Game sCP-IND-CPA

> *Initial.* The adversary commits to the challenge ciphertext policies $W_0^*, W_1^*$ before setup algorithm.
> *Setup.* Choose a sufficiently large security parameter $1^\lambda$, and run *Setup* to get a master secret key *sk* and public key *pk*. Retain *sk* and give *pk* to $\mathcal{A}$;
> *Phase 1.* $\mathcal{A}$ can perform a polynomially bounded number of queries to key generation oracle on attributes $L$, the only restriction on $L$ is that, $R(L, W_0^*) = R(L, W_1^*) = 0$ or $R(L, W_0^*) = R(L, W_1^*) = 1$;
> *Challenge.* $\mathcal{A}$ outputs two messages $M_0, M_1$ on which it wishes to be challenged with respect to $W_0^*$ and $W_1^*$. It requires that $M_0 = M_1$ if any attribute private key on $L$ satisfying $R(L, W_0^*) = R(L, W_1^*) = 1$ has been queried. The challenger randomly chooses a bit $b \in \{0, 1\}$, computes $C = Enc(M_b, W_b^*, pk)$ and sends $C$ to $\mathcal{A}$;
> *Phase 2.* $\mathcal{A}$ continues to issue queries to the key generation oracle, with the same restriction as before;
> *Guess.* Finally, $\mathcal{A}$ outputs a guess bit $b'$.

$\mathcal{A}$ wins the game if $b = b'$. The advantage of $\mathcal{A}$ in Game sCP-IND-CPA is defined as the probability that $\mathcal{A}$ wins the game minus $1/2$. This model can be considered to be analogous to the selective-ID model [16] utilized in IBE protocols. In their security model, the adversary should commit to the challenge identity ID before *Setup* phase.

**Definition 2.** *An anonymous CP-ABE satisfies sCP-IND-CPA if no polynomial time adversary can break the above game.*

In CP-A$^3$BE, as explained, we consider how to achieve user accountability in addition to fine-grained access-policy and user privacy. The system model for

CP-A$^3$BE is the same with anonymous CP-ABE, except here the algorithm for tracing is added.

Trace. *This algorithm is applied to trace an attribute private key in black-box to its original holder. It takes as input a pirate device, and outputs identity associated with this attribute private key in the pirate device.*

Because the CP-A$^3$BE is still one kind of anonymous CP-ABE, the adversary model and security requirement of sCP-IND-CPA are defined in the same way as anonymous CP-ABE. The only difference lies in the ciphertext-policy where it is defined by two parts $W = W' \vee \overline{W}$: The first part is the same as in the anonymous CP-ABE while the second part is for the identity. That is, $\overline{W}$ could be ∗ or specific $ID$. Accordingly, the challenge ciphertext would be $W_0^* = W_{0,1}^* \| W_{0,2}^*$ and $W_1^* = W_{1,1}^* \| W_{1,2}^*$. This kind of security implies that if a user has an attribute private key on attributes $L$ for identity $ID$, it cannot decrypt the ciphertext encrypted for the ciphertext-policy $W$ if $R(L \| ID, W) = 0$. Additionally, to trace the identity who shares the attribute private key, the tracing algorithm should be indistinguishable with the normal encryption algorithm to avoid detection by the pirate device.

## 2.2   Basic Mathematic Tools

We give a brief review on the property of pairings and some candidates of hard problem from pairings. Let $\mathbb{G}_1, \mathbb{G}_2$ be cyclic groups of prime order $p$, writing the group action multiplicatively. Let $g$ be a generator of $\mathbb{G}_1$, and $\hat{e} : \mathbb{G}_1 \times \mathbb{G}_1 \rightarrow \mathbb{G}_2$ be a map with the following properties. *Bilinearity*: $\hat{e}(g_1^a, g_2^b) = \hat{e}(g_1, g_2)^{ab}$ for all $g_1, g_2 \in \mathbb{G}_1$, and $a, b \in_R \mathbb{Z}_p$; *Non-degeneracy*: There exist $g_1, g_2 \in \mathbb{G}_1$ such that $\hat{e}(g_1, g_2) \neq 1$. In other words, the map does not send all pairs in $\mathbb{G}_1 \times \mathbb{G}_1$ to the identity in $\mathbb{G}_2$; *Computability*: There is an efficient algorithm to compute $\hat{e}(g_1, g_2)$ for all $g_1, g_2 \in \mathbb{G}_1$.

# 3   Improved Privacy-Aware CP-ABE Constructions

## 3.1   Anonymous CP-ABE with Short Public Parameters

First, we give a construction of anonymous CP-ABE with short public parameters. In this work, the ciphertext-policy has the same fine-grained access structure (ciphertext-policy) with CP-ABE scheme [8]. Details of the access structure in [8] are described below. Assume that the total number of attributes in the system is $n$ and the universal attributes set is $U = \{w_1, w_2, \cdots, w_n\}$. To encrypt a message, it specifies the ciphertext-policy $W = [W_1, W_2, \cdots, W_n]$. The notion of wildcard ∗ in the ciphertext policies means the value of "don't care". For example, let the ciphertext-policy $W = [1, 0, 1, *]$ when $n = 4$. This ciphertext-policy means that the recipient who wants to decrypt must have the value 1 for $W_1$ and $W_3$, the value 0 for $W_2$, and any possible values for $W_4$. Therefore, if the receiver has an attribute private key for $[1, 0, 1, 0]$, it can decrypt the ciphertext because the first three values for $W_1$, $W_2$ and $W_3$ are equivalent to the corresponding values in ciphertext-policy. Moreover, the fourth value 0 in the private key satisfies the ciphertext-policy because $W_4 = *$. If an attribute private key is associated with the attribute list $[1, 1, 1, 0]$, this attribute private key will not match the ciphertext-policy since $W_2 \neq 0$. To be more generalized, given an attribute list $L = [L_1, L_2, \cdots, L_n]$ and a ciphertext-policy $W = [W_1, W_2, \cdots, W_n]$, we say that $L$ matches $W$ if for all $i \in [1, n]$, $L_i \in W_i$, i.e., $L_i = W_i$ or $W_i = *$. In [8], each attribute can take two values 1 and 0. In our construction, we generalize the access structures such that each attribute can take two or more values. More formally, let $S_i = \{v_{i,1}, v_{i,2}, \cdots, v_{i,n_i}\}$ be a set of possible values for attribute $w_i$ where $n_i$ is the number of the possible values for $w_i$. Then the attribute list $L$ for a user is $L = [L_1, L_2, \cdots, L_n]$ where $L_i \in S_i$ for $1 \leq i \leq n$, and the generalized ciphertext policy $W$ is $W = [W_1, W_2, \cdots, W_n]$. The attribute list $L$ satisfies the ciphertext-policy $W$ (that is, $R(L, W) = 1$) if $L_i = W_i$ or $W_i = *$ for $1 \leq i \leq n$.

**Main Idea.** We use $H(i\|v_{i,k_i})$ to denote the $k_i$-th value $v_{i,k_i}$ for the $i$-th attribute. Instead, in [4], they used different public keys to denote the universal attributes, which makes the size of public parameters to be $O(N)$, where $N$ is the total number of all attribute values defined in the system. To keep the receiver-anonymity in ciphertext, we cannot just replace the public key $pk_{i,k_i}$ with $H(i\|v_{i,k_i})$ directly. The ciphertext of the $v_{i,k_i}$ is computed by splitting the random value used in encryption into two parts $H(1\|i\|v_{i,k_i}))$ and $H(0\|i\|v_{i,k_i})$, together with two different generators $g_1$ and $g_2$. The reason for choosing different generators is to prevent the public verifiability of the ciphertext's validity, which achieves hidden policy. User can only check whether his own attribute private key matches the ciphertext-policy. Furthermore, the user cannot check if the ciphertext is valid or not with respect to other attribute list he does not have, which keeps the ciphertext-policy hidden. The four algorithms of our scheme are defined as follows.

**Setup.** Let $\mathbb{G}_1, \mathbb{G}_2$ be cyclic groups of prime order $p$, and $\hat{e} : \mathbb{G}_1 \times \mathbb{G}_1 \to \mathbb{G}_2$ be a pairing defined in Section 2. Let $g_1, g_2$ be random elements from $\mathbb{G}_0$. Define a hash function $H : \{0, 1\}^* \to \mathbb{G}_0$. Assume there are $n$ attributes in universe. That is to say, let the universal attributes set be $U = \{\omega_1, \omega_2, \cdots, \omega_n\}$. And, each attribute has multiple values, where $S_i$ is the multi-value set for $\omega_i$ and $\mid S_i \mid = n_i$. This algorithm also chooses a random number $\alpha \in \mathbb{Z}_p$ and computes $T = \hat{e}(g_1, g_2)^\alpha$. The system public parameter is $para = (g_1, g_2, T, H)$. The system master secret key $msk$ is $\alpha$, which is only known to AC.

**KeyGen.** To generate an attribute private key for user with attribute list $L = [L_1, L_2, \cdots, L_n] = [v_{1,k_1}, v_{2,k_2}, \cdots, v_{n,k_n}]$, AC picks up random $s_1, s_2, \cdots, s_{n-1} \in \mathbb{Z}_p^*$ and computes $s_n = \alpha - \sum_{i=1}^{n-1} s_i \bmod p$. It also chooses $n$ random numbers $\{r_i\}_{1 \leq i \leq n} \in \mathbb{Z}_p^*$ and computes the attribute private key with respect to $L$ as

$sk_L = \{(d_{i0}, d_{i1}, d'_{i0}, d'_{i1})\} = \{(g_2^{s_i} H(1\|i\|v_{i,k_i})^{r_i}, g_1^{r_i}, g_1^{s_i} H(0\|i\|v_{i,k_i})^{r'_i}, g_2^{r'_i})\}_{1 \leq i \leq n}$. The validity of $sk_L = \{(d_{i0}, d_{i1}, d'_{i0}, d'_{i1})\}_{1 \leq i \leq n}$ can be verified through the following equation: $\prod_{i=1}^n \frac{\hat{e}(d_{i0}, g_1)\hat{e}(d'_{i0}, g_2)}{\hat{e}(d_{i1}, H(1\|i\|v_{i,k_i}))\hat{e}(d'_{i1}, H(0\|i\|v_{i,k_i}))} = T$.

**Enc.** To encrypt a message $M \in \mathbb{G}_2$ under ciphertext-policy $W = [W_1, W_2, \cdots, W_n]$, pick up a random value $z \in \mathbb{Z}_p$ and compute $C_0 = MT^z$. For each $1 \le i \le n$ and $1 \le t_i \le n_i$,

† if $v_{i,t_i} \in W_i$, choose $z_{i,t_i} \in \mathbb{Z}_p^*$ and compute $(C_{i,t_i,0}, C_{i,t_i,1}, C'_{i,t_i,0}, C'_{i,t_i,1})$
$$= ((H(1\|i\|v_{i,t_i}))^{z_{i,t_i}}, g_1^{z_{i,t_i}}, (H(0\|i\|v_{i,t_i}))^{z-z_{i,t_i}}, g_2^{z-z_{i,t_i}});$$

‡ if $v_{i,t_i} \notin W_i$, choose randomly $z_{i,t_i}, z'_{i,t_i} \in \mathbb{Z}_p^*$ and compute
$$(C_{i,t_i,0}, C_{i,t_i,1}, C'_{i,t_i,0}, C'_{i,t_i,1}) = ((H(1\|i\|v_{i,t_i}))^{z_{i,t_i}}, g_1^{z_{i,t_i}}, (H(0\|i\|v_{i,t_i}))^{z'_{i,t_i}}, g_2^{z'_{i,t_i}}).$$

Finally, output the ciphertext as $C = (C_0, \{(C_{i,t_i,0}, C_{i,t_i,1}, C'_{i,t_i,0}, C'_{i,t_i,1})\}$ for $1 \le t_i \le n_i$ and $1 \le i \le n$.

**Dec.** Assume a user has an attribute private key $sk_L = \{(d_{i0}, d_{i1}, d'_{i0}, d'_{i1})\}_{1 \le i \le n}$ on attribute list $L = [v_{1,k_1}, v_{2,k_2}, \cdots, v_{n,k_n}]$. To decrypt the ciphertext $C = (C_0, \{\{(C_{i,t_i,0}, C_{i,t_i,1}, C'_{i,t_i,0}, C'_{i,t_i,1})\}_{1 \le t_i \le n_i}\}_{1 \le i \le n})$ without the information of ciphertext-policy $W$, the user first computes $C' = \prod_{i=1}^{n} \frac{\hat{e}(C_{i,k_i,1}, d_{i0})\hat{e}(C'_{i,k_i,1}, d'_{i0})}{\hat{e}(C_{i,k_i,0}, d_{i1})\hat{e}(C'_{i,k_i,0}, d'_{i1})}$ and then decrypts the ciphertext as $M = C_0/C'$.

To check whether the decryption is correct or not, redundancy can be added in the plaintext such that the user knows if his attribute private key matches the ciphertext-policy. There are many ways to add redundancy, such as appending $0^\lambda$ to the message for security parameter $\lambda$. After decryption, the user can verify the correctness of decryption by checking whether the first $\lambda$ is $0^\lambda$.

### 3.1.1  Security Result

Before giving security result for the anonymous CP-ABE, we show definitions of the following problems and assumptions based on the bilinear groups.

**DBDH Problem.** The Decision Bilinear Diffie-Hellman (DBDH) problem is that, given $g, g^x, g^y, g^z \in \mathbb{G}_1$ for unknown random $x, y, z \in \mathbb{Z}_p^*$, $T \in \mathbb{G}_2$, to decide if $T = \hat{e}(g, g)^{xyz}$.

We say that a polynomial-time adversary $\mathcal{A}$ has advantage $\epsilon$ in solving the DBDH problem in groups $(\mathbb{G}_1, \mathbb{G}_2)$ if $| \, Pr[\mathcal{A}(g, g^x, g^y, g^z, \hat{e}(g, g)^{xyz}) = 1] - Pr[\mathcal{A}(g, g^x, g^y, g^z, \hat{e}(g, g)^r) = 1] \, | \ge 2\epsilon$, where the probability is taken over the randomly chosen $x, y, z, r$ and the random bits consumed by $\mathcal{A}$. $(t, \epsilon)$-DBDH assumption holds in $(\mathbb{G}_1, \mathbb{G}_2)$ if no $t$-time algorithm has the probability at least $\epsilon$ in solving the DBDH problem for non-negligible $\epsilon$.

**D-Linear Problem.** Let $z_1, z_2, z_3, z_4, z \in \mathbb{Z}_p$ be chosen at random and $g \in \mathbb{G}_1$ be a generator. The D-Linear problem is that given $g, g^{z_1}, g^{z_2}, g^{z_1 z_3}, g^{z_2 z_4}, T$, to decide if $T = g^{z_3 + z_4}$.

We say that a polynomial-time adversary $\mathcal{A}$ has advantage $\epsilon$ in solving the D-Linear Problem in groups $(\mathbb{G}_1, \mathbb{G}_2)$ if $| \, Pr[\mathcal{A}(g, g^{z_1}, g^{z_2}, g^{z_1 z_3}, g^{z_2 z_4}, T] - Pr[g, g^{z_1}, g^{z_2}, g^{z_1 z_3}, g^{z_2 z_4}, g^{z_3 + z_4}] \, | \ge 2\epsilon$, where the probability is taken over the randomly chosen $z_1, z_2, z_3, z_4$ and the random bits consumed by $\mathcal{A}$. $(t, \epsilon)$-D-Linear assumption holds in $(\mathbb{G}_1, \mathbb{G}_2)$ if no $t$-time algorithm has the probability at least $\epsilon$ in solving the D-Linear problem for non-negligible $\epsilon$. The D-Linear assumption was first proposed in [17] and one of its variants will be used in the proof. We have the following security result for the above construction:

**Theorem 1.** *The Anonymous CP-ABE construction is secure in sCP-IND-CPA model, under the DBDH and D-Linear assumption.*

*Proof.* Due to space limitations, the detailed proof is provided in the full version [18].

To achieve IND-sCP-CCA security in the standard model, we can use the technique of simulation-sound NIZK proofs [19]. The most efficient transformation from IND-sCP-CPA to IND-sCP-CCA is to use the Fujisaki-Okamoto technique [20], which adds only a little computation overhead on the original scheme. So, the resulted IND-sCP-CCA anonymous CP-ABE construction is very efficient.

### 3.2 Anonymous CP-ABE with Shorter Ciphertext

To further reduce the ciphertext size of the above scheme, we propose another construction by expressing the attribute values as bit pattern. The ciphertext-policy $W_i$ can be only one value or $*$. This technique, together with the above construction, will be applied to design the CP-A$^3$BE scheme in the next Section.

**Main Idea.** The value set $S_i$ for each attribute $\omega_i$ is expressed using bit pattern. Suppose the length of $\mid S_i \mid$ is $\rho_i$. Instead of computing the ciphertext for each value in $S_i$, we encrypt the message with respect to 0 or 1 for each bit by using the anonymous CP-ABE technique above. It is indistinguishable that some bit is encrypted for $0, 1$, or $*$, from the viewpoint of users. Without loss of generality, the values in set $S_i$ can be mapped to $\{1, 2, \cdots, \mid S_i \mid\}$ with some injective function. As a result, the ciphertext size can be reduced from $O(\mid S_i \mid)$ to $O(\log \mid S_i \mid)$. Here, for each $i$, the ciphertext policy $W_i$ can be some $v_{i,k_i}$ in $S_i$ or $*$. To encrypt a message, it specifies the ciphertext-policy $W = [W_1, W_2, \cdots, W_n]$ with AND gate as above.

**Setup.** Assume there are $n$ attributes in universe denoted by $U = \{\omega_1, \omega_2, \cdots, \omega_n\}$. Each attribute has multiple values. Let $S_i$ be the multi-value set for $\omega_i$ and $\mid S_i \mid = n_i$. Assume the length of $\mid S_i \mid$ is $\rho_i$. The system public parameter is the same as the above scheme $para = (g_1, g_2, T, H)$. The system master secret key $msk$ is $\alpha$.

**KeyGen.** To generate an attribute private key for user with attribute list $L = [L_1, L_2, \cdots, L_n] = [v_{1,k_1}, v_{2,k_2}, \cdots, v_{n,k_n}]$, AC picks up random $s_1, s_2, \cdots, s_{n-1} \in \mathbb{Z}_p^*$ and computes $s_n = \alpha - \sum_{i=1}^{n-1} s_i \bmod p$. For each $1 \leq i \leq n$, the following steps are taken:

1. AC picks up $s_{i,1}, s_{i,2}, \cdots, s_{i,\rho_i} \in Z_p^*$ such that $s_i = \sum_{k=1}^{\rho_i} s_{i,k} \bmod p$;
2. For each $1 \le t_i \le \rho_i$, AC chooses random numbers $(r_{i,t_i}, r'_{i,t_i})$ from $\mathbb{Z}_p^*$. Assume $v_{i,k_i} = (I_{i,1}, I_{i,2}, \cdots, I_{i,\rho_i}) \in \{0,1\}^{\rho_i}$. AC computes the attribute private key for $v_{i,k_i}$ as

$$D_i = \{(d_{i,t_i,0}, d_{i,t_i,1}, d'_{i,t_i,0}, d'_{i,t_i,1})\}_{1 \le t_i \le \rho_i}$$
$$= (g_2^{s_{i,t_i}} H(1\|i\|t_i\|I_{i,t_i})^{r_{i,t_i}}, g_1^{r_{i,t_i}}, g_1^{s_{i,t_i}} H(0\|i\|t_i\|I_{i,t_i})^{r'_{i,t_i}}, g_2^{r'_{i,t_i}})_{1 \le t_i \le \rho_i}.$$

The validity of $sk_L = \{D_i\}_{1 \le i \le n}$ can be also verified in a similar way as the construction in Section 3.1.

**Enc.** To encrypt a message $M \in \mathbb{G}_2$ under ciphertext-policy $W = [W_1, W_2, \cdots, W_n]$, pick up a random value $z \in \mathbb{Z}_p$ and compute $C_0 = MT^z$. For each $1 \le i \le n$,

1. If $W_i = v'_{i,k'_i} (= (I'_{i,1}, I'_{i,2}, \cdots, I'_{i,\rho_i}))$, choose $\{(z_{i,t_i}, z'_{i,t_i}, \bar{z}_{i,t_i})\}_{1 \le t_i \le \rho_i} \in \mathbb{Z}_p$. For $1 \le t_i \le \rho_i$, if $I'_{i,t_i} = 1$, compute $(C_{i,t_i,0}, C_{i,t_i,1}, C'_{i,t_i,0}, C'_{i,t_i,1}) = (H(1\|i\|t_i\|1)^{z_{i,t_i}}, g_1^{z_{i,t_i}}, H(0\|i\|t_i\|1)^{z-z_{i,t_i}}, g_2^{z-z_{i,t_i}})$ and $(\hat{C}_{i,t_i,0}, \hat{C}_{i,t_i,1}, \hat{C}'_{i,t_i,0}, \hat{C}'_{i,t_i,1}) = (H(1\|i\|t_i\|0)^{\bar{z}_{i,t_i}}, g_1^{\bar{z}_{i,t_i}}, H(0\|i\|t_i\|0)^{\bar{z}_{i,t_i}}, g_2^{\bar{z}_{i,t_i}})$; otherwise, compute $(C_{i,t_i,0}, C_{i,t_i,1}, C'_{i,t_i,0}, C'_{i,t_i,1}) = (H(1\|i\|t_i\|1)^{z_{i,t_i}}, g_1^{z_{i,t_i}}, H(0\|i\|t_i\|1)^{\bar{z}_{i,t_i}}, g_2^{\bar{z}_{i,t_i}})$, $(\hat{C}_{i,t_i,0}, \hat{C}_{i,t_i,1}, \hat{C}'_{i,t_i,0}, \hat{C}'_{i,t_i,1}) = (H(1\|i\|t_i\|0)^{z_{i,t_i}}, g_1^{z_{i,t_i}}, H(0\|i\|t_i\|0)^{z-z_{i,t_i}}, g_2^{z-z_{i,t_i}})$.
2. If $W_i = *$, choose $\{(z_{i,t_i}, z'_{i,t_i})\}_{1 \le t_i \le \rho_i}$ from $\mathbb{Z}_p$. For $1 \le t_i \le \rho_i$, compute $\{(C_{i,t_i,0}, C_{i,t_i,1}, C'_{i,t_i,0}, C'_{i,t_i,1})\} = \{H(1\|i\|t_i\|1)^{z_{i,t_i}}, g_1^{z_{i,t_i}}, H(0\|i\|t_i\|1)^{z-z_{i,t_i}}, g_2^{z-z_{i,t_i}}\}, (\hat{C}_{i,t_i,0}, \hat{C}_{i,t_i,1}, \hat{C}'_{i,t_i,0}, \hat{C}'_{i,t_i,1}) = (H(1\|i\|t_i\|0)^{z_{i,t_i}}, g_1^{z'_{i,t_i}}, H(0\|i\|t_i\|0)^{z''_{i,t_i}}, g_2^{z''_{i,t_i}})$, where $z'_{i,t_i} + z''_{i,t_i} = z$.

The ciphertext is $C = (C_0, \{(C_{i,t_i,0}, C_{i,t_i,1}, C'_{i,t_i,0}, C'_{i,t_i,1}), (\hat{C}_{i,t_i,0}, \hat{C}_{i,t_i,1}, \hat{C}'_{i,t_i,0}, \hat{C}'_{i,t_i,1})\}$ for $1 \le t_i \le \rho_i$ and $1 \le i \le n$.

**Dec.** Assume a user has an attribute private key $sk_L = \{D_i\}_{1 \le i \le n}$ for attribute list $L = [v_{1,t_1}, v_{2,t_2}, \cdots, v_{n,t_n}]$. To decrypt the ciphertext $C = (C_0, \{(C_{i,t_i,0}, C_{i,t_i,1}, C'_{i,t_i,0}, C'_{i,t_i,1}), (\hat{C}_{i,t_i,0}, \hat{C}_{i,t_i,1}, \hat{C}'_{i,t_i,0}, \hat{C}'_{i,t_i,1})\}_{1 \le t_i \le \rho_i}\}_{1 \le i \le n})$ without knowing ciphertext-policy $W$, the user first computes

$$C' = \prod_{i=1}^n (\prod_{t_i=1}^{\rho_i} \frac{\hat{e}(\widetilde{C}_{i,t_i,1}, d_{i,t_i,0})\hat{e}(\widetilde{C}'_{i,t_i,1}, d'_{i,t_i,0})}{\hat{e}(\widetilde{C}_{i,t_i,0}, d_{i,t_i,1})\hat{e}(\widetilde{C}'_{i,t_i,0}, d'_{i,t_i,1})}).$$

- If $I_{i,t_i} = 1$, $(\widetilde{C}_{i,t_i,b}, \widetilde{C}'_{i,t_i,b}) = (C_{i,t_i,b}, C'_{i,t_i,b})$ for $b \in \{0,1\}$;
- If $I_{i,t_i} = 0$, $(\widetilde{C}_{i,t_i,b}, \widetilde{C}'_{i,t_i,b}) = (\hat{C}_{i,t_i,b}, \hat{C}'_{i,t_i,b})$ for $b \in \{0,1\}$.

Finally, the user decrypts the ciphertext as $M = C_0/C'$.

The method given in Section 3.1 can be used here to check the correctness of decryption. We have the following security result for the construction:

**Theorem 2.** *The Anonymous CP-ABE construction is secure in sCP-IND-CPA model, under the DBDH and D-Linear assumption.*

*Proof.* The construction is similar to the construction in Section 3.1. The difference here is that the message is encrypted with respect to each bit, other than each value of the attribute. Therefore, the proof is easy to be derived from the proof for Theorem 1.

# 4  CP-A³BE: Privacy-Aware Attribute-Based Encryption with User Accountability

In this Section, we propose a CP-A³BE construction, that is, the anonymous CP-ABE with user accountability, which is based on the anonymous CP-ABE scheme in Section 3.1. In fact, to construct CP-A³BE, the technique can be also easily applied to the anonymous CP-ABE scheme [4].

**Main Idea.** In this scheme, user is issued an attribute private key for $L\|ID$, where $L$ is an attribute list and $ID$ is the user's identity. In a normal encryption algorithm, a message is encrypted under ciphertext-policy $W = W'\|*$ such that any user with $L\|ID$ satisfying $R(L\|ID, W) = 1$ is able to decrypt, regardless of the user's identity $ID$. This holds because the second part in the ciphertext-policy is "don't care" (This technique is used here to keep the one-to-many property in ABE, even though different identities have been inserted in the attribute private keys). In tracing algorithm, a message is encrypted with $W'\|ID^*$ to test whether the identity in the pirate device is $ID^*$. Due to the anonymity in CP-A³BE, the ciphertext is indistinguishable from other ciphertext under ciphertext-policy $W = W'\|*$. In this case, only user with private key on $L\|ID$ satisfying $R(L\|ID, W'\|ID^*) = 1$ can decrypt the ciphertext. As a result, the identity $ID^*$ can be determined in the pirate device. There are five algorithms of our CP-A³BE scheme, which are defined as follows:

**Setup.** Let $\mathbb{G}_1, \mathbb{G}_2$ be cyclic groups of prime order $p$, and $\hat{e} : \mathbb{G}_1 \times \mathbb{G}_1 \to \mathbb{G}_2$ be a pairing defined in Section 2. Let $g_1, g_2$ be random elements from $\mathbb{G}_0$. Define a hash function $H : \{0, 1\}^* \to \mathbb{G}_0$. Assume there are $n$ attributes in universe. That is to say, let the universal attributes set be $U = \{\omega_1, \omega_2, \cdots, \omega_n\}$. Each attribute has multiple values, where $S_i$ is the multi-value set for $\omega_i$ and $| S_i | = n_i$. This algorithm also chooses a random number $\alpha \in \mathbb{Z}_p$ and computes $T = \hat{e}(g_1, g_2)^\alpha$. The system public parameter is $para = (g_1, g_2, T, H)$. The system master secret key $msk$ is $\alpha$, which is only known to AC.

**KeyGen.** To generate an attribute private key for user with $ID = (I_1, I_2, \cdots, I_\rho) \in \{0, 1\}^\rho$ for attribute list $L = [L_1, L_2, \cdots, L_n] = [v_{1,k_1}, v_{2,k_2}, \cdots, v_{n,k_n}]$, AC picks up random $s_1, s_2, \cdots, s_n \in \mathbb{Z}_p^*$ and computes $s_{n+1} = \alpha - \sum_{i=1}^n s_i \bmod p$. AC also chooses $n + 1$ numbers $\{r_i\}_{1 \le k \le n} \in \mathbb{Z}_p^*$ and $\rho$ numbers $\{s_{n+1,k}\}_{1 \le k \le \rho}$

such that $s_{n+1} = \sum_{k=1}^{\rho} s_{n+1,k}$. Finally, it computes the attribute private key on $L$ as

$$sk_L = \{\{(d_{i0}, d_{i1}, d'_{i0}, d'_{i1})\}_{1 \leq i \leq n}, \{(d_{n+1,k,0}, d_{n+1,k,1}, d'_{n+1,k,0}, d'_{n+1,k,1})\}_{1 \leq k \leq \rho}\}$$

$$= \{(g_2^{s_i} H(1\|i\|v_{i,k_i})^{r_i}, g_1^{r_i}, g_1^{s_i} H(0\|i\|v_{i,k_i})^{r'_i}, g_2^{r'_i})\}, \{(g_2^{s_{n+1,k}} H(1\|n+1\|k\|I_k)^{r_{n+1,k}},$$

$$g_1^{r_{n+1,k}}, g_1^{s_{n+1,k}} H(0\|n+1\|k\|I_k)^{r'_{n+1,k}}, g_2^{r'_{n+1,k}})\}, 1 \leq i \leq n \wedge 1 \leq k \leq \rho.$$

**Enc.** To encrypt a message $M \in \mathbb{G}_2$ under ciphertext-policy $W = [W_1, W_2, \cdots, W_n] \vee W_{n+1}$ where $W_{n+1} = *$, this algorithm picks up a random value $z \in \mathbb{Z}_p$ and computes $C_0 = MT^z$.

1. For each $1 \leq i \leq n$,
   † if $v_{i,t_i} \in W_i$, choose $z_{i,t_i} \in Z_p^*$ and compute $(C_{i,t_i,0}, C_{i,t_i,1}, C'_{i,t_i,0}, C'_{i,t_i,1})$
   $= (H(1\|i\|v_{i,t_i})^{z_{i,t_i}}, g_1^{z_{i,t_i}}, H(0\|i\|v_{i,t_i})^{z-z_{i,t_i}}, g_2^{z-z_{i,t_i}})$;
   ‡ if $v_{i,t_i} \notin W_i$, choose randomly $z_{i,t_i}, z'_{i,t_i} \in Z_p^*$ and compute $(C_{i,t_i,0}, C_{i,t_i,1}, C'_{i,t_i,0}, C'_{i,t_i,1}) = (H(1\|i\|v_{i,t_i})^{z_{i,t_i}}, g_1^{z_{i,t_i}}, H(0\|i\|v_{i,t_i})^{z'_{i,t_i}}, g_2^{z'_{i,t_i}})$.
2. For $i = n+1$, this algorithm selects $z_{n+1,k}, z'_{n+1,k}$ from $\mathbb{Z}_p^*$. Then, for each $1 \leq k \leq \rho$, it computes

$$(C_{n+1,k,0}, C_{n+1,k,1}, C'_{n+1,k,0}, C'_{n+1,k,1}) = (H(1\|n+1\|k\|1)^{z_{n+1,k}}, g_1^{z_{n+1,k}},$$
$$H(0\|n+1\|k\|1)^{z-z_{n+1,k}}, g_2^{z-z_{n+1,k}})$$

$$(\hat{C}_{n+1,k,0}, \hat{C}_{n+1,k,1}, \hat{C}'_{n+1,k,0}, \hat{C}'_{n+1,k,1}) = (H(1\|n+1\|k\|0)^{z'_{n+1,k}}, g_1^{z'_{n+1,k}},$$
$$H(0\|n+1\|k\|0)^{z-z'_{n+1,k}}, g_2^{z-z'_{n+1,k}})$$

Finally, the ciphertext is computed as $C = (C_0, \{(C_{i,t_i,0}, C_{i,t_i,1}, C'_{i,t_i,0}, C'_{i,t_i,1})\}$ for $1 \leq t_i \leq n_i$ and $1 \leq i \leq n$, $\{(C_{n+1,k,0}, C_{n+1,k,1}, C'_{n+1,k,0}, C'_{n+1,k,1}), (\hat{C}_{n+1,k,0}, \hat{C}_{n+1,k,1}, \hat{C}'_{n+1,k,0}, \hat{C}'_{n+1,k,1})\}_{1 \leq k \leq \rho})$.

**Dec.** Assume a user has an attribute private key

$$sk_L = \{\{(d_{i0}, d_{i1}, d'_{i0}, d'_{i1})\}_{1 \leq i \leq n}, \{(d_{n+1,k,0}, d_{n+1,k,1}, d'_{n+1,k,0}, d'_{n+1,k,1})\}_{1 \leq k \leq \rho}\}$$

for $L = [v_{1,k_1}, v_{2,k_2}, \cdots, v_{n,k_n}]$. To decrypt the ciphertext $C$ without knowing ciphertext-policy $W$, he computes

$$C' = \prod_{i=1}^{n} \frac{\hat{e}(C_{i,k_i,1}, d_{i0})\hat{e}(C'_{i,k_i,1}, d'_{i0})}{\hat{e}(C_{i,k_i,0}, d_{i1})\hat{e}(C'_{i,k_i,0}, d'_{i1})} \prod_{k=1}^{\rho} \frac{\hat{e}(\widetilde{C}_{n+1,k,1}, d_{n+1,k,0})\hat{e}(\widetilde{C}'_{n+1,k,1}, d'_{n+1,k,0})}{\hat{e}(\widetilde{C}_{n+1,k,0}, d_{n+1,k,1})\hat{e}(\widetilde{C}'_{n+1,k,0}, d'_{n+1,k,1})}.$$

1. If $I_k = 1$, $(\widetilde{C}_{n+1,k,b}, \widetilde{C}'_{n+1,k,b}) = (C_{n+1,k,b}, C'_{n+1,k,b})$ for $b \in \{0,1\}$;

2. If $I_k = 0$, $(\widetilde{C}_{n+1,k,b}, \widetilde{C}'_{n+1,k,b}) = (\hat{C}_{n+1,k,b}, \hat{C}'_{n+1,k,b})$ for $b \in \{0,1\}$.

Finally, decrypt and output the ciphertext as $M = C_0/C'$.

**Trace.** Suppose a given pirate device can decrypt the ciphertext under ciphertext-policy $W$. AC extracts part of the attribute list $(L_{i_1}, L_{i_2}, \cdots, L_{i_k})$ out of $W$. The values in other positions except $\{i_1, i_2, \cdots, i_k\}$ in $W$ are $*$. AC checks the issuing record of attribute private key and determines the suspicious users set $S$, who have the attributes $(L_{i_1}, L_{i_2}, \cdots, L_{i_k})$. There are two ways to pinpoint the exact identity from $S$: If the size of set $S$ is not huge, then, AC just encrypts some message with respect to ciphertext-policy $W$ for each $ID \in S$ until the identity is found. To make the trace algorithm and encryption algorithm indistinguishable, the technique used in Section 3.2 is applied here.

AC picks up a random value $z \in \mathbb{Z}_p$ and computes $C_0 = MT^z$ to encrypt a message $M \in \mathbb{G}_2$ under ciphertext-policy $W = [W_1, W_2, \cdots, W_n] \vee W_{n+1}$ where $W_{n+1} = ID$,

1. For each $1 \leq i \leq n$,

   † if $v_{i,t_i} \in W_i$, AC picks $z_{i,t_i} \in Z_p^*$ and computes

$$(C_{i,t_i,0}, C_{i,t_i,1}, C'_{i,t_i,0}, C'_{i,t_i,1}) = (H(1\|i\|v_{i,t_i})^{z_{i,t_i}}, g_1^{z_{i,t_i}}, H(0\|i\|v_{i,t_i})^{z-z_{i,t_i}}, g_2^{z-z_{i,t_i}});$$

   ‡ if $v_{i,t_i} \notin W_i$, AC chooses $z_{i,t_i}, z'_{i,t_i} \in \mathbb{Z}_p^*$ and computes

$$(C_{i,t_i,0}, C_{i,t_i,1}, C'_{i,t_i,0}, C'_{i,t_i,1}) = (H(1\|i\|v_{i,t_i})^{z_{i,t_i}}, g_1^{z_{i,t_i}}, H(0\|i\|v_{i,t_i})^{z'_{i,t_i}}, g_2^{z'_{i,t_i}}).$$

2. For $i = n + 1$, assume $ID=(I_1, I_2, \cdots, I_\rho))$. AC chooses $\{(z_{n+1,k}, z'_{n+1,k}, \bar{z}_{n+1,k})\}$ for $1 \leq k \leq \rho$,

   † if $I_k = 1$, AC computes

$$(C_{n+1,k,0}, C_{n+1,k,1}, C'_{n+1,k,0}, C'_{n+1,k,1}) = (H(1\|n+1\|k\|1)^{z_{n+1,k}}, g_1^{z_{n+1,k}},$$
$$H(0\|n+1\|k\|1)^{z-z_{n+1,k}}, g_2^{z-z_{n+1,k}})$$

$$(\hat{C}_{n+1,k,0}, \hat{C}_{n+1,k,1}, \hat{C}'_{n+1,k,0}, \hat{C}'_{n+1,k,1}) = (H(1\|n+1\|k\|0)^{z'_{n+1,k}}, g_1^{z'_{n+1,k}},$$
$$H(0\|n+1\|k\|0)^{\bar{z}_{n+1,k}}, g_2^{\bar{z}_{n+1,k}})$$

   ‡ if $I_k = 0$, AC computes

$$(C_{n+1,k,0}, C_{n+1,k,1}, C'_{n+1,k,0}, C'_{n+1,k,1}) = (H(1\|n+1\|k\|1)^{z'_{n+1,k}}, g_1^{z'_{n+1,k}},$$
$$H(0\|n+1\|k\|1)^{\bar{z}_{n+1,k}}, g_2^{\bar{z}_{n+1,k}})$$

$$(\hat{C}_{n+1,k,0}, \hat{C}_{n+1,k,1}, \hat{C}'_{n+1,k,0}, \hat{C}'_{n+1,k,1}) = (H(1\|n+1\|k\|0)^{z_{n+1,k}}, g_1^{z_{n+1,k}},$$
$$H(0\|n+1\|k\|0)^{z-z_{n+1,k}}, g_2^{z-z_{n+1,k}})$$

It can be easily seen that the user is able to decrypt the ciphertext only when his identity is $ID$ and he has the attribute list $L=(L_{i_1}, L_{i_2}, \cdots, L_{i_k})$.

If $|S|$ is too huge, the tracing algorithm works in the following way: First, AC tries an attribute value $L_j$ from the position $j$ where $W_j = *$. Then, it encrypts a message as the normal encryption algorithm with respect to $W'$ such that all positions are set to be $*$, except the positions of $\{i_1, i_2, \cdots, i_k, j\}$ are set to be $L' = L \cup L_j$. The ciphertext is sent to the pirate device. If the ciphertext can be decrypted correctly, AC knows one of the users with $L'$ shares his attribute private key. The suspicious user set is of course not greater than $|S|$. AC continues the above procedure until the suspicious set $|S|$ is not too huge. Finally, the technique for small $|S|$ can be applied and the identity in the pirate device can be pinpointed. To verify the correctness of the decryption, we also use the method described in Section 3.1 by adding redundancy in the plaintext. Actually, based on the tracing algorithm, the scheme is secure against collusion attack, in which users with different attributes can collude to generate a pirate device. The tracing algorithm still works and at least one of the illegal users will be detected from the pirate device. Our definition and construction of tracing requires that the adversary produces a perfect pirate decoder device, namely a decoder that correctly decrypts all well-formed ciphertexts [21]. In reality, the pirate has a decoder that may work only a fraction of the time. When interact with such a decoder, just repeat the tracing algorithm for each suspicious identity such that the error-rate is lower than some predefined number. The tracing algorithm is indistinguishable from the normal encryption algorithm because of the anonymous CP-ABE. We have following security result for the construction of CP-A$^3$BE:

**Theorem 3.** *The CP-A$^3$BE construction is secure in sCP-IND-CPA model, under the DBDH and D-Linear assumptions.*

*Proof.* This construction is based on the construction in Section 3.1, with the technique of anonymous CP-ABE in Section 4.1. Therefore, the proof is easy to be derived from the proof for Theorem 1 and Theorem 2, and is omitted here.

# 5   Conclusion

Three requirements are desired in many secure access control systems, that is, 1) Fine-grained access policy, 2) User privacy, and 3) User accountability. ABE schemes are promising in providing fine-grained access policy, but no existing ABE schemes can achieve user accountability to prevent illegal key sharing while still maintaining user privacy. In this paper, we solved this problem by proposing the notion of accountable and anonymous CP-ABE (CP-A$^3$BE). We started by giving two improvements of privacy-aware CP-ABE. In the first improvement of anonymous CP-ABE, the size of public parameter is only $O(1)$, instead of $O(N)$ required in [4], where $N$ denotes the number of attributes in universe. In the second improvement, the size of public parameter and ciphertext is $O(1)$ and

$O(\log(N))$, respectively, while in [4], they are both $O(N)$. Based on the improvements, we presented a CP-A$^3$BE construction. The user accountability can be achieved in black-box model by embedding additional user-specific information into the attribute private key, while still maintaining hidden access policy. The construction of CP-A$^3$BE is provably secure.

## Acknowledgement

This work was supported in part by the US National Science Foundation under grant CNS-0831963 and the National Sciences and Engineering Research Council of Canada under Grant RGPIN/356059-2008. Thanks to Shucheng Yu, Cong Wang and Qian Wang for their helpful comments on this work.

## References

1. Sahai, A., Waters, B.: Fuzzy identity-based encryption. In: Cramer, R. (ed.) EUROCRYPT 2005. LNCS, vol. 3494, pp. 457–473. Springer, Heidelberg (2005)
2. Kapadia, A., Tsang, P.P., Smith, S.W.: Attribute-based publishing with hidden credentials and hidden policies. In: NDSS, pp. 179–192 (2007)
3. Yu, S., Ren, K., Lou, W.: Attribute-based content distribution with hidden policy. In: NPSEC 2008, pp. 39–44 (2008)
4. Nishide, T., Yoneyama, K., Ohta, K.: Attribute-based encryption with partially hidden encryptor-specified access structures. In: Bellovin, S.M., Gennaro, R., Keromytis, A.D., Yung, M. (eds.) ACNS 2008. LNCS, vol. 5037, pp. 111–129. Springer, Heidelberg (2008)
5. Goyal, V., Pandey, O., Sahai, A., Waters, B.: Attribute-based encryption for fine-grained access control of encrypted data. In: CCS 2006, pp. 89–98. ACM, New York (2006)
6. Ostrovsky, R., Sahai, A., Waters, B.: Attribute-based encryption with non-monotonic access structures. In: CCS 2007, pp. 195–203. ACM, New York (2007)
7. Bethencourt, J., Sahai, A., Waters, B.: Ciphertext-policy attribute-based encryption. In: IEEE Symposium on Security and Privacy 2007, pp. 321–334. IEEE, Los Alamitos (2007)
8. Cheung, L., Newport, C.: Provably secure ciphertext policy abe. In: CCS 2007, pp. 456–465. ACM, New York (2007)
9. Goyal, V., Jain, A., Pandey, O., Sahai, A.: Bounded ciphertext policy attribute based encryption. In: Aceto, L., Damgård, I., Goldberg, L.A., Halldórsson, M.M., Ingólfsdóttir, A., Walukiewicz, I. (eds.) ICALP 2008, Part II. LNCS, vol. 5126, pp. 579–591. Springer, Heidelberg (2008)
10. Boneh, D., Waters, B.: Conjunctive, subset, and range queries on encrypted data. In: Vadhan, S.P. (ed.) TCC 2007. LNCS, vol. 4392, pp. 535–554. Springer, Heidelberg (2007)
11. Katz, J., Sahai, A., Waters, B.: Predicate encryption supporting disjunctions, polynomial equations, and inner products. In: Smart, N.P. (ed.) EUROCRYPT 2008. LNCS, vol. 4965, pp. 146–162. Springer, Heidelberg (2008)
12. Li, J., Ren, K., Kim, K.: a$^2$be: Accountable attribute-based encryption for abuse free access control, http://eprint.iacr.org/2009/118

13. Hinek, M.J., Jiang, S., Safavi-Naini, R., Shahandashti, S.F.: Attribute-based encryption with key cloning protection, http://eprint.iacr.org/2008/478
14. Yu, S., Ren, K., Lou, W., Li, J.: Defending against key abuse attacks in kp-abe enabled broadcast systems. Accepted by SECURECOMM 2009 (to appear, 2009), http://eprint.iacr.org/2009/295
15. Abdalla, M., Catalano, D., Alexander, W., Dent, J.M.L., Neven, G., Smart, N.P.: Identity-based encryption gone wild. In: Bugliesi, M., Preneel, B., Sassone, V., Wegener, I. (eds.) ICALP 2006. LNCS, vol. 4052, pp. 300–311. Springer, Heidelberg (2006)
16. Boneh, D., Boyen, X.: Efficient selective-ID secure identity-based encryption without random oracles. In: Cachin, C., Camenisch, J.L. (eds.) EUROCRYPT 2004. LNCS, vol. 3027, pp. 223–238. Springer, Heidelberg (2004)
17. Boneh, D., Boyen, X., Shacham, H.: Short group signatures. In: Franklin, M. (ed.) CRYPTO 2004. LNCS, vol. 3152, pp. 41–55. Springer, Heidelberg (2004)
18. Li, J., Ren, K., Zhu, B., Wan, Z.: Privacy-aware attribute-based encryption with user accountability, Full version, http://eprint.iacr.org/2009/284
19. Sahai, A.: Non-malleable non-interactive zero knowledge and adaptive chosen ciphertext security. In: IEEE Symp. on Foundations of Computer Science (1999)
20. Fujisaki, E., Okamoto, T.: Secure integration of asymmetric and symmetric encryption schemes. In: Wiener, M. (ed.) CRYPTO 1999. LNCS, vol. 1666, pp. 537–554. Springer, Heidelberg (1999)
21. Chor, B., Fiat, A., Naor, M.: Tracing traitors. In: Desmedt, Y.G. (ed.) CRYPTO 1994, vol. 839, pp. 257–270. Springer, Heidelberg (1994)

# Hardware-Assisted Application-Level Access Control

Yu-Yuan Chen and Ruby B. Lee

Princeton University, Princeton NJ 08544, USA
{yctwo,rblee}@princeton.edu

**Abstract.** Applications typically rely on the operating system to enforce access control policies such as MAC, DAC, or other policies. However, in the face of a compromised operating system, such protection
mechanisms may be ineffective. Since security-sensitive applications are
most motivated to maintain access control to their secret or sensitive information, and have no control over the operating system, it is desirable
to provide mechanisms to enable applications to protect information with
application-specific policies, in spite of a compromised operating system.
In this paper, we enable application-level access control and information
sharing with direct *hardware* support and protection, bypassing the dependency on the operating system. We analyze an originator-controlled
information sharing policy (ORCON), where the content creator specifies who has access to the file created and maintains this control *after*
the file has been distributed. We show that this policy can be enforced
by the software-hardware mechanisms provided by the Secret Protection
(SP) architecture, where a Trusted Software Module (TSM) is directly
protected by SP's hardware features. We develop a proof-of-concept text
editor application which contains such a TSM. This TSM can implement many different policies, not just the originator-controlled policy
that we have defined. We also propose a general methodology for *trust-
partitioning* an application into security-critical and non-critical parts.

## 1    Introduction

Access control in a computer system mediates and controls accesses to resources.
It is an essential part of the security of a computer system, preventing illegitimate
access to sensitive or protected information. Various access control policies exist,
e.g. mandatory access control (MAC), discretionary access control (DAC), role-
based access control (RBAC), etc. One access control policy that has been hard
to achieve is ORCON [1,2], or originator-controlled access. This is neither a
MAC nor a DAC policy. It is not specified by a central authority (like DAC),
but its subsequent re-distribution by legitimate recipients must be controlled
(like MAC). While DAC allows individuals to specify the access policy for their
files, it cannot control how a legitimate user re-distributes those files. In this
paper, we propose a hardware-software mechanism for achieving flexible access
control and information sharing policies, including ORCON-like policies.

P. Samarati et al. (Eds.): ISC 2009, LNCS 5735, pp. 363–378, 2009.

Access control mechanisms are usually implemented by the operating system (OS), which also enforces the access control policies. However, if the OS is compromised, then the access control policy enforcement can also be compromised. Applications need some way to protect secret or sensitive information, in spite of a compromised OS, over which they typically have no control. Hence, we examine how an application can be provided a flexible mechanism to achieve an application-level access control or information sharing policy, without depending on the OS.

In addition to conventional access control mechanisms, cryptographic mechanisms have also been used to control access to protected information. For example, Digital Rights Management (DRM) systems[3,4,5] use cryptographic mechanisms for copy protection of digital media. Here, anyone can get access to the encrypted material, but only legitimate recipients may gain access to the plaintext material – at least, that is the goal of DRM systems. Strong cryptography can be used to protect the contents of sensitive files by encrypting them into an unintelligible mass, while decrypting them only when needed or authorized. However, two critical issues arise: how the keys are managed and how the decrypted plaintext is managed. Commercial DRM systems such as Advanced Access Content System (AACS) [3] are broken not because they use weak cryptography (as in the case of Content Scramble System (CSS) [4]), but because of the unsafe storage of the keys used by the application software [5]. Hardware protection mechanisms such as TPM [6] are designed to protect cryptographic keys by *sealing* them in the TPM hardware, and the keys are only retrieved when the system is running in a *verified* condition. TPM offers greater protection to the keys and includes the measurements of the integrity of the operating system in the trust chain to make sure that it has not been compromised. However, TPM's protection model does not consider how the keys are used and where they are stored after they are *unsealed*, therefore the access control of the decrypted sensitive information is still left to the application and the decrypted symmetric keys from the TPM chip can still be obtained by examining the memory contents[7,8]. Hence, it all boils down to the management of the keys and the decrypted plaintext.

The access control to keys is often delegated to the operating system, since it governs all accesses to resources. However, modern operating systems are large and complex, and hence more prone to software vulnerabilities. Further, in the monolithic kernel model of common operating systems such as Linux and Windows, all kernel modules have equal privilege, so that one compromised kernel module can access the memory of another kernel module, which may be security-critical. An attacker can gain control of the operating system by targeting one of the many device drivers, bypass the access control and retrieve the application's secret or sensitive information.

In this paper, we propose the following solution: *A small, verifiable application module that enforces its own policy with direct hardware protection that cannot be bypassed or manipulated by the operating system.* Implementing access control or information sharing policies in the application-space removes the dependency on the operating system and adds the flexibility of incorporating different policies.

Our solution architecture builds on top of the Secret Protection (SP) architecture [9,10], which requires a small addition to the processor hardware to protect a Trusted Software Module (TSM). We provide protection of the application by modifying it slightly to incorporate a TSM directly protected by the hardware to prevent any undesired information leakage. We implement an ORCON-like access control policy for protected documents that is designed to be enforced in a distributed manner.

The contributions of the paper are as follows:

- Proof-of-concept implementation of a distributed access control policy that is difficult to enforce, e.g. ORCON.
- Developing a methodology for *trust-partitioning* of an application.
- Demonstrating the versatility of the SP architecture for implementing different access control or information sharing policies.

Section 2 gives the detailed definition of our target access control policy. Section 3 describes the threat and trust models considered in this paper. Section 4 describes our solution architecture. Section 5 explains the methodology we developed to partition an application into a trusted and an untrusted part. Section 6 gives the security analysis of our solution. Section 7 describes related work in this area and Section 8 concludes the paper.

## 2    Problem Statement

Information sharing has different requirements in different contexts. For example, confidentiality is of top concern in a military system, whereas integrity is essential in commercial systems. We consider an information sharing policy that could be tailored to work in both environments, to meet the needs of both confidentiality and integrity. Consider the case where a secret document is to be distributed to selected recipients of different clearance levels, while the content of the original document cannot be modified. Further, the re-distribution of the content has to be approved by the content creator. This policy, previously known as Originator-Controlled policy (ORCON) [1,2], was proposed to address such a scenario. Since the control point of the policy is neither entirely centralized nor entirely distributed, it cannot be directly solved by applying Mandatory Access Control (MAC) or Discretionary Access Control (DAC).

In such an information sharing policy, the key players include the *content creator*, the *recipients* and the *trust group* (Figure 1). The recipients can be further categorized as *authorized* recipients, who are within the trust group and are allowed access to the content of the document by the content creator, and *unauthorized* recipients who are outside the trust group. Not all members of the trust group are authorized recipients of a given document. We formalize the problem statements of the information sharing policy as follows.

- **Problem 1: Dissemination to authorized recipients**
  The content creator wants to restrict access to the content to authorized recipients only. In other words, recipients who have not gained explicit approval from the content creator will not have access to the content.

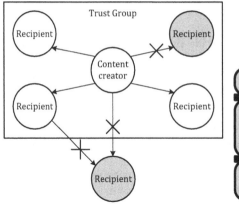

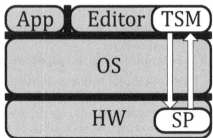

**Fig. 1.** Players in the information sharing policy. Gray circles represent unauthorized recipients, while white circles represent authorized recipients.

**Fig. 2.** Using SP architecture for flexible access-control enforcement. Grey parts are the untrusted system, white parts are trusted.

- **Problem 2: Prevent illegitimate re-dissemination**
  After the authorized recipients have gained access to the content, it should not be possible to redistribute or copy the original content to any unauthorized recipients. An unauthorized recipient must ask the content creator for explicit access rights in order to access the protected content.
- **Problem 3: Allow legitimate appending to the content**
  In the case where the content creator allows for appending extra information to the original content, an authorized recipient must be able to append to the original content, while preserving the authorized recipients of the original policy dictated by the content creator, i.e. the protected content may grow but the list of authorized recipients should remain unchanged.

To solve the above problems, we identify the requirements that must be met:
- The policy dictated by the content creator has to be tied to the corresponding protected content.
- The policy has to be enforced regardless of the presence of the content creator, i.e. the enforcement is distributed among the recipients.
- Updating (appending) the content should allow changes only in the content, not the policy. Therefore the policy should be physically separated from the content but logically tied to it.

Our solution architecture, as described in the following sections, adheres to these requirements and hence guarantees that the policy is never violated.

## 3    Threat and Trust Models

We assume that every recipient uses some type of computing device to access the content in the protected document, where each device has a Central Processing

Unit (CPU) that is trusted. Further, the content of the protected document is accessed by a piece of *editor* software that can read, display or modify the content in the document. For simplicity, we consider the protected content as digital text documents in this paper, although our proposed solution and methodology apply to any digital multimedia contents, e.g. digital photos, video or music.

The goal of an adversary is to gain access to the information in the protected content without explicit approval from the content creator. The adversary may have obtained the file of the protected document and have physical access to the computing device, and he can write his own software to run on the computing device to try to gain as much information as possible. Since the adversary has physical access, memory bus tapping or access to raw bits on disk are considered valid attacks. However, we *do not* consider any analog attacks, e.g. shoulder surfing or social engineering, since these attacks are out-of-band exploits that are not within the control of a computer system.

We divide the editor program into a *trusted* and an *untrusted* part, where the trusted part is guaranteed to perform the desired functions and any tampering with the trusted part will be detected, by means of our hardware protection mechanisms. However, the adversary can modify the untrusted part or the operating system to perform any malicious activities.

On the recipients' computing devices, we assume that a *trusted path* exists between the user input and the trusted CPU, and between the CPU and the display output. Hence, the device user can be assured that the input comes directly from him and that what is displayed is indeed that which is processed by the CPU. Various techniques exist [11,12,13] to support a trusted input path and a trusted display.

# 4   Architecture

Our solution consists of a combination of CPU hardware and application software, which builds upon the Secret Protection (SP) [9,10] architecture to provide direct hardware protection of the application. In essence, we partition the editor application into a trusted and an untrusted part and provide protection of the trusted part directly by the hardware, as shown in Figure 2.

## 4.1   SP Architecture

SP Architecture was first proposed [10] to protect the user's secret or sensitive information (*user mode*) and later modified [9] to protect a remote authority's and third parties' secret or sensitive information (*authority mode*). Our solution builds upon the authority mode SP [9]. We highlight the key architectural features of SP below.

The architecture consists of the Trusted Software Module (TSM) in the user-level application and the SP hardware in the microprocessor chip. There are two *hardware trust anchors* in the microprocessor chip: *Device Root Key* (DRK) and *Storage Root Hash* (SRH). The DRK is unique for each chip; it never leaves the chip and can not be read or written by any software. The only software that can use the DRK is the TSM, via a special instruction that can derive a new key from the DRK given nonces and/or constants. The SRH securely stores the root hash of a secure user-defined storage structure (on disk or on-line storage) accesssible only to the TSM. The SRH is accessible only to the TSM. Other software cannot read or write the SRH, including the operating system.

Hardware *Code Integrity Checking (CIC)* ensures the integrity of the TSM code while executing. Each instruction cache line embeds a MAC (a keyed hash), with the DRK as the key. The hash is verified before the instruction cache line is brought on-chip. Hardware *Concealed Execution Mode (CEM)* protects the TSM's data while it is executing, to guarantee confidentiality and integrity of any temporary data that the TSM uses, whether this is in on-chip registers or caches, or evicted to off-chip memory. During interrupt handling, hardware protects the contents of general registers and the interrupt return address from a potentially corrupted OS. All data cache lines containing protected data are encrypted and hashed when evicted from the microprocessor chip. A hardware encryption and hashing engine accelerates the automatic encryption (or decryption) and hash generation (or verification), reducing cryptographic overhead to the infrequent cache-miss handling of the last level of on-chip caches.

## 4.2  Distributed Access Control Architecture

The access control required by our information sharing policy is enforced by the new trusted part of the editor application, i.e. the TSM in the user-space. To guarantee the confidentiality and integrity of the protected document while it is opened by the editor, and to simplify the access control mechanism, we dedicate a special TSM buffer for use only by the TSM to store and manipulate any temporary data it uses. All the data in the TSM buffer are tagged as secure data in the processor's on-chip caches. When secure data cache lines are evicted from on-chip caches out to the main memory, the SP hardware mechanism will ensure that they are encrypted and hashed, by a key that is derived from the DRK. The TSM buffer does not interfere with the internal buffer structures of the editor program, so that the editor functions that do not involve the TSM are not modified at all. The TSM buffer is used by the TSM to hold temporary decrypted lines of the protected content. In other words, the protected content remains encrypted inside all internal buffers of temporary files used by the editor, only decrypted by the TSM in the TSM buffer when the TSM is active.

As mentioned in Section 2, the policy and the content should be physically separated but logically tied. We store the policy dictated by the content creator in the secure storage maintained by the SP hardware, and we tie together the policy and the content by a cryptographic hash that is also stored and protected in the secure storage. The root of trust of the secure storage (SRH) is protected on-chip and accessible only by the TSM, and only the TSM can legitimately access or modify the stored policies in the secure storage. Any illegitimate modifications to the stored policies will be caught by the TSM when checking the

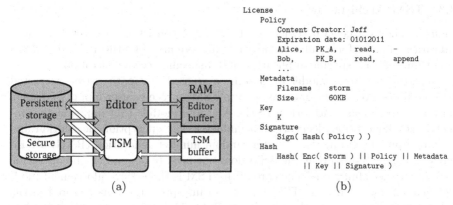

```
License
    Policy
        Content Creator: Jeff
        Expiration date: 01012011
        Alice,    PK_A,    read,    -
        Bob,      PK_B,    read,    append
        ...
    Metadata
        Filename    storm
        Size        60KB
    Key
        K
    Signature
        Sign( Hash( Policy ) )
    Hash
        Hash( Enc( Storm ) || Policy || Metadata
              || Key || Signature )
```

(a)                                        (b)

**Fig. 3.** (a): Partitioning the editor application and the system into untrusted (grey) and trusted (white) parts. The TSM gets its own buffer to work with temporary data, and it can access both the secure storage where the policies are stored and normal storage where the protected (encrypted) content is stored. (b): A license for the document. This contains a policy dictated by the content creator (PK_A represents the public key of Alice).

integrity of the secure storage. Figure 3(a) shows the interaction between the editor application (trusted and untrusted parts), the temporary buffers (SP-protected and unprotected), and the persistent storage (secure and unsecured). Physically separating the data and the policy reduces the amount of information that needs to be directly protected in the secure storage of SP, since a file can be very large.

The sensitive content is protected by encrypting the document with a key that is stored in the secure storage, along with the policy dictated by the content creator. Since the document is encrypted, it can be safely stored in any public storage without additional access control protection. The key to decrypt it is bound by the policy and the policy is enforced by the TSM. The TSM always controls the access to the decryption keys according to the corresponding policies. To ensure compliance with the requirements described in Section 2, in addition to the policy and the key, we store other pertinent information of the protected document in a data structure called a *license*, (see Figure 3(b)), which is stored in SP-protected secure storage. A license contains the access control policy, metadata, the key to decrypt the document, the originator's signature on the policy, and a hash over the encrypted document and all items in the license.

Before the user is allowed access to the content in the document, the TSM first checks the integrity (**Hash**) of the encrypted document and the license, to make sure they have not been tampered with. Then the TSM checks if the policy allows the particular recipient access to the content of the document. After all checks are successfully passed, the TSM decrypts the content of the document and stores it in the temporary TSM buffer and, through the trusted display link, displays the contents to the authenticated recipient.

### 4.3   TSM Architecture

Figure 4 shows a general structure of the TSM consisting of several modules (libraries) that perform different functionalities required by the TSM. The TSM is not limited to a specific application and a specific access control policy.

Since in our threat model we assume a trusted I/O path exists, a trusted I/O module serves as the gateway for the TSM to receive user input, to display output or to connect with other TSMs. A crypto module that implements symmetric key encryption/decryption, asymmetric key encryption/decryption and cryptographic hash functions, and a random number generation (RNG) are included in the TSM, so that the TSM does not need to depend on the operating system for these functions. The core of the TSM is the policy enforcement module that interacts with the TSM buffer and interprets the policy stored in the secure storage to mediate the I/O of the TSM. The policy enforcement module acts as the TSM resource manager that can be tailored to implement various access control policies. A user authentication module, along with a set of PKI interfaces is included in the TSM to take care of the user authentication required to guarantee that the owner of the public/private key pair specified in the policy is correctly authenticated. User authentication is described in Section 4.5.

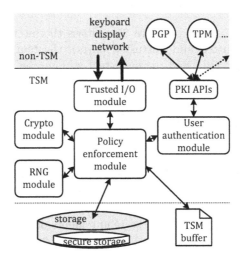

**Fig. 4.** TSM architecture. The trusted (white) parts of the system are the TSM and the SP-protected secure memory and secure storage.

### 4.4   Operation

We walk through an example to show how the TSM and the SP hardware protect and enforce the access control of the protected document.

1. The content creator creates the document containing sensitive data using any application he/she chooses.
2. The content creator dictates the policy he/she would like to enforce, e.g. *who* has *what* access to the content.

3. The content creator runs the editor application which contains the TSM, to turn the document into a protected document. A series of steps occur.
   (a) The TSM first randomly generates a new symmetric key.
   (b) The TSM encrypts the contents using the generated key and erases the plaintext.
   (c) The TSM calculates the hash of the policy and asks the content creator to sign the hash.
   (d) The TSM calculates the hash of the encrypted document, policy, metadata, key and the signature, and stores them in a newly created license in the secure storage.
4. The content creator can now distribute the encrypted document to all recipients he/she desires.
5. The TSM on the content creator side encrypts the license using the group encryption scheme [14] for the recipients (group encryption scheme and the trust group are described in Section 4.6).
6. The TSMs of the recipients' devices decrypt the license with their group decryption keys and securely store the license in the secure storage.
7. The TSM on the recipient side authenticates the recipient and checks the policy before granting access to the contents of the protected document.

## 4.5 User Authentication

User authentication is a difficult problem for the TSM, since we cannot rely on the operating system for existing user authentication mechanisms. To simplify the design of the TSM and not burden it with complex user authentication functions, we propose a public/private key authentication solution. We build a generic API interface that can interact with and make use of different public/private key applications, e.g. OpenPGP or GnuPG, that manage users' private keys. Below, we outline the protocol used by the TSM to authenticate a user utilizing other PKI applications.

When invoked by the user to read a policy-protected document, the TSM prompts the user for identity, for example, *Alice*. The TSM reads the corresponding policy in the secure storage to locate Alice's public key, PK_A. The TSM calls the RNG module to generate a new random number and uses PK_A to encrypt the random number as a challenge. The TSM sends the random challenge to the PKI application through the PKI interface and asks it to decrypt the random challenge.

The PKI application authenticates the user via its normal mechanisms, e.g. passphrase or TPM [6]. The PKI application returns the decrypted challenge to the TSM. The TSM checks for the validity of the random challenge to determine if the user has been successfully authenticated.

Ideally, the whole PKI application should be included in the TSM, since it is a security-critical function. If we consider the operating system as untrusted, the PKI application could also be compromised. However, our architecture still ensures that the keys that are used to decrypt the document, and the plaintext of the document, are never released outside the TSM.

### 4.6   Group Encryption and Trust Groups

We use group encryption [14] for distributing the protected license to the authorized recipients. Group encryption is the dual of the well-known group signature
scheme [15,16,17]. In a group signature scheme, a member of a group can anonymously sign a message on behalf of the group, without revealing his/her identity.
In a group encryption scheme, the sender can encrypt a piece of data and later
convince a verifier that it can be decrypted by the members of a group without
revealing the identity of the recipient. The authority in both cases is the only
entity that can reveal the identity of the signer in the group signature scheme
or the recipient of the group encryption scheme. One group in a group encryption scheme has one *group encryption key* and multiple *group decryption keys*
associated with it. The group encryption key is public and is used to encrypt
messages, while the group decryption keys are private.

In SP architecture [9], a trusted authority installs all TSMs and knows the
DRKs of all the SP devices. This is also the authority in the group encryption
scheme. In our architecture, the authority that initializes and installs the TSMs
creates a group that includes all SP devices, and assigns each SP hardware a
unique *group decryption key*, while publishing the *group encryption key* for that
group, such that in the secure storage of each SP device a pair of group encryption and decryption keys is stored and tied to the particular SP hardware.
Therefore the content creator can be assured that the license can only be decrypted by SP-enabled devices in the same group. For simplicity, we assume that
all SP-enabled devices are in the same group, although different groups of SP-
enabled devices can be established depending on application requirements. Note
that the authority that governs the SP-enabled devices and the trust groups
need not be the same as the certificate authority in the PKI systems for user
authentication.

In practice we may desire to have *multiple trust groups*, where each group may
contain an arbitrary number of SP devices. Since each originator may need to
distribute the protected document to a different set of authorized recipients, it
is desirable, although not necessary, to have separate groups for each originator.
This scheme can be easily incorporated in our solution since we can store multiple
group encryption-decryption key-pairs in the secure storage of each SP device,
and the TSM is responsible for distinguishing between different trust groups
and making sure that there is no information flow between trust groups, unless
it is explicitly allowed by the originator. Therefore, a recipient can belong to
multiple trust groups without the need to use multiple devices. However, since
there is only one authority that knows the DRKs of all devices and hence the
only authority that can properly insert group encryption-decryption key-pairs
into the devices, we cannot allow multiple authorities in a trust group without
extending the SP architecture [9].

## 5   Trust-Partitioning an Application

We developed a methodology for partitioning an existing application into a trusted
and an untrusted part. We chose *vi* [18] as our proof-of-concept application

to implement the application-level information sharing policy, since it is one of the most common text editors in the Unix operating system. Our methodology can also be applied to other applications.

To partition an application, we need to identify the entry and exit points into and out of the TSM. We first categorize the commands available in *vi*. Figure 5 shows the flow chart used to categorize the various commands of *vi* into 5 generic groups.

Table 1 shows the commands in each group. In particular, we are interested in the commands that are relevant to our information sharing policy, e.g. displaying the content of a file or appending new content to the original file, etc. The commands in **bold** (i.e., **ex** and **quit**) are modified *vi* commands and the commands in *italic* are new commands. These commands are the entry and exit points of the TSM and are the only commands that can legitimately manipulate

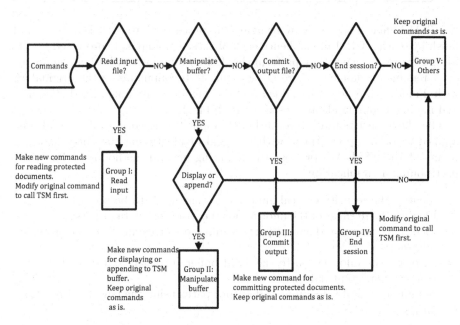

**Fig. 5.** Categorization of functions within an application for TSM protection

**Table 1.** The groups of *vi* commands after categorization

| Group I<br>Read input | Group II<br>Manipulate buffer | Group III<br>Commit output | Group IV<br>End session | Group V<br>Others |
|---|---|---|---|---|
| **ex** | print | write | **quit** | abbreviate |
| *tsm_ex* | read | *tsm_write* | | args |
| | *tsm_print* | | | cd |
| | *tsm_read* | | | delete |
| | | | | ... |
| | | | | *tsm_create* |

**Table 2.** New and modified vi commands

| | |
|---|---|
| *tsm_ex* filename | Open a protected document. |
| *tsm_print* line_number | Display the contents of a protected document. |
| *tsm_read* filename | Append the contents of filename to current protected document. |
| *tsm_write* | Automatically re-encrypt the protected document (with any appended data) and update the length and the hash stored in the license. $E_K$(document || appended_data) |
| *tsm_create* filename | Turn a document into a protected document. |
| **quit** & **ex** | Erase the plaintext in TSM buffer. |

the TSM buffer. They start by bringing the processor into SP CEM mode and finish by exiting CEM mode, hence each of these commands is protected by the SP hardware to ensure they perform the desired functions. All other commands of *vi* remain unchanged. There are a total of 70 commands in the original *vi*, with 2 modified, 5 new ones added and the remaining 68 unmodified. The new and modified commands are described in Table 2.

The above partitioning steps, although applied to *vi* specifically, can also be applied to other applications, with the goal of identifying the entry and exit points of the TSM. We propose the following general methodology for trust-partitioning an application:

1. Identify the security-critical information that needs to be protected.
2. Identify the *liveness* of the information, i.e. transient data or persistent data.
3. Identify the input and output paths leading to and leaving from the protected information.
4. Relocate the information to the TSM buffer (transient data) or the secure storage (persistant data).
5. Rebuild or modify the input and output paths using the new TSM functionalities.

## 6   Security Analysis

We analyze the security of our proposed solution according to three main security concerns: confidentiality, integrity and availability.

### 6.1   Confidentiality

In the information sharing policy, the content creator is most concerned with the confidentiality of the sensitive content in the protected document – only the authorized recipients can have access to the decrypted content.

We first consider the case where the adversary is outside the trust group, e.g., the adversary does not have a legitimate SP-enabled device. The adversary can try to attack the system by intercepting the communication (1) when the content creator is sending the encrypted document over to the recipients, or (2) when the content creator's device is sending the license to the recipients' devices. The adversary does not gain any information in the first case since the document sent over the communication is encrypted, and we assume the use of strong cryptography. Similarly, the communication channel intercepted in the second attack is also encrypted, using the group encryption key, which is known only by an SP device in the same group.

The attacker can also steal one of the recipients' devices and try to impersonate the authorized recipient. In this attack, in order for the adversary to successfully authenticate himself as the authorized recipient, he must know, or have access to, the private key of the authorized recipient.

We now consider the case where the adversary has a legitimate SP-enabled device and belongs to the correct trust group, but is not on the list of authorized recipients. The adversary now is also able to perform the previous three attacks. Further, the adversary can impersonate an authorized recipient and try to communicate with the content creator directly to ask for a legitimate license. However, the most that the adversary can do is to have both the encrypted document and the legitimate license stored in his/her device; the adversary still needs to have the private key of an authorized recipient to authenticate himself.

In the extreme case where the adversary is in the authorized recipient list – an *insider* attack – the adversary can access the contents of the document but has no way of digitally copying the contents to another file, since the plaintext document is only present in the TSM buffer during CEM mode and there is no command that allows direct memory copy of the plaintext from the TSM buffer to unprotected memory. The adversary can take pictures of the displayed content or memorize the content and later re-create it in another file. However, these attacks are not within the control of the computer system and hence, not in our threat model, as stated in Section 3.

Our solutions did not require the application used by the originator to create a new document to be trusted. Although the information could be stolen at this point, this is out-of-scope for this paper, since we are concerned not with the leaking of information when it is being created, but the leaking after it is recognized as important and being distributed.

## 6.2   Integrity and Availability

The integrity of the protected document and the corresponding policy is enforced by the Hash that ties together all the pertinent information of a policy-protected document. The Hash is stored in the secure storage, which is itself encrypted and integrity protected by the TSM using the keys accessible only by the TSM. The root of trust of the integrity of the secure storage is stored on the processor chip (SRH). Therefore, there is an integrity trust-chain from the protected content and license to the SRH, that does not depend on the potentially compromised OS.

SP architecture does not directly address denial-of-service attacks, therefore if the adversary modifies or completely deletes the document, or the license in the secure storage, any access to the protected information is lost. Although it is easy to achieve such denial-of-service attacks, they are not considered detrimental since no security-critical information is leaked by these attacks. In fact, these attacks show the fail-safe nature of the access control implementation. Nevertheless, SP architecture does provide intrinsic support for availability, in terms of the resiliency of the TSM to unrelated attacks. Since the trust chain consists only of the SP hardware and the TSM, attacks on the untrusted part of the application and the OS do not prevent the TSM from enforcing its access control functions.

## 7   Related Work

Several commercial solutions have been proposed to address the issue of information sharing, both in the context of digital media and digital documents. DRM solutions [3,4,5] focus on the copy-protection of the digital media, with a threat model that assumes that the whole *box* of the computing device is trusted, thus leading to the compromise of the encryption keys as described in Section 1. Cryptolope [19], known as cryptographic envelopes, also decouples the distribution of information and its license (called *superdistribution*) - similar to our solution. Cryptolope enables a commercial platform for the content creator and the publisher to license their content to the customers, by controlling the distribution of the decryption keys. However, Cryptolope assumes the same threat model as other DRM solutions – the device or the software on the device is trusted. Therefore, if an attacker can compromise the operating system or tap the memory bus, the attacker can have access to the decryption keys. Adobe Acrobat [20] has the ability to set permissions to protect sensitive files in the application level, including viewing, printing, changing, copying or commenting. But the password protection employed by Acrobat can be more easily defeated and is vulnerable to a malicious operating system as well. SISA [21] is a recent alliance of several industry companies, aiming to provide a secure end-to-end architecture for information sharing in a distributed environment. It involves several levels of access control, e.g. physical access control, network access control, storage access control, etc. Although the architecture provides extensive defense-in-depth, it still assumes the computing *box* as trusted. Also, the complexity of the architecture may make it more suitable only for large organizations.

Another area of related work is in hardware protection of application software. XOM [22] is another secure processor architecture that protects applications in an untrusted operating system environment. The protected applications running on XOM are kept in different *compartments*, each with its own compartment key. Like SP architecture, XOM has the ability to protect registers and encrypt memory traffic. Therefore, our application-level solution can also be mapped on the XOM processor by executing the TSM in a separate compartment. However, XOM is much more complicated than SP. TPM [6] is an industry solution to support trusted computing. Essentially TPM can be used to provide password protection, disk encryption and, most importantly, a trusted boot-chain. When employing TPM protection, applications can safely seal a piece of sensitive information inside the TPM chip. In other words, TPM can essentially *bind* a set of files to a particular host. However, since the TPM itself is not designed to provide protection of the decrypted plaintext once it leaves the TPM chip, a malicious operating system or hardware attacker can intercept the decrypted traffic in memory, although he/she cannot obtain the decryption keys in the TPM chip. Flicker [23] employs the newly introduced late launch instructions (both AMD and Intel) together with TPM to achieve a trusted execution environment for the protected part of an application. Like our proposal, it tries to minimize the trusted code base. However, unlike our proposal, Flicker does not consider hardware attacks. Also, our solution can achieve the same level of security without an external TPM chip. Overshadow [24] presented a framework for protecting applications without trusting the operating system. They do not require special hardware (like TPM, XOM or SP) but implement the protection mechanisms in the virtual machine monitor (VMM). They also do not consider hardware attacks and the TCB is larger since it has to include the entire VMM.

## 8  Conclusion

The SP security architecture provides a simple yet flexible software-hardware mechanism for protecting a Trusted Software Module (TSM) directly by SP hardware. This enables applications to express and enforce different security policies, without depending on the operating system over which they have no control. In this paper, we demonstrated the implementation of an originator-controlled (ORCON) distributed information sharing policy for documents. Such an access control policy is difficult to achieve with only MAC or DAC mechanisms. We achieve this in the user-space *vi* application, without relying on the operating system which can be compromised. The SP protection is rooted in the CPU hardware, defending against both software and hardware attacks. Our modified *vi* application is a proof-of-concept of the effectiveness of the SP hardware-software architecture. We also developed a general methodology for *trust-partitioning* an application, which is useful not only for our information sharing policy, but more generally for separating out the security-critical parts of applications.

## References

1. Graubart, R.: On The Need for A Third Form of Access Control. In: 12th National Computer Security Conference Proceedings, October 1989, pp. 296–303 (1989)
2. McCollum, C.J., Messing, J.R., Notargiacomo, L.: Beyond the Pale of MAC and DAC – Defining New Forms of Access Control. In: IEEE Computer Society Symposium on Research in Security and Privacy, pp. 190–200 (1990)
3. Advanced Access Content System (AACS), http://www.aacsla.com/home
4. Content Scramble System (CSS), http://www.dvdcca.org/css/
5. Leyden, J.: Blu-ray DRM Defeated: Copy-protection Cracked Again (January 23, 2007), http://www.theregister.co.uk/2007/01/23/blu-ray_drm_cracked/

6. Trusted Computing Group: Trusted Platform Module,
   https://www.trustedcomputinggroup.org/home
7. Halderman, J.A., Schoen, S.D., Heninger, N., Clarkson, W., Paul, W., Calandrino,
   J.A., Feldman, A.J., Appelbaum, J., Felten, E.W.: Lest We Remember: Cold Boot
   Attacks on Encryption Keys. In: SS 2008: Proceedings of the 17th Conference on
   Security Symposium, Berkeley, CA, USA, pp. 45–60. USENIX Association (2008)
8. Kumar, A.: Discovering Passwords in the Memory, White Paper, Paladion Networks (November 2003)
9. Dwoskin, J.S., Lee, R.B.: Hardware-rooted Trust for Secure Key Management and
   Transient Trust. In: Proceedings of the ACM Conference on Computer and Communications Security (CCS 2007), October 2007, pp. 389–400 (2007)
10. Lee, R.B., Kwan, P.C.S., McGregor, J.P., Dwoskin, J., Wang, Z.: Architecture for
    Protecting Critical Secrets in Microprocessors. In: ISCA 2005: Proceedings of the
    32nd Intl. Symposium on Computer Architecture, pp. 2–13 (2005)
11. Challener, D., Yoder, K., Catherman, R., Safford, D.: 15. In: A Practical Guide to
    Trusted Computing, pp. 271–276. IBM Press (2008)
12. Epstein, J.: Fifteen Years after TX: A Look Back at High Assurance Multi-Level
    Secure Windowing. In: ACSAC 2006, pp. 301–320 (2006)
13. Ocheltree, K., Millman, S., Hobbs, D., Mcdonnell, M., Nieh, J., Baratto, R.:
    Net2Display: A Proposed VESA Standard for Remoting Displays and I/O De-
    vices over Networks. In: Proceedings of the 2006 Americas Display Engineering
    and Applications Conference (ADEAC 2006) (October 2006)
14. Kiayias, A., Tsiounis, Y., Yung, M.: Group Encryption. In: Kurosawa, K. (ed.)
    ASIACRYPT 2007. LNCS, vol. 4833, pp. 181–199. Springer, Heidelberg (2007)
15. Camenisch, J., Stadler, M.: Efficient Group Signature Schemes for Large Groups
    (Extended Abstract). In: Kaliski Jr., B.S. (ed.) CRYPTO 1997. LNCS, vol. 1294,
    pp. 410–424. Springer, Heidelberg (1997)
16. Chaum, D., van Heyst, E.: Group Signatures. In: Davies, D.W. (ed.) EUROCRYPT
    1991. LNCS, vol. 547, pp. 257–265. Springer, Heidelberg (1991)
17. Chen, L., Pedersen, T.P.: New Group Signature Schemes. In: De Santis, A. (ed.)
    EUROCRYPT 1994. LNCS, vol. 950, pp. 171–181. Springer, Heidelberg (1995)
18. The Traditional vi, http://ex-vi.sourceforge.net/
19. Kohl, U., Lotspiech, J., Nusser, S.: Security for the Digital Library - Protectiong
    Documents Rather Than Channels. In: DEXA 1998: Proceedings of the 9th Inter-
    national Workshop on Database and Expert Systems Applications, p. 316 (1998)
20. Adobe Acrobat Family, http://www.adobe.com/products/acrobat
21. Secure Information Sharing Architecture (SISA) Alliance (2007),
    http://www.sisaalliance.com/
22. Lie, D., Thekkath, C.A., Horowitz, M.: Implementing an Untrusted Operating
    System on Trusted Hardware. In: SOSP 2003: Proceedings of the nineteenth ACM
    symposium on Operating systems principles, pp. 178–192 (2003)
23. McCune, J.M., Parno, B.J., Perrig, A., Reiter, M.K., Isozaki, H.: Flicker: an ex-
    ecution infrastructure for tcb minimization. In: Eurosys 2008: Proceedings of the
    3rd ACM SIGOPS/EuroSys European Conference on Computer Systems 2008, pp.
    315–328. ACM, New York (2008)
24. Chen, X., Garfinkel, T., Lewis, E.C., Subrahmanyam, P., Waldspurger, C.A.,
    Boneh, D., Dwoskin, J., Ports, D.R.: Overshadow: a virtualization-based approach
    to retrofitting protection in commodity operating systems. In: ASPLOS XIII, pp.
    2–13 (2008)

# Towards Trustworthy Delegation in Role-Based Access Control Model*

Manachai Toahchoodee, Xing Xie, and Indrakshi Ray

Department of Computer Science
Colorado State University
Fort Collins CO 80523-1873
{toahchoo,xing,iray}@cs.colostate.edu

**Abstract.** The need to delegate, which allows the temporary grant or transfer of access rights, arise in many applications. Although a lot of research appears in extending Role-Based Access Control (RBAC) to support delegation, not much appears on providing a formal basis for choosing delegatees. We provide an approach that allows one to assess the trustworthiness of potential delegatees in the context of the task that is to be delegated. It is also important to ensure that the choice of the delegatee does not cause any security policy violation. Towards this end, we show how to formally analyze the application using existing SAT solvers to get assurance that our choice of delegatee does not cause a security breach. Once the process of choosing delegatee can be formalized, it will be possible to automate delegation and use it for real-time applications.

## 1 Introduction

Role-Based Access Control (RBAC) is the de facto access control model for commercial organizations primarily because it is policy neutral and simplifies access control management. Since its conception, RBAC has evolved in various ways to meet the demands of various applications. One such extension is with regards to incorporating the notion of delegation. Delegation allows a user or a role to grant or transfer privileges to other users or roles. This makes it possible for organizations to continue functioning when some user is temporarily unavailable.

Although a lot of research appears in the area of delegation [1,2,3,4,5,6], not much appears in formalizing the basis on which a delegator selects a delegatee. The choice of a delegatee should be determined by two factors. First, the trustworthiness of an entity must be taken into account while considering it as a delegatee. This requires that the privileges should not be delegated to another user who the delegator does not consider trustworthy. This factor becomes even more critical in the presence of delegation chains. Second, choosing a delegatee should not introduce any security policy violation. For example, choosing a

---

* This work was supported in part by AFOSR under contract number FA9550-07-1-0042.

P. Samarati et al. (Eds.): ISC 2009, LNCS 5735, pp. 379–394, 2009.

specific delegatee may cause a violation in separation of duty constraints. We address these two factors.

The first factor requires evaluation of trustworthiness of candidates. Trust is a relationship between a truster and a trustee and it is dependent on a given *task*. The truster's trust for a trustee, with respect to a given task, depends on several factors, namely, *properties*, *experiences*, and *recommendations*. Properties are verifiable characteristics of the trustee. Experiences correspond to the past work experience of the trustee. Recommendations are the information that the truster obtains from reputable sources about the trustee. We show how to quantify these factors and assess the trustworthiness of an entity before designating him as the delegatee. We also show how trustworthiness of an entity can be used to decide and reason about delegation chains.

Sometimes a truster may not have enough information about a trustee with respect to a given task that will allow him to trust the trustee. However, information about related tasks may be available. We formalize the relationships among the various tasks using the concept of *task graphs*. Task graphs are directed acyclic graphs where the nodes correspond to the different tasks in an organization and the edges correspond to generalization/specialization and composition relationships. The labels on the edges give the degree of similarity between the different tasks. With the help of the task graphs, trust information of related tasks can be used to extrapolate the trust value for the given task.

Once a potential delegatee has been selected based on trustworthiness, we must ensure that this selection does not cause a security breach. We advocate the use of Alloy [7] for checking security policy violation. Alloy is a modeling language capable of expressing complex structural constraints and behavior. Alloy is supported by an automated constraint solver called Alloy Analyzer that searches instances of the model to check for satisfaction of system properties. The model is automatically translated into a Boolean expression, which is analyzed by SAT solvers embedded within the Alloy Analyzer. A user-specified scope on the model elements bounds the domain, making it possible to create finite Boolean formulas that can be evaluated by the SAT-solver. When a property does not hold, a counter example is produced that demonstrates how it has been violated. It has been successfully used in the modeling and analysis of real-world systems [8,9].

The paper is organized as follows. Section 2 describes some of the important work related to delegation and trust modeling. Section 3 presents our trust model and how to assess trustworthiness of entities with respect to a given task in a quantitative manner. Section 4 shows how trustworthiness of entities can be used to decide on the levels of delegation. Section 5 discusses how to compute trustworthiness of entities with respect to a given task when we do not have any information about the entity with respect to the given task. Section 6 illustrates how trust computation is performed for potential delegatees. Section 7 provides an approach using Alloy that evaluates the potential delegatees who satisfy the security policies. Section 8 concludes the paper with some pointers to future directions.

## 2    Related Work

One of the early works on delegation is by Barka and Sandhu [1] who proposed Permission-Based Delegation Model (PBDM) supporting permission and role delegations and revocations. The PBDM was refined subsequently by Zhang et al. [6] into three versions, namely, PBDM0, PBDM1, and PBDM2. RDM2000 [4] is an extension of PBDM0 which provides rules for delegation in the role hierarchy and identifies the prerequisites that must be satisfied by delegatees. Joshi et al. [2,3] focus on the relationship between delegation and role hierarchies in the context of Generalized Temporal Role-Based Access Control (GTRBAC) model. Crampton et al. [10] focussed on the workflow satisfiability problem (WSP) in the context of delegation. Workflow satisfiability in the context of delegation has also been addressed by Wang and Li [11]. The authors prove that WSP in the context of their $R^2BAC$ model is an NP-complete problem. In a subsequent work [5], Wang and Li formalize the notion of secure delegation.

One of the most important work in trust modeling is by Jøsang [12,13] where he claimed that trust is a relationship between two entities on a specific *statement* and is represented using degrees of *belief* b, *disbelief* d and *uncertainty* u. A statement describes a particular type of trust. Within a specific statement, Jøsang called the triple {b, d, u} as an opinion $\omega = \langle b, d, u \rangle$ representing the confidence to trust that declaration. Recommendation also plays a part in increasing the decision confidence. Jøsang utilized *subjective logic* to define *recommendation* and *consensus* formulae in order to take into account multiple subjective views on the same statement. Recommendation is when entity A asks another entity B for recommendation about how B trusts statement S, while consensus is the cumulative result caused by A and B in trusting S. Jøsang also formalized the notion of trust chains when recommenders indirectly provide input for a particular statement. In subsequent works[14], Jøsang added a new component *base rate a*, where $a \in [0, 1]$. The base rate gives the default trust value in the absence of information about a given entity. In another work [15], they show how to specify trust networks consisting of multiple paths between the trusted parties and provide a practical method for analyzing and deriving measures of trust in such environments.

Although a lot of research has been done in the context of access controls for open systems [16,17,18,19,20,21,22,23], we describe only the ones that are closely related to this work. Chakraborty et al. [17] proposed a Trust Based Access Control (TrustBAC) model where the assignment of users to roles depended on their trust values which can range from -1 to 1. The authors propose three factors, namely, *knowledge* $W_K$, *experience* $W_E$, and *recommendation* $W_R$, that impact a user's trustworthiness and show how to evaluate them. The trust value also takes into account history information in trustworthiness computation. Ray et al. [22] propose a trust model where the notion of trust contexts were formalized. The relationships among different contexts were represented using *context graph*. *Specialization* and *composition* are two kinds of relationships in context graph, where the labels on the graph indicate the degree of similarity between the contexts.

A lot of work also appears in the use of Alloy for analyzing security policies. Zao et al. [24] show how to model RBAC and Bell-Lapadula using Alloy. Schaad et al. [25] model user-role assignment, role-permission assignment, role hierarchy, and separation of duty features of RBAC extension using Alloy, and also describe how to detect conflicts.

# 3   Trust Modeling and Computation

*Delegator* refers to the role or user whose privileges are being transferred or granted to another role or user and the recipient of the privileges is termed *delegatee*. We show how the delegator can compute the trustworthiness of various entities in the context of the task that he is about to delegate.

Trust is a relationship between a truster and trustee with respect to a given context. The context in the case of delegation is the task for which delegation is needed. Trust relationship for a given context depends on three factors: *properties*, *experiences* and *recommendations*. Properties are verifiable characteristics of the trustee. For instance, it may be the role and credentials possessed by the trustee. Experiences are the past interactions that the truster had with the trustee. Recommendations are provided by third-parties whom the truster trusts about the capabilities of the trustee. In the following, we describe how the trust relationship is quantified.

## 3.1   Quantifying Properties

Properties depend on the attributes of the entity and also the role associated with it.

### Measuring Necessary Attributes $\mathcal{A}$

Every task in an organization requires some attributes of the user. For example, the task of performing surgery requires the user to be a certified surgeon. A task may require one or more attributes. The information about user attributes is contained in the credentials belonging to the user. Credentials are unforgeable and verifiable. Measuring necessary attributes requires evaluating what percentage of the necessary attributes are possessed by the user.

Let the set of attributes needed for task $T_i$ be denoted by $TA_i$ where $TA_i = \{a_{i1}, a_{i2}, \ldots, a_{in}\}$. Let $wa_{i1}$, $wa_{i2}$, ..., $wa_{in}$ be the weights of attributes $a_{i1}$, $a_{i2}$, ..., $a_{in}$ respectively. The weights of the attributes indicate their relative importance with respect to task $T_i$ and $\Sigma_{r=1}^{n} wa_{ir} = 1$. Each user profile contains the credentials possessed by the user. Let the set of all attributes possessed by the user $U_j$ be given by $UA_j$, where $UA_j = \{a_{j1}, a_{j2}, \ldots, a_{jm}\}$. Let $p = |TA_i \cap UA_j|$. The attribute value for user $j$ with respect to task $T_i$, denoted by $\mathcal{A}_{ij}$, is calculated as follows: $\mathcal{A}_{ij} = \Sigma_{k=1}^{p} wa_k$ where $wa_k$ $(1 \leq k \leq p)$ is the weight associated with attribute $a_k$ and $a_k \in TA_i \cap UA_j$.

### Measuring Role Attribute $\mathcal{R}$

The roles in the organization are arranged in the form of a hierarchy. The hierarchy can be represented as a labeled directed acyclic graph where the nodes represent the roles and the edges denote the hierarchical relationship. Note that, edges are drawn only for direct senior and junior relationship; transitive edges are not explicitly added. The edges in the hierarchy are labeled with a number in the range (0,1] which indicates the closeness relationship between the roles. A number close to 0 indicates that the two roles are very distant, whereas a number close to 1 denotes that the roles are very close. We assume that the assignment of the numbers is done by the system administrator who has knowledge about the relationships between roles. If there is a path between role $i$ and role $j$, the closeness relationship, denoted by $dist(r_i, r_j)$, is calculated by taking the product of all the edges constituting this path. Note that, if there are multiple paths connecting role $i$ and role $j$, both the paths should give the same value. Otherwise, the role graph is said to be inconsistent. The formal definition of the role graph appears below.

**[Weighted Role Hierarchy Graph]:** Weighted role hierarchy graph, denoted by $WRH = (V, A)$, is a weighted directed acyclic graph where $V$ is a set of nodes corresponding to the roles, and $A$ is a set of arcs corresponding to the hierarchical relationship; $(v_i, v_j) \in A$ indicates that role $v_j$ is directly senior to the role $v_i$. The weight of the edge $(v_i, v_j)$, denoted by $w(v_i, v_j)$, is a number in the range (0,1] that gives a measure of the closeness of the two roles.

Each task $T_i$ is associated with a set of roles $TR_i$ who are authorized to execute this task. The roles associated with a task include roles who have the direct permission to execute those tasks, as well as those authorized by virtue of role hierarchy. Each user $U_j$ also has a set of roles $UR_j$ assigned to him. We choose the role belonging to the user that is closest to some role associated with the task. The distance between these two roles gives the role attribute $\mathcal{R}_{ij}$ of user $U_j$ with respect to task $T_i$.

**Computing the Properties Value**
Some organizations may give greater importance to the role factor, whereas others may consider attribute factor to be more useful. Let $w_a$ and $w_r$ be the weights assigned to attributes and roles respectively, where $w_a, w_r \in [0, 1]$ and $w_a + w_r = 1$. The exact values of $w_a$ and $w_r$ will be decided by the organization's policies. We use these weights to compute the property value $\mathcal{P}_{ij}$ of user $U_j$ with respect to task $T_i$: $\mathcal{P}_{ij} = w_a * \mathcal{A}_{ij} + w_r * \mathcal{R}_{ij}$

### 3.2    Quantifying Experience

Experience constitutes an important factor in delegation. A delegator is more likely going to choose a candidate as a delegatee if the delegatee has prior experience of doing the task. Two factors contribute towards experience. One factor is when the task was performed, and the second factor is how well the task was performed. Note that, information about these factors is stored in the users' profile, $UP$. Events that have occurred in the recent past have more influence than that occurred in the distant past. To accommodate this, we give the most recent slot has the highest weight and the most distant slot has the lowest one.

---

**Algorithm 1.** Measuring Experience

---

**Input**: No. of slots $n$, User Profile $UP_j$
**Output**: $\mathcal{P}_{ij}$
**Procedure**:

   $performance = 0$
   **for all** $k : 1 \leq k \leq n$ **do**
     $weight\_slot_k = k$
   **end for**
   $total\_weight = n(n+1)/2$
   **for all** $k : 1 \leq k \leq n$ **do**
     $w_k = (2*k)/(n(n+1))$
   **end for**
   **for all** $k : 1 \leq k \leq n$ **do**
     $experience = experience + w_k * p_k$
   **end for**
   RETURN $experience$

---

For each time slot $t_k$, we get the value for performance $p_i$. Recall that, performance on the task measures how well the task has been performed. The performance on the task can be graded on a scale of $[0,1]$. A value closer to 0 indicates poor performance, while that closer to 1 indicates excellent performance. Not performing the task in a slot, gives a performance value equal to 0. Algorithm 1 shows how to assign weights to the various time slots and evaluate the experience. Sometimes the past experience may not exactly match the the task, but is related to it. We show how to extrapolate the trust value in such cases in Section 5.

### 3.3 Quantifying Recommendation

A truster may obtain recommendation from one or more recommenders about the trustee with respect to its ability to perform the given task. In order to quantify the recommendation obtained from each recommender, we need to evaluate two factors. First, we need to obtain the trust value that the truster has with respect to the recommender providing recommendation about the trustee with respect to the given task. If the recommender is sufficiently trusted, then we need to get from him the recommendation value for the trustee. Algorithm 2 shows how to compute the recommendation component.

### 3.4 Computing Trustworthiness

Trust, with respect to a given task $T_i$ for user $U_j$, denoted by $\mathcal{T}_{ij}$, depends on three factors, namely, properties $\mathcal{P}_{ij}$, experiences $\mathcal{E}_{ij}$, and recommendations, $\mathcal{R}_{ij}$. The exact weight assigned to each factor will be decided by the organization. Let $w_p$, $w_e$, and $w_r$ be the weights assigned to the three factors respectively where $w_p, w_e, w_r \in [0, 1]$ and $w_p + w_e + w_r = 1$. $\mathcal{T}_{ij}$ is given by, $\mathcal{T}_{ij} = w_p * \mathcal{P}_{ij} + w_e * \mathcal{E}_{ij} + w_r * \mathcal{R}_{ij}$. Note that $\mathcal{T}_{ij}$ will evaluate to some value in the range $[0,1]$. The

**Algorithm 2.** Measuring Recommendation

**Input:** Sequence of recommendations for user $U_j = < r_{1j}, r_{2j}, \ldots, r_{mj} >$, sequence of trust values for recommenders $= < t_1, t_2, \ldots, t_m >$
**Output:** $\mathcal{R}_{ij}$
**Procedure:**
    $reco = 0; total = 0$
    **for all** $k : 1 \leq k \leq m$ **do**
      $reco = reco + t_k * r_{kj}$
    **end for**
    **for all** $k : 1 \leq k \leq m$ **do**
      $total = total + t_k$
    **end for**
    $reco = reco/total$
    RETURN $reco$

delegator can choose a threshold value for trust $\mathcal{H}$. If $\mathcal{H} \leq \mathcal{T}_{ij}$, then user $U_j$ can be a potential delegatee.

# 4 Using Trust Values in Delegation Chains

The privilege that a user receives can be further delegated resulting in what is known as a delegation chain. In some cases, we may want to limit the level of delegation. This level of delegation can be decided by the trustworthiness of the users involved in the delegation chain. Thus, delegation chain is dependent on the concept of trust chains. Trust chains are formalized using the concept of trust graphs defined below.

[**Trust Graph**]: Let $TG = < N, E >$ be the directed acyclic graph that represents trust relationship for a given context. The set of nodes $N$ correspond to the entities in the system, and the set of edges $E$ represent the trust relationship between the nodes. The edge $(n_i, n_j)$ represents the trust relationship that node $n_i$ has for node $n_j$ with respect to the given task. The weight of the edge, denoted by $w(n_i, n_j)$, where $0 < w(n_i, n_j) \leq 1$, represents the trust value that node $n_i$ has with respect to node $n_j$. Note that, the absence of a trust relationship between nodes $n_r$ and $n_s$ is indicated by the missing edge $(n_r, n_s)$.

Given a trust graph, we define two types of operators to compute transitive trust. One is the sequential operator, and the other is the parallel operator. Sequential and parallel operators and their desirable properties have been proposed by Agudo et al. [26].

[**Sequential Operator**]: Sequential operator, denoted by $\otimes$, is a binary operator that takes as input two trust values and returns a trust value that is the product of the two input values. Formally, $\otimes : [0,1] \times [0,1] \to [0,1]$. The sequential operator is used for computing the transitive trust value in a single path in the trust graph. Algorithm 3 gives the description of how transitive trust is computed. For instance, to compute the transitive trust that $D$ has about $F$ with respect to the given context is the product of 0.2 and 0.6 which equals 0.12.

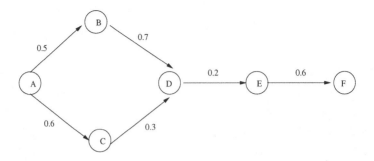

**Fig. 1.** Example of a Trust Graph

---

**Algorithm 3.** Computing Transitive Trust in a Single Path

---

**Input**: Trust Path $(n_1, n_2, \ldots, n_k)$
**Output**: Transitive trust between nodes $n_1$ and $n_k$
**Procedure**:

    $trans\_trust = 0$
    **for all** $i : 1 \leq i \leq (k-2)$ **do**
        $trans\_trust = trans\_trust * w(n_i, n_{i+1}) \bigotimes w(n_{i+1}, n_{i+2})$
    **end for**
    RETURN $trans\_trust$

---

The sequential operator is not adequate for calculating transitive trust when multiple paths are involved. For example, in Figure 1, computing transitive trust that $A$ has about $E$ using the path $(A, B, D, E)$ gives a different value than that obtained using the path $(A, C, D, E)$. The value is 0.07 for the path $(A, B, D, E)$ and it is 0.036 for the path $(A, C, D, E)$. Such differences are reconciled using the parallel operator. The parallel operator becomes useful when there are multiple paths from one node to another.

**[Parallel Operator]**: Parallel operator, denoted by $\bigoplus$, is a binary operator that takes as input two trust values and returns a trust value that is the minimum of the two input values. Formally, $\bigoplus : [0, 1] \times [0, 1] \rightarrow [0, 1]$. Algorithm 4 shows how to compute transitive trust when the source and destination are connected by parallel paths. The transitive trust that $A$ has for $D$, computed using this algorithm, equals 0.18.

The delegator can specify an acceptable level of trust to support delegation chains. Delegation is disallowed if the transitive trust value computed from the chain of delegation is below this minimum threshold.

## 5   Extrapolating Trust Values

Sometimes the delegator may not have enough information to assess the trustworthiness of a user with respect to some given task. Although the user is not

**Algorithm 4.** Computing Transitive Trust in the Presence of Multiple Paths

**Input:** Trust Paths $(n_1, n_{2_1}, \ldots, n_{(k-1)_1}, n_k)$, $(n_1, n_{2_2}, \ldots, n_{(k-1)_2}, n_k)$, $\ldots$, $(n_1, n_{2_j}, \ldots, n_{(k-1)_j}, n_k)$

**Output:** Transitive trust between nodes $n_1$ and $n_k$

**Procedure:**

    min = 1;
    **for all** $l : 1 \leq l \leq j$ **do**
       $trans\_trust_l = 0$
    **end for**
    **for all** $l : 1 \leq l \leq j$ **do**
       **for all** $i : 1 \leq i \leq (k-2)$ **do**
          $trans\_trust_l = trans\_trust_l + w(n_i, n_{i+1}) \bigotimes w(n_{i+1}, n_{i+2})$
       **end for**
    **end for**
    **for all** $l : 1 \leq l \leq j$ **do**
       **if** $trans\_trust_l < min$ **then**
          $min = trans\_trust_l$
       **end if**
    **end for**
    RETURN $min$

associated with a given task, it is possible that he has done some related tasks. To handle such scenarios, we define the different relationships that can exist among the tasks in an organization.

### Specialization Relation

Different tasks may be related by the generalization/specialization relationship which is anti-symmetric and transitive. We use the notation $T_i \subset T_j$ to indicate that task $T_i$ ($T_j$) is a generalization (specialization) of task $T_j$ ($T_i$). For instance, *Surgery Treatment* $\subset$ *Heart Bypass Surgery Treatment* and *Heart Treatment* $\subset$ *Heart Bypass Surgery Treatment*. However, the degree of specialization is different in the two cases. The *degree of specialization* captures this difference. The degree of specialization is denoted as a fraction whose value is determined using domain knowledge.

### Composition Relation

Sometimes tasks can be linked together using the composition relation. A task can either be *elementary* or *composite*. An elementary task is one which cannot be subdivided into other tasks, whereas a composite task is one that is composed from other tasks. The individual tasks that form a composite one are referred to as the *component* tasks. A component task can either be composite or elementary. We use the notation $T_i \ll T_j$ to indicate that the task $T_i$ is a component of task $T_j$. For instance, we may have the component tasks *operation* and *medication* that are part of the composite task *Catheter-assisted Procedures*. This is denoted as *operation* $\ll$ *Catheter-assisted Procedures*.

Sometimes a composite task $T_i$ may be composed from the individual tasks $T_j$, $T_k$ and $T_m$. All these tasks may not contribute equally to form $T_i$. The *degree of composition* captures this idea. A degree of composition is associated with each composition relation. Since two tasks related by composition will not be exactly identical, the degree of composition is denoted as a fraction. The sum of all these fractions equals one if $T_i$ is composed of $T_j$, $T_k$, and $T_m$ only. If $T_i$ is composed of $T_j$, $T_k$, and $T_m$ and also other component contexts, then the sum of fractions associated with $T_j$, $T_k$, and $T_m$ must be equal to or less than one. The exact value of the fraction representing the degree of composition will be determined by domain knowledge.

The generalization/specialization and composition relations are formally specified using the notion of *task graphs* defined below.

[**Task Graph**]: A task graph $TG = \langle N, \mathcal{E}_c \cup \mathcal{E}_s \rangle$ is a weighted directed acyclic graph satisfying the following conditions.

- $N$ is a set of nodes where each node $n_i$ is associated with a task $T_i$.
- The set of edges in the graph can be partitioned into two sets $\mathcal{E}_c$ and $\mathcal{E}_s$. For each edge $(n_i, n_j) \in \mathcal{E}_c$, the task $T_i$ corresponding to node $n_i$ is a component of the task $T_j$ corresponding to node $n_j$. The weight of the edge $(n_i, n_j)$, denoted by $w(n_i, n_j)$, indicates the percentage of component task that makes up the composite one. For each edge $(n_i, n_j) \in \mathcal{E}_s$, the task $T_i$ corresponding to node $n_i$ is a specialization of task $T_j$ corresponding to node $n_j$. The weight of the edge $(n_i, n_j)$, denoted by $w(n_i, n_j)$, indicates the degree of specialization.

## 5.1   Computing the Degree of Specialization and Composition

Consider two tasks $T_i$ and $T_j$ where $T_i \subset T_j$, that is, $T_j$ is a specialization of $T_i$. The degree of specialization is computed as follows. Let $n_i$, $n_j$ be the nodes corresponding to tasks $T_i$ and $T_j$ in the weighted graph. Let the path from $n_i$ to $n_j$ consisting of specialization edges be denoted as $(n_i, n_{i+1}, n_{i+2}, \ldots, n_{j-1}, n_j)$. The degree of specialization $= \Pi_{p=i}^{j-1} w(n_p, n_{p+1})$. This corresponds to our notion that the similarity decreases as the length of the path from the generalized node to the specialized node increases. Note that, in real world there may be multiple paths from $T_i$ to $T_j$. In such cases, it is important that the degree of specialization yield the same values when any of these paths are used for computation.

Consider two tasks $T_i$ and $T_j$ such that $T_j$ is a component of $T_i$. Degree of composition captures what portion of $T_i$ is made up of $T_j$. The degree of composition is computed as follows. Let $n_i$, $n_j$ be the nodes corresponding to contexts $T_i$ and $T_j$ in the task graph. Let there be $m$ paths consisting of composition edges from $n_i$ to $n_j$. Let the $q$th path ($1 \leq q \leq m$) from $n_i$ to $n_j$ be denoted as $(n_i, n_{i_q+1}, n_{i_q+2}, \ldots, n_{j_q-1}, n_j)$. The degree of composition $= \Sigma_{q=1}^{m}(w(n_i, n_{i_q+1}) \times w(n_{j_q-1}, n_j) \times \Pi_{p=i_q+1}^{j_q-2} w(n_p, n_{p+1}))$.

## 6   Trust Computation for Example Application

Consider a small healthcare organization that has six roles, namely, *senior doctor, junior doctor, cardiologist, surgeon, physician's assistant* and *patient. senior doctor* is senior to *junior doctor*, and *junior doctor* is senior to *cardiologist* and *physician's assistant*. Allen and Miller are assigned to *senior doctor*, Bell and Nelson are assigned to *junior doctor*, Cox is assigned to *cardiologist*, and Davis is assigned to *physician's assistant*. Allen is also assigned to *surgeon* and Evans is assigned to *patient*. Allen is the assigned surgeon for performing Coronary Artery Disease Angioplasty (CAD type A) surgery on patient Evans. Since Allen has to leave town for family emergency, he must delegate the surgeon role to another doctor. He cannot delegate the surgeon role to his two trusted colleagues, Miller and Nelson, because they will be on vacation. The hospital policy requires that a person assigned to a doctor role or senior can be delegated the role of surgeon. This rules out Davis. Thus, he computes trust values for the only two viable candidates, Bell and Cox.

**Quantifying Properties:** To perform the CAD type A surgery, the hospital requires the following attributes from the candidates. First, the candidate should be a doctor ($a_1 = doctor$) and he should be able to perform a CAD type A surgery ($a_2 = Surgery_A$). So, $TA = \{doctor, Surgery_A\}$. The hospital policy ranks the ability to perform a CAD type A surgery higher than the doctor position, so the policy administrator assigned $w_{Surgery_A} = 0.7$ and $w_{doctor} = 0.3$. The hospital administrator assigned the value of closeness equal to 0.6 between roles *Senior Doctor* and *Junior Doctor* ($dist(Senior\ Doctor,\ Junior\ Doctor)$=0.6), and that between roles *Junior Doctor* and *Cardiologist* equals 0.3 ($dist(Junior\ Doctor,\ Cardiologist)$=0.3). Hence, by using the computation method explained in Section 5, we get the value of closeness between role *Senior Doctor* and *Cardiologist* equals to $0.6*0.3 = 0.18$ ($dist(Senior\ Doctor,\ Cardiologist)$=0.18). The hospital policy ranks the importance of necessary attributes and role attributes equally, hence $w_a = w_r = 0.5$.

Now, we quantify the properties of both candidates. Bell is a doctor who can perform the CAD type A surgery ($UA_{Bell} = \{doctor, Surgery_A\}$), and Cox is a cardiologist who can perform a bypass surgery ($UA_{Cox} = \{cardiologist, Surgery_B\}$). So, $A_{Bell} = w_{Surgery_A} + w_{doctor} = 0.7 + 0.3 = 1$ and $A_{Cox} = w_{doctor} = 0.3$. Since Bell is a junior doctor, $R_{Bell} = dist\ (Senior\ Doctor,\ Junior\ Doctor)$=0.6. Since Cox is a cardiologist, $R_{Cox} = dist\ (Senior\ Doctor,\ Cardiologist)$=0.18.

Using this information, we calculate the properties value of the candidates:

$P_{Bell} = w_a * A_{Bell} + w_r * R_{Bell} = 0.5 * 1 + 0.5 * 0.6 = 0.8$, and
$P_{Cox} = w_a * A_{Cox} + w_r * R_{Cox} = 0.5 * 0.3 + 0.5 * 0.18 = 0.24$.

**Quantifying Experience:** Here the experience is quantify based on the number of heart operations the candidates have done in the past five years and the unit of the slot of the time period is equal to one year. The weight for each time slot where $slot_1$ represents the time period closest to the present time is defined by policy as follow: $w_{slot_1} = 1, w_{slot_2} = 0.8, w_{slot_3} = 0.6, w_{slot_4} = 0.4$, and $w_{slot_5} = 0.2$. Bell has performed surgery once 300 days ago ($slot_1$) with performance 0.7 ($p_{Bell_{slot_1}} = 0.7$) and Cox has performed surgery once 700 days ago ($slot_2$ ) with performance 0.8 ($p_{Cox_{slot_2}} = 0.8$). Thus, the experience value of both candidates can be calculated as follow:

$\mathcal{E}_{Bell} = \sum_{i=1}^{5} w_{slot_i} * p_{Bell_{slot_i}} = 1 * 0.7 + 0 + 0 + 0 + 0 = 0.7$, and
$\mathcal{E}_{Cox} = \sum_{i=1}^{5} w_{slot_i} * p_{Cox_{slot_i}} = 0 + 0.8 * 0.8 + 0 + 0 + 0 = 0.64$.

**Quantifying Recommendation:** Here, we have two recommenders–Miller and Nelson. According to hospital policy, the recommendation coming from senior doctor is more trustworthy than the one coming from junior doctor. So, the administrator set the trust value that hospital has about Miller ($t_{Miller}$) to 0.8 and the trust value that hospital has about Nelson ($t_{Nelson}$) to 0.2. Miller recommendation for Bell ($r_{MillerBell}$) and Cox ($r_{MillerCox}$) are 0.4 and 0.6, respectively. Nelson recommendation for Bell ($r_{NelsonBell}$) and Cox ($r_{NelsonCox}$) are 0.9 and 0.2, respectively. The computation results yield the recommendation for Bell and Cox as follow:

$$\mathcal{R}_{Bell} = \frac{t_{Miller} * r_{MillerBell} + t_{Nelson} * r_{NelsonBell}}{t_{Miller} + t_{Nelson}} = \frac{0.8 * 0.4 + 0.2 * 0.9}{0.8 + 0.2} = 0.5,$$

and

$$\mathcal{R}_{Cox} = \frac{t_{Miller} * r_{MillerCox} + t_{Nelson} * r_{NelsonCox}}{t_{Miller} + t_{Nelson}} = \frac{0.8 * 0.6 + 0.2 * 0.2}{0.8 + 0.2} = 0.52.$$

**Computing Trustworthiness:** Allen prefers the delegatee with more experience. So, he set the weights for properties ($w_p$), experience ($w_e$), and recommendation ($w_r$) to 0.2, 0.6, and 0.2, respectively. The trustworthiness of Bell and Cox can be computed as follow:

$$\mathcal{T}_{Bell} = w_p * \mathcal{P}_{Bell} + w_e * \mathcal{E}_{Bell} + w_r * \mathcal{R}_{Bell} = 0.2 * 0.8 + 0.6 * 0.7 + 0.2 * 0.5 = 0.68,$$

and

$$\mathcal{T}_{Cox} = w_p * \mathcal{P}_{Cox} + w_e * \mathcal{E}_{Cox} + w_r * \mathcal{R}_{Cox} = 0.2 * 0.24 + 0.6 * 0.64 + 0.2 * 0.52 = 0.54$$

Bell is selected to be the delegatee after comparing the trustworthiness values between both candidates.

## 7    Model Analysis

Once we have determined the most trustworthy candidate, we need to formally ensure that the choice of this delegatee does not cause a security breach. We do the formal analysis using the Alloy Analyzer. An Alloy model consists of *signature* declarations, *fields*, *facts* and *predicates*. Each signature consists of a set of *atoms* which are the basic entities in Alloy. Atoms are *indivisible* (they cannot be divided into smaller parts), *immutable* (their properties do not change) and *uninterpreted* (they do not have any inherent properties). Each field belongs to a signature and represents a relation between two or more signatures. A relation denotes a set of tuples of atoms. Facts are statements that define constraints on the elements of the model. Predicates are parameterized constraints that can be invoked from within facts or other predicates.

The basic types in the access control model, such as, *User*, and *Role* are represented as signatures. For instance, the declarations shown below define a set named *User*, and a set named *Role* that represents the set of all users, and roles in the system. Note that we use the *abstract* signature to represent these sets, and the different of users, and roles are modeled as the subsignatures of each signature. The analyzer will then recognize that users, and roles consist of only these different types, and nothing else.

```
abstract sig User{}
one sig Allen, Bell, Cox, Davis, Evans,
         Miller, Nelson extends User{}

abstract sig Role{}
one sig SeniorDoctor, JuniorDoctor, Assistant,
         Cardiologist, Surgeon, Patient extends Role{}
```

The different relationships between the RBAC components are also expressed as signatures. Signature *UserRoleAssign* which represents the roles assigned to user has a field called *URAsmember* that maps to a cartesian product of *User* an d *Role*. Signature *UserRoleAcquire* which represents the roles user can acquire through the assignment and role hierarchy has a field called *URAcqmember* that maps to a cartesian product of *User* and *Role*. We use the signature *RoleHierarchy* to represent role hierarchy relationship.

```
one sig UserRoleAssign{URAsmember: User -> Role}
one sig UserRoleAcquire{URAcqmember: User -> Role}
one sig RoleHierarchy{RHmember : Role -> Role}
```

The various invariants in the RBAC model are represented as facts in Alloy. For instance, the fact *URAcq* states that user can acquire all roles assign ed to him together with all of his junior roles. This is specified in Alloy as shown below. Other invariants are modeled in a similar manner.

```
fact URAcq{
UserRoleAcquire.URAcqmember =
UserRoleAssign.URAsmember +
(UserRoleAssign.URAsmember).^(RoleHierarchy.RHmember)}
```

The policy constraints are modeled as predicates. First, consider the cardinality constraint. The following constraint says that role $r$ can be assigned to only one user.

```
pred Cardinality(r: Role, uracq: User->Role){
       (#((uracq).r) >= 1) &&
       (#((uracq).r) <= 1)}
```

Next, consider the prerequisite constraint that says that if a user $u$ can acquire role $r1$, then he can also acquire role $r2$. The other forms are modeled in a separate manner.

```
pred Prerequisite(u:User, r1, r2: Role,
uracq: User->Role){
    (u->r2 in uracq) => (u->r1 in uracq)}
```

The separation of duty constraint says that if a user $u$ can acquire role $r1$, then he cannot acquire the conflicting role $r2$.

```
pred SoD(u:User, r1, r2: Role, uracq: User->Role){
    (u->r1 in uracq) => not (u->r2 in uracq)}
```

The different types of delegation are also modeled as predicates. The grant and transfer operation can be modeled as follows:

```
pred Grant[u: User, r: Role,
   uracq, uracq': User->Role]{
     uracq' = uracq + (u->r)}

pred Transfer[u1, u2: User, r: Role,
   uracq, uracq': User->Role]{
     uracq' = uracq + (u2->r) - (u1->r)}
```

Finally, we need to verify whether the selected delegatee could cause any security policy violation. We create an *assertion* that specifies the properties we want to check. After we create the assertion, we will let ALLOY analyzer validate the assertion by using *check* command. If our assertion is wrong in the specified scope, ALLOY analyzer will show the counterexample. For example, suppose we want to check whether separation of duty constraint is violated when Allen delegates his role to Bell. The assertion below will check whether the separation of duty constraint is violated after the transfer operation. The separation of duty constraint says that user cannot be assigned both *Assistant* and *Surgeon* roles. The counterexample illustrates that even though user *Bell* is not assigned to *Assistant* role, he can still acquire it from the effect of role hierarchy.

```
assert TestConflict3{
    all u1, u2: User, r: Role, uracq, uracq': User->Role|
        ((u1 = Allen) && (u2 = Bell) && (r=Surgeon) &&
        (uracq = UserRoleAcquire.URAcqmember) &&
        (u1->r in UserRoleAcquire.URAcqmember) &&
        (u2->Assistant not in UserRoleAssign.URAsmember) &&
        Transfer[u1, u2, r, uracq, uracq']) =>
           SoD[u2, r, Assistant, uracq']}
check TestConflict3
```

The result shown that, although Bell is the most trustworthy candidate, we cannot choose him as Allen's delegatee. Next, we verify the situation where Cox, another candidate with the lower trustworthiness, is chosen as the delegatee. The assertion below will check whether the separation of duty constraint is violated after the transfer operation.

```
assert TestConflict4{
    all u1, u2: User, r: Role, uracq, uracq': User->Role|
        ((u1 = Allen) && (u2 = Cox) && (r=Surgeon) &&
```

```
(uracq = UserRoleAcquire.URAcqmember) &&
(u1->r in UserRoleAcquire.URAcqmember) &&
(u2->Assistant not in UserRoleAssign.URAsmember) &&
Transfer[u1, u2, r, uracq, uracq']) =>
      SoD[u2, r, Assistant, uracq']}
check TestConflict4
```

Here, the analyzer cannot find the counterexample, which means the separation of duty constraint defined in the model is not violated. This indicates that Cox is a more suitable delegatee for Allen.

## 8 Conclusion and Future Work

Delegation gives temporary privilege to one or more users, that allows critical tasks to be completed. We provide a formal approach for choosing delegatees. The approach evaluates the trustworthiness of candidates, and then ensures that the chosen candidate does not cause a security breach. We also illustrate how trustworthiness can be used to decide on the length of the delegation chain. A lot of work remains to be done. The first work is with regards to implementing the model such that trust computation can be done in an efficient manner. The second is with respect to validating the model in the context of real-world applications. The results of this validation can be further used to refine the model.

## References

1. Barka, E., Sandhu, R.S.: A Role-Based Delegation Model and Some Extensions. In: Proceedings of the 23rd National Information Systems Security Conference (2000)
2. Joshi, J., Bertino, E.: Fine-grained role-based delegation in presence of the hybrid role hierarchy. In: Proceedings of the 11th ACM Symposium on Access Control Models and Technologies, pp. 81–90 (2006)
3. Joshi, J., Bertino, E., Latif, U., Ghafoor, A.: A generalized temporal role-based access control model. IEEE Trans. Knowl. Data Eng. 17(1), 4–23 (2005)
4. Zhang, L., Ahn, G.J., Chu, B.: A rule-based framework for role based delegation. In: Proceedings of the 6th ACM Symposium on Access Control Models and Technologies, pp. 153–162 (2001)
5. Wang, Q., Li, N., Chen, H.: On the security of delegation in access control systems. In: Jajodia, S., Lopez, J. (eds.) ESORICS 2008. LNCS, vol. 5283, pp. 317–332. Springer, Heidelberg (2008)
6. Zhang, X., Oh, S., Sandhu, R.S.: PBDM: A Flexible Delegation Model in RBAC. In: Proceedings of the 8th ACM Symposium on Access Control Models and Technologies, pp. 149–157 (2003)
7. Jackson, D.: Alloy 3.0 reference manual (2004),
   http://alloy.mit.edu/reference-manual.pdf
8. Georg, G., Bieman, J., France, R.B.: Using Alloy and UML/OCL to Specify Run-Time Configurati on Management: A Case Study. In: Practical UML-Based Rigorous Development Methods - Countering or Integrating the eXtremists. LNI, vol. P-7, pp. 128–141. German Informatics Society (2001)

9. Taghdiri, M., Jackson, D.: A lightweight formal analysis of a multicast key management scheme. In: König, H., Heiner, M., Wolisz, A. (eds.) FORTE 2003, vol. 2767, pp. 240–256. Springer, Heidelberg (2003)
10. Crampton, J., Khambhammettu, H.: Delegation and satisfiability in workflow systems. In: Proceedings of the 13th ACM Symposium on Access Control Models and Technologies, pp. 31–40 (2008)
11. Wang, Q., Li, N.: Satisfiability and resiliency in workflow systems. In: Proceedings of the 12th European Symposium on Research in Computer Security, pp. 90–105 (2007)
12. Jøsang, A.: Artificial reasoning with subjective logic. In: Proceedings of the 2nd Australian Workshop on Commonsense Reasoning (1997)
13. Jøsang, A.: An algebra for assessing trust in certification chains. In: Proceedings of the Network and Distributed Systems Security Symposium (1999)
14. Jøsang, A., Bhuiyan, T.: Optimal trust network analysis with subjective logic. In: Proceedings of the Second International Conference on Emerging Security Information, Systems and Technologies, pp. 179–184 (2008)
15. Jøsang, A., Gray, E., Kinateder, M.: Simplification and analysis of transitive trust networks. Web Intelligence and Agent Systems 4(2), 139–161 (2006)
16. Agudo, I., Lopez, J., Montenegro, J.A.: Enabling Attribute Delegation in Ubiquitous Environments. Mobile Networks and Applications 13(3-4), 398–410 (2008)
17. Chakraborty, S., Ray, I.: TrustBAC: integrating trust relationships into the RBAC model for access control in open systems. In: Proceedings of the 11th ACM Symposium on Access Control Models and Technologies, pp. 49–58 (2006)
18. Cruz, I.F., Gjomemo, R., Lin, B., Orsini, M.: A location aware role and attribute based access control system. In: Proceedings of the 16th ACM SIGSPATIAL International Symposium on Advances in Geographic Information Systems, p. 84 (2008)
19. Damiani, E., di Vimercati, S.D.C., Samarati, P.: New paradigms for access control in open environments. In: Proceedings of the 5th IEEE International Symposium on Signal Processing and Information Technology, pp. 540–545 (2005)
20. Li, N., Mitchell, J.C., Winsborough, W.H.: Design of a role-based trust-management framework. In: Proceedings of the 2002 IEEE Symposium on Security and Privacy, pp. 114–130 (2002)
21. Priebe, T., Dobmeier, W., Kamprath, N.: Supporting attribute-based access control with ontologies. In: Proceedings of the 1st International Conference on Availability, Reliability and Security, pp. 465–472 (2006)
22. Ray, I., Ray, I., Chakraborty, S.: An interoperable context sensitive model of trust. Journal of Intelligent Information Systems 32(1), 75–104 (2009)
23. Wang, L., Wijesekera, D., Jajodia, S.: A logic-based framework for attribute based access control. In: Proceedings of the 2004 ACM Workshop on Formal Methods in Security Engineering, pp. 45–55 (2004)
24. Zao, J., Wee, H., Chu, J., Jackson, D.: RBAC Schema Verification Using Lightweight Formal Model and Constraint Analysis (2002), http://alloy.mit.edu/publications.php
25. Schaad, A., Moffett, J.D.: A Lightweight Approach to Specification and Analysis of Role-Based Access Control Extensions. In: Proceedings of the 7th ACM Symposium on Access Control Models and Technologies, pp. 13–22 (2002)
26. Agudo, I., Gago, M.C.F., Lopez, J.: A model for trust metrics analysis. In: Furnell, S.M., Katsikas, S.K., Lioy, A. (eds.) TrustBus 2008. LNCS, vol. 5185, pp. 28–37. Springer, Heidelberg (2008)

# Secure Interoperation in Multidomain Environments Employing UCON Policies*

Jianfeng Lu[1], Ruixuan Li[1,**], Vijay Varadharajan[2], Zhengding Lu[1], and Xiaopu Ma[1]

[1] Intelligent and Distributed Computing Lab, College of Computer Sci. and Tech. Huazhong University of Sci. and Tech., Wuhan 430074, P.R. China
lujianfeng@smail.hust.edu.cn, rxli@hust.edu.cn,
{zdlu,xpma}@smail.hust.edu.cn
[2] Department of Computing, Macquarie University, NSW 2109, Australia
vijay@ics.mq.edu.au

**Abstract.** Ensuring secure interoperation in multidomain environments based on role based access control (RBAC) has drawn considerable research works in the past. However, RBAC primarily consider static authorization decisions based on subjects' permissions on target objects, and there is no further enforcement during the access. Recently proposed usage control (UCON) can address these requirements of access policy representation for temporal and time-consuming problems. In this paper, we propose a framework to facilitate the establishment of secure interoperability in multidomain environments employing Usage Control (UCON) policies. In particular, we propose an attribute mapping technique to establish secure context in multidomain environments. A key challenge in the establishment of secure interoperability is to guarantee security of individual domains in presence of interoperation. We study how conflicts arise and show that it is efficient to resolve the security violations of cyclic inheritance and separation of duty.

**Keywords:** Multidomain, interoperation, usage control, cyclic inheritance, separation of duty.

## 1 Introduction

Ensuring secure interoperation in multidomain environments based on role based access control (RBAC) has drawn considerable research works in the past [1]. Although RBAC [2] has become widely accepted as the principal type of access control model in theory and in practice, it primarily considers static authorization decisions based on subjects' permissions on target objects, and there is no further enforcement during the access. In recent information systems, the interactive and concurrent concepts should be introduced to access control. Obviously, RBAC model and other traditional access control model have difficulties,

---

* This work is supported by National Natural Science Foundation of China under Grant 60873225, 60773191 and 60403027, National High Technology Research and Development Program of China under Grant 2007AA01Z403.
** Corresponding author.

P. Samarati et al. (Eds.): ISC 2009, LNCS 5735, pp. 395–402, 2009.
© Springer-Verlag Berlin Heidelberg 2009

or lack the flexibility to specify these requirements. Recently proposed usage control (UCON) [3, 4] offers a promising approach for the next generation of access control, it can address these requirements of access policy representation for temporal and time-consuming problems.

The above observations motivate us to consider new secure Interoperation policy. In this paper, we employ attribute mapping techniques to propose an interoperation policy framework in multidomain environments based on UCON model. In this policy framework, parts of foreign subject attributes will be mapped to local attributes, once these associations are set up, all required foreign attributes are dynamically mapped to local attributes, and the authorization can be made based on these local attributes. A key challenge in the establishment of secure interoperability is to guaran-tee security of individual domains in presence of interoperation. This paper focuses on two types of security violations of cyclic inheritance and separation of duty (SoD). We study how these security violations arise and show that it is efficient to resolve them.

The rest of this paper is organized as follows. Section 2 proposes the attribute mapping technique for secure interoperation framework. Section 3 studies how security violations arise and show that it is efficient to resolve these security violations. Some related work in interoperation are reviewed in Section 4. Finally, Section 5 concludes this paper.

## 2    Attribute Mapping Technique for Interoperation Policy

Zhang et al. [5] present an example motivating the new features of UCON. As the access control of this motivating example is not a simple action, the authorization decisions are not only based on subjects' permissions on target objects, but also need further enforcement during the access. In this way, traditional access control models lack the flexibility to specify policies in these scenarios, UCON is the preferred policy. However, this example does not fit in with multidomain environments. In multidomain environments, many types of user attributes' semantics cannot be interpreted across multiple domains, the first and foremost problem is to interpret these attributes across multiple domains. We now identify a complete taxonomy of attributes.

Attributes can be classified into different categories based on different items. Firstly, we classify attributes based on available scope as follows.

***Localdomain attributes:*** This type of attributes is defined in a domain whose semantics can be interpreted only within local domain, but has no meaning or visibility in other domains.

***Multidomain attributes:*** Comparing with localdomain attributes, multidomain attributes' semantics can be interpreted across multiple domains.

Secondly, we classify attributes based on liveness as follows.

***Temporary attributes:*** Temporary attributes are created at the time a usage is started and deleted at the end of a single usage.

*Persistent attributes:* Persistent attributes live longer for multiple usage decisions.

Thirdly, we classify attributes based on whether the attributes can be updated during the usage process as follows.

*Mutable attributes:* Mutable attributes can be modified by the system automatically and do not require any administrative actions for update.

*Immutable attributes:* Immutable attributes cannot be changed by the subject's activity. Only administrative actions can change it.

The secure interaction between two or more administrative domains motivates the need for attributes translations that foreign attributes can be interpreted and understandable to local entities. In multidomain interaction scenario, only parts of attributes need to be translated. Firstly, multidomain attributes' semantics can be interpreted across multiple domains. Secondly, temporary attributes are alive only for a single usage. Therefore, we only need to establish a flexible policy for dynamic LPM (*localdomain persistent mutable*) and LPI (*localdomain persistent immutable*) attributes mapping to make interoperation in two domains employing UCON policies, and then the communications between two domains are mainly created by attribute mapping technique. The characterize definition about attribute mapping is as follows.

**Definition 1.** *Attribute Mapping: The attribute mapping is formalized as a 5-tuple: $< a_1, D_1, a_2, D_2, m >$, $a_1$ is an attribute in domain $D_1$, and $a_2$ is an attribute in domain $D_2$ respectively. In general, $D_1$ is the foreign domain, and $D_2$ is the local domain. The fifth parameter $m$ is the mapping modes $\mapsto_{LPM}$ or $\mapsto_{LPI}$, which denotes the association of the two attributes $a_1$ and $a_2$. $\mapsto_{LPM}$ denotes that LPM attributes from the foreign domain $D_1$ will be translated to local domain $D_2$. $\mapsto_{LPI}$ implies that LPI attributes from the foreign domain $D_1$ will be mapped to local domain $D_2$.*

For the $\mapsto_{LPM}$ mappings, let $\Gamma$ be a set which includes the LPM attributes mapped from foreign domain to local domain, let $\Gamma'$ be a set which includes the LPM attributes from local domain associated with $\Gamma$, $\mapsto_{LPM} : \Gamma \to \Gamma'$ is a function from $\Gamma$ to $\Gamma'$, then $\mapsto_{LPM}$ obviously is a monotone increasing function. And for the $\mapsto_{LPI}$ mappings, let $\Gamma$ be a set which includes the LPI attributes from two interoperate domains, and $\mapsto_{LPI}$ be a binary relation on $\Gamma$. Obviously, $\mapsto_{LPI}$ associates the LPI attributes, and these associations form a combined hierarchy that is partially ordered on $\Gamma$.

## 3   Security Issues for Attribute Mappings

A key challenge in the establishment of secure interoperability is to guarantee security of individual domains in presence of interoperation. There are many types of security violations leaded by establishing an interoperation policy among heterogeneous systems. These violations may arise because different domains may adopt different models, semantics, schema format, data labeling schemes, and constraints for representing their access control policies [6, 7, 8]. This section focuses on two types of security violations: cyclic inheritance, and SoD. We study how these security violations arise and show that it is efficient to resolve them.

### 3.1 Violations of Cyclic Inheritance

Violations of cyclic inheritance mainly occur in interoperation of systems employing multilevel security policies, such as lattice-based access control (LBAC) and role-based access control (RBAC) [8, 9]. The cross-domain hierarchy relationship may introduce a cycle in the interoperation lattice enabling a subject lower in the access control hierarchy to assume the permissions of a subject higher in the hierarchy.

**Definition 2.** *A cyclic inheritance violation is expressed as*

$$\exists (a_i, a_j) \in A \times A, (b_k, b_l) \in B \times B \left( (a_j \mapsto_{\text{LPI}} b_k) \wedge (b_l \mapsto_{\text{LPI}} a_i) \right) \Rightarrow (a_j, a_i)$$

*where $A = a_1, \ldots, a_m$, $B = b_1, \ldots, b_n$, $i$, $j$, $k$, $l$, $m$ and $n$ are integers, such that $1 \leq i \neq j \leq m$, $1 \leq k \neq l \leq n$. Each $a_i$ is an attribute in attribute set $A$, and $b_k$ is an attribute in attribute set $B$. $A$ and $B$ are two different domains. The notation $(a_i, a_j)$ is a two-tuples, which means that the attribute $a_i$ is the ancestor of $a_j$.*

Cyclic inheritance usually arises from the circulation in the $I - hierarchy$. There are four cases of cyclic inheritance: $a_i$ is a direct or indirect ancestor of $a_j$, and $b_k$ is a direct or indirect ancestor of $b_l$. It is noted that $a_i$ is an ancestor of itself. Combine with the above cases also can generate other sub cases. A cyclic inheritance causes an attribute to inherit its senior attribute, as get all senior attributes and all junior attributes of a given attribute is tractable.

**Theorem 1.** *The checking problem for violations of cyclic inheritance is in* **P**.

*Proof.* One algorithm for detecting problem for cyclic inheritance violations is as follows. For each attribute $a$ in a domain A, one first computes all senior attributes and junior attributes of $a$, includes the $I - hierarchies$ and LPI attribute mappings. Then compares these two sets of attributes, if the intersection is not empty, there exists at least one cyclic inheritance violations, otherwise not. This algorithm has a time complexity of $O(Na(Na + Nm).$, where $Na$ is the number of LPI attributes with $I - hierarchy$, $Nm$ is the number of LPI attribute mappings. □

### 3.2 Violations of Separation of Duty

SoD is widely considered to be a fundamental principle in computer security [10]. Violations of SoD constraints may occur in an interoperation policy because of the interplay of various policy constraints across domains. When a sensitive task is comprised of $m$ permissions, an SSoD policy requires the cooperation of at least $n$ (for some $2 \leq\leq m$) different users to complete the task. In other words, there shouldn't exist a set of fewer than $n$ users that together have all the $m$ permissions to complete the sensitive task. We now formally define the SSoD violation.

**Definition 3.** *An SSoD violation is expressed as*

$$\forall \{u_1, \ldots, u_{n-1}\} \subseteq U \left( \bigcup_{i=1}^{n-1} Auth_P(u_i) \supseteq \{p_1, \ldots, p_m\} \right)$$

*where m and n are integers, such that $1 \leq n \leq m$, $\{p_1, \ldots, p_m\}$ is the set of all possible permissions. $Auth_P : U \rightarrow 2^P$ is a function, where $U$ is the user set, and $2^P$ is the power set of permissions.*

In the literature on RBAC, statically mutually exclusive roles (SMER) constraints are used to enforce SSoD policies [2]. In RBAC model, permissions are assigned to roles. But the role is only a special type of subject attribute in UCON model, and there are many types of subject attributes in UCON model as shown in section 2.2, which play a very important role on the authorization based decision, that makes SMER constraints not suit to UCON policy. Consequently, we formally define two types of statically mutually exclusive attributes (SMEA) in UCON.

**Definition 4.** *An SD-SMEA (single-dimensional statically mutually exclusive attributes) constraint is expressed as*

$$\forall u \in U \left( |ATT(u) \cap AS| < n \right)$$

*where $AS = a_1, \ldots, a_m$ be the set of all mutually exclusive attributes where each $a_i$ is an attribute, m and n are integers, such that $2 \leq n \leq m$, each $a_i$ is an attribute, $ATT : U \rightarrow 2^A$ is a function, where $U$ is the user set, and $2^A$ is the power set of attributes. SD-SMEA constraint means that no user is a member of n or more attributes in AS.*

The SD-SMEA is a general form, SMER is an instance of SD-SMEA where the $ATT(u)$ is a role set that assigned to a user: $ATT(u) = \{r \in R[(u, r_1) \in UA \wedge (r_1, r) \in RH]\}$

For example, as shown in Fig.1, $AS = \{Junior - Member, Rookie\}$, $n = 2$ in $(AS, n)$, from the example,

$ATT(u_1) = \{Administrator, Teacher, Student, Junior - Member, Rookie\}$,
$ATT(u_2) = \{Teacher, Junior - Member\}, ATT(u_3) = \{Student, Rookie\}$,
$|ATT(u_1) \cap AS| = 2 = n, |ATT(u_2) \cap AS| = 1 < n, |ATT(u_3) \cap AS| = 1$

Therefore, this example violates the SD-SMEA constraint as $|ATT(u_1) \cap AS| = 2 = n$ violates Definition 4.

**Definition 5.** *A MD-SMEA (multi-dimensional statically mutually exclusive attributes) constraint is expressed as*

$$\forall u \in U \{\forall as_i \in AS \left( |ATT(u) \cap as_i| < n_i \right)\}$$

*where $AS = \{as_1, \ldots, as_m\}$ be the set of mutually exclusive attributes sets where each $as_i = \{as_{i1}, as_{i2}, as_{i3} \ldots\}$, each element in $as_i$ is an attribute, the corresponding $n_i$ is integer in the integer array $N = \{n_1, \ldots, n_m\}$, m is an integer,* such that $1 < n_i \leq |AS_i|$, Each $a_i$ is an attribute, $(AS, N)$ is a two-tuples, it means that no user is a member of $n_i$ or more attributes in $as_i$ for every $1 \leq i \leq m$. ATT is the same meaning with Definition 4. MD-SMEA also includes mutable attributes. Let $ATT(u)$ is the value of mutable attributes, $as_i$ is an interval, and $n_i = 1$ ($n_i$ can be any integer which larger than zero).

Assume that no user can be assigned to both $Junior - Member$ and $Rookie$, and his virtual-money exceeds \$1000. Then we can generate a MD-SMEA constraint to enforce this SoD policy:

$$as_1 = \{Junior - Member, Rookie\}, as_2 = (1000, +\infty), n_1 = 2, n_2 = 1.$$

We assume user $u_1$, $u_2$ and $u_3$ be assigned corresponding roles as shown in Fig.1. And the virtual-money of $u_1$ is \$500, the virtual-money of $u_2$ is \$900, the virtual-money of $u_3$ is \$1200. We now use the definition of MD-SMEA to verify whether the above example enforces MD-SMEA policy.

For $u_1$ : $ATT(u_1) = \{\{Administrator, Teacher, Student, Junior - Member, Rookie\}, \{virtual - money = \$1200\}\}$, $|ATT(u_1) \cap as_1| = |\{Junior - Member, Rookie\}| = 2 = n_1, |ATT(u_1) \cap as_2| = 0 < n_2 = 1$; For $u_2$ : $ATT(u_2) = \{\{Teacher, Junior - Member\}, \{virtual - money = \$1200\}\}$, $|ATT(u_2) \cap as_1| = |\{Junior - Member\}| = 1 < n1 = 2, |ATT(u_2) \cap as_2| = 0 < n_2 = 1$; For $u_3$ : $ATT(u_3) = \{\{Student, Rookie\}, \{virtual - money = \$1200\}\}$, $|ATT(u_3) \cap as_1| = |\{Rookie\}| = 1 < n_1 = 2, |ATT(u_3) \cap as_2| = |1200| = 1200 > n_2 = 1$. From the above analysis, the user $u_1$ and $u_3$ violate MD-SMEA constraints because they don't satisfy all of the restriction of the MD-SMEA constraints. It is significant that MD-SMEA constraint can't be regarded as the combination of many SD-SMEA constraints since they are different conceptions. When we make the form definition of SD-SMEA and MD-SMEA, the verification problem is urgent: "do the satisfaction checking problems for SD-SMEA and MD-SMEA constraints can be done?" Both SD-SMEA and MD-SMEA constraints restrict the attribute memberships of a single user in order to enforce SSoD policies. Therefore, checking whether an UCON state satisfies a set of SD-SMEA and MD-SMEA constraints is efficient.

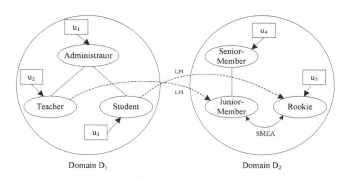

**Fig. 1.** An example of SD-SMEA constraint

**Theorem 2.** *The satisfaction checking problem for SD-SMEA constraints is in **P**.*

*Proof.* One algorithm for solving SD-SMEA constraints is as follows. For each user in $U$, one first computes the set $UA$ of all attributes in which the user is a member of, and then counts how many attributes in this set also appear in the set of attributes in the SMEA constraint, and finally compares this number with $n$. This algorithm has a time complexity of $O(Nu \times Na \times M)$, where $Nu$ is the number of users in $U$, $Na$ is the number of attributes, and $M$ is the number of SD-SMEA constraints.                                                                            $\square$

**Theorem 3.** *The satisfaction checking problem for MD-SMEA constraints is in **P**.*

*Proof.* The proof is essentially the same as that for Theorem 2: The satisfaction checking problem for a MD-SMEA constraint can be regarded as $N$ SD-SMEA constraints, where $N$ is the number of mutually exclusive attributes sets.    $\square$

## 4    Related Work

Ensuring secure interoperation in multidomain environments has drawn considerable research work in the past. Kapadia et al. [1] proposed a secure interoperability using dynamic role translation to implement access control across domains in the form of role mappings among individual domains. In [11], M. Shehab et al. proposed a distributed secure interoperability protocol that ensures secure interoperation of the multiple collaborating domains without compromising the security of collaborating domains. Shafiq et al. [12] extended the IRBAC model by proposing a secure interoperation framework in which all roles in the interacting domains are matched and policies are integrated to form a global RBAC policy.

The first and foremost challenge in establishing secure interoperation is the composition of a consistent and conflict-free interoperation policy. Several research efforts have been devoted to the resolution of the conflicts among role mappings. E. C. Lupu et al. [13] focused on the problems of conflict detection and resolution for policy conflicts, including authorization policies and obligation policies. Cyclic inheritance and separation of duties may appear in an interoperation policy [8]. The resolution of interoperation inconsistencies related to SoD constraint has not been adequately investigated and the existing approaches rely on manual intervention of policy administrators to resolve SoD conflicts [13]. In this paper, we give new definition of the violation of cyclic inheritance and SoD, and show that there exist efficient algorithms to resolve these violations.

However, RBAC primarily consider static authorization decisions based on subjects' permissions on target objects, and there is no further enforcement during the access. Recently proposed usage control [3] models extend traditional access control models for next generation access control by integrating obligations, conditions as well as authorizations, and by including continuity and mutability properties, which make UCON have strong expressive power and policy specification flexibility. Role mappings are the basic approach for the interoperation among multiple individual domains. The attribute mapping technique can be regarded as the extended of role mapping technique.

## 5   Conclusion

This paper presents an attribute mapping technique which can establish a secure interoperation in multidomain environments based on usage control policies. In order to ensure the security of individual domains in presence of interoperation, we study how conflicts arise and show that it is efficient to resolve the security violations of cyclic inheritance and SoD.

## References

1. Kapadia, A., AlMuhtadi, J., Campbell, R., et al.: IRBAC 2000: Secure Interoperability using Dynamic Role Translation. University of Illinois, Technical Report: UIUCDCS-R-2000-2162 (2000)
2. ANSI. American National Standard for Information Technology-Role Based Access Control. ANSI INCITS 359-2004 (2004)
3. Park, J., Sandhu, R.: The UCONABC Usage Control Model. ACM Transactions on Information and System Security 7(1), 128–174 (2004)
4. Zhang, X., Parisi-Presicce, F., Sandhu, R., Park, J.: Formal Model and Policy Specification of Usage Control. ACM Transactions on Information and Systems Security 8(4), 351–387 (2005)
5. Zhang, X., Park, J., Parisi-Presicce, F., Sandhu, R.: A Logical Specification for Usage Control. In: 9th ACM Symposium on Access Control Models and Technology, pp. 1–10. ACM Press, New York (2004)
6. Bonatti, P., Vimercati, S.D.C., Samarati, P.: An Algebra for Composing Access Control Policies. ACM Transaction on Information and System Security 5(1), 409–422 (2002)
7. Dawson, S., Qian, S., Samarati, P.: Providing Security and Interoperation of Heterogeneous Systems. Distributed and Parallel Databases 8, 119–145 (2000)
8. Gong, L., Qian, X.: Computational Issues in Secure Interoperation. IEEE Transactions on Knowledge and Data Engineering 22(1), 14–23 (1996)
9. Dawson, S., Qian, S., Samarati, P.: Providing Security and Interoperation of Heterogeneous Systems. Distributed and Parallel Databases 8(1), 119–145 (2000)
10. Clark, D., Wilson, D., Kuhn, D.R.: A comparison of Commercial and Military Computer Security Policies. In: IEEE Symposium on Security and Privacy, pp. 184–195. IEEE Press, Los Alamitos (1987)
11. Shehab, M., Bertino, E., Ghafoor, A.: SERAT: Secure Role Mapping Technique for Decentralized Secure Interoperability. In: 10th ACM Symposium on Access Control Models and Technologies, Stockholm, pp. 159–167. ACM Press, Sweden (2005)
12. Shafiq, B., Joshi, J.B.D., Bertino, E.: Secure Interoperation in a Multidomain Environment Employing RBAC Policies. IEEE Transactions on Knowledge and Data Engineering 17(11), 1557–1577 (2005)
13. Lupu, E., Sloman, M.: Conflicts in Policy-Based Distributed Systems Management. IEEE Transactions on Software Engineering 25(6), 852–869 (1999)

# Specification and Enforcement of Static Separation-of-Duty Policies in Usage Control*

Jianfeng Lu, Ruixuan Li**, Zhengding Lu, Jinwei Hu, and Xiaopu Ma

Intelligent and Distributed Computing Lab, College of Computer Sci. and Tech.,
Huazhong University of Sci. and Tech., Wuhan 430074, P.R. China
lujianfeng@smail.hust.edu.cn, {rxli,zdlu,jwhu}@hust.edu.cn,
xpma@smail.hust.edu.cn

**Abstract.** Separation-of-Duty (SoD) policy is a fundamental security principle for prevention of fraud and errors in computer security. The research of static SoD (SSoD) policy in recently presented usage control (UCON) model has not been explored. Consequently, this paper attempts to address two important issues: the specification and enforcement of SSoD in UCON. We give a set-based specification scheme, which is simpler and more general than existing approaches. As for the enforcement, we study the problem of determining whether an SSoD policy is enforceable, and show that directly enforcing an SSoD policy is a coNP-complete problem. In indirect enforcement, we generate the least restrictive static mutually exclusive attribute (SMEA) constraints to enforce SSoD policies, by using the attribute level SSoD requirement as an intermediate step. The results are fundamental to understanding the effectiveness of using constraints to enforce SSoD policies in UCON.

**Keywords:** Separation-of-Duty, usage control, constraint.

## 1   Introduction

Separation-of-duty (SoD) is widely considered to be a fundamental security principle for prevention of fraud and errors in computer security, and widely applied in business, industry, and government [1,2]. Although SoD has been studied extensively in the information security, and it has been recognized that "one of RBAC's great advantages is that SoD rules can be implemented in a natural and efficient way" [3], as a related and fundamental problem, research of SoD policy in recently presented usage control (UCON) [4] model has not been explored. UCON has been considered as the next generation access control model with distinguishing properties of decision continuity and attribute mutability [4,5]. Consequently, this paper focuses on static SoD (SSoD) policies in UCON$_A$

---

* This work is supported by National Natural Science Foundation of China under Grant 60873225, 60773191 and 60403027, National High Technology Research and Development Program of China under Grant 2007AA01Z403.
** Corresponding author.

P. Samarati et al. (Eds.): ISC 2009, LNCS 5735, pp. 403–410, 2009.

which is a sub-model of UCON only considering authorizations. Since an authorization decision is determined by subject's and object's attributes, and these attribute values can be updated as side-effects of the authorization, the study of SSoD policies in authorization models is more pressing than that in obligation and condition models. In this paper, we provide a set-based specification scheme for SSoD policies. Furthermore, we study a number of problems related to generating SMEA constraints for enforcing SSoD policies in UCON$_A$ systems. We study the problem of determining whether an SSoD policy is enforceable, and generate SMEA constraints to indirect enforce SSoD policies, by using attribute level SSoD requirements (ASSoD) as an intermediate step from SSoD policies to SMEA constraints. The research of the SSoD policy in UCON$_A$ is important for emerging applications as usage control scenarios, and it can also increase UCON's strengths in that it enables the use of constraints to support SSoD policies.

The rest of this paper is organized as follows. Section 2 describes related works. Section 3 gives the specification of SSoD policies. Section 4 studies the problem of determining whether an SSoD policy is enforceable. Section 5 uses SMEA constraints to indirect enforce SSoD policies. Section 6 concludes this paper.

## 2    Related Work

The concept of SoD can be traced back to 1975 when by Saltzer and Schroeder [6] took it as one of the design principles for protecting information, under the name "separation-of-privilege". The research community has taken an active interest in incorporating SoD controls into computer systems since the late 1980s, Clark and Wilson [1] applied SoD principle to data objects to ensure integrity and to control frauds along with well-formed transactions as two major mechanisms for controlling fraud and error. Later on, SoD has been studied by various researchers as a principle to avoid frauds. One of the best known requirements for SoD is embodied in the Chinese Wall model [7], in which access to documents that could result in a commercial conflict of interest is strictly controlled.

In this paper, the specification scheme of the SSoD policy we propose has its basis in our set-based approach to conflict of interest, and it is considerably simpler syntactically than other schemes because the SSoD policy is expressed in terms of restrictions on permissions other than attributes, such as roles, and we make no attempt to define the conditions that must be met for the constraint to be satisfied. As for the enforcement, Sandhu presented transaction control expressions, a history based mechanism for dynamically enforcing SoD policies [8,9]. Simon and Zurko combined the Object SoD and Operational SoD and introduced a notion of history based SoD [10]. Crampton [11] employed blacklist to enforce historical constraints, it does not need to keep a historical record. However, since these approaches for SoD only consider constraint sets with a few elements, they will have unacceptable overheads to support large range of constraints. UCON$_A$ includes RBAC which increases the difficulty of enforcement

SSoD policies in UCON$_A$. Motivated by the SMER constraints [12],we enforce SSoD policies in UCON$_A$ by using SMEA constraints. For a more detailed description of UCON, the reader can refer to [4,5].

## 3    The Specification of SSoD Policies

We now give a formal basis for representing SSoD policies in UCON$_A$ models based on the following requirements. (1) An SSoD policy must be a high-level requirement. Clark et al. identified SoD as a high-level mechanism that is "at the heart of fraud and error control" [13]. It states a high-level requirement about the task without the need to refer to individual steps in the task. (2) An SSoD policy must be described in terms of restrictions on permissions. In the ANSI RBAC standard [14], the distinction between SSoD policies as objectives and static mutually exclusive role (SMER) constraints as a mechanism is not clear. One problem is that the SMER constraints may be specified without a clear specification of what objectives they intend to meet; consequently, it is unclear whether the higher-level objectives are met by the constraints or not. Another problem is that even though when SMER constraints are specified there exist a clear understanding of what SSoD policies are desired, when the assignment of permissions to roles changes, the SMER constraints may no longer be adequate for enforcing the desired SSoD policies [12]. (3) An SSoD policy must capture restrictions on user set involved in the task. In practice, the number of users in any organization is bounded. It needs to consider the SSoD policies with an upper bound on the number of users in an access control state.

**Definition 1.** *An SSoD policy ensures that at least $k$ users from a user set are required to perform a task that requires all these permissions. Formally,*

- *$P$ and $U$ denote the set of permissions, the set of users;*
- *$UP \subseteq U \times P$, a user-permission assignment relation;*
- *$auth\_P_\varepsilon[u] = \{allowed(u,p) \Rightarrow preA(ATT(u),p)\};$*
- *$\forall (P,U,k) \in SSoD, \forall U' \subseteq U : |U'| < k \Rightarrow \bigcup_{u \in U'} auth\_p_\varepsilon(u) \not\supseteq P.$*

*where $P = \{p_1,\ldots,p_m\}$, $U = \{u_1,\ldots,u_n\}$, $m$, $n$, and $k$ are integers, such that $2 \leq k \leq min(m,n)$, min returns the smaller value of the two. $ATT(u)$ denotes the user's attributes, preA is the pre-authorizations in UCON, and $allowed(u,p)$ indicates that user $u$ is assigned permission $p$. $\varepsilon$ is a UCON$_A$ state which constituted by the set of assignments for all objects' attributes. We write an SSoD policy as ssod $< P,U,k >$. We say that a UCON$_A$ state $\varepsilon$ is safe with respect to an SSoD policy $e$, if in state $\varepsilon$ no $k-1$ users from $U$ together have all the permissions in $P$, and we write it as $safe_e(\varepsilon)$. An UCON$_A$ state $\varepsilon$ is safe with respect to a set $E$ of SSoD policies, which we denote by $safe_E(\varepsilon)$, if and only if $\varepsilon$ is safe with respect to every policy in the set $E$.*

## 4    Enforceability of SSoD Policies

In a UCON$_A$ system, not all SSoD policies are enforceable. For example, given an SSoD policy $e = \{ssod < \{p_1,p_2\}, \{Alice, Bob, Carl\}, 2 >\}$, which ensures that at least two users together from $\{Alice, Bob, Carl\}$ are allowed to have $\{p_1, p_2\}$. Assume that $allowed(u, p_1) \Rightarrow ATT(u) = \{engineer, student, 50\}$, where engineer is a role, student denotes the identity of user, and 50 is a trust value. It means that a user who must cover all these attributes can be allowed to have $p_1$. Where $allowed(u, p_2) \Rightarrow ATT(u) = \{programmer, student, 75\}$ has the similar meaning. Suppose that $ATT(Alice) = \{supervisor, student, 100\}$, where $supervisor$ is a senior role to both $engineer$ and $programmer$. Obviously, $safe_e(\varepsilon)$ is false, because $Alice$ can be a member of both $p_1$ and $p_2$. In order to address this, forbid $Alice$ from having the attribute set $\{supervisor, student, 100\}$. This is undesirable, if an attribute can not be assigned to a user, then the attribute value should not be included in the domain of the user attribute.

**Definition 2.** $(I, M)$ *is a attribute set, where $I$ is the set of immutable attributes, and $M$ is the set of mutable attributes. We say $(I_1, M_1) \leq (I_2, M_2)$ if and only if for each attribute $a \in I_1$, there exists an attribute $a' \in I_2$ such that $a \leq a'$; and for each attribute $b \in M_1$ there exists an attribute $b' \in M_2$ such that $b \leq b'$.*

Obviously, $\leq$ associates the user attribute sets, and these associations form a combined hierarchy that is partially ordered. If $(I, M)$ satisfies exactly the requirement of $allowed(u, p)$, we say $(I, M)$ is the threshold attribute set of $p$.

**Definition 3.** *Given an SSoD policy ssod $< P, U, k >$, let $(I_{p_i}, M_{p_i})$ is the threshold attribute set of each $p_i$ in $P$, and $(I_t, M_t)$ is an attribute set. If $\forall (I_{p_i}, M_{p_i})$ $(I_{p_i} \leq I_t \Rightarrow M_{p_i} \leq M_t)$, we say $(I_t, M_t)$ is an ancestor attribute set. Assuming that $(I_i, M_i)$ is an ancestor attribute set, there does not exist another ancestor attribute set $(I_j, M_j)$ that $(I_j, M_j) \leq (I_i, M_i)$, we say $(I_i, M_i)$ is an least ancestor attribute set.*

**Theorem 1.** *An SSoD policy $e = ssod < P, U, k >$ is not enforceable if and only if the number of ancestor attribute sets for $e$ is less than $k$.*

*Proof.* For the "if" part, we assume that if the condition in the theorem exists. Then one can construct a UCON$_A$ state in which there are $k - 1$ users and each of the users is assigned one of the $k - 1$ ancestor attribute set $(I, M)$ for $e$. Thus these $k - 1$ users together cover all $m$ permissions, and result in an unsafe state. For the "only if" part, we show that if the condition in the theorem does not exist, then the SSoD configuration is enforceable. Consider that the number of ancestor attribute sets is $k$. we can declare every pair of $(I, M)$ to be mutually exclusive, which forbids any user to cover any two of them, this makes $safe_e(\varepsilon)$ true. $\qquad\square$

## 5   Enforcing SSoD Policies by SMEA Constraints

We now show that directly enforcing SSoD policies is intractable (coNP-complete).

**Theorem 2.** *The verification problem of $safe_e(\varepsilon)$ is coNP-complete.*

*Proof.* In ANSI RBAC model [14], a role is a collection of users and a collection of permissions, and the permission is a collection of object-right pairs. The UCON$_A$ model can support RBAC in its authorization process, in UCON$_A$, user-role assignment can be viewed as subject attributes and permission-role assignment as attributes of object and rights [5]. And let $U$ in an SSoD policy $e = ssod < P, U, k >$ be the user set of all possible users in an RBAC state $\varepsilon_0$, then the verification problem of $safe_e(\varepsilon)$ is equivalent to the one in [12] for the theorem that checks whether an RBAC state is safe or not with respect to an SSoD policies, which is NP-complete. □

In RBAC, constraints such as mutually exclusive roles (SMER) are introduced to enforce SSoD policies [12]. We present a generalized form of the SMEA in this paper, which is directly motivated by SMER constraint.

**Definition 4.** *A statically mutually exclusive attribute (SMEA) constraint is expressed as*

$$smea < \{(I_1, M_1), \ldots, (I_m, M_m)\}, \{u_1, \ldots, u_n\}, k >$$

*where each $(I_i, M_i)$ is an attribute set and $m$ and $n$ are integers such that $2 \leq k \leq min(m, n)$.*

**Definition 5.** *A UCON$_A$ state $\varepsilon$ is safe with respect to a SMEA constraint when*

$$\forall u_i \in \{u_1, \ldots, u_n\}(|(ATT_\varepsilon(u_i) \cap \{(I_1, M_1), \ldots, (I_m, M_m)\})| < k)$$

*which means that no user from $\{u_1, \ldots, u_n\}$ is a member of $k$ or more attribute sets in $\{(I_1, M_1), \ldots, (I_m, M_m)\}$, and we write it as $safe_c(\varepsilon)$. A UCON$_A$ state $\varepsilon$ is safe with respect to a set $C$ of SMEA constraints if it is safe with respect to every constraint in $C$, and we write it as $safe_C(\varepsilon)$.*

As each SMEA constraint restricts the attribute set memberships of a single user, it is efficient to check whether an UCON$_A$ state satisfies a set of SMEA constraints, and thus provides a justification for using SMEA constraints to enforce SSoD policies.

## 5.1   Translating SSoD Policies to ASSoD Requirements

SMEA constraints are expressed in term of restrictions on attribute memberships, but SSoD policies are expressed in terms of restrictions on permissions. In order to generate SMER constraints for enforcing SSoD policies, the first step is to translate restrictions on attribute sets other than on permissions for SSoD policies. For each permission $p_i$ in $\{p_1, \ldots, p_m\}$, there exists a $(I_{p_i}, M_{p_i})$ which is the threshold attribute set of $p_i$. In this way, we can define the attribute level SSoD requirement, and translate an SSoD policy to ASSoD requirements.

**Definition 6.** *An attribute level Static Separation-of-Duty (ASSoD) require-ment is expressed as*

$$assod < \{(I_1, M_1), \ldots, (I_m, M_m)\}, \{u_1, \ldots, u_n\}, k >$$

*where each $(I_i, M_i)$ is an attribute set, $m$ and $n$ are integers such that $2 \le n \le m$.*

**Definition 7.** *A UCON$_A$ state $\varepsilon$ is safe with respect to ASSoD requirement when*

$$\forall \{u'_1 \ldots u'_{k-1}\} \subseteq \{u_1, \ldots, u_n\}(\cup_{i=1}^{k-1})ATT_\varepsilon(u'_i) \not\supseteq \{(I_1, M_1), \ldots, (I_m, M_m)\}$$

*It means that there should not exist a set of fewer than $k$ users from $\{u_1, \ldots, u_n\}$ that together have memberships in all the $m$ attribute sets in the requirement. A UCON$_A$ state $\varepsilon$ is safe with respect to a set $A$ of ASSoD requirements if it is safe with respect to every requirement in $A$, and we write it as $safe_A(\varepsilon)$.*

Let $A$ denote the set of ASSoD requiremets derived from $< E, \varepsilon >$, if $< E, \varepsilon >$ is not enforceable, then it can not be translated to any ASSoD requirements, let $A = \emptyset$. Else if $< E, \varepsilon >$ is enforceable, then for each $e \in E$, and for each permission $p_i$ in $\{p_1, \ldots, p_m\}$, there exist many attribute sets corresponding to it. Assume that each permission in $\{p_1, \ldots, p_m\}$ relates to the number of attribute sets is $\{k_1, \ldots, k_m\}$, then the total number of elements in $< A, \varepsilon >$ is $k_1 \times k_2 \times \ldots \times k_m$. Theorem 3 shows that for the ASSoD configuration $< A, \varepsilon >$ derived from an enforceable SSoD configuration $< E, \varepsilon >$ captures the same security requirement.

**Theorem 3.** *Given an SSoD configuration $< E, \varepsilon >$, and the ASSoD configu-ration $< R, \varepsilon >$ derived from $< E, \varepsilon >$, then $safe_A(\varepsilon) \Leftrightarrow safe_E(\varepsilon)$.*

*Proof.* Firstly, we show that if $safe_R(\varepsilon)$ is false, then $safe_E(epsilon)$ is also false. If $safe_R(\varepsilon)$ is false, then there exist $r = assod < \{(I_1, M_1), \ldots, (I_m, M_m)\}, U, n >$ and $k - 1$ users that together cover all attribute sets in $\{(I_1, M_1), \ldots, (I_m, M_m)\}$. Given the way in which $ASSoD < R, \varepsilon >$ is derived from $< E, \varepsilon >$, there exists an SSoD policy in $E$ such that the at-tribute set in $R$ together have all the permissions in it, therefore $safe_E(\varepsilon)$ is also false. Secondly, we show that if $safe_E(\varepsilon)$ is false, then $safe_A(\varepsilon)$ is also false. If $safe_E(\varepsilon)$ is false, then there exist $e = ssod < \{p_1, \ldots, p_m\}, \{u_1, \ldots, u_n\}, k >$ and $k - 1$ users together cover all permissions in $\{p_1, \ldots, p_m\}$. For each per-mission $p_i$ in the permission set, there exists an attribute set $(I_i, M_i)$ covering $p_i$, if it contains some sub attribute set, then we divide it, then there exists $r =< \{(I_1, M_1), \ldots, (I_m, M_m)\}, U, n >$ derived from $e$. Given the way in which $ASSoD < R, preA >$ is derived from $< E, \varepsilon >$, then $r \in R$, therefore $safe_A(\varepsilon)$ is also false. $\square$

**Theorem 4.** *Given a UCON$_A$ state $\varepsilon$, and a set $A$ of ASSoD requirements, determine if $safe_A(\varepsilon)$ is coNP-complete.*

*Proof.* The proof is similar to the one in Theorem 2: let each attribute set in AS-SoD requirement map to a permission. Then the ASSoD requirement is mapped to an SSoD policy. $\square$

## 5.2   Generating SMEA Constraints to Enforce ASSoD Requirements

**Definition 8.** *Let $C$ be a set of SMEA constraints, and $R$ be a set of ASSoD requirement, $C$ enforces $R$ if and only if $safeC(\varepsilon) \Rightarrow safe_R(\varepsilon)$.*

**Theorem 5.** *The ASSoD requirement $a = assod < \{(I_1, M_1), \ldots, (I_m, M_m)\}, \{u_1, \ldots, u_n\}, k >$ can be enforced by the following SMEA constraint*

$$c = \bigcup_{\substack{i \neq j}}^{i,j \in [1,m]} \{c = smea < \{(I_i, M_i), (I_j, M_j)\}, \{u_1, \ldots, u_n\}, 2 >\}$$

*Proof.* The ASSoD requirement means that $k$ users are required to cover all $m$ attribute sets. The constraint set $C$ means that every two attribute sets in $\{(I_1, M_1), \ldots, (I_m, M_m)\}$ are mutually exclusive, then m users are needed to cover the $m$ attribute sets, as $2 \leq n \leq m$, Thus $safe_{\{C\}}(\varepsilon)$ is true.    □

Although the above SMEA constraints ($k = 2$) can enforce any ASSoD requirement, this may result in constraints that are more restrictive than necessary. Ideally, we want to generate SMEA constraints that can enforce the ASSoD requirement, and avoid generating constraints that are overly restrictive. For this, we prefer to use the less restrictive constraint set.

**Definition 9.** *Let $C_1$ and $C_2$ be two sets of SMEA constraints, $C_1$ is more restrictive than $C_2$ if $safe_{C_1}(\varepsilon) \Rightarrow safe_{C_2}(\varepsilon) \wedge safe_{C_2}(\varepsilon) \not\Rightarrow safe_{C_1}(\varepsilon)$, and we write it as $C_1 \succ_\varepsilon C_2$.*

We now give an algorithm to generate the relatively less SMEA constraints to enforce ASSoD requirement. Given an ASSoD requirement $a$, the first step is to compute the most restrictive SMEA constraint set $C$ by enumerating all possible SMEA constraints (where $m=k=2$) ; the second step is to remove any constraint in $C$ that the remainders can also enforce a; the third step is to weaken the more restrictive constraint in the set. This algorithm can be used by the step 1 and 2, which we only try to remove the constraints in $C$ is efficient, the output $C$ will be a relative less restrictive SMEA constraints to enforce $a$. Although by systematically enumerating all the cases in step 3 will generate the least restrictive SMEA constraints, the runtime will be expensive. We generally prefer to use step 1 and 2, and return the output $C$.

## 6   Conclusion

This paper presents two main contributions to the research of SoD policy in $UCON_A$: the specification and enforcement of SSoD in UCON. The specification is set-based and we show that it has simpler syntax than existing approaches. For the enforcement aspect, we have studied a number of problems related to generating SMEA constraints for enforcing SSoD policies in $UCON_A$ system. We show that directly enforcing SSoD policies in $UCON_A$ system is intractable (coNP-complete), study the problem how to verify whether a given SSoD configuration is enforceable, translate the SSoD policy to ASSoD requirement which be used as an intermediate step, and generate the least restrictive SMEA constraints from a set of ASSoD requirements. The results are fundamental to understanding the effectiveness of using constraints to enforce SSoD policies in UCON.

# References

1. Clark, D., Wilson, D., Kuhn, D.R.: A Comparison of Commercial and Military Computer Security Policies. In: 8th IEEE Symposium on Security and Privacy, pp. 184–195. IEEE Press, Los Alamitos (1987)
2. Clark, D., Wilson, D., Kuhn, D.R.: Evolution of a Model for Computer Integrity. Technical Report, Invitational Workshop on Data Integrity, Section A2, pp. 1–3 (1989)
3. Ferraiolo, D.F., Kuhn, D.R., Chandramouli, R.: Role-Based Access Control. Artech House, 47–63 (April 2003)
4. Park, J., Sandhu, R.: The UCONABC Usage Control Model. ACM Transactions on Information and System Security 7(1), 128–174 (2004)
5. Zhang, X., Parisi-Presicce, F., Sandhu, R., Park, J.: Formal Model and Policy Specification of Usage Control. ACM Transactions on Information and Systems Security 8(4), 351–387 (2005)
6. Saltzer, J.H., Schroeder, M.D.: The Protection of Information in Computer Systems. Proceed Communications of the ACM 63(9), 1278–1308 (1975)
7. Brewer, D., Nash, M.: The Chinese Wall security policy. In: 10th IEEE Symposium on Security and Privacy, pp. 206–214. IEEE Press, California (1989)
8. Sandhu, R.: Transaction Control Expressions for Separation of Duties. In: 4th Annual Computer Security Applications Conference, pp. 282–286. IEEE Press, Orlando (1988)
9. Sandhu, R.: Separation of Duties in Computerized Information Systems. In: The IFIP WG11.3 Workshop on Database Security, pp. 18–21. IEEE Press, Halifax (1990)
10. Schaad, A., Lotz, V., Sohr, K.: A Model-checking Approach to Analyzing Organizational Controls in a Loan Origination Process. In: 11th ACM Symposium on Access Control Models and Technologies, pp. 139–149. ACM Press, California (2006)
11. Crampton, J.: Specifying and Enforcing Constraints in Role-based Access Control. In: 8th ACM Symposium on Access Control Models and Technologies, pp. 43–50. ACM Press, New York (2003)
12. Li, N., Tripunitara, M., Bizri, Z.: On Mutually Exclusive Roles and Separation-of-Duty. ACM Transactions on Information and System Security 10(2), 1–35 (2007)
13. Li, N., Mitchell, J.C., Winsborough, W.H.: Beyond Proof-of-Compliance: Security Analysis in Trust Management. Journal of the ACM 52(3), 474–514 (2005)
14. ANSI. American National Standard for Information Technology-Role Based Access Control. ANSI INCITS 359-2004 (2004)

# Nonce Generators and the Nonce Reset Problem

Erik Zenner

Department of Mathematics,
Technical University of Denmark
e.zenner@mat.dtu.dk

**Abstract.** A nonce is a cryptographic input value which must never repeat within a given context. Nonces are important for the security of many cryptographic building blocks, such as stream ciphers, block cipher modes of operation, and message authentication codes. Nonetheless, the correct generation of nonces is rarely discussed in the cryptographic literature.

In this paper, we collect a number of nonce generators and describe their cryptographic properties. In particular, we derive upper bounds on the nonce collision probabilities of nonces that involve a random component, and lower bounds on the resulting nonce lengths.

We also discuss an important practical vulnerability of nonce-based systems, namely the nonce reset problem. While ensuring that nonces never repeat is trivial in theory, practical systems can suffer from accidental or even malicious resets which can wipe out the nonce generators current state. After describing this problem, we compare the resistance of the nonce generators described to nonce resets by again giving formal bounds on collision probabilities and nonce lengths.

The main purpose of this paper is to provide a help for system designers who have to choose a suitable nonce generator for their application. Thus, we conclude by giving recommendations indicating the most suitable nonce generators for certain applications.

**Keywords:** Cryptography, Security Engineering, Nonce, Nonce Reset, Nonce Generator.

## 1 Introduction

Nonces are cryptographic inputs with the property that each value only occurs once within a given context[1]. Many modern cryptographic algorithms require a key and a nonce as input, and as long as the key is unchanged, the nonce must not repeat. Examples for cryptographic solutions that require nonces are stream ciphers, certain block cipher modes of operation, some message authentication codes (in particular Wegman-Carter based codes [2]), and certain entity authentication solutions.

---

[1] The term "nonce" is sometimes understood by cryptographers to be an abbreviation for "number used once". Even though this etymology incorrect [1], it is a useful mnemonic for cryptographic purposes.

P. Samarati et al. (Eds.): ISC 2009, LNCS 5735, pp. 411–426, 2009.

One possibility of generating nonces is the use of a random number generator (RNG). However, in order to avoid collisions, the nonce length has to be large, which may be problematic particularly in light-weight cryptographic systems with limited memory or bandwidth. In addition, a cryptographically strong RNG is not always available. Thus, a popular solution is to use a deterministic, stateful generator that keeps track of the nonces already used. The most obvious candidate for such a generator is a simple counter. As long as the generator does not "wrap around" (i.e. reaches a value that is longer than the nonce length, forcing it to start from 0 again), such a generator is good enough for most practical purposes.

However, this is only the case as long as the generator actually maintains its inner state. While this seems trivial in theory, it can not be taken for granted in practice. An unexpected power-down can mean the loss of all information that was not stored in non-volatile memory, and for many applications, constantly storing the nonce to Flash memory or a hard disk is not an option. For such systems, solutions are required that guarantee the nonce property also after a system reset.

*Prior Art:* Even though nonces play a prominent role in cryptography, hardly any literature exists on the issue. An overview of some known nonce techniques can be found in a discussion threat in the CFRG mailing list from early 2007 [3]. The use of nonces in security proofs was modeled by Rowaway [4]. In addition, a number of practical cryptosystems have been broken due to errors in the nonce handling [5,6,7,8].

*Contribution:* While the nonce generators described in this paper have been used in practice, our main contribution is the derivation of concrete bounds for collision probability and the nonce length. To the best of our knowledge, this is the first time that a full formal treatment of popular nonce generator techniques is given. We also give the first scientific discussion of the nonce reset problem and analysis of the techniques used to address it. Using the results of this paper, system designers can compare the suitability of the different nonce generators for their target application and make choices based on mathematical bounds.

*Paper Structure:* In Section 2, we review a number of nonce generators, most of which are well-known in the literature. We derive formal upper bounds on the nonce collison probabilities and lower bounds on the nonce lengths. Then we proceed to describe the nonce reset problem in Section 3. Here we also introduce a number of solutions to the problem, again giving formal bounds on collision probability and nonce length. Finally, in Section 4, we compare the nonce generators proposed and conclude the paper.

*Notation:* Throughout this paper, we will use the following variables. The maximum number of nonces produced by a generator is denoted by $\theta$. The maximum number of nonce resets is denoted by $r - 1$, i.e. $r$ is the maximum number of (re-)initialisations. The collision probability is denoted by $p_c$, and the maximum allowable collision probability is denoted by $p_{max}$. Finally, the nonce length is denoted by $l$. If a nonce consists of a counter and a random part, then the lengths of these parts are denoted by $l_1$ and $l_2$, respectively.

In the algorithmic descriptions of Section 3, $a \leftarrow b$ denotes the assignment of value $b$ to variable $a$, while $a = b$ denotes the logical comparison between $a$ and $b$.

## 2   Standard Nonce Generators

Before discussing the nonce reset problem, we briefly review three basic types of nonce generators (NGs) and give bounds for the corresponding collision probabilities and nonce lengths.

*Choosing the Right Nonce Length:* It is important upon designing a nonce-based system to pick the right nonce length $l$. An *upper bound* for $l$ often results from application limititations such as expensive bandwidth or storage. While it may be possible to choose any desired nonce length on e.g. a desktop or laptop computer, limitations may exist for resource-restricted devices. In light-weight systems such as smart cards, sensor nodes, RFID chips etc., non-volatile memory as well as transmission bandwidth is limited and expensive. Thus, a solution that simply chooses a large nonce to elimiate all potential problems is not an option in such a scenario – the nonce length has to be optimised as far as possible.

To this end, a *lower bound* for $l$ is required. As it turns out, this lower bound depends on the type of NG used, as well as the maximum number $\theta$ of nonces required within one context. In Section 3, we will see that additional factors play a role if we also want the NG to address the reset problem.

*Deterministic vs. Probabilistic NGs:* Nonce generators can be either deterministic or probabilistic.

- Deterministic NGs use some kind of inner state to keep track of the values already used as nonces, ensuring that the same value never gets used again. Such generators have two functions: Init() is executed upon setting up the NG, while Next() outputs the next nonce value and updates the inner state.
- Probabilistic NGs use some kind of external randomness source to generate nonces. While all of them have a Next() function, some of them are stateless and do thus not require an Init() function. We denote them as "probabilistic" NGs because the sequences produced by them can in theory contain collisions. In practice, however, the collison probability $p_c$ can be kept arbitrarily small by making the nonce length $l$ large. Note that probabilistic NGs require a good RNG to function properly. Implementing a good RNG, however, is one of the hardest tasks in practical cryptography (see, e.g., [9]), and if the RNG is faulty, the true collision probability may be much higher than expected.

*Additional Constraints:* In some protocols in the literature, the NG is expected to have additional properties, such as unpredictability or pseudo-randomness. However, in this paper, we follow the cryptographically more rigorous view presented by Rogaway [4], namely that the role of a NG should be limited to guaranteeing collision-freeness. If additional properties are required, they have to be made explicit ("The protocol requires an pseudo-random nonce") and should be provided by the appropriate cryptographic primitives (e.g. a pseudo-random function) in a separate step. Thus, no such additional constraints are considered here.

## 2.1   Counter-Based Generator

The most widespread deterministic generator is a simple counter. The `Init()` function consists of setting the counter `cnt` to 0 or to a random value, and the `Next()` function outputs `cnt` and increases it by 1 (modulo $2^l$)[2].

Note that this type of generator is well-known and well-understood; we only repeat some known facts for completeness sake.

*Nonce length:* As long as the number $\theta$ of nonces drawn is at most $2^l$, the output of a counter-based generator is guaranteed to be collision-free. This yields the trivial condition on the nonce length that $l \geq \log_2(\theta)$.

## 2.2   RNG-Based Generator

The most common probabilistic NG simply outputs an $l$-bit random number `rnd` every time a nonce is requested. This NG does not maintain an inner state and thus, does not need an `Init()` function. As with the counter-based generator, the RNG-based generator is well-known and well-understood in the literature. Note that if a pseudo-RNG is used instead (as is often the case in practice), it should be a cryptographically secure one as formalised e.g. in [9]. If this security advice is heeded, it should not be possible to distinguish between the pseudo-RNG and a real RNG. Thus, in the following, the following facts for a real RNG can be applied to a cryptographically sound pseudo-RNG just as well.

*Nonce length:* The birthday bound (see e.g. [12, Section 6.6]) states that if $\theta$ out of $2^l$ elements are drawn in a mutually independent way, the collision probability $p_c$ is upper bounded by $\frac{\theta^2 - \theta}{2 \cdot 2^l}$.

---

[2] Another wide-spread type of deterministic NG is the use of the system clock [10,11]. If there is at least one clock tick between two accesses to the `Next()` function, and if the clock is never reset or wrapped around, then this can be seen as a special case of a counter, where not every available nonce is actually used. However, there are additional problems, such as synchronisation problems or the possibility that someone (even inadvertently) resets the system clock, thus creating a nonce re-use that goes unnoticed by the application.

This formula can be used to calculate the minimum length of a nonce. If $p_{max}$ denotes the highest acceptable collision probability, we have:

$$\frac{\theta^2 - \theta}{2 \cdot 2^l} \leq p_{max} \quad \Leftrightarrow \quad 2^l \geq \frac{\theta^2 - \theta}{2 \cdot p_{max}}$$

*Example:* If we need at most $\theta = 2^{20}$ nonces and a collision probability of at most $p_{max} = 2^{-20}$, then we get $2^l \geq 2^{59}$, meaning that the nonce has to have a minimum length of 59 bit. Thus, compared to the counter solution, the nonce has to be almost three times as long[3].

## 2.3   Mixed Solution

Another possibility is to combine the two approaches above by concatenating an $l_1$-bit counter cnt and an $l_2$-bit random number rnd into one nonce of length $l = l_1 + l_2$. For every call to the Next() function, cnt will be increased by one, and a new random number rnd will be generated. Obviously, this has the disadvantages of the RNG-based solution, namely that an RNG is required and that there is a risk for collisions. However, the collision probability and thus the nonce length is reduced by the counter part, and the solution offers some advantages in the case of nonce resets (see Section 3).

While the mixed solution is used in practice, we are not aware of a thorough discussion in the cryptographic literature. Thus, we give a more detailed analysis of its properties in the rest of this section.

*Nonce length:* As for the RNG-based generator, the nonce length depends on the collision probability. Thus, we start by giving a general collision bound for the mixed solution.

**Lemma 1.** *Assume that the number $2^{l_1}$ of possible counter values divides the maximum number $\theta$ of required nonces. Then the collision probability for the mixed solution is 0 if $0 \leq \theta \leq 2^{l_1}$, and*

$$p_c \leq \frac{\theta^2 - \theta \cdot 2^{l_1}}{2 \cdot 2^l}$$

*otherwise.*

*Proof.* Let us simplify notation by writing $S = 2^l$, $S_1 = 2^{l_1}$, and $S_2 = 2^{l_2}$. Note that for $\theta \leq 2^{l_1}$, no collision can occur, since we are guaranteed to use a new counter cnt each time. Beyond that point, a collision can occur if for two nonces with the same counter cnt, the random part rnd also collides. If a total of $\theta$ nonces is output by this NG, then for each value of cnt, we have $\frac{\theta}{S_1}$ calls to

---

[3] In many cryptographic texts, the simplified rule $2^l \approx \theta^2$ is used, meaning that nonces are chosen to be *exactly* twice as long as in the counter case. This, however, ignores the influence of the acceptable collision probability $p_{max}$.

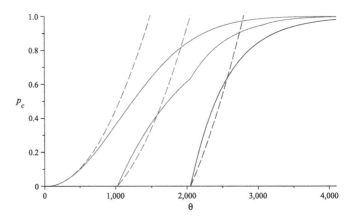

**Fig. 1.** Mixed Solution: Collision probabilities for $l = 20$ and $l_1 \in \{6, 10, 11\}$

the RNG, generating an $l_2$-bit random part. Thus, for each counter, the collision probability $p'_c$ is bounded by the birthday bound as

$$p'_c \leq \frac{\left(\frac{\theta}{S_1}\right)^2 - \left(\frac{\theta}{S_1}\right)}{2 \cdot S_2} = \frac{\theta^2 - \theta S_1}{2 \cdot S_1^2 \cdot S_2}.$$

In total, a collision occurs if there is a collision for any of the $S_1$ counters, i.e. the total collision probability is bounded by

$$p_c \leq S_1 \cdot p'_c = S_1 \cdot \left(\frac{\theta^2 - \theta S_1}{2 \cdot S_1^2 \cdot S_2}\right) = \frac{\theta^2 - \theta S_1}{2 \cdot S_1 \cdot S_2} = \frac{\theta^2 - \theta S_1}{2S}.$$

Resubstituting $S_1$ and $S$, we obtain the desired bound. □

**Corollary 1.** *Assume that the number $2^{l_1}$ of possible counter values divides the maximum number $\theta$ of required nonces. If the maximum acceptable collision probability is $p_{\max}$, then the nonce length for the mixed solution has to be at least*

$$l \geq \log_2 \left(\frac{\theta^2 - \theta \cdot 2^{l_1}}{2 \cdot p_{\max}}\right).$$

*A Cautionary Note:* Note that the above estimate is only correct if $S_1$ divides $\theta$. In situations where this is not the case and where $\theta$ is small compared to $S_1$, the bound on $p_c$ may be too low. Figure 1 illustrates this problem for three sample setups, namely for $l_1 = 6, 10, 11$ (from left to right) and a total nonce length of $l = 20$. The correct value for $p_c$ is shown with a solid line, while the above bound is shown with a dashed line[4]. As can be seen, the error gets larger with

---

[4] Note that the "break" in the curve for $l_1 = 10$ is not a plotting error, but a property of the probability function, which is always convex with the exception of the points where $S_1$ divides $\theta$.

increasing $l_1$. In fact, the error is bounded by $\frac{S_1}{8S_2}$, i.e. it is rather insignificant for small values of $l_1$ but must not be ignored for large $l_1$. This result is proven in Appendix A.

A simple way of solving this problem when designing a system is to choose $S_1$ and $\theta$ such that $S_1$ divides $\theta$. If $S_1$ is small compared to $\theta$, this should not be a problem. If, on the other hand, $S_1$ is close to $\theta$, it is probably worth increasing $l_1$ by a few bits such that $S_1 \geq \theta$, thus achieving a collision probability of 0.

## 3   The Nonce Reset Problem

In actual implementations, the inner state of a deterministic NG has to be stored between two calls to the Next() function. Basically, there are two possibilities:

- **Volatile memory (VM):** This type of memory requires power to maintain its state. Examples are various types of RAM, but also CPU registers. The problem with using this kind of memory is that the NG state will be lost when the system suffers a (planned or accidential) power-down.
- **Non-volatile memory (NVM):** This type of memory maintains its state even if not powered. There are two types of solutions:
  - **Electronically addressed:** This includes technologies like EEPROM or Flash. They are rather expensive and slow compared to VM. As a result, on most platforms, designers will try to use as little electronically addressed NVM as possible.
  - **Mechanically addressed:** This includes typical "secondary" storage like magnetical or optical storage media (e.g. hard disks or DVDs). They have the disadvantage of being very slow compared to VM.

Long-term cryptographic keys are typically stored in NVM, and while they are in use, they are also loaded into VM. This way, they can be accessed fast and will nonetheless survive a system crash. For NGs, however, this solution is not always feasible. Electronically addressed NVM is often not available, and mechanically addressed NVM would slow down the system performance considerably due to the frequent changes of the NG state. Thus, practical solutions often store the NG state in volatile memory only. If, however, the key survives a system crash while the NG state does not, then some way of re-setting the NG is required.

It turns out that this reset function is often forgotten by NG designers. The classical mistake is to re-use the old key, but to start a new instance of the NG [5]. This means that the Init() function is called for the second time, which leads to a nonce re-use for deterministic NGs.

If the solution is built such that the cryptographic key survives a system crash, then the NG should have a Reset() function, which may or may not be identical to the Init() function. If Reset() and Init() are different, then it is important to always remember the following master rule for nonce initialisation upon system start-up: *If no key exists, run Init(). If a key exists, run Reset().*

In the following, we will discuss a number of proposals for how to add a Reset() function to a counter-based NG. In addition, note that the simple RNG-based NG also solves the reset problem, albeit not in an optimal way.

### 3.1   Randomised Reset

A wide-spread solution is to reset the counter to a random $l$-bit value. Note that this solution requires an RNG, which has the disadvantages already discussed.

In addition, the solution is no longer deterministic and opens up for the possibility of nonce collisions. Note that since the `Next()` function computes the new counter as $i \leftarrow i + 1 \bmod 2^l$, counters will "wrap" if they get larger than $2^l - 1$. Thus, we can imagine the set of counters to be a cycle of length $2^l$. Each sequence of counters between two resets marks a segment on this cycle, as illustrated in Figure 2. A collision between two sequences of counters occurs if those segments overlap, also shown in Figure 2.

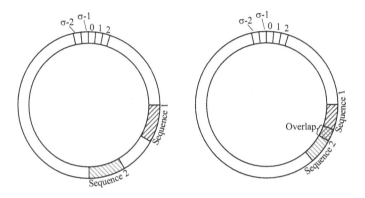

**Fig. 2.** The cycle of counter values

*Nonce length:* A first intuition is that choosing the same nonce length as for an RNG-based NG would be save. But in this case, our new solution would not offer any advantages compared to an RNG-based NG. Thus, we are interested in showing that the required nonce length can be made smaller, as follows.

**Lemma 2.** *After $r - 1$ resets, the probability for at least one collision in a randomised reset solution is at most $\frac{r-1}{2^l}\left(\theta - \frac{r}{2}\right)$.*

*Proof.* We number the nonce sequences by $1, 2, \ldots, r$ and denote their respective lengths by $s_1, s_2, \ldots, s_r$. Before the first reset, there is only sequence 1, i.e. there can not be any collisions unless $s_1 \geq 2^l$.

Now consider the drawing of the starting point for sequence 2. Obviously, it must not collide with any of the $s_1$ points on sequence 1. In addition, it must not coincide with any of the $s_2 - 1$ points before sequence 1 either, since otherwise, the sequences will overlap (see Figure 3). Thus, a collision occurs with a probability of $\frac{1}{2^l}(s_1 + s_2 - 1)$.

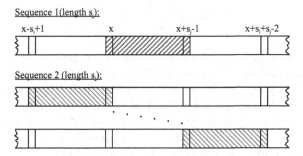

If sequence 1 starts with nonce $x$, then sequence 2 will overlap if its starting nonce lies between $x - s_i + 1$ and $x + s_j - 1$ (both inclusive).

**Fig. 3.** Overlapping sequences

For sequence 3, we already have to take the sequences 1 and 2 into account, and so on. In general, the probability $p_i$ ($i \geq 2$) for a new sequence to overlap with an already existing one is upper bounded as follows:

$$p_i \leq \frac{1}{2^l} \cdot \sum_{j=1}^{i-1} (s_j + s_i - 1).$$

The overall probability that at least one collision has occured after $r$ sequences (i.e., $r - 1$ resets) is then upper bounded by

$$
\begin{aligned}
p_c &\leq \sum_{i=2}^{r} p_i \leq \frac{1}{2^l} \cdot \sum_{i=2}^{r} \sum_{j=1}^{i-1} (s_j + s_i - 1) \\
&= \frac{1}{2^l} \cdot \left( (r-1) \sum_{i=1}^{r} s_i - \sum_{i=1}^{r-1} i \right) \\
&= \frac{1}{2^l} \cdot \left( (r-1) \cdot \theta - \frac{r \cdot (r-1)}{2} \right) \\
&= \frac{r-1}{2^l} \cdot \left( \theta - \frac{r}{2} \right) \qquad\qquad \square
\end{aligned}
$$

Note that this can be considered as a generalisation of the bound for purely random nonces. Purely random nonces correspond to random reset system with one reset after each output nonce, meaning $r = \theta$. In this case, the above collision bound becomes $\frac{\theta-1}{2^l} \cdot \left( \theta - \frac{\theta}{2} \right)$. This is the same as $\frac{\theta^2 - \theta}{2 \cdot 2^l}$, which is the bound we already knew for purely random nonces.

**Corollary 2.** *Assume that a counter-based NG with randomised reset suffers at most $r - 1$ resets during the lifetime of one key. If the maximum acceptable collision probability is $p_{\max}$, then we need*

$$l \geq \log_2 \left( \frac{r-1}{p_{\max}} \cdot \left( \theta - \frac{r}{2} \right) \right).$$

*Example:* Consider the case of a resource-restricted device[5] with a built-in key and a maximum lifetime of 5 years ($2^{27.23}$ seconds). After each power-down, the device needs 30 seconds ($2^{4.91}$) to re-boot, which limits the number of possible resets to $r = 2^{22.32}$. On the other hand, if the system is running, it can send (due to bandwidth restrictions) at most 100 nonces ($2^{6.64}$) per second, i.e. up to $\theta = 2^{33.87}$ nonces in its lifetime. Thus, a naive application of the above corollary yields a minimum nonce length of $56.19 - \log_2(p_{\max})$ bit.

However, this approach overestimates the required nonce length. The reason is that the system can not be busy re-booting all the time while at the same time producing nonces all the time. In fact, the number $r - 1$ of calls to the Reset() function and the number $\theta$ of calls to the Next() function depend on each other. An analysis of the function $f(r) = \log_2(\frac{r}{p_{\max}} \cdot (\theta - \frac{r}{2}))$ where $\theta$ is written as a function of $r$ shows that the function is constantly increasing in the interval $[1, \theta]$. Accordingly, the maximum is reached for $r = \theta$. Since $r$ has a known upper bound, we have $\theta = r = 2^{22.32}$, proving that a minimum nonce length of $43.64 - \log_2(p_{\max})$ bit is in fact sufficient.

## 3.2   Mixed Solution 1

The mixed solution described in Section 2.3 can also be used to solve the nonce reset problem. Again, an RNG is required, which may induce new problems into the solution.

*Nonce Length:* A general bound for the collision probability of this mixed solution can be given as follows.

**Lemma 3.** *After $r - 1$ resets, the probability for at least one collision in the mixed solution from section 2.3 is bounded by*

$$p_c \leq \frac{\theta \cdot (\theta + 2^{l_1}(r - 1))}{2 \cdot 2^l}.$$

*Proof.* We write again $S_1 = 2^{l_1}, S_2 = 2^{l_2}$ and $S = 2^l$. We start our analysis by observing that the worst case occurs if for each of $(r - 1)$ resets, the same value is assigned to the counter part (as is the case in a counter solution without randomised reset). In this case, we have $r$ sequences, each of which has a length of $\frac{\theta}{r}$ nonces. Thus, for no value of cnt we can have more than $\frac{\theta}{S_1 \cdot r} + 1$ assignments to rnd between two resets, and the total number $a$ of rnd values for each cnt is bounded by

$$a \leq r \cdot \left( \frac{\theta}{S_1 \cdot r} + 1 \right) \leq \frac{\theta}{S_1} + r.$$

---

[5] The example is taken from a real-world solution for intelligent homes.

For each value of `cnt`, this means that the collision probability is upper bounded using the birthday bound by

$$p'_c[j] \leq \frac{a_j^2 - a_j}{2S_2}.$$

Consequently, the total collision probability for all $S_1$ values of `cnt` is bounded by

$$p_c \leq \sum_{j=1}^{S_1} \frac{a_j^2 - a_j}{2S_2} = \frac{1}{2S_2} \left( \sum_{j=1}^{S_1} a_j^2 - \sum_{j=1}^{S_1} a_j \right).$$

The sum $\sum_{j=1}^{S_1} a_j^2$ with a term sum of $\theta$ and terms in an interval $[0, \ldots, \frac{\theta}{S_1} + r]$ can be shown to be upper bounded by $\theta \cdot \left( \frac{\theta}{S_1} + r \right)$. In addition, it holds that $\sum_{j=1}^{S_1} a_j = \theta$. Thus, the bound can be computed to be

$$p_c \leq \frac{1}{2S_2} \left( \theta \cdot \left( \frac{\theta}{S_1} + r \right) + \theta \right) = \frac{\theta \cdot (\theta + S_1(r-1))}{2S}.$$

By resubstituting $S_1$ and $S$, we obtain the desired result.    □

Note that this bound is a special case of the bound for the mixed solution without nonce reset, since for $r = 0$, we obtain $p_c \leq \frac{\theta \cdot (\theta - 2^{l_1})}{2 \cdot 2^l} = \frac{\theta^2 - \theta 2^{l_1}}{2 \cdot 2^l}$, which is exactly the bound for the mixed solution from Section 2.3.

**Corollary 3.** *If the maximum acceptable collision probability is $p_{max}$, then the minimum nonce length for the mixed solution from section 2.3 is*

$$l \geq \log_2 \left( \frac{\theta \cdot (\theta + 2^{l_1}(r-1))}{2 \cdot p_{max}} \right).$$

### 3.3  Mixed Solution 2

An alternative is to modify the mixed solution described in Section 2.3 as follows: For every call to the `Next()` function, only the `cnt` part is updated. On the other hand, for every call to the `Init()` or `Reset()` function, the `cnt` part is set to 0, and the `rnd` part is set to a random value which is preserved until the next reset.

*Nonce Length:* This solution can be made very resistant against nonce resets by choosing the parameters as follows:

- If $2^{l_1} \geq \theta$, the construction is completely resistant against collisions as long as no nonce resets occur.
- If $2^{l_2} \geq \frac{r^2 - r}{2 \cdot p_{max}}$, the probability for a collision in case of a reset will be less than $p_{max}$.

Thus, the recommended nonce length for this solution is $l \geq \log_2 \left( \theta \cdot \frac{r^2 - r}{2 \cdot p_{max}} \right)$ for $r > 1$. Note that if the maximum number $r - 1$ of expected resets is small compared to the total number of nonces, this solution is superior to mixed solution 1.

| function Init()      | function Reset()          | function Next()        |
|----------------------|---------------------------|------------------------|
| 1. $i \leftarrow 0$  | 1. retrieve $p$ from NVM  | 1. if $i = p$          |
| 2. $p \leftarrow u$  | 2. $i \leftarrow p$       | 2.   $p \leftarrow p + u$ |
| 3. store $p$ in NVM  | 3. $p \leftarrow p + u$   | 3.   store $p$ in NVM  |
|                      | 4. store $p$ in NVM       | 4. output $i$          |
|                      |                           | 5. $i \leftarrow i + 1$ |

**Fig. 4.** Counter-based NG using reset points

## 3.4   Reset Points

A completely different solution is to use reset points. This means that instead of using random start values after a reset, deterministic values are used in such a way that collision-freeness can be guaranteed. This is achieved by occasionally storing a safe reset point to non-volatile memory[6]. If this is done only rarely, the slow hardware access has little impact on the overall system performance.

To this end, we choose an interval size $u$ which defines the distance between two reset points. If no reset occurs after $u$ calls to the Next() function, a new reset point is stored. If, on the other hand, a reset occurs, the counter is set to the last stored reset point. Figure 4 describes this solution.

*Nonce length:* Note that the last nonce value ever to be produced by the system reaches its maximum if for each of the $r - 1$ calls to the function Reset(), a full $u$ nonce values go unused. This means that even the largest nonce will be less than $\theta + (r-1) \cdot u$, and that the nonce length has to be at least $\log_2(\theta + (r-1) \cdot u)$.

Note that this is a generalisation of the nonce length given for simple counter-based NGs. If the system suffers no resets, then $r - 1 = 0$, and the above formula yields the well-known nonce length of $\log_2(\theta)$.

## 4   Comparison and Conclusions

### 4.1   Comparison

Table 1 compares the collision bounds for the solutions described above. All probabilities are given under the assumption that the actual number of nonces produced and actual number of nonce resets occuring do not exceed the anticipated values $\theta$ and $r - 1$, respectively. The table also indicates whether an RNG is required. Remember that if this is the case, there is a probability (albeit low when choosing the right parameters) of producing a nonce collision. For these cases, the table indicates whether one collision significantly increases the risk of getting a whole sequence of collisions.

---

[6] This technique is mentioned in passing by Bernstein in [13], where he writes: "Store a safe nonce value – a new nonce larger than any nonce used – on disk alongside the key."

Table 1. Comparison of nonce generators

| | coll. prob. without reset | coll. prob. with reset | RNG required? | colliding sequences |
|---|---|---|---|---|
| Counter w. rand. reset | $p_c = 0$ | $p_c \leq \frac{r-1}{2^l}\left(\theta - \frac{r}{2}\right)$ | yes | yes |
| RNG-based nonce | $p_c \leq \frac{\theta^2-\theta}{2\cdot 2^l}$ | $p_c \leq \frac{\theta^2-\theta}{2\cdot 2^l}$ | yes | no |
| Mixed solution 1 | $p_c \leq \frac{\theta^2-\theta\cdot 2^{l_1}}{2\cdot 2^l}$ | $p_c \leq \frac{\theta\cdot(\theta+2^{l_1}(r-1))}{2\cdot 2^l}$ | yes | no |
| Mixed solution 2 | $p_c = 0$ | $p_c \leq \frac{r^2-r}{2\cdot 2^l}$ | yes | yes |
| Couter w. reset points | $p_c = 0$ | $p_c = 0$ | no | n.a. |

If no nonce resets are to be expected, the simple counter-based NG (not contained in the table) is the optimal strategy, yielding a minimum nonce length and a collision probability of zero.

If, however, nonce resets can happen, then the choice of the optimal NG and its parameters depends on the application situation. However, it seems that for many applications, the use of nonce reset points offers an optimal strategy. If storing a reset point at regular intervals is an option, this solution gives a guarantee for collision-freeness while having the shortest nonce length ($\log_2(\theta + (r - 1) \cdot u)$ bit) of all solutions discussed. In addition, it does not require a random-number generator, thus removing an often vulnerable component from the solution.

Where regular storing of reset points is not an option, the mixed solution 2 will often give good results. Note that for most systems, a nonce reset is a rare event, and for small values of $r$, mixed solution 2 provides a low collision probability and a low nonce length.

### 4.2  Conclusions

In this paper, we have collected and described a number of nonce generators that are used in practice. For all of these generators, we have derived formal bounds on the collision probabilities and nonce lengths. In addition, we have described the nonce reset problem, given a theoretical analysis of suitable nonce generators and discussed their resistance against nonce resets. To the best of our knowledge, this is the first time that a full formal treatment of popular nonce generator techniques is given. In particular, we hope to have given system designers a toolbox for choosing the right nonce generator and parameters for their target application.

## Acknowledgements

The author wishes to thank G. Leander and L.R. Knudsen for inspiring discussions, D. Wagner for helpful comments on an early draft of this paper, and several anonymous reviewers for proposed improvements.

# References

1. Wiktionary: Nonce (2009), http://en.wiktionary.org/wiki/nonce
2. Wegmann, M., Carter, J.: New hash functions and their use in authentication and set equality. Journal of Computer and System Sciences 22, 265–279 (1981)
3. List, C.M.: Consequences of nonce reuse (2007), http://www1.ietf.org/mail-archive/web/cfrg/
4. Rogaway, P.: Nonce-based symmetric encryption. In: Roy, B., Meier, W. (eds.) FSE 2004. LNCS, vol. 3017, pp. 348–359. Springer, Heidelberg (2004)
5. Borisov, N., Goldberg, I., Wagner, D.: Intercepting mobile communications: The insecurity of 802.11. In: Proc. 7th International Conference on Mobile Computing and Networking, pp. 180–189. ACM, New York (2001)
6. Kohno, T.: Attacking and repairing the WinZip encryption scheme. In: Proc. 11th ACM Conference on Computer and Communications Security (CCS 2004), pp. 72–81. ACM Press, New York (2004)
7. Sabin, T.: Vulnerability in Windows NT's SYSKEY encryption. BindView Security Advisory (December 16, 1999), http://marc.info/?l=bugtraq&m=94537756429898&w=2
8. Wu, H.: The misuse of RC4 in Microsoft Word and Excel (2005), http://eprint.iacr.org/2005/007
9. Barak, B., Halevi, S.: A model and architecture for pseudo-random generation with applications to /dev/random. In: Proc. 12th ACM Conference on Computer and Communications Security (CCS 2005), pp. 203–212. ACM Press, New York (2005)
10. Gong, L.: A security risk of depending on synchronized clocks. ACM Operating Systems Review 26, 49–53 (1992)
11. Neuman, B., Stubblebine, S.: A note on the use of timestamps as nonces. Operating Systems Review 27, 10–14 (1993)
12. Shoup, V.: A Computational Introduction to Number Theory and Algebra. Cambridge University Press, Cambridge (2005)
13. Bernstein, D.: The Poly1305-AES message-authentication code. In: Gilbert, H., Handschuh, H. (eds.) FSE 2005. LNCS, vol. 3557, pp. 32–49. Springer, Heidelberg (2005), http://cr.yp.to/mac.html#papers

# A    Detailed Analysis of the Mixed Solution

In Section 2.3, the collision probability for the mixed solution was upper bounded by $\frac{\theta^2-\theta\cdot S_1}{2S}$. However, this bound only holds if $S_1$ divides $\theta$; otherwise, the bound is too low. In the following, we derive a universal bound.

**Theorem 1.** *The collision probability for the mixed solution in Section 2.3 is upper bounded by* $\frac{\theta^2-\theta\cdot S_1}{2S} + \frac{S_1}{8S_2}$.

*Proof.* Let us start by introducing the following notation. We write $\theta = q\cdot S_1 + r$, where $q$ and $r$ are the unique quotient and remainder, resp., when dividing $\theta$ by $S_1$.

The exact collision probability for the mixed solution can be modelled as follows. Imagine that there are $S_1 = 2^{l_1}$ containers and $S_2 = 2^{l_2}$ balls. With each call $i$ to the NG, one ball is drawn at random (with replacement) and thrown into $i$-th container. This means that after $\theta$ iterations, all $S_1$ containers contain $q$ balls, and $r$ containers contain one additional ball. Thus, the total exact collision probability can be described by the following formula:

$$p_c = 1 - \left(\prod_{i=1}^{q-1}\left(1-\frac{i}{S_2}\right)\right)^{S_1}\cdot\left(1-\frac{q}{S_2}\right)^r.$$

By using the approximation that $1 - \prod(1-p_i) \leq \sum p_i$ for $0 < p_i \leq 1$, we can upper bound this probability as follows:

$$p_c \leq S_1\cdot\sum_{i=1}^{q-1}\frac{i}{S_2}+r\cdot\left(\frac{q}{S_2}\right) = S_1\cdot\frac{q\cdot(q-1)}{2\cdot S_2}+\frac{rq}{S_2}.$$

Substituting $S_2$ by $S/S_1$, we obtain:

$$p_c \leq S_1^2\cdot\frac{q\cdot(q-1)}{2\cdot S}+\frac{rqS_1}{S} = \frac{S_1^2(q^2-q)+2rqS_1}{2S}.$$

This bound is a correct bound in the sense that it is always larger than the correct probability function. Now let us consider the estimate given in Section 2.3:

$$\frac{\theta^2-\theta\cdot S_1}{2S} = \frac{(q\cdot S_1+r)^2-(q\cdot S_1+r)\cdot S_1}{2S}$$
$$= \frac{S_1^2(q^2-q)+2rqS_1+(r^2-S_1r)}{2S}.$$

As we can see, this bound differs from the above by the term $\frac{r^2-S_1r}{2S}$. Since $r \leq S_1$ by definition of $r$, this term is always $< 0$ with the exception of $r = 0$, in which case both functions are identical. Figure 5 illustrates this by showing the correct probability (dotted), the simplified bound (solid), and the correct bound (dashed).

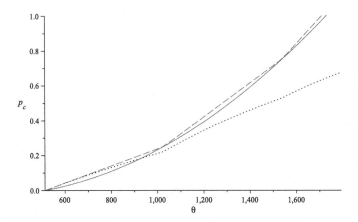

**Fig. 5.** Comparing correct and simplified bound for mixed solution ($l = 20, l_1 = 9$)

Analysis of the error function $\frac{r^2 - S_1 r}{2S}$ shows that it achieves its maximum for $r = \frac{S_1}{2}$, yielding a maximum error of $-\frac{S_1}{8S_2}$. By adding this maximum error to the simplified bound, we obtain a bound that is always correct and prove the theorem.    $\square$

# MAC Precomputation
# with Applications to Secure Memory

Juan Garay[1,*], Vladimir Kolesnikov[2], and Rae McLellan[2]

[1] AT&T Labs – Research, 180 Park Ave., Florham Park, NJ 07932
`garay@research.att.com`
[2] Bell Labs, 600 Mountain Ave., Murray Hill, NJ 07974, USA
`{kolesnikov,rae}@research.bell-labs.com`

**Abstract.** We present *ShMAC* (Shallow MAC), a fixed input length message authentication code that performs most of the computation *prior* to the availability of the message. Specifically, ShMAC's message-dependent computation is much faster and smaller in hardware than the evaluation of a pseudorandom permutation (PRP), and can be implemented by a small *shallow* circuit, while its precomputation consists of one PRP evaluation.

A main building block for ShMAC is the notion of *strong differential uniformity* (SDU), which we introduce, and which may be of independent interest. We present an efficient SDU construction built from previously considered differentially uniform functions.

Our motivating application is a system where a hardware-secured processor uses memory controlled by an adversary. We present in technical detail a novel, more efficient approach to encrypting and authenticating memory and discuss the associated trade-offs, while paying special attention to minimizing hardware costs and the reduction of DRAM latency.

## 1 Introduction

With the publicized attacks on consumer computer products, such as the iPhone [1] and Xbox, security of computing has become a topic of widespread commercial interest. Broadly speaking, security of computing can be divided into two main areas — hardware and software security. Software security is concerned with integrity of the software and prevention of control or compromise by an outside attacker. Hardware security, on the other hand, assumes that the adversary has full physical access to the device and may use oscilloscopes and logic analyzers to observe and compromise the computing system. This paper focuses on ways to efficiently provide hardware security. For that purpose, we present a new MAC technique, and discuss its application in securing memory.

Recent VLSI advances have provided strongly tamper-resistant hardware computing platforms by integrating complete Systems on a Chip, through SoC technology. It is considered infeasible to all but government-scale attackers to perform

---

* Work partly done while the author was at Bell Labs.

P. Samarati et al. (Eds.): ISC 2009, LNCS 5735, pp. 427–442, 2009.

meaningful analysis of the internals of production SoC. Ideally, we would store and execute the entire computation on a SoC, eliminate external DRAM (Dynamic Random Access Memory), and encrypt all off-chip communication. However, this is not possible in most practical scenarios, due to prohibitive costs of such large SoC. In this paper we consider the question of how to encrypt off-chip DRAM transactions with minimal performance degradation and cost increase. Note, such transactions occur much more frequently than network messages and have much more stringent latency requirements. Since processor performance is so tightly dependent on off-chip memory latency, speeding up the encryption/authentication process is of primary importance.

For many on-chip bus protocols, e.g., [2,3], the address is available early in the bus transaction between the processor and memory controller, while the larger-size data follows later and is composed of multiple transfers of sub-units. Such serialization of data transfers in on-chip buses is an engineering trade-off between performance and the number of wires required for a wider bus. Therefore, an encryption/authentication algorithm which can postpone data-dependent computation, can start earlier in the memory transaction and potentially reduce the performance impact of an encrypted memory system. This paper describes an efficient way to encrypt off-chip memory transactions and provide data authentication that takes advantage of the early arrival of the memory address.

**Our Contributions.** Our main contribution is a new fixed input length Message Authentication Code (MAC) construction which allows the bulk of the MAC computation to be performed *before* the message $m$ is available. The computation dependent on $m$ is the evaluation of (a new variant of) an $\epsilon$-*differentially uniform* ($\epsilon$-DU) function [4,5] (cf. Section 2) and an XOR operation, which is much simpler and faster than a typical MAC implementation via a block cipher. In envisioned instantiations, MAC precomputation is a PRP (e.g., a full 10-round AES) evaluation, and the remaining computation (dependent on $m$) is an evaluation of 2- or 4-round AES.

As a second contribution, we present a secure DRAM architecture, discussing at length security/efficiency trade-offs and underlying design choices.

**Related Work.** As our work consists of two relatively independent (but complementary) contributions – a cryptographic construction and a secure DRAM design – we separate the discussions of related work accordingly below. First, we discuss Wegman-Carter [6] and related constructions, followed by an overview of previous work on secure DRAM.

*On precomputation in Fixed Input Length MACs.* Some details and applications of the specific property of MAC precomputation have been discussed in the literature (e.g., [7]), although, to our knowledge, not in the severely restricted environments (with respect to both data-dependent computation and precomputation time and chip surface area) that we consider. In this section we overview previous work on message authentication, with emphasis on precomputation. We discuss the relationships between the building blocks, clarify the terminology and review some efficient constructions.

We are mainly interested in validating 256-bit data blocks. The direct approach is to simply encrypt (e.g., with a blockcipher, such as AES) the data concatenated with the address, and possibly some redundancy. However, this solution is unsatisfactory since it does not allow for precomputation, and, further, requires both encryption and decryption hardware.

Before discussing previous work in more detail, we recall some definitions. Let $H : K \times X \to Y$ be a function family, indexed by the key $k \in K$. A Universal Hash Function (UHF), or universal$_2$, $H$ guarantees that $\forall x_1 \neq x_2 \in X, \Pr_k[H_k(x_1) = H_k(x_2)] \leq \frac{1}{|Y|}$. That is, no pair of preimages is mapped into the same value by more than one $|Y|$-th of the functions. A stronger notion of Strongly Universal (SU) $H$ requires that $\forall x_1 \neq x_2 \in X, \forall y_1, y_2 \in Y, \Pr_k[H_k(x_1) = y_1 \wedge H_k(x_2) = y_2] = \frac{1}{|Y|^2}$. In other words, $H_k$ maps all distinct $x_1, x_2$ independently and uniformly.

One of the most celebrated MAC schemes, and also one that naturally allows precomputation, was proposed by Wegman and Carter [6]. Extending the authors' previous work on UHF families, in [6] they introduced the notion of SU hash families, and showed that $MAC_{k,r}(m) = H_k(m) \oplus r$ is an unconditionally secure MAC, where $H$ is an SU function, $r$ is a one-time pad, and $k$ is a random index into the family $H$.

Stinson [8] formalized the notion of $\epsilon$-Almost SU (ASU), a more general class of functions usable with the Wegman-Carter MAC construction. As the name suggests, ASU functions simply allow less strict bounds on the probability guaranteed by SU. Stinson also showed how to combine a (faster) UHF with an ASU function to obtain a faster ASU function. Brassard [9] pointed out that a pseudorandom generator could be used in place of one-time pad. Krawczyk [10] noticed that $\epsilon$-Almost XOR Universal (AXU) function families, weaker than ASU's, are sufficient for Wegman-Carter MAC. (Recall that $H$ is $\epsilon$-AXU, if $\forall x_1 \neq x_2 \in X, \forall c \in Y, \Pr_k[H_k(x_1) \oplus H_k(x_2) = c] \leq \epsilon$. Krawczyk called this notion *otp-secure*, but AXU is the more frequently used term today.)

Following these fundamental results, a lot of work went into the design of efficient universal, AU, ASU and AXU functions. Most of the research concentrated on software-efficient functions, i.e., those that take advantage of CPU's instruction sets which, in particular, include multiplication. Unfortunately, algebraic solutions are not acceptable in our setting, due to the latency and cost of hardware implementation of multiplication.

In fact, acceptable solutions would only be those that reuse the circuitry of the PRFG to generate the pad $r$ and to evaluate $H$. Our solution does just that. Alternatively, a MAC scheme with a similar performance can be extracted from a large volume of previous work. Several papers contribute pieces of the total solution, but, to our knowledge, none explicitly states it; further, several sources use conflicting terminology.

Firstly, we point out that neither UHF nor AU functions are sufficient for Wegman-Carter MAC security. This is because they do not guarantee that an offset in the argument will not result in an unpredictable offset in the value of $H_k$. For example, the identity function is a UHF, but clearly a Wegman-Carter

MAC based on it is easily forged. We note, however, that UHF and AU are often used in MACs for efficiency reasons, but only as *part* of the function $H$; a stronger ASU component is additionally required in $H$ [8]. Further, some sources (e.g., [11]) "blend" the notions of UHF and SU, in fact defining UHF as SU.

Therefore, although it was previously observed, in [5], for example, that it is possible to obtain AU functions from four-round AES, such results are not applicable for our uses of Wegman-Carter MAC. To our knowledge, the only explicit AXU construction from an $\epsilon$-DU function recently appeared in [12]. In particular, it uses AXU derived from a 4-round AES in the Wegman-Carter MAC. That MAC construction, however ([12], Algorithm 1), generates fresh keys for $H$ and the pad $r$ for each MAC evaluation, which is an unacceptable overhead for our setting. We observe, however, that in our setting, the keys of $H$ could be reused, which would bring the resource requirements down to those in our proposed construction.

*Related Work on Secure Memory.* There is a vast amount of work on securing memory. One direction uses smart cards or other separate adjunct chips such as TPMs (Trusted Platform Module) [13]. These methods are usually limited; for example, they do not protect intellectual property contained in the software running on an (insecure) host, but only secure execution of small parts of it by running it on the smart card/TPM. An interesting use of a smart card processor was proposed in the X$\mu$P system [14]. X$\mu$P allows the ROM-less smart card to execute signed code, using the terminal as a (cheap) storage. [14] describes ways of securing the computation, including a public key and symmetric key-based authentication of the executing code. At a high level, the symmetric-key case resembles our setting; however, X$\mu$P is not as severely restricted, uses computationally expensive hash functions, and does not attempt MAC precomputation.

Another system is XOM [15], which provides architectural support for software licensing and allows code to be authenticated and run even under untrusted operating systems. XOM requires a significant modification of the processor's instruction cache, the addition of special instructions, and operating system (OS) support. Our system is more general, is independent of the instruction set, and supports any processor architecture.

Closer to our setting, securing memory in a SoC system was announced by IBM [16] and considered academically (e.g., AEGIS [17,18], CryptoPage [19], TEC-Tree [20]). These systems validate memory by maintaining a hash tree of the entire DRAM, each transaction requiring 20-30 DRAM accesses and hash evaluations. Caching part of the tree somewhat reduces the performance impact [21] at the cost of on-chip resources. Solutions to the similar problem of "online memory checking" (see [22,23] and references therein), where the checker (processor) ensures (only) the integrity of adversarially controlled storage (RAM), also incur a logarithmic overhead. Our MAC approach is an order of magnitude faster (but with weaker replay protection). We believe such compromise is well suited for many industrial applications.

Other systems such as the one presented in [24], PE-ICE [25] or TEC-Tree [20], forgo Merkle trees but require significant on-chip storage for nonce or checksum values updated on each memory write. While the amount of on-chip storage can be as small as a byte for each encrypted off-chip storage block, this method doesn't scale to to support the desired gigabytes of off-chip DRAM. Further, it can be shown that "natural" CRC-based integrity checking mechanisms (e.g., [24]) have critical vulnerabilities [26].

Given these overheads, we choose to forgo replay attack protection, but instead mitigate the threat by changing the encryption keys at reasonably frequent intervals. In encryption and authentication, we focus on efficiency and minimal additional on-chip resources. In our system, authenticating a memory access takes slightly more than a PRP evaluation, and is effectively further reduced by the precomputation of the MAC.

**Organization of the Paper.** In Section 2 we introduce the necessary notation, definitions and building blocks that we will be using. Section 3 is the cryptographic core of this work. We first discuss the intuition behind, and then formally present our MAC construction — *ShMAC*, together with an evaluation of its performance and instantiation considerations. In Section 4 we present the system aspects of our secure memory architecture. In particular, we discuss the assumptions, security objectives, and restrictions of our system, and its use of ShMAC. Due to space limitations, proofs, as well as additional system design considerations are presented in the full version of the paper [26].

## 2   Preliminaries

We denote the security parameter by $k$, keys by $\ell \in \{0,1\}^k$, a pseudorandom permutation generator by PRPG, and a pseudorandom permutation by PRP. The constructions in this work assume the existence of PRPGs.

### 2.1   Message Authentication Code (MAC)

A MAC is a tool for ensuring data integrity. It is most commonly used in authenticating communication, and we use it in a similar setting. In our setting, the data is stored in an untrusted location and MAC is used to ensure its integrity.

In a traditional MAC, the tag generation function is stateless and deterministic, and verification is done by applying the tagging function to compute the correct tag of the given message, and comparing it with the candidate tag. We need a slightly more general notion, which we call a *nonce-based MAC*, and which allows the generation function to use nonces. More formally:

**Definition 1.** *A nonce-based message authentication code is a stateless algorithm* $MAC : \{0,1\}^k \times \{0,1\}^k \times \{0,1\}^* \to \text{TAG}$, *which on input key* $\ell \in \{0,1\}^k$, *nonce* $r \in \{0,1\}^k$ *and message* $m \in \{0,1\}^*$, *outputs a tag* $t \in \text{TAG}$. *(Here* TAG *is the domain of tags, which depends on $k$.) We will sometimes write* $MAC_{\ell,r}(m)$ *to mean* $MAC(\ell, r, m)$; *we will also sometimes omit $r$ and just write* $MAC_\ell(m)$ *for simplicity.*

*Now let $\ell \in_R \{0,1\}^k$, and $\mathcal{A}$ be a nonce-respecting polynomial-time adversary with access to oracle $\mathcal{O}(r,m) = MAC_{\ell,r}(m)$. $\mathcal{A}$ outputs a message $m'$ and its alleged signature (i.e., a nonce-tag pair) $\tau' = (r', t')$, subject to the condition that it never received $t'$ from $\mathcal{O}(r', m')$. We say that $MAC$ is secure if for every such $\mathcal{A}$, $\Pr[MAC_{\ell,r'}(m') = t'] < 1/k^c$ for every $c$ and sufficiently large $k$.*

In the above definition, by "nonce-respecting adversary" we mean an adversary who never queries the MAC oracle with the same nonce twice. We give $\mathcal{A}$ the freedom to choose his nonces at will with the single above restriction. Throughout the paper, all our adversaries are nonce-respecting.

We remark that, although we define MAC in its commonly encountered general form, in our application we will use the fixed-length variant of this definition, and specifically for messages of length $k$, i.e., $m \in \{0,1\}^k$ rather than $m \in \{0,1\}^*$. Further, it will be convenient for us to use keys longer than $k$ bits, and thus we allow $\ell \in \{0,1\}^{ck}$, where $c$ is a small constant (e.g., $c = 2$).

We note that Definition 1 imposes a *strong unforgeability* property [27], which enforces that $\mathcal{A}$ cannot create new valid tags on the old (i.e., already tagged) messages. In contrast, "regular" message authentication schemes often do not consider a forgery a valid message-signature pair $(m, \tau')$ when the oracle was queried on $m$ and returned $\tau \neq \tau'$. In our application, however, strong unforgeability is essential. We note that, as a side effect, strong unforgeability allows us to avoid the introduction and discussion of verification oracles in the definition of MAC. (See [27] for further discussion on this topic.)

We remark that, in practice, MAC schemes are built directly from PRPGs. Similarly to PRPGs, practical MAC schemes are not defined for all $k$, but rather, for some fixed but sufficiently large $k$. Our MAC construction will follow the latter paradigm, but we will perform the analysis in the asymptotic setting.

## 2.2   $\epsilon$-Differential Uniformity and Properties of AES Rounds

A main building block for our MAC construction is *Strongly Differentially Uniform* (SDU) functions, introduced in Section 3.2. An SDU function family is a stronger version of a *Differentially Uniform* (DU) family, which is widely used in block cipher design and which we now present as background.

For our application we will use the sub-class of *keyed $\epsilon$-DU permutations*, due to their efficiency. Therefore, for simplicity, we do not discuss here $\epsilon$-DU functions in their full generality. However, we note that unkeyed permutations or functions [4] could also be used in our constructions, and our analysis (appropriately modified for indices, etc.) equally applies.

Let $\mathcal{G} : \{0,1\}^k \times \{0,1\}^k \to \{0,1\}^k$ be a keyed permutation family. Let $\Delta x, \Delta y \in \{0,1\}^k$ be fixed and let $X \in \{0,1\}^k$ be a uniformly distributed random variable. Let $G_\ell \in \mathcal{G}$. (In the sequel, we may sometimes omit index $\ell$ and write $G \in \mathcal{G}$, when $\ell$ is clear from the context or where it does not play a role.) The *differential probability* $DP(\Delta x, \Delta y, \ell)$ is defined as

$$DP(\Delta x, \Delta y, \ell) = \Pr_X[G_\ell(X) \oplus G_\ell(X \oplus \Delta x) = \Delta y]. \tag{1}$$

Here $\Delta x$ and $\Delta y$ are viewed as input/output differences. The *expected* differential probability $EDP(\Delta x, \Delta y)$ is the expectation of $DP(\Delta x, \Delta y, \ell)$, over all keys $\ell$. We are interested in the *maximum EDP (MEDP)*:

$$MEDP(\mathcal{G}) = max_{\Delta x, \Delta y \in \{0,1\}^k \backslash 0} EDP(\Delta x, \Delta y). \qquad (2)$$

Informally, a small *MEDP* value corresponds to good bit mixing by $\mathcal{G}$ — indeed, small *MEDP* means that any change in the (randomly chosen) input of the cipher results in an unpredictable output. However, small *MEDP* does not necessarily imply "security under multiple queries," since the *MEDP* experiment is defined over all keys $\ell$.

**Definition 2.** *We say that a permutation family $\mathcal{G}$ as defined above is $\epsilon$-Differentially Uniform ($\epsilon$-DU), if $MEDP(\mathcal{G}) \leq \epsilon$.*

It is well known [28,29] that the *MEDP* of two-round AES (AES2) is at most $1.6 \cdot 2^{-28}$, and the *MEDP* of four-round AES (AES4) is at most $1.8 \cdot 2^{-110}$. Thus, AES2 is a $1.6 \cdot 2^{-28}$-DU permutation, and AES4 is a $1.8 \cdot 2^{-110}$-DU permutation.

## 3   *ShMAC*: MAC with Precomputation

In this section we present *Shallow MAC* (ShMAC), a MAC scheme which takes advantage of precomputation. The required precomputation essentially consists of one PRP evaluation, while the message-dependent portion is a small shallow circuit, which can be evaluated in a fraction of the time required for a PRP evaluation. (As a bonus, in our envisioned instantiation, precomputation can share hardware gates with the rest of MAC computation. This is a critical advantage in cases where chip area is restricted, as it is in FPGAs.)

Recall that we require a low-latency MAC scheme, simultaneously "cheap" to implement in hardware, and faster than the evaluation of a PRP (e.g., AES) or a hash function. This requirement precludes many standard MAC solutions, such as AES-based, which require availability of the message at the beginning of the computation. (To be concrete about the involved latencies, recall that AES requires the sequential evaluation of at least 10 rounds[1]. Further, many (but not all) Universal Hash Function (UHF)-based constructions require expensive group arithmetic and additional hardware, and thus are unacceptable in this setting. See Section 1 for more details.)

However, as also discussed in Section 1, in many systems the address of the memory transaction arrives before the data, and thus the hardware MAC unit is idle waiting for the data. We explore the possibility of using these idle cycles to perform *precomputation* to speed up MAC generation.

---

[1] For our application, fewer rounds (e.g., 8) would provide an adequate level of security, because the keys are refreshed frequently, and $\mathcal{A}$ would only have on the order of seconds or minutes to "crack" the MAC.

### 3.1   The Intuition behind ShMAC

An $\epsilon$-DU permutation family $\mathcal{G}$ (e.g., 2-round AES), an object with much weaker security properties than a PRPG, can in principle be the core of a MAC, with appropriate pre- and post-computation. Indeed, $G \in_R \mathcal{G}$ provides good bit mixing, but only on random inputs. We satisfy this by using (nonce-based new and secret) precomputed randomness to mask the data $d$ prior to each application of $G$. This masking of the inputs additionally prevents adversary $\mathcal{A}$ from collecting any information on (the key of) $G$, even if $\mathcal{A}$ sees MAC evaluated on messages of his choice (i.e., queries the MAC oracle adaptively).

Note that even though $\mathcal{A}$ has no knowledge of the random mask $mask_r$ derived from nonce $r$, he can attempt a forgery using the same $r$ (and thus the same $mask_r$). The output unpredictability guarantees of DU functions are too weak to protect against this attack, since in our scenario $\mathcal{A}$ knows $G(d \oplus mask_r)^2$. We strengthen the notion of DU to preserve its guarantees even after one query to $G$ – see Definition 3 below. In terms of implementation, it turns out that masking the output of $G$ with fixed secret randomness (which can be viewed as part of $G$'s key) is sufficient to satisfy the stronger requirements, and results in a secure MAC.

### 3.2   $\epsilon$-Strongly Differentially Uniform Functions

In this section we introduce the notion of *Strong Differential Uniformity* (SDU), discuss its relationship with DU, and present an efficient construction. The new notion is a natural building block in MAC constructions, including ours, and may have applications in other areas. For simplicity, we give an asymptotic notion of $\epsilon$-Strongly Differentially Uniform ($\epsilon$-SDU) permutations, by allowing $\epsilon$ to be a function of the security parameter $k$.

**Definition 3.** *Let $\mathcal{G} : \{0,1\}^k \times \{0,1\}^k \to \{0,1\}^k$ be a permutation family indexed by security parameter $k$, and $\mathcal{A}$ be a computationally unbounded TM. Consider the following experiment* $\mathsf{SDU}_{\mathcal{A},\mathcal{G}}(k)$:

1.   $G \leftarrow \mathcal{G}$ *is selected at random by choosing the key. Further, a random $R \in \{0,1\}^k$ is chosen.*
2.   $\mathcal{A}$ *provides $d$, and receives $G(d \oplus R)$. $\mathcal{A}$ outputs $\Delta x, \Delta y \in \{0,1\}^k \setminus 0$.*
3.   *The output of the experiment is defined to be 1 if $G(d \oplus R) \oplus \Delta y = G(d \oplus R \oplus \Delta x)$, and 0 otherwise.*

*We say that $\mathcal{G}$ is $\epsilon(k)$-Strongly Differentially Uniform ($\epsilon$-SDU for short), if for all $\mathcal{A}$, $\Pr[\mathsf{SDU}_{\mathcal{A},\mathcal{G}}(k) = 1] \leq \epsilon(k)$, where the probability is taken over the random choices used in the experiment.*

It is easy to see how the $\epsilon$-SDU notion is derived from $\epsilon$-DU's. Indeed, Definition 2 can be cast asymptotically and in game style, resulting in exactly Definition 3,

---

[2] It is easy to see that if $G$ is unkeyed, $\mathcal{A}$ can easily construct a forgery.

with the exception that in the corresponding experiment $DU_{\mathcal{A},\mathcal{G}}(k)$, $\mathcal{A}$ is not given $G(d \oplus R)$. Note that the notion of $\epsilon$-SDU is strictly stronger than that of $\epsilon$-DU. Indeed, while unkeyed $\epsilon$-DU functions exist [4], unkeyed $\epsilon$-SDU functions don't. (This is because $\forall \Delta x$, an $\epsilon$-SDU $\mathcal{A}$ can output a winning $\Delta y$ since he can invert the received $G(d \oplus R)$.)

We now show how to construct an efficient $\epsilon$-SDU permutation from any $\epsilon$-DU permutation, such as AES, at additional negligible cost.

**Lemma 1.** *Let $k$ be a security parameter, and $\mathcal{G}'$ be a keyed (or unkeyed) $\epsilon$-DU permutation family. Let $\mathcal{G} = \{G = G' \oplus \ell_1 | G' \in \mathcal{G}', \ell_1 \in \{0,1\}^k\}$ be a family additionally keyed by uniformly chosen $\ell_1 \in_R \{0,1\}^k$. Then $\mathcal{G}$ is an $\epsilon$-SDU permutation family, for the same $\epsilon$.*

### 3.3 ShMAC Construction

Let $d$ be a data block, and $r \in \{0,1\}^k$ be a nonce; in practice $r$ may be a counter or chosen randomly for each MAC evaluation. Let $\mathcal{G}$ be a $\epsilon$-SDU permutation family (Definition 3), where $\epsilon = \epsilon(k)$ is negligible in $k$. Let $G$ be a random member of $\mathcal{G}$, selected by randomly choosing the key $\ell$. Let $F : \{0,1\}^k \to \{0,1\}^k$ be chosen at random, and unknown to the adversary; in practice $F$ is implemented by a PRPG, such as AES. In Construction 1, we use a PRPG as a source of indexed secret fresh randomness for each evaluation of MAC.

**Construction 1** *Let $F, G, r, d$ be as above. Shallow MAC is the algorithm:*

$$\text{ShMAC}_\ell(r, d) = (r, G(d \oplus F(r))) \qquad (3)$$

**Theorem 1.** *Construction 1 is a nonce-based MAC as defined in Definition 1.*

Note that ShMAC can be executed on multiple data blocks by simple concatenation of MACs of individual blocks. This observation is motivated by the fact that efficient $\epsilon$-SDU functions may not be readily available from the literature for larger data blocks. For simplicity, we state the following lemma for the case of two blocks; it can be naturally extended to any number of blocks.

**Lemma 2.** *Let $F, G, r$ be as above, and let $d_0$ and $d_1$ be data blocks. Then*

$$MAC(d_0, d_1) = (r, G(d_0 \oplus F(r, 0)), G(d_1 \oplus F(r, 1))) \qquad (4)$$

*is a nonce-based MAC, as defined in Definition 1.*

### 3.4 ShMAC Instantiation Considerations

Theorem 1 is stated with respect to an ideal object — a randomly chosen function. In practice, this is implemented by means of a PRPG, and therefore Theorem 1 becomes conditional on the existence of PRPGs. Of course, this transition into the computational model improves the chances of $\mathcal{A}$ to forge the MAC, but it can be easily shown that this improvement is negligible. Note that $\epsilon$-DU functions (and thus $\epsilon$-SDU functions) are known to exist, and their use does not constitute an assumption.

As noted previously, "shortcut" 2- or 4-round versions of AES are $\epsilon$-DU permutations. Further, *AddRoundKey*, the final phase of each AES round, implements the transformation of Lemma 1. At the same time, in many hardware implementations, the AES key schedule is precomputed, with round keys being randomly chosen. Such shortcut AES implementations satisfy the stronger $\epsilon$-SDU requirements and are sufficient for security of MAC. In our implementation, we follow this approach.

Depending on the application, the desired input length of MAC may vary. In our encrypted memory system, for example, we operate on 256-bit blocks. We wish to point out several observations that apply to such usage scenarios. First, it is not necessary to use a "wider" (e.g., 256-bit) block cipher as the PRPG $F$. Wider block ciphers are more expensive, since they aim to achieve strong bit mixing on the full block. For example, 256-bit Rijndael requires 14 rounds, vs. the 10 rounds of its 128-bit AES sibling. In our application $F$ is only a source of randomness, and it is sufficient to execute AES twice with corresponding adjustment of the nonce $r$ to $(r, 0)$ and $(r, 1)$. Second, $\mathcal{G}$ must be chosen properly as well. Similarly to AES, 256-bit Rijndael achieves good bit mixing after only 4 rounds [30,31]. Alternatively, we could apply Lemma 2 and execute 128-bit $\mathcal{G}$ (e.g., AES2 or AES4) on each of the two 128-bit halves of masked data.

In our secure memory application, we choose the nonce $r$ to be a concatenation of the address of the memory location and a global RAM transaction counter (the latter may have to be wrapped around for efficiency). This provides a simple and efficient way of generating nonces. Further, this method allows binding the memory value to the memory location, preventing replay of valid data at wrong locations[3]. Another advantage of this nonce choice is that the bulk of the nonce, the memory address, need not be written to memory, as it is managed by underlying subsystems. Further, this method ensures that the nonce is available before the data arrives, thus allowing precomputation.

## 4   Applications: Secure DRAM

We now give an overview of a SoC-based secure system which makes use of ShMAC. While we are mainly interested in integrity checking, for completeness we also discuss a (weak) form of memory encryption. As noted in Section 1, all system operations (with the exception of memory transactions) take place inside the presumably secure and tamper-resistant chip. Therefore, securing the memory, which might be adversarially controlled, closes the main avenue of attack.

---

[3] The binding between the data and the nonce is guaranteed by the strong definition of MAC that we use. Indeed, it disallows a poly-time $\mathcal{A}$ to generate new nonce-tag pairs even on previously signed data. Note that this does not prevent replay at the same memory location. We discuss this trade-off in Section 4.

We now discuss the hardware aspects of an encrypted memory implementation using ShMAC for authentication, associated trade-offs and improvements; we view this technical discussion as an additional contribution of this paper.

In our discussion, we omit some of the aspects of the system, such as secure-boot procedures, the design of which is not related to MAC. We start by presenting the Encryption/Authentication Unit (EAU), its on-chip location, connectivity and relationship with other units.

### 4.1    Overview of Memory Encryption and Authentication

As shown in the conceptual block diagram, Figure 1(a), the EAU is interposed between a conventional DRAM controller and the interface logic that allows potential bus masters, such as CPUs and DMA engines, to access secured off-chip memory. DRAM write transactions are encrypted on the way out to DRAM and read transactions are decrypted coming back from DRAM to the SoC. During the encryption process, a MAC is generated and stored with each encrypted block of memory. During subsequent DRAM read operations, the stored MAC is compared with a newly recomputed MAC to detect corrupted off-chip memory contents.

Each MAC is associated with a fixed number of data bytes, called an *encryption block*, which is the minimal unit of data. That is, the EAU supports only block-size read or write DRAM transactions (and transparently handles creation and verification of the associated MACs). The bus interface logic handles transactions of all sizes. If a bus write transaction affects only a portion of an encryption block, the EAU first needs to read, decrypt and verify the unavailable bits (if any) of the encryption block from off-chip DRAM. Then, it merges the bits to create a full updated encryption block, before it is re-encrypted and written to DRAM.

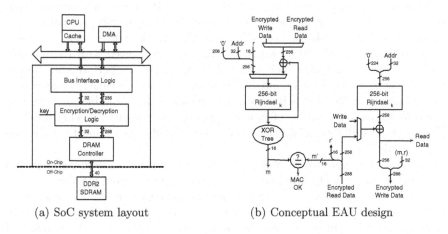

(a) SoC system layout          (b) Conceptual EAU design

**Fig. 1.** SoC-based Encrypted DRAM

Encryption block size and the number of bits in the MAC is a complex engineering trade-off. Clearly, each bit of MAC stored in DRAM is unavailable for user data and therefore represents overhead in an encrypting memory system. Short MAC may not not offer sufficient protection against forgers. Since MACs are stored in the same DRAM as the encryption blocks, there is also the physical and costs constraints of the DRAM data width. For most SoC systems, the DRAM is usually 16 or 32-bits wide. Therefore, MACs with that granularity are preferred.

Similarly, the size of the encryption block is determined by the range of data sizes expected in typical SoC bus transactions. DMA transfers generate bus transactions of size from single bytes to multi-word IP packets, but CPUs present a characteristic bus transaction width that corresponds to their cache line size. Choosing an encryption block size the same as the cache line will efficiently support the most frequent bus transactions.

Our encrypted memory supports a physical 32-bit wide DRAM system. Encryption blocks are 256-bit wide and the associated MAC can be as short as 32 bits, while providing reasonable security. Each DRAM transaction is therefore eight 32-bit words of data followed by one or two words of MAC. This way, the memory overhead is as low as 12.5% and up to 8/9-ths of the DRAM contents is available for user storage.

*Stateless vs. stateful integrity checks.* In our design, the EAU is stateless. This is necessary due to severe on-chip resource restrictions. It is not hard to see that encryption and authentication process as described above exposes the system to replay attacks. For example, an adversary can replace the current contents of memory with a value that was stored in that same location previously. Similarly, an adversary can simply not update the DRAM as required by a write transaction. It is easy to see that the system will decrypt and mistakenly accept such data as valid.

Stateful operation is *necessary* to prevent such attacks. Keeping on-chip state per each memory location, however, is prohibitively expensive. A natural solution is to build a Merkle tree [32] of MACs for the entire memory space, as proposed and deployed in, e.g., [33,17,18]. However, even with the possible optimizations, maintaining such a tree of MAC values is a performance bottleneck (20-30 memory accesses and hash evaluations for each DRAM transaction) and requires significant on-chip resources, which is unacceptable in many settings, including ours. Instead of expensive tree-based integrity checking, we use a much faster method to achieve a level of security sufficient for most commercial applications.

To limit the exposure to replay attacks, we propose periodic refreshing of encryption keys so as to invalidate sufficiently stale encrypted memory state. It is easy to implement, e.g., by maintaining two memory regions, each encrypted with its own key, and growing one region at the expense of the other. If the keys are refreshed often enough, say, every two minutes, then the window of vulnerability to replay attacks is fairly narrow.

This idle-time key refreshment is much more efficient than maintaining a Merkle tree. We believe that frequent key expiration, and a single MAC per encryption block affords practical levels of security with much less mechanism and performance penalty, and thus is a better security/performance trade-off, suitable for most industrial applications.

## 4.2   EAU Implementation Using ShMAC

The most direct method to encrypt and authenticate off-chip memory transactions, would be to encrypt the concatenated address and data[4]. This would produce a 288-bit encrypted memory write value, as shown in Figure 2(a). However, this method serializes the encryption process with memory write operations and, more importantly, adds decryption delay to the already performance-limiting DRAM read latency. Additionally, this scheme requires both encrypt logic and decrypt logic, which is unacceptable for FPGA implementations.

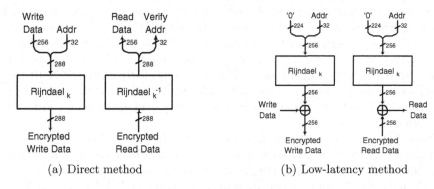

(a) Direct method          (b) Low-latency method

**Fig. 2.** Encryption/authentication methods for off-chip transactions

Motivated by low-latency and small footprint requirements, we prefer a different encryption approach, shown on Figure 2(b), and separate MAC generation from the data encryption process. Figure 1(b) illustrates a conceptual EAU design, described below[5]. A 256-bit pad is generated by Rijndael encryption of the address[6]. The pad is then XORed with the Write data to produce the encrypted result. Since XOR is its own inverse, the same encryption function can be used for both encryption as well as decryption. While encryption remains serialized with the DRAM write operation, the pad calculation can start as soon as the address is available early in the bus transaction. More importantly, for

---

[4] Additional redundant data can be added under the encryption, if stronger integrity checks are desired.

[5] Reasonable security parameter sizes were included in Figure 1(b) for concreteness, however, their values should be evaluated for specific instantiations.

[6] We note that the 256-bit pad can be more efficiently generated by two parallel 128-bit AES encryptions in fewer rounds. We omit this, as well as other natural optimizations, for the sake of clarity.

performance-critical read operations, the pad calculation can occur in parallel with DRAM read latency. Once encrypted DRAM data is available, a single XOR operation is the only additional delay incurred by decryption. As a result, decrypted data is returned to the processor with negligible delay[7].

MAC generation proceeds as follows. The PRP $F$ of Construction 1 is achieved by running full Rijndael on the address concatenated with a nonce $r$. The nonce value can be a global counter that increments with each memory write transaction. We stress that there is no need for expensive pseudorandom generation of the nonce. Note that this first step of the MAC calculation can start as soon as an address is available, simultaneously with the encryption process. The Rijndael output is then XORed with the encrypted data and the same Rijndael data path is reused to compute $G$ of $\epsilon$-SDU family $\mathcal{G}$. In our implementation, $G$ is a four-round Rijndael evaluation[8]. The output of $G$ is collapsed via an XOR tree to a value $m$, which is concatenated with the original unmodified nonce $r$ to form the MAC written to DRAM — this is the ShMAC output.

In contrast with the decryption process, the MAC verification for memory read operations must first wait for the DRAM latency in order to acquire the original nonce $r$, which is stored off-chip. Once data and MAC arrive, $F$ is computed on the address appended with $r$ (14 rounds of 256-bit Rijndael). This value is then XORed with the encrypted read data and the same Rijndael data path is reused to compute $G$, which consists of four rounds of Rijndael. The XOR tree collapses the result to generate $m$, which is compared with $m'$, the value of the just-read, off-chip MAC. If they match, the memory read operation is considered uncorrupted.

Note that MAC verification can only start after the original MAC value is read and much later than the actual decryption process, which means that data would have already been returned to the processor before the MAC is verified. We can afford this delay because in our application we consider MAC failure to be so dire that the system effectively resets and discards any use of the corrupted data. Thus, we do not need to implement any recovery mechanisms, such as rollbacks.

*Trade-offs and design choices.* Due to the unacceptable cost of tree-based integrity checking, it was our decision to use weaker but much more efficient authentication, which allows replay attacks within a small window (e.g., one to several minutes). We believe this is a reasonable compromise. Next, we argue that our authentication approach effectively limits the forger to replay attacks.

Performance considerations require use of short MACs. We first argue that even 16-bit security is sufficient in many practical security applications[9]. (Of

---

[7] Admittedly, reusing the pad for the same DRAM location results in a weakness of the encryption process. However, varying the pad, for example, based on a counter, would preclude pad precomputation for read transactions, or require significant on-chip storage.

[8] As discussed in Section 3.4, we alternatively could use parallel execution of two instances of 2- or 4-round AES.

[9] Of course, by 16-bit security we mean that the probability of a polynomial-time adversary forging a MAC is $(\frac{1}{2})^{16}$, and not that it takes $2^{16}$ operations to break it.

course, this parameter would need to be evaluated for each concrete system instantiation, using the following discussion as a guideline.) Indeed, on average, it would take the adversary $2^{15}$ attempts to forge just one memory block. Note that in our system each unsuccessful attempt would be followed by a forced reboot (a natural reaction to a break-in attempt), which might take around a minute to complete. This means that forging a single block would take an expected 20 days of continuous attacks; forging even two blocks (expected $2^{30}$ attempts) is infeasible. Thus, attackers are likely to use other attack avenues, such as exploiting the replay permissiveness.

Achieving 16-bit security requires the use of MACs of greater length, since the ShMAC output includes a nonce. In our system, the ShMAC nonce consists of the concatenation of the address and $r$. We first observe that nonces for different memory locations would never collide; however, nonces may collide within the same memory locations. If many collisions occur, the adversary may eventually accumulate some useful information about $G$. We mitigate this threat with periodically refreshing $F$ and $G$ (by changing their keys). As an additional disadvantage to the adversary, he does not learn the full value of $G$'s, but only a fraction of it. Thus, we believe that a choice of length for $r$ in the 16–48 bits range would be appropriate for most applications.

Refer to [26] for additional design considerations and trade-offs.

# References

1. Gonsalves, A.: Hackers report breaking Phone 2.0. InformationWeek (03.12.2008)
2. ARM: ARM advanced microcontroller bus architecture rev 2.0 (1999)
3. IBM: IBM 128-bit processor local bus version 4.7 (2007)
4. Nyberg, K.: Differentially uniform mappings for cryptography. In: Helleseth, T. (ed.) EUROCRYPT 1993. LNCS, vol. 765, pp. 55–64. Springer, Heidelberg (1994)
5. Minematsu, K., Tsunoo, Y.: Provably secure MACs from differentially-uniform permutations and AES-based implementations. In: Robshaw, M.J.B. (ed.) FSE 2006. LNCS, vol. 4047, pp. 226–241. Springer, Heidelberg (2006)
6. Wegman, M.N., Carter, J.L.: New hash functions and their use in authentication and set equality. J. Comput. System Sci. 22, 265–279 (1981)
7. Krawczyk, H., Bellare, M., Canetti, R.: RFC2104 - HMAC: Keyed-hashing for message authentication, http://www.faqs.org/rfcs/rfc2104.html
8. Stinson, D.R.: Universal hashing and authentication codes. In: Feigenbaum, J. (ed.) CRYPTO 1991. LNCS, vol. 576, pp. 74–85. Springer, Heidelberg (1992)
9. Brassard, G.: On computationally secure authentication tags requiring short secret shared keys. In: Advances in Cryptology – CRYPTO 1982, pp. 79–86 (1982)
10. Krawczyk, H.: LFSR-based hashing and authentication. In: Desmedt, Y.G. (ed.) CRYPTO 1994. LNCS, vol. 839, pp. 129–139. Springer, Heidelberg (1994)
11. Weisstein, E.W.: Universal hash function. From MathWorld–a Wolfram web resource, http://mathworld.wolfram.com/UniversalHashFunction.html
12. Jakimoski, G., Subbalakshmi, K.P.: On efficient message authentication via block cipher design techniques. In: Kurosawa, K. (ed.) ASIACRYPT 2007. LNCS, vol. 4833, pp. 232–248. Springer, Heidelberg (2007)
13. Trusted Computing Group: TCG Specification Architecture Overview. Revision 1 edn. (July 2007)

14. Chevallier-Mames, B., Naccache, D., Paillier, P., Pointcheval, D.: How to disembed a program? Cryptology ePrint Archive, Report 2004/138 (2004)
15. Lie, D., Thekkath, C.A., Mitchell, M., Lincoln, P., Boneh, D., Mitchell, J.C., Horowitz, M.: Architectural support for copy and tamper resistant software. In: ASPLOS, pp. 168–177. ACM, New York (2000)
16. Hall, W.E., Jutla, C.S.: Parallelizable authentication trees. In: Preneel, B., Tavares, S. (eds.) SAC 2005. LNCS, vol. 3897, pp. 95–109. Springer, Heidelberg (2006)
17. Suh, G.E.: AEGIS: A Single-Chip Secure Processor. PhD thesis, MIT (2005)
18. Suh, G., O'Donnell, C., Devadas, S.: Aegis: A single-chip secure processor. IEEE Design and Test of Computers 24(6), 570–580 (2007)
19. Duc, G.: Cryptopage. Master's thesis, ENST, Bretagne (June 2004)
20. Elbaz, R., Champagne, D., Lee, R.B., Torres, L., Sassatelli, G., Guillemin, P.: Tec-tree: A low-cost, parallelizable tree for efficient defense against memory replay attacks. In: Paillier, P., Verbauwhede, I. (eds.) CHES 2007. LNCS, vol. 4727, pp. 289–302. Springer, Heidelberg (2007)
21. Gassend, B., Suh, G.E., Clarke, D., Dijk, M.V., Devadas, S.: Caches and hash trees for efficient memory integrity verification. In: 9th Intl. Symp. on High Performance Computer Architecture (2003)
22. Blum, M., Evans, W., Gemmell, P., Kannan, S., Naor, M.: Checking the correctness of memories. In: FOCS 1991, pp. 90–99 (1991)
23. Dwork, C., Naor, M., Rothblum, G.N., Vaikuntanathan, V.: How efficient can memory checking be? In: TCC 2009 (2009)
24. Vaslin, R., Gogniat, G., Netto, E.W., Tessier, R., Burleson, W.P.: Low latency solution for confidentiality and integrity checking in embedded systems with off-chip memory. In: ReCoSoC, pp. 146–153 (2007)
25. Elbaz, R., Torres, L., Sassatelli, G., Guillemin, P., Bardouillet, M., Martinez, A.: A parallelized way to provide data encryption and integrity checking on a processor-memory bus. In: DAC 2006, pp. 506–509 (2006)
26. Garay, J., Kolesnikov, V., McLellan, R.: MAC precomputation with applications to secure memory. Cryptology ePrint Archive (2009)
27. Bellare, M., Goldreich, O., Mityagin, A.: The power of verification queries in message authentication and authenticated encryption. Cryptology ePrint Archive, Report 2004/309 (2004), http://eprint.iacr.org/
28. Keliher, L., Sui, J.: Exact maximum expected differential and linear cryptanalysis for two-round Advanced Encryption Standard. IET Information Security 1(2), 53–57 (2007)
29. Daemen, J., Rijmen, V.: Understanding two-round differentials in AES. In: De Prisco, R., Yung, M. (eds.) SCN 2006. LNCS, vol. 4116, pp. 78–94. Springer, Heidelberg (2006)
30. Daemen, J., Rijmen, V.: AES proposal: Rijndael, http://www.iaik.tugraz.at/Research/krypto/AES/
31. Daemen, J.: Annex to AES proposal Rijndael. Chapter 5. Propagation and correlation, http://www.iaik.tugraz.at/Research/krypto/AES/
32. Merkle, R.: Secrecy, authentication, and public key systems. PhD thesis, Stanford Univeristy (1979)
33. Hunt, G.D.H.: Secure processors for secure devices and secure end-to-end infrastructure, http://www.research.ibm.com/jam/secure-processors5-30-06.pdf

# HMAC without the "Second" Key

Kan Yasuda

NTT Information Sharing Platform Laboratories, NTT Corporation, Japan
yasuda.kan@lab.ntt.co.jp

**Abstract.** We present a new secret-prefix MAC (Message Authentication Code) based on hash functions. Just like the well-known HMAC algorithm, the new MAC can utilize current hash functions without modifying their Merkle-Damgård implementations. Indeed, the new MAC is almost the same as HMAC except that the *second call* to the secret key, which is made at the finalization stage, is *omitted*. In this way we not only increase efficiency over HMAC but also reduce the cost of managing the key, as the new MAC invokes a key only once at the initialization stage, and the rest of the process depends solely on incoming data. We give a rigorous security proof of the new MAC algorithm. Like HMAC, our new MAC is proven to be a secure PRF (Pseudo-Random Function) based on a reasonable assumption about the underlying compression function. In theory our assumption is neither stronger nor weaker than the PRF-type compression-function requirement for the PRF security of HMAC. In practice our assumption looks somewhat similar to the PRF-type requirement for the security of HMAC.

**Keywords:** Cascade construction, prefix-free PRF, hybrid argument, multi-oracle family, affix.

## 1 Introduction

HMAC [1] is a commonly-used, widely-standardized [2,3] MAC (Message Authentication Code) algorithm. The virtues of HMAC are twofold. First, HMAC can make use of current hash functions without making any modifications. Recall that most of the modern cryptographic hash functions are based on the Merkle-Damgård construction [4,5] with fixed-IV (Initial Value) usage, with 10*-type padding, and with 64-bit length encoding at the end. Such implementations of hash functions are fully compatible with the HMAC algorithm. Second, HMAC is provably secure. The newer proof [6] shows that HMAC is a secure PRF (Pseudo-Random Function)—hence a secure MAC—based on a PRF-type assumption about the underlying compression function. The assumption is "close enough" to the standard keyed-via-IV PRF property, which is a well-established requirement imposed on a compression function [1,7].

However, HMAC has a disadvantage of managing its secret key. That is, HMAC makes a call to the secret key *twice* in the process. The first call is to produce a secret IV at the initialization stage, and the second one is to envelope the last chaining variable at the finalization stage. The second call causes a

P. Samarati et al. (Eds.): ISC 2009, LNCS 5735, pp. 443–458, 2009.

considerable inconvenience, because the system must either keep the secret key all the time during the operation or access its "secured" storage again at the final stage of the process in order to obtain the secret key.

This problem was partially resolved by the MDP [8] scheme at the cost of losing the original advantages that HMAC had. Specifically, the MDP scheme provided a secret-prefix MAC algorithm, invoking a secret key only once at the first stage of the process, but at the same time MDP produced both design and security problems. The design problem lay in the iteration method. That is, MDP introduced modifications to the Merkle-Damgård construction, requiring direct access to the underlying compression function—*i.e.*, a direct call to the *compression* function rather than a black-box call to the *hash* function. The other problem lay in the compression-function assumption. Namely, the PRF security proof of MDP had to make a *related-key* PRF assumption about the compression function. Such an assumption seems indeed more demanding than the standard PRF requirement from both the theoretical [9] and the practical [10,11] aspects.

Therefore, we would like to resolve the "second-key" problem in HMAC without counterbalancing the original advantages of HMAC. More specifically, our goal is to come up with a new MAC algorithm that acquires the following three desirable characteristics:

1. **Secret-Prefix.** The new MAC should invoke its secret key only once. We prefer especially a secret-prefix MAC for ease of key management.
2. **Merkle-Damgård.** The new MAC should retain Merkle-Damgård hash functions. Most of the modern hash functions are of this type. In particular, the SHA-2 family of hash functions [12] employs the Merkle-Damgård construction, and this family of hash functions is expected to be widely used at least until the year 2012 (*i.e.*, the expected publication of SHA-3 [13]).
3. **PRF Assumption.** Like the case of HMAC, the security of the new MAC should be based on a reasonable assumption about the underlying compression function. Specifically, we want our assumption to be as close to the standard (non-related-key) PRF property as possible.

**Our Results.** The above goal is achieved by our new MAC algorithm, which we call $H^2$-MAC. The $H^2$-MAC algorithm is very simple to define: Given a Merkle-Damgård hash function $H : \{0,1\}^* \to \{0,1\}^n$ and a secret key $K \in \{0,1\}^n$, the tag $T$ for a message $M \in \{0,1\}^*$ is computed as

$$T = H\big(H(K\|pad\|M)\big) = H^2(K\|pad\|M),$$

where $pad \in \{0,1\}^{m-n}$ is a fixed constant and $m$ the block size of the underlying compression function. In this way $H^2$-MAC transforms a Merkle-Damgård hash function into a secret-prefix MAC algorithm.

The $H^2$-MAC algorithm has a beneficial side effect of improving efficiency over HMAC. Namely, $H^2$-MAC skips over the key derivation at the finalization stage, which helps reduce the number of compression-function calls by 1. The improvement is beneficial especially to short messages.

**Table 1.** Comparison between the well-known HMAC algorithm and our new $H^2$-MAC. The quantity $\ell$ represents the length, in blocks, of the input message $M$ after being padded, *i.e.*, $\ell = \lceil (|M| + 65)/m \rceil$.

|  | HMAC | $H^2$-MAC |
|---|---|---|
| # of compression-function calls | $\ell + 3$ | $\ell + 2$ |
| # of secret-key calls | twice (start and end) | once (start only) |
| Compression-function assumptions | PRF (keyed via IV) with key derivation | PRF (keyed via IV) with an affix |

Our $H^2$-MAC is provably secure. We prove that $H^2$-MAC is a secure PRF under the assumption that the compression function satisfies a property we call PRF-AX (Pseudo-Random Function with an AffiX). PRF-AX makes only a slight modification to the standard PRF property and remains nearly the same as PRF. In theory, PRF-AX is neither stronger nor weaker than the assumption made in the PRF security proof of HMAC [6]—we call the assumption in [6] PRF-KD (Pseudo-Random Function with Key Derivation)—but we believe that in practice PRF-AX is comparable to PRF-KD. See Table 1 for a summary of our results.

**Intuitive Reasoning behind Our New Construction.** Coron *et al.* [14] introduce the "HMAC" construction $H^2(0^m \| M)$, which looks similar to our $H^2$-MAC. The "HMAC" construction is a mode of operation for keyless hash functions and is proven to be *indifferentiable* from a random oracle. The indifferentiability implies that the "HMAC" construction, when combined with a secret prefix (which is then almost identical to our $H^2$-MAC), can be used as a secure MAC algorithm.[1]

The above plausible argument, however, is based on an informal assumption that the underlying compression function is a *random oracle*. We need to give a separate treatment in the *standard model* in order to ensure the PRF security of the secret-prefix MAC algorithm based on some formal (*e.g.*, PRF-like) assumption about the compression function. This issue has been already addressed by Bellare and Ristenpart in their multi-property-preserving EMD construction [16].

In this paper we perform a formal analysis of the $H^2$-MAC algorithm in the standard model. We adopt the powerful techniques of multi-oracle families [7] in order to study the PRF security of our construction.

**Organization of the Paper.** In Sect. 2 we briefly mention previous constructions of similar MAC algorithms. Section 3 defines symbols and notions necessary for presenting the paper. Section 4 gives a formal definition of our new $H^2$-MAC algorithm, followed by its security proof in Sect. 5. In Sect. 6 we open up a general discussion on the design strategy employed by our $H^2$-MAC algorithm. Section 7 concludes the paper.

---

[1] This idea of obtaining a secret-prefix MAC from an indifferentiable construction also appears in the sponge construction [15].

## 2   Related Work

The secret-prefix method applied to a plain Merkle-Damgård hash function is obviously insecure due to the "extension" attack [17,18]. The ENMAC algorithm [19] increases efficiency over HMAC by providing a secret-prefix algorithm for short messages, but ENMAC still requires the second key application for long messages. The MDP scheme [8] operates as a secret-prefix MAC algorithm for messages of any length, but MDP has the disadvantages that we have already pointed out in Sect. 1.

There are other types (other than secret-prefix) of MACs that aim to improve efficiency over HMAC, such as the Sandwich construction [20] and the BNMAC algorithm [21]. These two schemes, however, pursue different aims and do not avoid the second-key problem.

## 3   Preliminaries

**Bit-String Operations.** Given a finite bit string $x \in \{0,1\}^*$, we write $|x|$ for the length in bits of the string $x$. Given two strings $x, y \in \{0,1\}^*$, the notation $x \| y$ represents the concatenation of $x$ and $y$, whereas $x \oplus y$ the exclusive OR of $x$ and $y$. The symbol $\|$ is often omitted; e.g., we simply write 10 in place of $1\|0$. The symbol $0^n$ denotes the $n$-bit string $00 \cdots 0 \in \{0,1\}^n$. We use the wild card $*$ and write $0^*$ to make the value $n$ implicit. Given a non-negative integer $n$ (less than $2^{64} - 1$), we let $\langle n \rangle_{64}$ denote the big-endian 64-bit binary representation of the integer $n$, so that we have $\langle n \rangle_{64} \in \{0,1\}^{64}$. We use the abbreviation symbol $\ldots$ and write $\langle \ldots \rangle_{64}$ to make the value $n$ implicit.

**Compression Functions.** Throughout the paper we fix a compression function $F : \{0,1\}^{n+m} \to \{0,1\}^n$, where $n$ and $m$ are positive integers. We call $m$ the *block size*. We impose a requirement $n + 65 \le m$, which is not a severe restriction as it is satisfied by most of the modern cryptographic hash functions [12,13]. The time complexity is the sum of running time and code size. We fix a model of computation and a method of encoding. We write $\text{Time}_F(q)$ for the time complexity necessary for computing the function $F$ $q$-many times.

**Merkle-Damgård Hash Functions.** We iterate the compression function $F$ to obtain a function $F^*$ as follows:

---

**Algorithm.** $F_V^*(x[1] \cdots x[\ell])$

---

Input: Chaining variable $V \in \{0,1\}^n$, blocks $x[i] \in \{0,1\}^m$ for $i = 1, \ldots, \ell$
  $v[0] \leftarrow V$
  **For** $i = 1, \ldots, \ell$ **do**
    $v[i] \leftarrow F(v[i-1] \| x[i])$
  **endfor**
Output: Final value $v[\ell] \in \{0,1\}^n$

---

This defines a new function $F_V^* : \{0,1\}^{m*} \to \{0,1\}^n$, where the domain $\{0,1\}^{m*}$ is the set of bit strings whose length is a multiple of $m$. We follow the convention $F_V^*(\varepsilon) = V$ on the null input $\varepsilon$.

Now let $IV \in \{0,1\}^n$ be a fixed constant. Using the constant $IV$ and 64-bit length encoding $\langle \ldots \rangle_{64}$ we obtain a hash function $H : \{0,1\}^* \to \{0,1\}^n$ as follows:[2]

$$H(M) := F_{IV}^*(M\|10^*\|\langle|M|\rangle_{64}),$$

where the wild card $*$ is the minimum number of zeros necessary to make the length of the resulting string a multiple of $m$ bits.

Whenever we write

$$X\|10^*\|\langle\ldots\rangle_{64},$$

it is understood that the input "..."to the encoding function is the length $|X|$ of the string $X$. For example, we write $X\|Y\|Z\|10^*\|\langle\ldots\rangle_{64}$ as a shorthand for $X\|Y\|Z\|10^*\|\langle|X|+|Y|+|Z|\rangle_{64}$. Here, also recall that the length of the resulting string $X\|Y\|Z\|10^*\|\langle\ldots\rangle_{64}$ is a multiple of $m$ bits.

We use the following system of notation for dividing a message $M \in \{0,1\}^*$ into blocks. We write

$$M[1]\cdots M[\ell] \leftarrow M$$

to mean that each block value is assigned to $M[i]$, satisfying the following three conditions:

1. $M[1]\cdots M[\ell] = M$,
2. $|M[i]| = m$ for $i = 1, \ldots, \ell - 1$, and
3. $1 \leq |M[\ell]| \leq m$.

Note that given a message $M$ its block decomposition is uniquely determined.

**HMAC.** Using a Merkle-Damgård hash function $H : \{0,1\}^* \to \{0,1\}^n$ constructed as above, the well-known HMAC algorithm is defined as follows:

---

**Algorithm.** $HMAC_K(M)$

---

Input: Key $K \in \{0,1\}^n$, message $M \in \{0,1\}^*$
$\quad \bar{K} \leftarrow K\|0^{m-n}$
$\quad Y \leftarrow H\big((\bar{K} \oplus ipad)\|M\big)$
$\quad T \leftarrow H\big((\bar{K} \oplus opad)\|Y\big)$
Output: Tag $T \in \{0,1\}^n$

---

In the above definition $ipad, opad \in \{0,1\}^m$ are two different constants ($ipad$ is a repetition of the byte 0x36, whereas $opad$ a repetition of 0x5c). See also Fig. 1 for an illustration of the HMAC algorithm.

---

[2] Strictly speaking, we note that inputs to $H$ are restricted to $2^{64} - 1$ bits.

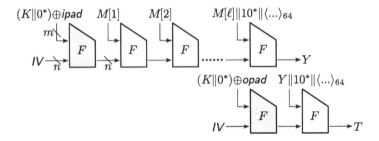

**Fig. 1.** The well-known HMAC algorithm. The diagram describes only the case $|M[\ell]| \leq m - 65$. The shaded boxes correspond to key derivation.

**The Notion of PRFs.** We consider a distinguisher $\mathcal{A}$, which is an oracle machine that outputs either 1 or 0. We let $\mathcal{A}^{\mathcal{O}}$ denote the value returned by $\mathcal{A}$ after interacting with the oracle $\mathcal{O}$. The oracle $\mathcal{O}$ is either a "real" oracle $G_K : \{0,1\}^* \rightarrow \{0,1\}^n$ or an "ideal" oracle $\mathcal{R} : \{0,1\}^* \rightarrow \{0,1\}^n$. The real oracle picks a key $K$ uniformly at random from its key space $\{0,1\}^n$. We write $K \xleftarrow{\$} \{0,1\}^n$ for such a sampling operation. The ideal oracle picks a function $\mathcal{R}$ uniformly at random from the space of functions mapping $\{0,1\}^*$ onto $\{0,1\}^n$.[3] We then define the *advantage* function as

$$\mathrm{Adv}_G^{\mathrm{prf}}(\mathcal{A}) := \Pr\left[\mathcal{A}^{G_K(\cdot)} = 1\right] - \Pr\left[\mathcal{A}^{\mathcal{R}(\cdot)} = 1\right],$$

where the probabilities are defined over the choice of $K$, the choice of $\mathcal{R}$, and the internal coins of $\mathcal{A}$. We use the notation

$$\mathrm{Adv}_G^{\mathrm{prf}}(t, q, \ell) := \max_{\mathcal{A}} \mathrm{Adv}_G^{\mathrm{prf}}(\mathcal{A}),$$

where the maximum runs over all adversaries $\mathcal{A}$ whose time complexity is at most $t$, each making at most $q$ queries in total to its oracles, each query being at most $\ell$ blocks (Recall that the block size is $m$).

**The Notion of Secure MACs.** The required property for a secure MAC is so-called existential unforgeability under chosen-message attacks. It is well-known that this property is implied by the notion of a PRF (*e.g.*, [6]). In the current paper the notion of a secure MAC itself is not used, as we prove that our construction is indeed secure as a PRF.

**Multi-oracle Families (for PRFs).** The techniques of multi-oracle families were developed in [7] for analyzing the cascade construction. It is a general principle which can be applied to any kind of indistinguishability, including the notion of PRF just defined and that of PRF-AX to be defined in Sect. 5.1.

---

[3] Formally speaking, we restrict the domain to $\{0,1\}^{2^{64}-1}$, so that the function $\mathcal{R}$ is chosen from the corresponding restricted space.

For PRF it works as follows. Suppose that we key our compression function $F$ as $F_K(\cdot) = F(K\|\cdot)$, obtaining $F_K : \{0,1\}^m \to \{0,1\}^n$ with keys $K \xleftarrow{\$} \{0,1\}^n$, and that we have an ideal function $R : \{0,1\}^m \to \{0,1\}^n$. We consider an oracle $F \otimes \cdots \otimes F$ ($q$-fold) which picks independent $q$-many keys $K_1, \ldots, K_q \xleftarrow{\$} \{0,1\}^n$ and upon a query $(i,x)$ returns $F_{K_i}(x)$. An ideal oracle $R \otimes \cdots \otimes R$ picks independent $q$-many functions $R_1, \ldots, R_q$ from the function space and upon a query $(i,x)$ returns $R_i(x)$. We define

$$\mathrm{Adv}^{\mathrm{prf}}_{F \otimes \cdots \otimes F}(\mathcal{A}) := \Pr\big[\mathcal{A}^{F \otimes \cdots \otimes F} = 1\big] - \Pr\big[\mathcal{A}^{R \otimes \cdots \otimes R} = 1\big].$$

The symbol $\mathrm{Adv}^{\mathrm{prf}}_{F \otimes \cdots \otimes F}(t,q)$ is similarly defined. Note that the quantity $q$ denotes the *total* number of queries made across the different indices $i$.

**Lemma 1.** *If $F$ is a secure PRF, then so is $F \otimes \cdots \otimes F$ ($q$-fold). Specifically, we have*

$$\mathrm{Adv}^{\mathrm{prf}}_{F \otimes \cdots \otimes F}(t,q) \leq q \cdot \mathrm{Adv}^{\mathrm{prf}}_{F}(t',q),$$

*where the time complexity $t'$ is about $t + \mathrm{Time}_F(q)$.*

*Proof.* The proof is a very standard hybrid argument and can be found in [7]. □

## 4 Specification of the $H^2$-MAC Algorithm

Our $H^2$-MAC algorithm is defined as follows:

---

**Algorithm.** $H^2MAC_K(M)$

---

Input: Key $K \in \{0,1\}^n$, message $M \in \{0,1\}^*$
  $Y \leftarrow H(K\|pad\|M)$
  $T \leftarrow H(Y)$
Output: Tag $T \in \{0,1\}^n$

---

See also Fig. 2 for an illustration of the $H^2$-MAC algorithm. Note that in the definition we introduce a fixed constant *pad*, which is exactly defined as

$$pad = 1\|0^{m-n-65}\|\langle n\rangle_{64} \in \{0,1\}^{m-n},$$

so that $Y\|pad$ coincides with the standard Merkle-Damgård strengthening, *i.e.*, $Y\|pad = Y\|10^*\|\langle\ldots\rangle_{64}$.

  Our scheme takes a secret key $K \in \{0,1\}^n$. We assume that the key length is always equal to $n$ bits. This is for simplicity of our analysis. The case $|K| \neq n$ can be treated in a similar way, requiring some modifications to the PRF-AX assumption, but it would only add unnecessary complication. In practice the $H^2$-MAC algorithm works fine with a key length different from $n$.

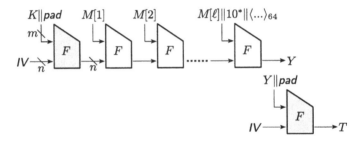

**Fig. 2.** Our $H^2$-MAC algorithm. The diagram describes only the case $|M[\ell]| \leq m - 65$. Note that we set $pad = 10^{m-n-65} \| \langle n \rangle_{64}$, so that it is in accordance with the standard Merkle-Damgård padding rule. The shaded boxes correspond to affixes.

## 5    Security of the $H^2$-MAC Algorithm

In this section we prove our main security result. Before going into the proof details, we start with defining our PRF-AX assumption. After proving the main theorem, we review the notion of PRF-AX by making a comparison with that of PRF-KD for HMAC.

### 5.1    Our Assumption PRF-AX

The notion of PRF-AX is almost the same as that of PRF except that an adversary is allowed to obtain a piece of additional information called *affix*. Consider keying a compression function $F : \{0,1\}^{n+m} \to \{0,1\}^n$ via its IV as $F_K(\cdot) := F(K\|\cdot)$, where as usual $K \xleftarrow{\$} \{0,1\}^n$. In the standard PRF setting, an adversary has access to the $F_K(\cdot)$ oracle and tries to distinguish it from the random oracle $R(\cdot)$. On the other hand, in our PRF-AX setting, an adversary is given access not only to the $F_K(\cdot)$ oracle but also to the *affix oracle*,[4] which upon request returns $F(IV\|K\|pad)$. The adversary's goal is to distinguish such a pair of oracles from the pair of a random oracle $R : \{0,1\}^m \to \{0,1\}^n$ and a random string $r \xleftarrow{\$} \{0,1\}^n$ (in place of the affix oracle), the string $r$ being independent from the choice of $R$. We define

$$\mathrm{Adv}_F^{\mathrm{prf\text{-}ax}}(\mathcal{A}) := \Pr\left[\mathcal{A}^{F_K(\cdot), F(IV\|K\|pad)} = 1\right] - \Pr\left[\mathcal{A}^{R(\cdot), r} = 1\right].$$

The symbol $\mathrm{Adv}_F^{\mathrm{prf\text{-}ax}}(t, q)$ is defined in the same way as before.

We admit that PRF-AX is a requirement *strictly* stronger than the standard PRF. For example,[5] consider a block cipher $E_K : \{0,1\}^n \to \{0,1\}^n$ with keys $K \xleftarrow{\$} \{0,1\}^n$. For the moment ignore the constant *pad* (or set it null). If $E$ is a secure block cipher, then $E_K(\cdot)$ should be a good PRF, but it becomes

---

[4] The term "affix oracle" might be somewhat misleading, as it returns a single fixed value. The adversary is allowed to make only one "request" (*i.e.*, a query without a value) to the affix oracle.

[5] This instructive example of a block cipher was given by one of the referees.

completely insecure as soon as the affix oracle $E_{IV}(K)$ is given to the adversary (The adversary can recover the secret key $K$ using the public value $IV$). This counterexample exploits the fact that $E_{IV(\cdot)}$ is invertible. We believe that a well-designed compression function should not have such characteristics.

## 5.2  PRF-AX Multi-oracle Families

As to PRF-AX, the notion of multi-oracle families is as follows. We key our compression function $F$ as $F_K(\cdot) = F(K\|\cdot)$, obtaining $F_K : \{0,1\}^m \to \{0,1\}^n$ with keys $K \xleftarrow{\$} \{0,1\}^n$ as usual. We consider an oracle $F \otimes \cdots \otimes F$ ($q$-fold) which picks independent $q$-many keys $K_1, \ldots, K_q \xleftarrow{\$} \{0,1\}^n$ and upon a (PRF) query $(i, x)$ returns $F_{K_i}(x)$. Upon an affix query $i$, the oracle returns $F(IV\|K_i\|pad)$. An ideal oracle $R \otimes \cdots \otimes R$ picks independent $q$-many functions $R_1, \ldots, R_q$ from the function space $R : \{0,1\}^m \to \{0,1\}^n$ and upon a query $(i, x)$ returns $R_i(x)$. The oracle also picks $q$-many random values $r_1, \ldots, r_q$ from the space $\{0,1\}^n$ and upon an affix query $i$ returns $r_i$. The symbols $\mathrm{Adv}^{\mathrm{prf\text{-}ax}}_{F \otimes \cdots \otimes F}(\mathcal{A})$ and $\mathrm{Adv}^{\mathrm{prf\text{-}ax}}_{F \otimes \cdots \otimes F}(t, q)$ are defined in the same way as before.

## 5.3  Main Security Result

We now state our main theorem:

**Theorem 1.** *The $H^2$-MAC algorithm is a secure PRF if the underlying compression function $F$ is a secure PRF-AX. Specifically, we have*

$$\mathrm{Adv}^{\mathrm{prf}}_{H^2MAC}(t, q, \ell) \leq (\ell q + q + 1) \cdot \mathrm{Adv}^{\mathrm{prf\text{-}ax}}_F(t', q),$$

*where the time complexity $t'$ is about $t + \mathrm{Time}_F(\ell q + q)$.*

The proof of Theorem 1 is divided into three parts. The first part is to replace the initial key derivation with a secret key itself. The last two parts are to implement the techniques of multi-oracle families.

**Part 1: From $H^2$-MAC to $\tilde{H}^2$.** We introduce a slightly modified MAC algorithm $\tilde{H}^2$. The difference arises from the first stage: The derivation of a secret chaining variable $F(IV\|K\|pad)$ is simply replaced with the key $K$ itself.

---

**Algorithm. $\tilde{H}^2_K(M)$**

---

Input: Key $K \in \{0,1\}^n$, message $M \in \{0,1\}^*$
$\quad x[1] \cdots x[\ell] \leftarrow K\|pad\|M\|10^*\|\langle \ldots \rangle_{64}$
$\quad Y \leftarrow F^*_K(x[2] \cdots x[\ell])$
$\quad T \leftarrow H(Y)$
Output: Tag $T \in \{0,1\}^n$

---

In the following lemma we reduce the security of $H^2$-MAC to that of $\tilde{H}^2$ under the assumption that $F$ is a secure PRF-AX. The affix oracle plays an essential role in the proof.

**Lemma 2.** *If the compression function $F$ is a secure PRF-AX and the $\tilde{H}^2$ algorithm is a secure PRF, then the $H^2$-MAC algorithm is also a secure PRF. Specifically, we have*

$$\mathrm{Adv}^{\mathrm{prf}}_{H^2MAC}(t,q,\ell) \leq \mathrm{Adv}^{\mathrm{prf\text{-}ax}}_{F}(t',1) + \mathrm{Adv}^{\mathrm{prf}}_{\tilde{H}^2}(t,q,\ell),$$

*where the time complexity $t'$ is about $t + \mathrm{Time}_F(\ell q)$.*

*Proof.* Let $\mathcal{A}$ be a distinguisher attacking the $H^2$-MAC algorithm. Assume that the time complexity of $\mathcal{A}$ is at most $t$, that $\mathcal{A}$ makes at most $q$ queries, and that the length of each query is at most $\ell$ blocks. Construct an adversary $\mathcal{B}$ that attacks $F$ in the PRF-AX sense by utilizing $\mathcal{A}$ as a black-box, as follows:

---

**Adversary $\mathcal{B}$**

---

Make a query to its affix oracle and receive $V \in \{0,1\}^n$
Run $\mathcal{A}$ and answer each query $M$ as follows:
  $x[1] \cdots x[\lambda] \leftarrow 0^m \| M \| 10^* \| \langle \ldots \rangle_{64}$
  Return $H\big(F_V^*(x[2] \cdots x[\lambda])\big)$ to $\mathcal{A}$
Output whatever $\mathcal{A}$ outputs

---

Observe that $\mathcal{B}^{F_K(\cdot), F(IV\|K\|pad)}$ is exactly the same as running $\mathcal{A}^{H^2(K\|pad\|\cdot)}$. Similarly, running $\mathcal{B}^{R(\cdot),r}$ coincides with running $\mathcal{A}^{\tilde{H}^2_K(\cdot)}$ (Actually, $\mathcal{B}$ never makes a query to its $R : \{0,1\}^m \to \{0,1\}^n$ oracle). Therefore, we have

$$\begin{aligned}
\mathrm{Adv}^{\mathrm{prf\text{-}ax}}_{F}(t',1) &\geq \mathrm{Adv}^{\mathrm{prf\text{-}ax}}_{F}(\mathcal{B}) \\
&= \Pr\big[\mathcal{B}^{F_K(\cdot), F(IV\|K\|pad)} = 1\big] - \Pr\big[\mathcal{B}^{R(\cdot),r} = 1\big] \\
&= \Pr\big[\mathcal{A}^{H^2(K\|pad\|\cdot)} = 1\big] - \Pr\big[\mathcal{A}^{\tilde{H}^2_K(\cdot)} = 1\big] \\
&= \Pr\big[\mathcal{A}^{H^2(K\|pad\|\cdot)} = 1\big] - \Pr\big[\mathcal{A}^{\mathcal{R}(\cdot)} = 1\big] \\
&\quad - \Pr\big[\mathcal{A}^{\tilde{H}^2_K(\cdot)} = 1\big] + \Pr\big[\mathcal{A}^{\mathcal{R}(\cdot)} = 1\big] \\
&= \mathrm{Adv}^{\mathrm{prf}}_{H^2MAC}(\mathcal{A}) - \mathrm{Adv}^{\mathrm{prf}}_{\tilde{H}^2}(\mathcal{A}),
\end{aligned}$$

where $t'$ is about $t + \mathrm{Time}_F(\ell q)$ and $\mathcal{R}(\cdot) : \{0,1\}^* \to \{0,1\}^n$ is an ideal oracle. This proves Lemma 2. $\qquad\square$

**Part 2: From $\tilde{H}^2$ to $F \otimes \cdots \otimes F$.** The next step is to reduce the security of $\tilde{H}^2$ to that of multi-oracle family. We prove the following lemma:

**Lemma 3.** *The MAC algorithm $\tilde{H}^2$ is a secure PRF if $F \otimes \cdots \otimes F$ is a secure PRF-AX multi-oracle family. Specifically, we have*

$$\mathrm{Adv}^{\mathrm{prf}}_{\tilde{H}^2}(t,q,\ell) \leq (\ell+1) \cdot \mathrm{Adv}^{\mathrm{prf\text{-}ax}}_{F \otimes \cdots \otimes F}(t',q),$$

*where the product $F \otimes \cdots \otimes F$ is $q$-fold and the time complexity $t'$ is about $t + \mathrm{Time}_F(\ell q)$.*

*Proof.* The proof is based on a hybrid argument. Our proof is conducted along the lines of [7], except that in our case adversaries are no longer prefix-free. This is another place where we need the affix oracle.

We introduce intermediate functions $G_i$ for $i = 0, 1, \ldots, \ell+1$. These functions take a random function $g : \{0,1\}^* \to \{0,1\}^n$ as a "key."

---

**Algorithm.** $G_i(M)$

---

Input: Function $g : \{0,1\}^* \to \{0,1\}^n$, message $M \in \{0,1\}^*$
   $x[1] \cdots x[\lambda] \leftarrow M \| 10^* \| \langle \ldots \rangle_{64}$
   **If** $\lambda < i$ **then**
     $T \leftarrow g(x[1] \cdots x[\lambda])$
   **endif**
   **If** $\lambda = i$ **then**
     $Y \leftarrow g(x[1] \cdots x[\lambda])$
     $T \leftarrow H(Y)$
   **endif**
   **If** $\lambda > i$ **then**
     $V \leftarrow g(x[1] \cdots x[i])$
     $T \leftarrow H\big(F_V^*(x[i+1] \cdots x[\lambda])\big)$
   **endif**
Output: Tag $T \in \{0,1\}^n$

---

In the above definition of $G_i$, the cases $i = 0$ and $i = \ell + 1$ should be regarded as follows. If $i = 0$, then the last condition $\lambda > i$ is always satisfied. The input value $x[1] \cdots x[i] = x[1] \cdots x[0]$ should be treated as the null string $\varepsilon$, at which the function $g$ returns a random value $V = g(\varepsilon)$. On the other hand, if $i = \ell+1$, then only the first condition $\lambda < i$ is satisfied. This means that the function $G_{\ell+1}$ behaves like a random function.

Now let $\mathcal{A}$ be a distinguisher attacking $\tilde{H}^2$, whose time complexity is at most $t$, each making at most $q$ queries, each query being at most $\ell$ blocks. We define probabilities $P_i := \Pr[\mathcal{A}^{G_i(\cdot)} = 1]$ for $i = 0, 1, \ldots, \ell+1$. Note that the probabilities are taken over the choice of $g : \{0,1\}^* \to \{0,1\}^n$.[6] Now we can rewrite the advantage as

$$\mathrm{Adv}_{\tilde{H}^2}^{\mathrm{prf}}(\mathcal{A}) = \Pr\big[\mathcal{A}^{\tilde{H}_K^2(\cdot)} = 1\big] - \Pr\big[\mathcal{A}^{\mathcal{R}(\cdot)} = 1\big]$$

$$= P_0 - P_{\ell+1}$$

$$= \sum_{i=0}^{\ell}(P_i - P_{i+1}).$$

So it remains to bound each term $P_i - P_{i+1}$. For this, we construct adversaries $\mathcal{B}_i$ for $i = 0, 1, \ldots, \ell$. Each $\mathcal{B}_i$ is a distinguisher which attacks the multi-oracle

---

[6] Formally, the domain of $g$ should be restricted to $\{0,1\}^{2^{64}-1}$.

family $F \otimes \cdots \otimes F$ by having black-box access to the adversary $\mathcal{A}$, where $\mathcal{A}$ is now considered as an adversary trying to distinguish between $G_i$ and $G_{i+1}$.

---

**Adversary $\mathcal{B}_i$**

---

Set a counter $c \leftarrow 0$
Run $\mathcal{A}$ and answer its $\alpha$-th query $M_\alpha$ as follows:
    $x_\alpha[1] \cdots x_\alpha[\ell] \leftarrow M_\alpha \| 10^* \| \langle \ldots \rangle_{64}$
    **If** $\ell < i$ **then** $T \xleftarrow{\$} \{0,1\}^n$; return $T$ to $\mathcal{A}$
    **else** (*i.e.*, $\ell \geq i$)
        **If** $x_\alpha[1] \cdots x_\alpha[\ell] = x_\beta[1] \cdots x_\beta[\ell]$ for some previous $\beta < \alpha$ **then** $s_\alpha \leftarrow s_\beta$
        **else** $c \leftarrow c + 1$; $s_\alpha \leftarrow c$ **endif**
        **If** $\ell = i$ **then** make an affix query to its $s_\alpha$-th oracle and receive $T$
        return $T$ to $\mathcal{A}$ **endif**
        **If** $\ell = i + 1$ **then** make a query $x[\ell]$ to its $s_\alpha$-th oracle and receive $Y$
        return $H(Y)$ to $\mathcal{A}$ **endif**
        **If** $\ell \geq i + 2$ **then** make a query $x[i+1]$ to its $s_\alpha$-th oracle and receive $V$
        return $H\big(F_V^*(x_\alpha[i+2] \cdots x_\alpha[\ell])\big)$ to $\mathcal{A}$ **endif**
    **endif**
Output whatever $\mathcal{A}$ outputs

---

Observe that running $\mathcal{B}_i^{F \otimes \cdots \otimes F}$ is the same as running $\mathcal{A}^{G_i(\cdot)}$. Similarly, running $\mathcal{B}_i^{R \otimes \cdots \otimes R}$ coincides with running $\mathcal{A}^{G_{i+1}(\cdot)}$. Therefore, we have

$$P_i - P_{i+1} = \Pr\big[\mathcal{A}^{G_i(\cdot)} = 1\big] - \Pr\big[\mathcal{A}^{G_{i+1}(\cdot)} = 1\big]$$
$$= \Pr\big[\mathcal{B}_i^{F \otimes \cdots \otimes F} = 1\big] - \Pr\big[\mathcal{B}_i^{R \otimes \cdots \otimes R} = 1\big] \leq \mathrm{Adv}_{F \otimes \cdots \otimes F}^{\mathrm{prf\text{-}ax}}(\mathcal{B}_i),$$

where each $\mathcal{B}_i$ makes at most $q$ queries, and each $\mathcal{B}_i$'s time complexity is at most $t' \approx t + \mathrm{Time}_F(\ell q)$. Hence we get

$$\mathrm{Adv}_{\tilde{H}^2}^{\mathrm{prf}}(\mathcal{A}) \leq \sum_{i=0}^{\ell} \mathrm{Adv}_{F \otimes \cdots \otimes F}^{\mathrm{prf\text{-}ax}}(\mathcal{B}_i) \leq (\ell + 1) \cdot \mathrm{Adv}_{F \otimes \cdots \otimes F}^{\mathrm{prf\text{-}ax}}(t', q),$$

as desired. This proves Lemma 3.         $\square$

**Part 3: From $F \otimes \cdots \otimes F$ to $F$.** The last step is an adaptation of Lemma 1 to the PRF-AX setting. It translates into the following:

**Lemma 4.** *If $F$ is a secure PRF-AX, then so is $F \otimes \cdots \otimes F$ ($q$-fold). Specifically, we have*
$$\mathrm{Adv}_{F \otimes \cdots \otimes F}^{\mathrm{prf\text{-}ax}}(t, q) \leq q \cdot \mathrm{Adv}_F^{\mathrm{prf\text{-}ax}}(t', q),$$
*where the time complexity $t'$ is about $t + \mathrm{Time}_F(q)$.*

*Proof.* The proof is essentially the same as that of Lemma 1.         $\square$

**Putting Them Together.** Given Lemmas 2, 3 and 4, it is now easy to prove Theorem 1. We successively compute as:

$$
\begin{aligned}
\mathrm{Adv}_{H^2MAC}^{\mathrm{prf}}(t, q, \ell) &\leq \mathrm{Adv}_F^{\mathrm{prf\text{-}ax}}(t', 1) + \mathrm{Adv}_{\tilde{H}^2}^{\mathrm{prf}}(t, q, \ell) \\
&\leq \mathrm{Adv}_F^{\mathrm{prf\text{-}ax}}(t', 1) + (\ell + 1) \cdot \mathrm{Adv}_{F \otimes \cdots \otimes F}^{\mathrm{prf\text{-}ax}}(t', q) \\
&\leq \mathrm{Adv}_F^{\mathrm{prf\text{-}ax}}(t', 1) + (\ell + 1)q \cdot \mathrm{Adv}_F^{\mathrm{prf\text{-}ax}}(t'', q) \\
&\leq (\ell q + q + 1) \cdot \mathrm{Adv}_F^{\mathrm{prf\text{-}ax}}(t'', q),
\end{aligned}
$$

where $t' \approx t + \mathrm{Time}_F(\ell q)$ and $t'' \approx t + \mathrm{Time}_F(\ell q + q)$. This proves Theorem 1.  □

### 5.4   Tightness of the (Birthday) Bound

Note that the generic birthday attacks [22] are applicable to our $H^2$-MAC algorithm. Hence the bound obtained in Theorem 1 is essentially tight, as it ensures security up to the birthday limit (*i.e.*, the scheme remains secure for $\ell q \ll 2^{n/2}$). This bound is the same as the one for the cascade construction [7] and is essentially comparable to the one for HMAC [6].

### 5.5   Comparison between PRF-AX and PRF-KD

We compare our PRF-AX assumption with the assumption made in the PRF security proof of HMAC [6], PRF-KD. The PRF-KD property consists of two conditions:

1. The function $F_K(\cdot) := F(K\|\cdot)$ is a secure PRF (against $q$-many queries), and
2. The function $F'_{K'}(\cdot) := F(IV\|K'\|\cdot)$ is a secure PRF against two constant queries (related to the two constants *ipad* and *opad*), under a fixed related-key attack (coming from the two constants).

Therefore, just like the PRF-AX property, PRF-KD is a slight strengthening of the standard PRF property, allowing an adversary to have two pieces of additional information (*i.e.*, key derivation). However, there exists a subtle difference between the two properties. Namely, in the case of PRF-AX, the *same* key $K$ is used for both $F_K(\cdot)$ oracle and the affix oracle, allowing an adversary to access the two oracles concurrently. On the other hand, in the case of PRF-KD, the above two conditions are independent requirements; the key $K$ for the $F_K(\cdot)$ oracle has no relation to the key $K'$ for key derivation. Neither the assumption implies the other; our PRF-AX property is neither stronger nor weaker than PRF-KD. See Appendix for examples of exhibiting separation between the two notions.

## 6   Discussion on the Design Principle

In this section we discuss the feasibility of the design principle (*i.e.*, the removal of the second key) adopted by the $H^2$-MAC algorithm. At first glance the absence of the second key seems to make the construction less secure. There are both the positive and the negative sides to this issue.

**Positive: Multi-property Preservation.** The $H^2$ construction preserves several properties. It is easily seen that $H^2(\cdot)$ is collision-resistant if $H(\cdot)$ is (and hence if the underlying compression function $F$ is [4,5]). Coron *et al.* [14] proved that $H^2(0^m\|\cdot)$ is an indifferentiable construction. Furthermore, we have proven that the construction is "almost" PRF-preserving. Therefore, the $H^2$ construction is secure against these kinds of attacks under respective assumptions about the compression function.

**Negative: Security under Weakened Assumption.** The security of HMAC does not necessarily require its compression function be a secure PRF. The MAC security (rather than the PRF security) of HMAC can be proven under weaker-than-PRF assumptions about the compression function. For example, Bellare [6] provides a security proof based on a "privacy-preserving MAC" property of the compression function (together with some other weak assumptions). Also, Fischlin [23] proves the MAC security of HMAC under a non-malleability condition of the compression function (together with some other weak assumptions). Currently we do not have a proof to ensure the MAC security of the $H^2$-MAC algorithm based on a weaker-than-PRF property of the compression function.

It is worth pointing out that the second key serves a crucial role in Carter-Wegman MACs based on universal hash functions [24]. The second key enables us to use "weak" hashing for the inner part, by utilizing a "strong" outer finalization with the second key.

**Negative: Secret Storage vs. Secret Computation.** The critical aspect of the $H^2$-MAC (in fact, any secret-prefix MAC) algorithm is that as soon as one of the intermediate values is leaked, the adversary can easily compute any extension of that message locally. Hence, the evaluation of these values must be done securely.

This problem does not seem to exist in HMAC. In HMAC, such leakage of the intermediate values would lead to existential forgery after off-line collision search with one additional query, but such an attack is less powerful and more costly. Thus, secure computation appears to be a price to pay for relaxing the storage condition of the secret key.

It remains as interesting future work to investigate further the role of the second key. Hopefully such a study gives a better insight about the gap in security between the two types of constructions.

## 7    Conclusion

We have presented a new MAC algorithm called $H^2$-MAC. The new algorithm removes the second-key application from HMAC, improves efficiency over HMAC and achieves similar PRF-based provable security. In this sense, we have shown that, at least for PRF-based security, the second key is redundant. The role of the second key for other types of security needs further investigation.

# Acknowledgments

The author wishes to thank the ISC 2009 anonymous reviewers for making helpful comments and pointing out some errors in the proofs.

# References

1. Bellare, M., Canetti, R., Krawczyk, H.: Keying hash functions for message authentication. In: Koblitz, N. (ed.) CRYPTO 1996. LNCS, vol. 1109, pp. 1–15. Springer, Heidelberg (1996)
2. Krawczyk, H., Bellare, M., Canetti, R.: HMAC: Keyed-hashing for message authentication, RFC2104 (1997)
3. JTC1: Information technology—Security techniques—Message Authentication Codes (MACs)—Part 2: Mechanisms using a dedicated hash-function, ISO/IEC 9797-2:2002 (2002)
4. Merkle, R.C.: One way hash functions and DES. In: Brassard, G. (ed.) CRYPTO 1989. LNCS, vol. 435, pp. 428–446. Springer, Heidelberg (1990)
5. Damgård, I.: A design principle for hash functions. In: Brassard, G. (ed.) CRYPTO 1989. LNCS, vol. 435, pp. 416–427. Springer, Heidelberg (1990)
6. Bellare, M.: New proofs for NMAC and HMAC: Security without collision resistance. In: Dwork, C. (ed.) CRYPTO 2006. LNCS, vol. 4117, pp. 602–619. Springer, Heidelberg (2006)
7. Bellare, M., Canetti, R., Krawczyk, H.: Pseudorandom functions revisited: The cascade construction and its concrete security. In: The 37th FOCS, pp. 514–523. IEEE Computer Society, Los Alamitos (1996)
8. Hirose, S., Park, J.H., Yun, A.: A simple variant of the Merkle-Damgård scheme with a permutation. In: Kurosawa, K. (ed.) ASIACRYPT 2007. LNCS, vol. 4833, pp. 113–129. Springer, Heidelberg (2007)
9. Bellare, M., Kohno, T.: A theoretical treatment of related-key attacks: RKA-PRPs, RKA-PRFs, and applications. In: Biham, E. (ed.) EUROCRYPT 2003. LNCS, vol. 2656, pp. 491–506. Springer, Heidelberg (2003)
10. den Boer, B., Bosselaers, A.: Collisions for the compressin function of MD5. In: Helleseth, T. (ed.) EUROCRYPT 1993. LNCS, vol. 765, pp. 293–304. Springer, Heidelberg (1994)
11. Contini, S., Yin, Y.L.: Forgery and partial key-recovery attacks on HMAC and NMAC using hash collisions. In: Lai, X., Chen, K. (eds.) ASIACRYPT 2006. LNCS, vol. 4284, pp. 37–53. Springer, Heidelberg (2006)
12. NIST: Secure Hash Standard (SHS), FIPS 180-2 (2004)
13. NIST: Announcing request for candidate algorithm nominations for a new cryptographic hash algorithm (SHA 3) family, Federal Register Notice, vol. 72(212) (2007)
14. Coron, J.S., Dodis, Y., Malinaud, C., Puniya, P.: Merkle-Damgård revisited: How to construct a hash function. In: Shoup, V. (ed.) CRYPTO 2005. LNCS, vol. 3621, pp. 430–448. Springer, Heidelberg (2005)
15. Bertoni, G., Daemen, J., Peeters, M., Van Assche, G.: On the indifferentiability of the sponge construction. In: Smart, N.P. (ed.) EUROCRYPT 2008. LNCS, vol. 4965, pp. 181–197. Springer, Heidelberg (2008)
16. Bellare, M., Ristenpart, T.: Multi-property-preserving hash domain extension and the EMD transform. In: Lai, X., Chen, K. (eds.) ASIACRYPT 2006. LNCS, vol. 4284, pp. 299–314. Springer, Heidelberg (2006)

17. Tsudik, G.: Message authentication with one-way hash functions. In: INFOCOM 1992, pp. 2055–2059 (1992)
18. Kaliski, B., Robshaw, M.: Message authentication with MD5. CryptoBytes 1(1), 5–8 (1995)
19. Patel, S.: An efficient MAC for short messages. In: Nyberg, K., Heys, H.M. (eds.) SAC 2002. LNCS, vol. 2595, pp. 353–368. Springer, Heidelberg (2003)
20. Yasuda, K.: "Sandwich" is indeed secure: How to authenticate a message with just one hashing. In: Pieprzyk, J., Ghodosi, H., Dawson, E. (eds.) ACISP 2007. LNCS, vol. 4586, pp. 355–369. Springer, Heidelberg (2007)
21. Yasuda, K.: Boosting Merkle-Damgård hashing for message authentication. In: Kurosawa, K. (ed.) ASIACRYPT 2007. LNCS, vol. 4833, pp. 216–231. Springer, Heidelberg (2007)
22. Preneel, B., van Oorschot, P.C.: MDx-MAC and building fast MACs from hash functions. In: Coppersmith, D. (ed.) CRYPTO 1995. LNCS, vol. 963, pp. 1–14. Springer, Heidelberg (1995)
23. Fischlin, M.: Security of NMAC and HMAC based on non-malleability. In: Malkin, T. (ed.) CT-RSA 2008. LNCS, vol. 4964, pp. 138–154. Springer, Heidelberg (2008)
24. Wegman, M.N., Carter, L.: New hash functions and their use in authentication and set equality. J. Comput. Syst. Sci. 22(3), 265–279 (1981)

# A    Separation between PRF-AX and PRF-KD

Assume we have a secure PRF-AX function $F : \{0,1\}^{n+m} \to \{0,1\}^n$. We shall construct a new function $F'$ which is also secure in the sense of PRF-AX but completely insecure in the sense of PRF-KD. For this, recall that we have $ipad = 0x36\cdots$ and $opad = 0x5c\cdots$. Now write $\mathrm{msb}_2(X)$ for the second most significant bit of a given string $X \in \{0,1\}^*$. Then observe that $\mathrm{msb}_2(ipad \oplus opad) = 1$. For $X \in \{0,1\}^n$ and $Y \in \{0,1\}^m$ define

$$F'(X\|Y) := \begin{cases} F\big(X\|(Y \oplus ipad \oplus opad)\big) & \text{if } X = IV \text{ and } \mathrm{msb}_2(Y) = 1 \\ F(X\|Y) & \text{otherwise.} \end{cases}$$

It can be seen that $F'$ is a secure PRF-AX function. However, $F'$ is totally insecure in the sense of PRF-KD, as the two values $F'\big(IV\|(\bar{K} \oplus ipad)\big)$ and $F'\big(IV\|(\bar{K} \oplus opad)\big)$ are always the same.

Conversely, assume we have a secure PRF-KD function $F : \{0,1\}^{n+m} \to \{0,1\}^n$. We shall construct a new function $F'$ which is secure in the sense of PRF-KD but not in the sense of PRF-AX. For $X \in \{0,1\}^n$, $Y \in \{0,1\}^n$ and $Z \in \{0,1\}^{m-n}$ define

$$F'(X\|Y\|Z) := \begin{cases} F(Y\|0^m) & \text{if } X = IV \text{ and } Z = pad \\ F(X\|Y\|Z) & \text{otherwise.} \end{cases}$$

We see that $F'$ remains secure in the sense of PRF-KD but becomes insecure in the sense of PRF-AX, since the value returned by the affix oracle coincides with the value returned by the $F'_K(\cdot)$ oracle upon the query $0^m$.

# Adding Trust to P2P Distribution of Paid Content

Alex Sherman[1], Angelos Stavrou[2], Jason Nieh[1], Angelos D. Keromytis[1],
and Cliff Stein[1]

[1] Computer Science Department, Columbia University
[2] Computer Science Department, George Mason University

**Abstract.** While peer-to-peer (P2P) file-sharing is a powerful and cost-effective content distribution model, most paid-for digital-content providers (CPs) use direct download to deliver their content. CPs are hesitant to rely on a P2P distribution model because it introduces a number of security concerns including content pollution by malicious peers, and lack of enforcement of authorized downloads. Furthermore, because users communicate directly with one another, the users can easily form illegal file-sharing clusters to exchange copyrighted content. Such exchange could hurt the content providers' profits. We present a P2P system TP2P, where we introduce a notion of trusted auditors (TAs). TAs are P2P peers that police the system by covertly monitoring and taking measures against misbehaving peers. This policing allows TP2P to enable a stronger security model making P2P a viable alternative for the distribution of paid digital content. Through analysis and simulation, we show the effectiveness of even a small number of TAs at policing the system. In a system with as many as 60% of misbehaving users, even a small number of TAs can detect 99% of illegal cluster formation. We develop a simple economic model to show that even with such a large presence of malicious nodes, TP2P can improve CP's profits (which could translate to user savings) by 62% to 122%, even while assuming conservative estimates of content and bandwidth costs. We implemented TP2P as a layer on top of BitTorrent and demonstrated experimentally using PlanetLab that our system provides trusted P2P file sharing with negligible performance overhead.

## 1 Introduction

While P2P presents a powerful and cost-effective file-sharing model due to its ability to leverage the participating users' uplink bandwidth, most paid-content providers (CPs) typically rely on direct download methods to distribute their paid content. For example, Apple iTunes [1], Amazon [2] and Sony distribute content either directly from their website or via contracted content delivery networks (CDNs) such as Akamai [3]. The cost of content delivery, which involves either building infrastructure or paying CDN fees, is quite significant. While some CPs, such as Warner Bros. and AOL, have begun to experiment with limited P2P deployment [4] based on proprietary technology most CPs are reluctant to embrace the cheaper P2P content delivery model. Their worry is that unlike direct download, P2P introduces a number of security concerns such as unauthorized downloads of paid content and increased illegal content sharing that could reduce CPs' profits. In this paper, we introduce TP2P – an architecture that augments P2P to address these security concerns. We hope that our extension can help promote a

P. Samarati et al. (Eds.): ISC 2009, LNCS 5735, pp. 459–474, 2009.

wider adoption of P2P by content providers and that the content delivery cost savings could benefit end-users via lowered content prices.

In a P2P model where peers download content from one another rather than from a centrally managed system, a number of security issues arise. These security threats include content pollution (as malicious peers could be serving garbage data to one another), unauthorized download of paid content (as it can not be enforced by a CP server or a CDN), and increased illegal file-sharing of copyrighted content by the P2P peers. Peers can protect against content pollution via standard hash-checking of the file chunks that they receive. However, in a P2P system peers do not have an incentive to enforce exclusively authorized downloads by other peers or to abstain from forming illegal file-sharing clusters with their neighbors. One reason that some CPs are hesitant to rely on a P2P model for content distribution is that it can easily deteriorate into a free file-sharing community similar to Xbox-sky [5] and Red Skunk Tracker [6]). Since P2P users communicate with one another directly during file distribution, it is easy for them to form clusters for an illegal file-sharing. To form a cluster, malicious users can use a simple protocol to "signal" one another as an invitation to join a cluster. Members of a formed cluster can exchange content that they have purchased or that they will purchase in the future from the CP. Thus they reduce CP's profits, as each member of the cluster only pays for a fraction of content that they obtain.

In order to protect against the threat of illegal file-sharing cluster formation and unauthorized downloads, we present a new technique that we call "trusted auditing". Trusted auditors (or TAs) are a new class of peers that police the P2P system by covertly monitoring other peers for any sign of misbehavior, such as admitting unauthorized users or protocol "signaling" that may lead to illegal sharing cluster formation. The TAs help detect and stop such cluster formation and thus protect the CPs profits. We model the behavior of TAs and show analytically and with simulations that TAs can effectively thwart the formation of illegal file-sharing clusters. By introducing this type of policing by trusted auditors we can provide security guarantees that are similar to those of direct download systems. We show via an economic model that since the cost of using the TAs is small TP2P can yield significant profits for the CPs.

When TAs detect misbehavior by a peer, a variety of countermeasures may be taken. For example, offending peers can be banned from the P2P system to a direct download system where they cannot exfiltrate any peer information. **We stress that we do not address illegal content sharing over out-of-band channels**. Sharing of files after they have been downloaded can happen regardless of the file distribution mechanism used by the CP: a user can download a movie via a CDN and then post it for free download on PirateBay [7]. However, it is important to address the threat of additional illegal sharing that may occur over the P2P delivery system used by the CP. Imagine the effects of having millions of iTunes users connected to a P2P system. Many of the users have demonstrated the willingness to purchase content. Regular users are reluctant to visit illegal pirated content sites because of the potential legal consequences as these sites are policed by RIAA, MPAA and third-party companies such as Media Defender [8]. At the same time, simple software can help iTunes users probe their P2P peers and invite them to share media libraries. Users know that their iTunes peers already have high quality purchased content and probably share similar interests since they have learned one another's IPs by P2P sharing of the same content. With the use of TAs, TP2P prevents such additional illegal file-sharing when CP switches to the P2P model.

The contributions of our work are as follows:

• We introduce *automated trusted auditors* as a controlled and inexpensive way to monitor and detect certain types of misbehavior in a P2P system.

• We present an analytical model that shows how TAs effectively thwart malicious users from forming illegal file-sharing clusters. Our analysis shows that even where TAs are but a small fraction of all peers, they are sufficient at protecting the P2P system against unauthorized file sharing.

• Using a simple economic model we further show that TP2P provides a more cost-effective solution than direct download. This results in higher profits for a CP even in the presence of a large percentage of malicious users.

• Finally, we implement TP2P security elements on top of BitTorrent to demonstrate that our system can provide its functionality in an existing, widely-used P2P system with only modest modifications.

## 2  Related Work

As broadband Internet access becomes more prevalent, digital content stores such as Apple Itunes and Amazon have begun to distribute richer digital content over the Internet, such as TV series episodes and full-length movies. Since each download requires significant bandwidth, these stores typically contract Content Delivery Networks (CDNs) to distribute their content. Commercial CDNs include Akamai [3], Limelight [9] and VitalStream [10]. Since CDNs are centrally managed they can enforce appropriate security measures on behalf of a digital store, such as authorization of customers and encryption of served content. However, the price paid to CDNs for their services is quite high. Market research [11] suggests that digital media vendors spend 20% of their revenue on infrastructure costs for serving content. *While free academic alternative CDNs such as Coral [12] and CoDeeN [13] exist, these systems are typically limited in their deployment and the amount of bandwidth they are allowed to use.*

An alternative powerful distribution model is Peer-to-Peer (P2P) systems such as BitTorrent [14], Napster [15] and Kazaa [16]) among others. No extra contracted bandwidth is required as users leverage one another's upload links to "share" content. Bit-Torrent is perhaps the most popular of these systems, and many analytical works [17, 18, 19, 20] have shown the high efficiency and scalability characteristics of BitTorrent.

Some companies have begun to adopt the P2P model with some security measures. MoveDigital [21] implements a gateway in front of a P2P system to allow only authorized users access. However, once inside, users can leverage the system for further illegal sharing without limitations. For example, if a user can learn the IP addresses of other users inside the system, she can start sharing content with those users directly for free, bypassing the up-front payment. Moreover, users might choose to participate in the P2P system and pay to download files to gain knowledge about other participants that have similar interests. Then, they can easily form another, private P2P community, a darknet [22], for free future exchange of similar content

In contrast, TP2P is designed explicitly to guard against such free file sharing using an open system architecture that is resistant to exploitation even in the presence of malicious nodes. TAs used in TP2P are owned and managed by the content provider, and are unlike reputation-based systems [23] where users simply rate each other such that the resulting ratings may not be trustworthy.

An additional problem for efficient P2P distribution of content is "free-riding" by users who do not upload to their neighbors [24]. This problem can be partially addressed by BitTorrent's tit-for-tat mechanism [25] which was found to be fairly robust [26]. Additional solutions that consider incentives in P2P systems have also been proposed [27,28,29,30]. We believe, that our technique of using TAs could also be used to solve this problem. We leave this idea as an item for future work and focus here on using TAs to prevent illegal cluster formation.

## 3   Architecture

The TP2P architecture is designed as an additional layer for common P2P systems. This layer consists of components that enforce stronger security and trust in the P2P system: the authenticator service and trusted auditors. While TP2P layer can be applied to virtually any common P2P system, we use BitTorrent as the underlying P2P system as a proof of concept. We selected BitTorrent given its popularity, open implementation, and its very efficient file-swarming mechanism where users share individual blocks of a given file.

The goal of BitTorrent is to distribute a file as fast as possible to all connected peers. BitTorrent splits the file (such as a digital movie) into a number of chunks. Participating *peers* exchange individual chunks of the file using a *file swarming* approach. The swarming algorithm is fully distributed and nodes use it to decide from which peers they are going to request their missing chunks. In addition, in each file-sharing instance there are one or more *Seeds* present. Seeds are peers that have all the chunks of the given file. The party that advertises the content typically initializes one or more *Seeds* with the full content of the file. A file-sharing instance also contains a *Tracker* that tracks all participating peers. A peer joins the system by contacting the Tracker. It receives a set of usually up to 50 IP addresses of other participating Peers. The Peer then exchanges chunks of the file with the other Peers and periodically updates its progress to the Tracker via *announce* messages.

### 3.1   System Overview and Usage

When the user decides to purchase content for the first time, she registers at the content provider's portal. She picks a username and a password and enters her payment information (*i.e.,* credit card number). She then downloads a software client that allows her to browse for files, purchase content and perform P2P downloads. For each purchase at the CP's portal she obtains a verifiable token (signed credential) that authorizes her to download the purchased content file. The authenticator also generates credentials for her to be used for secure communication during the download *session.* (We occasionally refer to the file-sharing instance as a download session.) We describe these parameters in Section 3.2.

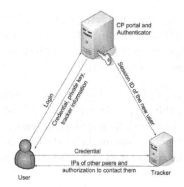
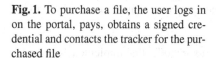

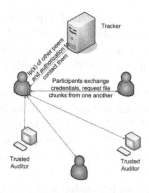

**Fig. 1.** To purchase a file, the user logs in on the portal, pays, obtains a signed credential and contacts the tracker for the purchased file

**Fig. 2.** Users authenticate one another and request file pieces. A fraction of trusted auditors is mixed in among the file-sharing peers.

The user is then directed to a tracker that manages a file-sharing instance for the purchased file. The tracker validates that the user is authorized to perform the download by verifying her credentials. The user's interaction with the authenticator and the tracker is depicted in Figures 1 and 2. As in BitTorrent, the tracker *assigns* a set of other clients or *peers* to the new client. The client shares pieces of the purchased file with her assigned peers using BitTorrent's *file-swarming* approach. TP2P differs significantly from Bit-Torrent in the assignment of the peers. The TP2P tracker ensures that a certain fraction of the peers that it assigns are *trusted auditors* (TAs), as shown in Figure 2. TAs are special peers who, in addition to participating in a download session, detect misbehaving peers. The detected malicious peers are identified and "banished" from the system.

### 3.2 Authorization

We first describe the authenticator and other modules which enforce strong authorization and authentication. When a user purchases the content at the CP's portal, their credit card is charged the cost of the content. At that point the authenticator running on the CP's portal generates authorization credentials for the user (that will authorize her to participate in a download session for this content) and sends them to the user over a secure connection (using SSL). CP also stores the purchase record in case the user loses her credential due to reboot or another failure and needs to come back to the authenticator.

**Authorization Credentials.** The authorization credentials given to the user include a temporary public/private key pair and a signed credential (akin to a public-key certificate) signed by the authenticator, whose public key is implicitly trusted by all participating users (*i.e.*, it is distributed along with the software, or is otherwise well known). More specifically, we use public-key-signed policy statements (similar in form to public-key certificates [31]) issued by the content provider as the basis for authorization in our system. These credentials are given to authorized users after a purchase is made, and can be used as proof to both the Tracker and the other participants in a P2P download session. The credential includes a Session ID that identifies the user's download session, an expiration time, the user's IP address and public key, and an Instance ID (a unique identifier of the file-sharing instance managed by the Tracker).

**Verification by Tracker.** Following the previous step, the user establishes an encrypted TCP connection to the Tracker using the Tracker's public key and sends its signed credential. The Tracker validates the digital signature of the credential against the authenticator's public key, confirms that the user's IP matches the one in the credential, and that the credential has not expired and that the Instance ID refers to a valid download instance.

If all the parameters are confirmed, the Tracker assigns and sends a list of other peers to the new user, along with a new credential that lets the new user contact other nodes of the same session.

**Peer Verification.** When establishing communication, nodes that implement the correct protocol verify their peers using the tracker-issued credentials: the signature, IP address, public/private key binding, expiration, and instance ID. After verification, they negotiate a symmetric session key for their encrypted TCP connection using their public/private keys.

**Certificate Revocation.** If a peer loses a credential, due to hardware down-time or local network down-time, she will re-login to the authenticator with her username and password and obtain new credentials. However, before the new credentials are issued the authenticator revokes the old credentials. The authenticator contacts the tracker previously assigned to this user and invalidates the old session ID. In response the tracker sends out new ACLs to the peers assigned to the peer with revoked session ID. The credentials and ACL revocation prevent a user from having multiple simultaneous identities in the system, thereby avoiding a scenario where a malicious user may attempt to steal and reuse the identity of an authorized user. Observe, that if the user machine is assigned a new IP by a local DHCP server after a network down-time, the new IP will be included with the new signed credentials.

### 3.3  Detecting Malicious Behavior with Trusted Auditors

The TAs imitate other peers through their participation in the P2P file exchange. In addition, they passively and actively detect malicious nodes that either allow download of unauthorized content or signal one another to form illegal file-sharing clusters. After malicious nodes are detected they are "banished" to an isolated direct-download system for future downloads. There they can no longer exploit the P2P system to form new content-sharing clusters. As a deterrent, the banished nodes may also be charged a penalty of the bandwidth cost for their future downloads. Alternatively, they may be warned with a temporary fine or threatened with legal action. The deterrent may vary according to the policy chosen by the CP as we discuss in Section 4.3. Since TAs also consume bandwidth and require a maintenance cost the relative number of the TAs must remain small.

**Defining Malicious Behavior.** The system maintains and updates a definition list of malicious behavior (MDL) that can lead to unauthorized downloads and establishment of covert channels. The TAs can easily monitor the behavior described in the MDL. TAs can steer clear of false positives by following the full protocol that the malicious users use to exchange illegal content and incriminate them with evidence of such transaction. Initially the MDL includes unauthorized or unencrypted connections, connections to a non-protocol port and connections to a proper port that is not formatted according to TP2P protocol.

The CP employs two strategies in updating the MDL. One strategy involves actively searching, studying and running the software that malicious users use. The second strategy is learning the malicious probing format and pattern on the fly. This approach is based on recent work done at UC Berkeley on the RolePlayer system [32]. RolePlayer installed on a TA machine can quickly learn and replay various network communication patterns.

In order to form an illegal cluster malicious users attempt to discover one another by either establishing a covert channel or by accepting one. With high probability, the malicious node will probe a TA or reply to a probe from a TA and thus be detected and banished from the P2P system. Of course, here is also a small probability that a malicious node will find other malicious users by such probing and form a file-sharing *cluster* which diminishes with the size of the cluster. Thus, a more aggressive malicious node who may attempt to probe more neighbors aiming in forming a bigger malicious cluster, runs a higher risk of being detected by a TA and being banished to a direct download system. We explore and model the optimal strategy for a malicious node and the detection probability in detail in Section 4.

**Behavior of Trusted Auditors.** TAs act as hidden "sentinels" in the system to prevent malicious probing, and therefore significantly limit the ability of illegal cluster formation. To stay hidden, TAs mimic different roles: regular or "neutral" nodes and malicious nodes. In their "neutral" role, TAs mimic the behavior of P2P peers by implementing the same discovery and download protocols, exhibit similar download speeds, arrival and departure rates as the regular clients. In their "malicious" role, TAs mimic the behavior of malicious nodes by sending out probes to their neighbors at the same rate as other malicious nodes.

### 3.4  Security Analysis

TP2P architecture was designed to ensure that a P2P content delivery system could exhibit similar security properties to a direct download system. In particular, we consider threats where users may attempt to exploit the P2P system by attempting illegal cluster formation and unauthorized downloads. We further classify the former threat into an *insider* and *outsider* attacks. In insider attacks a P2P participant may contact another participant during a download session. For outsider attacks, a node records the IP of it's peers and tries contact them from another IP address either during or after the download session. In the next session, we discuss how TP2P addresses those threats.

**Insider Attacks.** One class of attacks against TP2P can stem from malicious users who purchase content and thus obtain the proper authorization to join a file-sharing instance.

Such *insider* malicious users can then attempt to discover other malicious users among the file-sharing peers and form a collaborative network for future unauthorized sharing. For instance, if five malicious users with similar interests discover each another during a file-sharing instance, then in the future only one out of five will need to purchase new content and share it with the rest. TP2P offers protection against this abuse by including TAs in the file-sharing network. The role of the TAs is to detect any malicious user attempting to scavenge information for future sharing. There are two ways in which TAs can detect malicious users: either because the malicious user contacts the TA and attempts to share unauthorized content, or because she allowed a TA to contact her and share content without proper authorization.

*But how can we make sure that the identities of TAs are not exposed to the malicious nodes rendering them ineffective?* There are two ways in which a TA can be exposed over time: either by learning the TA network locations (IP addresses) or by observing their behavior in the P2P system (*i.e.,* when they perform active probing or detect a malicious node). To avoid simple detection of the TAs' IP address pool based on their location, we can rent IP address space from Internet Service Providers based on their user population [33]. Moreover, for more sophisticated attacks that can learn even those IPs over time, we can request the TAs' IPs to be given via the same DHCP servers that the ISPs use for their own users. This will make the tracing of the TAs IPs futile since their IPs do not only change over time but are also shared with regular Internet users.

The second way to expose a TA is to learn to identify its behavior, in particular as it pretends to be malicious and probes other nodes. However, this is only true if malicious nodes already have the knowledge of what is deemed a "normal" probing rate or they don't probe at all (thus exposing the TAs). In both cases, this requires some sort of previous shared knowledge among malicious nodes about the malicious behavior that they should exhibit. However even in the extreme case that malicious users have pre-agreed on a way of probing, the TAs can mimic such behavior because they are also receiving a fraction of the malicious probes. Thus, the TAs can adjust their behavior based on the probes that they themselves receive (remember that TAs communicate with one another their common knowledge about the received probing rates).

*How can we protect the system from Denial of Service (DoS) attacks?* Since TAs mimic the malicious node probing behavior, the increased rate of probing may cause TAs to amplify their probing and thus cause a DoS attack. To avoid this, we use randomized traffic thresholds for the probing rates received from the attackers. TAs do not probe beyond those rates. At the same time, malicious nodes that use DoS run the risk of being easily detected by the TAs. Thus, a DoS to scan for other malicious nodes in the P2P, even a short one, represent a prohibitive cost for the malicious user since the probability of being detected and shut down is high.

**Outside Probing.** In this type of attack, an insider participates legitimately in a download session and collects the list of Peer IPs. It attempts to contact these IPs in search of other malicious nodes from an external IP either during or after the download session. Observe that contacts from outsiders who learn these IPs from a third party also fall under this type of attack. To address such outside scanning we use TAs who are not part of the P2P network to mimic the behavior of the outside scanning. Note that for a malicious node inside the P2P network there is little incentive to answer an outside probe. The reason for this is that outsiders are less likely to have content for trading. On the other hand, nodes inside the P2P are far more likely to have content worth trading since they have proven that they are actually willing to buy such content. All things being equal in terms of scanning, by replying to outside probes malicious insiders run the same risk of detection with uncertain gains. In practice, there is no incentive for a malicious insider to respond to outside probes. The TAs prevent a possible DoS behavior by setting high random thresholds in the traffic they receive. Furthermore, as we show in our analysis in Section 4, the mere knowledge that TAs are present in the network causes rational malicious nodes to behave more cautiously and thus less dangerously towards the CP. TAs help to set the bar of malicious exploitation high by detecting malicious users who have purchased content and thus have gained authorized entry into the system. Furthermore, TAs detect users that do not honor (enforce) the authorization credentials generated by the authenticator.

**Unauthorized Downloads.** Similarly, TAs probe the peers to check whether they allow unauthorized or unauthenticated downloads, by attempting to connect to them without proper credentials. Peers that deviate from the protocol by not enforcing the security checks are banned to the direct download system and may be selectively warned or penalized as a deterrent.

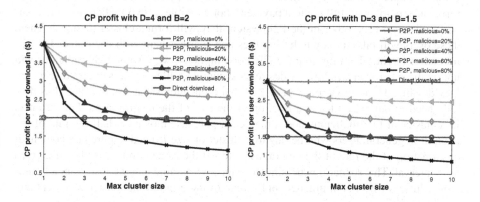

**Fig. 3.** CP profit per user download. Distinct combinations of $D$ (profit before bandwidth) and $B$ (bandwidth cost) capture variations in possible royalties and bandwidth agreements.

## 4   Analysis

We have discussed two threats that may exist in the P2P framework for paid content distribution: unauthorized downloads and illegal file-sharing cluster formation. In the case of the first threat the user that allows an unauthorized download has nothing to gain by deviating from the TP2P protocol. In the case of the second threat each user that joins a cluster gets some content for free from the other cluster members. In this section, we focus our analysis on the threat of cluster formation and effectiveness of

TAs in thwarting such activity. We analyze the strategy of malicious nodes and show that even a small number of the TAs can effectively curb the growth of clusters and successfully protect the CP's profits.

## 4.1   Economic Impact

We propose a simple economic model to quantify the impact that malicious nodes have on the CP's profit. We assume that the average price of digital content sold by the CP is $S$ dollars. The CP pays a large part of that price as royalties $R$ to the content owner (a movie studio for example), and retains $D. (D = S − R)$. In a direct download system the CP also pays $B$ for the bandwidth required to serve a file of average size to the end user. Thus the CP's profit per movie purchase is, on average, $(D − B)$. The market research in [11] shows that digital movie and audio stores pay roughly $60 − 70\%$ of end price ($S$) in royalties and the cost of bandwidth amounts to about $20\%$. Using a store similar to Apple Itunes as an example, one can purchase standard length (1GB) digital movies for \$10. We assume that $D$, the store's profit before bandwidth cost is \$3 to \$4 and $B$, the cost of bandwidth is roughly \$2 per download. We experiment with these assumptions in this section, but our results hold for wider ranges of values.

Using a P2P download approach the CP saves on most of the bandwidth cost and claims a full $D$ as profit. Unfortunately, in the presence of malicious users the CP collects smaller amount of revenue, and thus smaller profit since the malicious nodes form clusters to avoid full content payment. For example, if two malicious users manage to discover each other in the P2P system they will form a cluster of size 2. Then, these users will take turns purchasing files and sharing them with each other for free instead of buying them through the CP. For simplicity, we assume that malicious and non-malicious (or *neutral*) users desire to accumulate files at the same rate (e.g. say they download one movie per week), and that their interests are similar and thus they only need to purchase files at a fraction of the rate of the neutral users. For instance, in a cluster of two malicious nodes they each purchase movies at half the rate of the neutral. In general, users who belong to a cluster of size $K$ need to purchase content at a $\frac{1}{K}$ fraction of the rate of the neutral users to get the same number of files in a given time interval. This scenario is pessimistic, since we assume that we lose from all malicious clusters whereas in practice, only some of the users in the cluster will want any particular file.

A single download session consists of up to $N_s$ nodes that are all assigned to one another by a tracker. For a popular file, the system runs multiple download sessions of up to $N_s$ nodes each. We assume that a single session contains at most $M$ malicious nodes, $T$ TAs and $Q$ neutral nodes with $N_s = Q+M+T$. In a BitTorrent network a typical value of $N_s$ is around $50 − 60$ nodes, thus in our system we will assume a maximum bound of $N_s = 100$. Let $M_i$ be the number of users in the system who are malicious and who belong to clusters of size $i$. Then $M = \sum_{i=1} M_i$. We define $m_i = M_i/(M + Q)$ and $m = M/(M + Q)$ to denote the ratio of malicious users to the total number of malicious and neutral nodes. The amortized profit received by the CP for each file is:

$$\text{Profit} = D \cdot (1 - m) + D \cdot \sum_{i \geq 1} \frac{m_i}{i}. \tag{1}$$

The first term in Equation 1 is the CP profit from neutral users who pay a full price. The second term is from malicious users who pay only a fraction $\frac{1}{i}$ of the price based on their cluster size $i$ (assuming multiple downloads). On the other hand, the profit of the CP in a direct-download system per download is $D-B$. As a reminder, we do not attempt to solve *out-of-band* sharing that can exist with both direct and P2P systems. Rather, we are interested in curbing file-sharing from clusters formed by malicious exploitation of the P2P distribution system itself.

Using Equation 1, we produce the CP profit plots varying $D$ and $B$. Figure 3 depicts CP profit curves for the P2P and the direct download systems for values of $m$ ranging from 0 to 80%. Each plot uses different values for $D$ and $B$ to allow for variations in the cost of royalties and bandwidth. The x-axis shows the maximum size of a malicious cluster, $K$. The y-axis shows the average profit claimed by the CP user download. Each plot contains two horizontal lines: the top one representing a profit of a P2P system assuming no malicious nodes and the bottom one representing profit of a direct download system. The difference between the two plots is exactly $B$, the cost of bandwidth per download. The non-linear curves plot Equation 1 and represent the profit of a P2P system with various fractions $m$ of malicious users. The plots show that as the fraction of malicious nodes and the file-sharing clusters that they form grow the profits for the P2P system dwindle. In fact, as the malicious nodes' fraction approaches 80% and for malicious clusters of > 20 nodes, the CP collects less than half the profits of a direct download approach. Even for less aggressive collections of malicious users, we see that most of the economic advantage of P2P rapidly diminishes.

## 4.2   Probing Game

We model the interaction between the malicious nodes and TAs as a *probing game*: malicious nodes probe and reply to probes from other malicious nodes in order to form and grow a malicious cluster. To detect malicious nodes, TAs also pretend to be malicious. They actively send probes and reply to probes of malicious nodes. To avoid being detected a malicious node must not probe all of its neighbors. Instead, she chooses a finite strategy that we call a growth factor (GF) which reflects the minimum cluster size that she aims to belong to at the end of the download session. The malicious node probes and replies to probes until she discovers at least $GF - 1$ other malicious nodes which may include a TA pretending to be malicious. For $GF = 1$ malicious nodes behave as a neutral nodes. If $GF > M$ the malicious nodes are certain to hit a TA and thus become detected before they can grow into a cluster of size $GF$. Thus, $1 < GF \leq M$.

In general, we make the following assumptions: malicious nodes remain "active" (i.e. they send and reply to probes) until they reach their growth factor of $GF$. Each malicious node knows both $M$ and $T$ in a download session, and based on that picks the most profitable value of $GF$. We suggest a good value for $GF$ later in the section based on a simulation of multiple games. In the end of the session if a cluster formed during the session includes a TA (that pretended to be malicious) all the malicious nodes in the cluster are assumed to be "detected" and they are warned and "banished" by the CP. Such nodes still do not know which of the cluster nodes was trusted and thus cannot assume that they can share with the nodes they already discovered. Both malicious nodes and TAs send probes to randomly chosen neighbors at the same probing rate per node. TAs send probes at the same rate to be indistinguishable from malicious nodes. Otherwise, collaborating malicious nodes could easily pick out TAs in the system.Upon receiving probes, neutral nodes simply ignore them.

Figure 4 shows the probability that a malicious node *succeeds* in forming the desired cluster size. Here the fraction of malicious nodes in the download session $m$ is fixed at 50% and the number of trusted $T$ is varied over different ratios of $M/T$. The x-axis shows the strategy (i.e. growth factor) chosen by the malicious nodes in the game. The y-axis gives the probability that a node succeeds in achieving reaching its selected growth factor. As an example, the scenario of $M/T = 1$ (the number of malicious nodes and TAs is the same) and a target $GF = 2$, shows that the probability of a node succeeding in forming a cluster of size 2 is about 25%. Thus there is a 3/4 chance that a node gets detected in such a game. An important observation about this plot is that all curves are decreasing monotonically. That means that as the malicious nodes become more aggressive by picking larger growth factors, they are also more likely to be *detected*. Interestingly, even for the top curve ($M/T = 10$) and the least aggressive target of $GF = 2$, there is only a 77% chance that such a node succeeds (*i.e.,* there is a 23% chance that it becomes detected). So, even in a favorable scenario, the probability that the node does not become detected in $k$ independent games is roughly only $.77^k$.

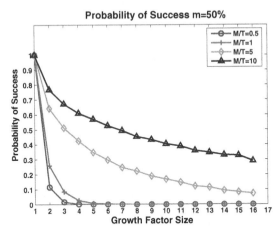

**Fig. 4.** For a single game, probability that a malicious node succeeds in forming a cluster of at least its growth factor with 50% of malicious users

### 4.3   Simulation and Results

We used MatLab to simulate the overall behavior of a BitTorrent-like P2P system with neutral, malicious and trusted nodes. We varied the overall system size, ranging from $10^5$ to $10^7$ participants. Our results remain consistent for all sizes. The plots presented in this paper are obtained using a population of $2 \cdot 10^6$ nodes. Our aim was to examine the performance limits of our system under diverse operating conditions by varying both the fraction of the malicious nodes $M$ and their relative ratio to the TAs $M/T$. In addition, we wanted to find which growth factor is more beneficial for the malicious nodes across multiple downloads. We picked 30 downloads as the number we use for the multiple plots, because at 30 downloads we have detected the overwhelming majority of the formed clusters for all $M/T$ ratios we consider. In addition, after 30 downloads, we notice that new clusters are formed almost exclusively by the new malicious arrivals and thus we consider the distribution to be stable. We assumed a *renewal rate* (departure and arrival of new users between downloads ) of 5%. (Higher renewal rates result in even less effective cluster formation for malicious nodes).

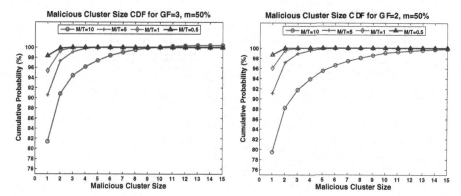

**Fig. 5.** Cumulative probability of forming clusters for growth factor $GF = 2$ and $GF = 3$ for multiple games. Notice that both plots look similar and that for $M/T = 10$, $GF = 2$ results in slightly larger cluster sizes.

In Figure 5, we present results from multiple downloads and for growth factors $GF = 2$ and $GF = 3$. The depicted results indicate that there is very little difference in the malicious cluster size distribution (CDF) when comparing $GF = 2$ and $GF = 3$ with the first having slightly better results. Therefore, the malicious users should select $GF = 2$ as their growth factor if they want to optimize their probability of being in a larger cluster over multiple downloads.

The main result of the system with TAs is a high detection rate of the malicious nodes. In fact, in our simulation even starting with $m = 60\%$ of malicious nodes and $M/T = 10$ with $GF = 2$ after the multi-game simulation reaches steady state we observed that more than 99% of the malicious nodes in the system have been "detected". 80% of the malicious nodes failed to form clusters of even a small size prior to detection.

With the *conservative* policy, the CP warns the detected malicious users but leaves them in the P2P system. The CP threatens a fine or court action for illegal activity and forces them to re-download a new software client. If the CP believes that almost all such users will behave neutrally then it continues to make $\$D$ in profit from these users. Equation 2 presents the amortized profit per download under this policy.

$$AP = D \cdot (1 - m) + (D \cdot \sum_{i \geq 1} \frac{m_i}{i}) - B \cdot \frac{T}{M + Q} \qquad (2)$$

This is an extension of equation 1 with the additional term: $-B \cdot \frac{T}{M+Q}$ that accounts for the bandwidth used by the TAs normalized by the total number of malicious and neutral users in the system. Figure 6 compares the profits of an unprotected system with that of TP2P based on a multi-game simulation (with the parameters as describe above) when it reaches steady state. TP2P shows much higher profits. For instance with $m = 60\%$, $M/T = 10$, $D = 4$ and $B = 2$ the profit is 122% higher for TP2P. Observe, that if instead the CP decides to move the detected nodes to a direct download system and charge them a penalty of their bandwidth cost the equation 2 also describes the resulting profit. In this situation the CP still makes $\$D$ from each download.

With an *aggressive* policy the CP does not trust the detected users to behave neutrally after a warning. The CP moves the detected users to a direct download system but does

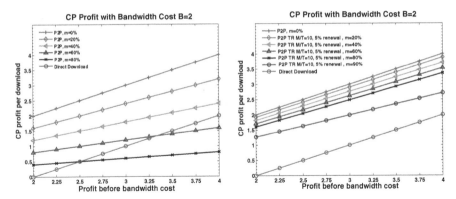

**Fig. 6.** Comparison of CP profits between a protected and an unprotected system for bandwidth cost of 2. On the left we have a system without TAs and on the right a system with a ratio of TAs to malicious users being 10. The protected system yields more profits that are comparable to a P2P system without malicious peers.

not charge them a bandwidth penalty. These users are no longer a threat but the CP now loses $B of bandwidth cost for their downloads. Equation 3 shows the profit based on this policy where the CP loses $B on $m_1$ fraction of nodes - the singleton malicious nodes that are detected.

$$AP = D \cdot (1 - m) + (D \cdot \sum_{i \geq 2} \frac{m_i}{i}) + (D - B) \cdot m_1 - B \cdot \frac{T}{M + Q} \tag{3}$$

Even with this assumption in the case of $m = 60\%$, $M/T = 10$, $D = 4$ and $B = 2$ the steady state profit is 62% greater even with very few TAs. For a very high initial value of $m = 90\%$, the profit curve under this policy overlaps with direct download. The CP can improve the profit by moving detected nodes only temporarily until they can gain higher reputation. We leave this item for future study.

## 5   Implementation and Performance

We implemented an TP2P prototype by adding modifications to the existing BitTorrent client and Tracker (ver. 3.9.1) written in Python. Our modest modifications included adding secure channel communication using RC4 encryption, assignment of trusted auditors by the Tracker, and the distribution of credentials by the tracker to the peers. We conducted our experiments using PlanetLab [34] to compare the download speed of TP2P clients compared to BitTorrent clients on a set of geographically distributed machines given the overhead of secure communication and credentials distribution and verification in TP2P. Most machines used were equipped with 3GHz processors and ran the Linux 2.6.12 kernel.

For our first test, we deployed 41 BitTorrent clients randomly distributed in the continental US. A node was designated as the Seed client and initialized it with a 512MB movie file. To stress our system, we stored no parts of the file on the rest of the clients before the test. We ran the Tracker process on a machine outside of PlanetLab, a blade server with 3.06GHz processors, running a Linux 2.6.11 kernel, and a 10Mbit/sec up-load bandwidth link. We ran the test both with the unmodified BitTorrent code and with TP2P. The BitTorrent download times were only 0.8% faster on average, showing that TP2P adds negligible performance overhead.

For our second test, we performed a similar experiment as the first test but using a more dynamic scenario where peers join the download system at staggered times. We began with one Seed and 76 clients. The 76 clients joined the system at 2 minute intervals. By the time the later peers start, more clients in the system already have partial data sets. Therefore, newer clients have more sources to download the data from and thus their download times are generally faster. For this test, TP2P clients on average slightly outperformed BitTorrent by about 0.5%. This was due to the fact that the TP2P nodes contact the Tracker more frequently and receive new connection assignments at a faster rate at startup. As a result, initially they have slightly more choices for selecting faster sources. The CPU overhead on the TP2P clients was also minimal as RC4 encryption is a fast stream cipher. Average CPU utilization on the TP2P and BitTorrent clients was almost identical at roughly 1.3% and 1.23% respectively.

## 6    Conclusions

We introduced the concept of TAs to a P2P setting: by policing the system, TAs are able to enforce TP2P protocols and guarantee security properties that are similar to those of a direct download system. We have analyzed TP2P by modeling it as a game between malicious users who try to form free file sharing clusters and trusted auditors who curb the growth of such clusters. We have combined this analysis with a simple economic model to quantify the cost-effectiveness of our approach in the presence of malicious users. Our analysis shows that even when 60% of the participants in a system are malicious users, our system can detect 99% of malicious users and prevent them from forming large clusters, thereby providing strong protection of the P2P system against unauthorized file sharing. For most configurations, our analysis shows that TP2P yields profits that are between 62% and 122% higher than a direct download system based on conservative profit and bandwidth cost models. We demonstrate that TP2P can be implemented on top of BitTorrent with modest modifications, and provides its content protection and economic benefits with negligible performance overhead compared to vanilla BitTorrent.

## Acknowledgements

This work was supported by NSF Grants CNS-07-14277, CNS-04-26623, and Google Inc.

## References

1. Apple itunes, http://www.apple.com/itunes
2. Amazon, http://www.amazon.com/
3. Akamai, http://www.akamai.com/

4. Warner bros and p2p,
   http://arstechnica.com/news.ars/post/20060130-6080.html
5. Xbox-sky, http://bt.xbox-sky.com/
6. Red Skunk Tracker, http://www.inkrecharge.com/ttrc2/
7. Piratebay, http://thepiratebay.org/
8. About media defender,
   http://rss.slashdot.org/~r/Slashdot/slashdot/~3/157024087/article.pl
9. Limelight, http://www.limelightnetworks.com/
10. Vitalstream, http://www.vitalstream.com/
11. Aguilar, J.G.: Personal communication (February 2006)
12. Freedman, M.J., et al.: Coral, http://www.coralcdn.org/
13. Pai, V., et al.: Codeen, http://codeen.cs.princeton.edu/
14. Bittorrent, http://www.bittorrent.com
15. Napsterm, http://www.napster.com
16. Kazaa, http://www.kazaa.com
17. Yang, X., de Veciana, G.: Service capacity of peer to peer networks. In: INFOCOM (2004)
18. Arthur, D., Panigrahy, R.: Analyzing the efficiency of bit-torrent and related peer-to-peer networks. In: SODA (January 2006)
19. Massoulie, L., Vojnovic, M.: Coupon replication systems. In: SIGMETRICS (2005)
20. Qiu, D., Srikant, R.: Modeling and performance analysis of bittorrent-like peet-to-peer networks. In: SIGCOMM (2004)
21. Movedigital, http://www.movedigital.com/
22. Biddle, P., England, P., Peinado, M., Willman, B.: The Darknet and the Future of Content Distribution. In: Proceedings of the 2nd ACM Workshop on Digital Rights Management (November 2002)
23. Xiong, L., Liu, L.: Peertrust: Supporting reputation-based trust in peer-to-peer communities. IEEE TKDE, Special Issue on Peer-to-Peer Based Data Management 6 (2004)
24. Locher, T., Moor, P., Schmid, S., Wattenhofer, R.: Free riding in bittorrent is cheap. In: 5th Workshop on Hot Topics in Networks (2006)
25. Cohen, B.: Incentives build robustness in bittorrent. In: Workshop on Economics of P2P Systems (2003)
26. Liogkas, N., Nelson, R., Kohler, E., Zhang, L.: Exploiting BitTorrent for Fun (but not Profit). In: International Workshop on P2P Systems (2006)
27. Vishnumurthy, V., Chandrakumar, S., Sirer, E.G.: Karma: A virtual Currency for Peer-To-Peer Systems. In: ACM Workshop on the Economics of Peer-to-Peer Systems (June 2003)
28. Ngan, T., Nandi, A., Singh, A., Wallach, D., Druschel, P.: Designing Incentives-Compatible Peer-to-Peer Systems. In: Proceedings of the 2nd Workshop on Future Directions in Distributed Computing (2004)
29. Piatek, M., Isdal, T., Anderson, T., Krishnamurthy, A., Venkataramani, A.: Do Incentives Build Robustness in BitTorrent? In: Proceedings of the 3rd USENIX Symposium on Networked Systems Design and Implementation (NSDI) (April 2007)
30. Piatek, M., Isdal, T., Anderson, T., Krishnamurthy, A., Venkataramani, A.: Building Bit-Tyrant, a (more) strategic BitTorrent client. USENIX;login: 32(4), 8–13 (2007)
31. CCITT: X.509: The Directory Authentication Framework. International Telecommunications Union, Geneva (1989)
32. Cui, W., Paxson, V., Weaver, N., Katz, R.H.: Protocol-independent adaptive replay of application dialog. In: Proceedings of the 13th Annual Network and Distributed System Security Symposium (NDSS) (February 2006)
33. Top 22 U.S. ISPs by Subscriber: Q3 2006: Market Research,
    http://www.isp-planet.com/research/rankings/usa.html
34. Planetlab, http://www.planetlab.org/

# Peer-to-Peer Architecture for Collaborative Intrusion and Malware Detection on a Large Scale

Mirco Marchetti, Michele Messori, and Michele Colajanni

Department of Information Engineering
University of Modena and Reggio Emilia
{mirco.marchetti,michele.messori,michele.colajanni}@unimore.it

**Abstract.** The complexity of modern network architectures and the epidemic diffusion of malware require collaborative approaches for defense. We present a novel distributed system where each component collaborates to the intrusion and malware detection and to the dissemination of the local analyses. The proposed architecture is based on a decentralized, peer-to-peer and sensor-agnostic design that addresses dependability and load unbalance issues affecting existing systems based on centralized and hierarchical schemes. Load balancing properties, ability to tolerate churn, self-organization capabilities and scalability are demonstrated through a prototype integrating different open source defensive software.

## 1   Introduction

Distributed and collaborative systems are emerging as the most valid solutions to face modern threats coming from multiple sources. To identify network intrusions and new malware as soon as possible, hierarchical architectures for intrusion detection have been proposed, such as [1]. They are able to gather information from a wide network space and allow early detection of emerging threats because they are based on multiple sensors placed in different network segments and on a hierarchical collaboration scheme. This approach allows administrators to deploy timely countermeasures because all the network segments hosting at least one sensor can be alerted about new threats as soon as they are detected in any part of the collaborative system.

The problems affecting existing collaborative solutions based on hierarchical or centralized architectures are well known: peer dependability issues, limited scalability and load unbalance. We present a distributed collaborative architecture that aims to address these main issues through a cooperative peer-to-peer scheme based on a Distributed Hash Table (DHT).

The peer-to-peer architecture proposed in this paper aims to capture and analyze real malware specimens and propose countermeasures instead of just recognizing that a malware is spreading. Moreover, it disseminates network activity reports on the basis of a behavioral analysis of the captured payload, thus being able to provide a description of the malware behavior. The communication

P. Samarati et al. (Eds.): ISC 2009, LNCS 5735, pp. 475–490, 2009.

model is different from existing systems because it adopts a publish/subscribe scheme as an option for the distribution of the result of the analysis, while each alert is inserted into an ad-hoc message persistently stored in PAST [2]. This solution guarantees privacy and availability of the information. Other proposals based on peer-to-peer defensive schemes (e.g., [3, 4]) differ from this paper because their focus is on novel algorithms for anomaly detection that should be facilitated by cooperation. On the other hand, our focus is on the software architecture that is flexible enough to work with different algorithms.

Other peer-to-peer schemes (e.g., [5, 6, 7]) are used to disseminate information about malicious IP addresses through some publish/subscribe model. Our architecture uses a publish/subscribe scheme only for the communication of the analysis results, while events are persistently stored and can be retrieved successively. Other proposals have some peculiarity that is not addressed in this paper. For example, DOMINO [6] is an interesting architecture because its overlay network combines peer-to-peer and hierarchical components and uses Chord [8] to distribute alert information. The main goal of Worminator [9] is to guarantee a high level of privacy of the shared information. It extracts relevant information from alert streams and encodes it in Bloom Filters. This information forms the basis of a distributed watchlist and includes IP addresses and ports. The watchlist can be distributed through various mechanisms, ranging from a centralized trusted third party to a decentralized peer-to-peer overlay network. However, we should be aware that the main goal of these architectures is to compile an updated blacklist of the IP addresses at the origin of some attacks. On the other hand, this paper has a broader scope: our architecture manages IP addresses and other important information, such as binary code of malware, signature of IDS and malware behavior. Moreover, our architecture is sensor agnostic, and is able to support heterogeneous algorithms and techniques for intrusion detection and malware analysis. For these reasons, it differs from Indra [7] that is a distributed intrusion detection architecture that relies on custom sensors.

This paper is organized as follows. Section 2 describes the design of the proposed architecture. Section 3 highlights its main benefits with respect to hierarchical and centralized architectures. Section 4 details the main features of the prototype that is based on open source software. Section 5 reports the experimental results achieved through the prototype. Scalability, load balancing, robustness and self-organization properties at a larger scale with thousands of collaborative nodes are demonstrated through simulation. Section 6 outlines main conclusions and future work.

## 2   Architecture Design

The main goal of this paper is to design a distributed architecture where each component collaborates to the intrusion and malware detection and to the dissemination of the local analyses including:

- malware behavior,
- malware diffusion,

- network-based attacks,
- diffusion of intrusions,
- identification of suspicious IP addresses,
- identification of the servers from which the malware is downloaded.

The novel architecture should address the main issues of hierarchical collaborative schemes in order to guarantee high scalability, fault tolerance, dependability and self-organization.

To accomplish these goals we propose a flat distributed architecture composed by several cooperating nodes that communicate through a DHT overlay network. An overview of this architecture is shown in Figure 1. Each node, called *collaborative alert agregator*, accomplishes the same high level functions: generation of local security alerts, forwarding of relevant alerts to the other collaborative nodes, analysis of received events and communications of the analysis results. All the communications among the collaborative alert agregators are carried out through a peer-to-peer overlay providing a fully connected mesh network. This solution does not require centralized coordination nor supernodes.

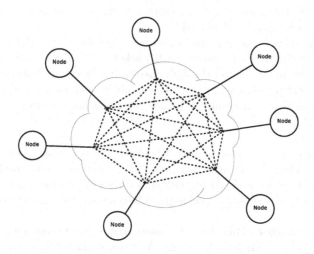

**Fig. 1.** Node connections to the DHT

The design of a collaborative alert agregator is represented in Figure 2. It is possible to identify three layers: the *sensor layer*, the *local aggregation layer*, and the *collaboration layer*. Each layer is described in the following sections.

## 2.1   Sensor Layer

The sensor layer provides a node with intrusion detection alerts and malware samples. It consists of one or multiple types of sensors. For example, the current version relies on four classes of sensors.

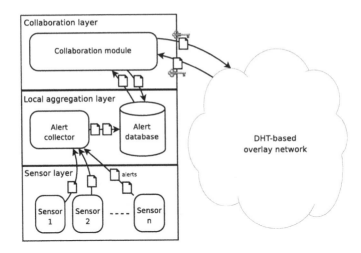

**Fig. 2.** Design of a collaborative alert aggregator

- **Host IDS.** Intrusion detection system monitoring activities on the same host in which they are deployed.
- **Network IDS.** Sensors placed in strategic positions in order to capture all network traffic and to analyze each packet looking for malicious content. They can be implemented through custom hardware appliances or installed on a general purpose computer. When illicit activities are detected, they generate an alert containing information on malicious network packets (such as a TCP/IP header and a signature identifier) and a description of the attack.
- **Honeypot.** These tools are able to collect malware and to trace malicious activities. They consist of server processes hosted on a vulnerable computer connected to a network. As an honeypot does not provide any useful service, any attempt of reaching it and logging into its services can be considered an attack.
- **Sensor manager.** This class of sensors represents a component of a multi-tier, hierarchical NIDS architecture. A sensor manager forwards information to other managers belonging to upper tiers. It can also aggregate and filter data, thus reducing the number of forwarded alerts and decreasing the network traffic and the computational load of the other components of the hierarchical architecture.

It is not necessary to install the sensors on the same physical machine hosting the local aggregation and the collaboration layers. Sensors can interface with alert collectors installed on remote hosts.

### 2.2   Local Aggregation Layer

The local aggregation layer is responsible for collecting, filtering and aggregating all the heterogeneous alerts received from the sensors of the lower layer. While it is possible for the *alert collector* to execute arbitrarily complex aggregation and correlation algorithms, its fundamental task is to pre-process all the received alerts that may be syntactically and semantically heterogeneous. All the alerts are classified and stored in the local *alert database* that is used by the upper layer as the only sensor-independent interface to store and retrieve heterogeneous events.

### 2.3   Collaboration Layer

The collaboration layer is the only component connected to the collaboration overlay network that is based on DHT. Being part of the overlay network, we can assign a unique node identifier *nodeID* to each *collaboration module*. The collaboration module has three main purposes.

– It is responsible for retrieving new events that have been stored in the local alert database. These events are submitted to the DHT-based overlay network through a key (*messageId*) that is computed over a significant field of the event. As an example, the key used to submit a malware specimen can be computed by applying a hash function to the malware, while a NIDS alert can be submitted to the overlay network twice, using as keys the signature ID and the IP address from which the illicit network packet originated. The strategy used to determine which fields are involved in key computation can be configured to fulfill specific analysis goals.
– It receives messages submitted by other collaboration modules that are connected to the same overlay network. Each collaboration module is responsible for a portion of the hash space (determined by the *nodeID* and by the implementation details of the DHT algorithm), and receives all the messages whose *messageId* fall within that hash space. This design choice allows each collaboration module to receive all the messages that are relevant to one scenario. As an example, this approach allows a single collaboration module to receive all the events caused by the same source, thus achieving an effective network-based and distributed alert aggregation and correlation scheme.
– Each collaboration module is responsible for the dissemination of the analysis results to the other nodes connected to the overlay network. This guarantees the timely disseminations of network activity reports to all the collaborative nodes without any centralized bottleneck.

There are two ways to retrieve new alerts from the database: the collaboration layer reads the collected data at regular interval (pull mode) or it is driven by external calls (push mode). In the implementation presented in Section 4 we use the pull mode.

### 2.4   Event Processing

The collaboration layer gets new events from the database and processes them sequentially. For each event, it sends a number of messages depending on the type of event and analysis goals. In the current version of the architecture, each event retrieved from the database may have up to four interesting fields: the malware's binary code, the IP address of the server from which the malware has been downloaded, the IDS signature ID and the IP address of the attacker. For each of these fields a message is created using the hash of its value as *messageId*. This message is received by the node with the *nodeID* closer to the *messageId*. Before the insertion of a new message, the collaboration module monitors the pre-existence of its *messageId* within the DHT. If there is not that value, the message is inserted, and the sender node signals the arrival of a new message to the receiver node. Otherwise, if the *messageId* already exists, the sender node contacts the receiver node and it informs it about the generation of a new event with the same *messageId*.

The receiver node behaves differently depending on the message type. If the message contains a new malware, the receiver node takes care of its behavioral analysis by relying on a local or remote sandbox. In all the other instances, the node executes some anomaly detection strategies that are based on the frequency of received events. It is important to observe that in this paper we do not focus on specific event analysis algorithms, but on the architecture which permits to collaborate and to share information. In particular, the proposed architecture is algorithm agnostic and flexible enough to adopt several different analysis strategies. Finally, the event analysis results are disseminated to all the interested collaborative alert aggregators following a publish/subscribe paradigm [10].

## 3   Peer-to-Peer vs. Hierarchical Architecture

### 3.1   Fault Tolerance

The completely distributed nature of the proposed architecture is inherently fault tolerant, and lacks single points of failure that are typical of hierarchical and centralized architectures, where alert aggregation, correlation and analysis functions are aggregated in the root node of the architecture [1]. This node represents a single point of failure and when it is unreachable the hierarchical architecture is unable to complete any collaborative task. The effectiveness of hierarchical architectures can be impaired even by failures of the nodes belonging to intermediate layers of the tree. As an example, a failure of one of the tier-1 nodes causes the isolation of the complete sub-tree having the faulty node as its root.

On the other hand, the proposed architecture leverages a completely distributed, network-driven aggregation and correlation technique. Each node is responsible for aggregating and correlating only a small subset of alerts and malware samples. If a node becomes unreachable, only the messages that would have been handled by the faulty node are lost, while all the other nodes are not influenced by the failure. Moreover, depending on the implementation of DHT routing of the overlay network, the collaborative nodes can detect the failure of a peer, and autonomously modify their local overlay routing tables accordingly.

Hence, the proposed architecture is able to autonomously reorganize itself and restoring its efficiency with no human intervention.

Message replication schemes can also be used to reduce the (minimal and transitory) message losses due to the failure of a collaborative node. In the current version, it is possible to set a *replication constant* $k$ denoting that, for each message, $k$ copies are created and maintained by the DTH overlay. One message is sent to the node whose unique identifier $nodeID$ is responsible for the message key. The other $k-1$ messages are sent to the $k-1$ nearest neighbors, thus guaranteeing full reliability for up to $k-1$ failures, because the network would maintain constant the number of replicas through periodic inspections (experimental evaluation of message loss probability for higher number of concurrent faults are presented in Section 5). By tuning the value of $k$, it is possible to achieve the desired trade-off between overhead and fault tolerance.

## 3.2   Load Balancing

Hierarchical architectures, such as [1], concentrate malware analysis and alert correlation tasks on the root node, so that they can avoid replicated analyses. As a bad consequence, the computational load on the root is significantly higher than the load of the intermediate nodes, to the extent that much more powerful hardware is necessary to host the root services.

Another advantage of the proposed DHT-based distributed architecture is represented by its intrinsic load balancing properties. Let us consider a scenario in which a network participating to a collaborative hierarchical architecture is targeted by an attacker, while the other participating networks are not. In a similar situation, the load related to alert management and correlation is unevenly distributed because only the nodes related to the attacked network are involved in alert management. Hence, an attacker could easily overload the path connecting the attacked networks to the hierarchy root by attacking few selected networks connected to the same higher-level node.

Uneven load distribution and overload risks are mitigated by the proposed distributed alert aggregation and correlation scheme. As we avoid one centralized aggregator, then there is no single path through which all the alerts generated by a sensor (or a set of sensors in the same network) are transmitted. Alerts gathered by one node in an attacked network are routed to multiple nodes, based on the messageId characterizing each alert. Even if one network (leaf) is heavily attacked, this scenario is well managed and the load is automatically distributed among many nodes through different branches.

## 3.3   Scalability

Hierarchical architectures are based on a multi-tier management infrastructure connecting the lowest layer alert managers (the leaves of the management tree) to the root manager. Each manager node in the hierarchical architecture is able to aggregate alerts generated by a finite number $n$ of lower-layer sensors or managers on the basis of computational complexity of alert management operations and on bandwidth constraints. Hence, a hierarchical architecture can be modeled as an n-ary tree, whose number of intermediate elements grows logarithmically with the number of the leaves.

On the other hand, in the proposed architecture, all the alert management operations are distributed among the leaves and there is no need for a separate management infrastructure. This is a huge advantage in terms of scalability and management, because it is not necessary to reconfigure the architecture hierarchy, possibly by adding new layers to the management tree, whenever the number of leaves increases.

### 3.4    Number of Stored Copies

Another advantage of the peer-to-peer architecture is represented by the smaller number of copies of individual alerts and malware specimens. In a hierarchical architecture, a copy of each different alert and malware specimen is maintained in each node of the management tree by which the alert has been received. Let us consider a tree of managers having order $n$, $l$ leaves and height $h = \log_n(l)$. If $c$ represents the number of copies of each alert stored by the nodes belonging to the hierarchical architecture, then we have:

$$h \leq c \leq \sum_{i=0}^{h-1} n^i$$

This means that the number of copies $c$ is comprised between the number of manager nodes in the path between the leaf generating the alert and the root of the tree ($h$) and the total number of manager nodes when the same alert has been issued by all the leaves in the tree ($\sum_{i=0}^{h-1} n^i$). As $h$ grows proportionally to the logarithm of the number of leaves, then the number of copies of each alert (and malware specimen) that a hierarchical architecture needs to maintain grows logarithmically with the number of leaves.

On the other hand, in the peer-to-peer architecture the number of copies of each different alert is determined by the replication factor $k$, which is a configurable parameter independent of the number of nodes connected to the overlay.

A comparison between the number of stored copies is presented in Table 1. The first row of this table represents the number of copies stored in the peer-to-peer architecture having a replication factor $k = 5$. The number of copies does not depend on any other parameter. The other rows contain the number of copies stored in a hierarchical architecture characterized by a different number of nodes and order. As an example, the second row shows that the number of copies of the same alert in a hierarchical architecture with $l = 1000$ nodes and order $n = 10$ is between 3 and 111, depending on how many leaves issue the same alert.

## 4    Prototype

The viability of the proposed architecture has been demonstrated through a prototype. In compliance with the architecture description in Section 2, each

**Table 1.** Number of store copies stored in the peer-to-peer and hierarchical architectures

| Architecture | Minimum | Maximum |
|---|---|---|
| DHT overlay, $k = 5$ | 5 | 5 |
| Hierarchical, $l = 1000$, $n = 10$ | 3 | 111 |
| Hierarchical, $l = 10000$, $n = 10$ | 4 | 1111 |
| Hierarchical, $l = 100000$, $n = 10$ | 5 | 11111 |
| Hierarchical, $l = 8000$, $n = 20$ | 3 | 421 |
| Hierarchical, $l = 160000$, $n = 20$ | 4 | 8421 |

collaborative alert aggregator consists of different software modules that can be divided in three classes. The first two classes include typical network defense items. The third class includes communication software. The entire prototype is based on open source software.

The current implementation of the collaborative alert aggregator can rely upon heterogeneous sensors, thus being able to detect a wide range of threats. In particular, we used Snort [11] (standard de-facto for signature based network intrusion detection) as a NIDS sensor, and Nepenthes [12] as a low-interaction honeypot.

The alert collector is implemented through the Prelude software [13]. All the communications between the Prelude manager and the sensors is based on the Intrusion Detection Message Exchange Format (IDMEF) [14], which is an IETF standard for the generation and management of messages related to security incidents.

The alert collector is configured to store all the collected alerts and malware samples in the local alert database, implemented with MySQL [15].

The collaboration layer is implemented in Java and guarantees a platform independent application. The DHT-based overlay network relies on the FreePastry libraries, a Java implementation of Pastry network [16, 17, 18, 19, 20]. These libraries guarantee a useful emulation environment, and implementation of two applications based on Pastry: PAST [2, 21], that is the persistent peer-to-peer storage utility used to store information, and Scribe [10, 22], used for multicast communications.

The collaboration module can be configured by editing an XML [23] file. The interface with the alert database is implemented through JDBC drivers [24], thus guaranteeing a high interoperability with the most common DBMS.

Events retrieved from the database are classified according to their type, and managed by different classes. The current version of the prototype implements four classes managing four heterogeneous types of event: malware samples, IP addresses related to the server from which the malware is downloaded, IP addresses related to hostile activity and signatureId of alerts generated by a NIDS sensor. The collaboration module is modular and its functions can be easily extended by adding new classes for the management of other types of event.

After a message has been received, the collaboration module is responsible for its storage, its analysis (possibly leveraging external services, such as Norman sandbox [25] and CWSandbox [26]), and communication of the analysis results to the other nodes. Each node can subscribe to specific areas of interests, thus receiving only a subset of analysis results. In the current implementation, alert storage is handled by PAST, while multicast dissemination of the analysis results relies on Scribe. Finally, a custom application based on FreePastry provides a one-to-one communication service.

PAST is a persistent storage utility distributed peer-to-peer network. It is based on a Pastry layer for routing messages and for network management. We adopt it because it is able to balance the storage space between all the nodes and guarantees high fault tolerance. Each PAST node, identified by an ID of 128 bits ($nodeID$), acts both as a storage repository and as a client access. Any file inserted into PAST uses a 160-bit key, namely $fileId$ ($messageId$ in our case). When a new file is inserted into PAST, Pastry directs the file to the nodes whose $nodeID$ is numerically closer to the 128 most significant bits of $messageId$, each of which stores a copy of the file. The number of involved nodes depends on the replica factor chosen for availability and persistence requirements.

## 5    Validation and Performance Evaluation

Viability and performance of the proposed architecture have been demonstrated through extensive experiments and simulations. Small scale experiments, through the prototype comprising few tens of nodes, have been carried out by executing several instances of the collaboration module in few hosts and by binding each instance on different port numbers. Tests for large number of nodes include the network emulation environment provided by the FreePastry libraries. This solution allows us to launch up to one thousand nodes on each Java Virtual Machine.

Very large scale simulation involving up to ten thousand nodes are carried out through an ad-hoc simulator. It considers the high-level behavior of the hierarchical and DHT-based architectures and omits the low level details related to the transmission of messages over the network and their storage within the local alert database. Although simplified, the simulator uses the same routing schemes of the prototype, and it has been validated by executing the same (reduced scale) experiments on the prototype and on the simulators and by verifying the complete agreement of the results.

### 5.1    Dependability and Fault Tolerance

The completely distributed nature of the proposed architecture is inherently fault tolerant, and lacks single points of failure. While the failure of few nodes does not impair the architecture dependability, it is possible that an alert message sent to a faulty node can be lost. To minimize the chances of loosing alerts, the proposed architecture relies on the message replication scheme described in Section 3. It is possible to configure a replication factor $k$, so that each message is sent to the $k$ collaborative alert aggregators whose $nodeID$ is nearest to the message key.

The ability of the proposed architecture to sustain faults and node churn is demonstrated through several simulations. In each run we simulate an overlay network consisting of a variable number of collaborative nodes (from 1000 to 10000), and we randomly generate a set of messages. Once guaranteed that each node is responsible for at least one message, we simulate the concurrent failure of a percentage of collaborative alert aggregators, ranging from 1% to 10% of the nodes. Then, we wait for PAST to run a scheduled update, thus restoring $k$ copies of each message not lost due to the concurrent node failures, and we check whether all messages created at the beginning of the simulation are still available. The results are presented in Tables 2 and 3.

In Table 2 we compare networks of different sizes (between 1000 and 5000) where nodes use a replica factor $k = 5$ (5 copies of each message). The number of faulty nodes is denoted as a percentage of the total number of nodes. For each failure rate and for each network size, we run the simulation 100,000 times. The message loss rate is expressed as a percentage of the 100,000 runs in which *at least* one message has been lost. As an example, the cell in the fourth row and second column in Table 2 denotes that a network of 1000 collaborative nodes with 40 concurrent failures (4% for the number of nodes) lost at least one message in only 0.01% of the 100,000 tests.

**Table 2.** Message loss probability for $k = 5$ and for different numbers of nodes and faults

| Concurrent fault rate (%) | 1000 nodes | 2000 nodes | 5000 nodes |
|:---:|:---:|:---:|:---:|
| 1 | 0 | 0 | 0 |
| 2 | 0 | 0 | 0 |
| 3 | 0 | 0 | 0,01 |
| 4 | 0,01 | 0,02 | 0,04 |
| 5 | 0,02 | 0,04 | 0,12 |
| 6 | 0,06 | 0,12 | 0,24 |
| 7 | 0,11 | 0,22 | 0,6 |
| 8 | 0,19 | 0,42 | 1,04 |
| 9 | 0,35 | 0,76 | 1,8 |
| 10 | 0,58 | 1,16 | 2,99 |

In Table 3 we report the results about the influence of the replica factor $k$ on the probability of losing a message. In these simulations we use a network size of 10,000 collaborative nodes, and we vary both the concurrent failure rate (expressed as percentage of the number of nodes) and the replica factor (with values of $k$ ranging from 4 to 6). As in the previous set of experiments, for each combination of fault rate and replica factor we run 100,000 simulations. The packet loss probability is expressed as the percentage of simulations in which at least one message has been lost.

These experiments demonstrate that by using appropriate values of the replica factor $k$, it is possible to achieve negligible message loss probability even for

**Table 3.** Message loss probability for a network of 10,000 nodes and for different $k$

| Concurrent fault rate (%) | k=4 | k=5 | k=6 |
|---|---|---|---|
| 1 | 0,009 | 0 | 0 |
| 2 | 0,16 | 0,003 | 0 |
| 3 | 0,735 | 0,019 | 0,001 |
| 4 | 2,117 | 0,075 | 0,002 |
| 5 | 5,022 | 0,219 | 0,015 |
| 6 | 9,732 | 0,542 | 0,037 |
| 7 | 16,64 | 1,186 | 0,081 |
| 8 | 25,859 | 2,172 | 0,159 |
| 9 | 36,682 | 3,774 | 0,315 |
| 10 | 48,685 | 5,904 | 0,529 |

large networks and (unrealistic) concurrent failures of hundreds of geographically distributed and independent nodes.

## 5.2   Load Balancing and Scalability

In this section we compare the load of the proposed peer-to-peer architecture against that of the lowest layer of alert manager in the hierarchical architecture presented in [1].

Experiments are carried out by simulating a network of 5000 collaborative nodes, and a hierarchical architecture with the same number of leaves. Each intermediate manager node of the hierarchical network is connected to a random number of elements in the lower level, uniformly chosen between 5 and 20. The resulting tree has 4 layers and 420 low-level managers, directly connected to the leaf sensors. Figures 3 and 4 compares the load distribution among the 5000 collaborative alert aggregators in the DHT-based overlay and the load distribution among the 420 manager nodes of the hierarchical architecture.

Figure 3 represents the best-case scenario for the hierarchical architecture, in which all the 500,000 messages (100 messages for each leaf, on average) are uniformly distributed among the leaves. The uniform distribution among all the leaves simulates an unrealistic scenario in which network attacks are uniformly distributed among all the sensors that generate alerts at the same rate. The two lines of Figure 3 represent the cumulative distribution function (CDF) of the number of messages that each node in the collaborative architecture (line *P2P*) and each low-level manager in the hierarchical architecture (line *Hierarchical*) has to manage. The X axis represents the ratio between the number of messages handled by a node and the expected average value (that is, $500,000/5000 = 100$ messages per collaborative node in the collaborative architecture, and $500,000/420 = 1190.48$ messages per manager in the hierarchical architecture). The vertical line represents the ideal load distribution, in which all the nodes handle a number of messages that is equal to the expected average (hence the ratio between the handled messages and the expected ratio is 1).

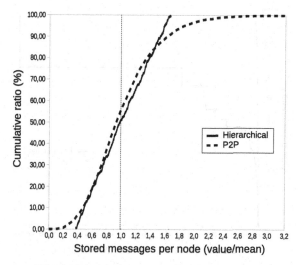

**Fig. 3.** Load balancing with random inserts

As shown in Figure 3, characterized by a uniform distribution of alerts among sensors, the load of the hierarchical architecture is better distributed than that of the collaborative overlay. Indeed, all the nodes in the hierarchical architecture handle a number of messages between 40% and 150% of the expected average. However, even in the best-case scenario for the hierarchical architecture, the load distribution of the two collaborative architectures is comparable. Indeed, the two distributions behave similarly for about 70% of the nodes; 20% of the nodes in the peer-to-peer architecture are more loaded than the most loaded nodes of the hierarchical architecture, but the highest load is still manageable and limited to 3.2 times than the expected load.

In Figure 4 we report the results of a more realistic scenario, in which 500,000 messages are not generated uniformly by all the collaborative alert aggregators, but they follow a Zipf distribution[1] (known to be a realistic representation of traffic distribution and load of nodes in the Internet). The two cumulative distributions in this figure show the benefits of the hash-based event distribution algorithm implemented by the DHT-based overlay network. The node managing an alert is determined by the alert content and not by the network generating it, hence the load distribution of the proposed architecture in this scenario is identical to that presented in Figure 3. On the other hand, the load distribution in the hierarchical architectures is highly unbalanced. For example, the most loaded manager handles a number of messages equal to 103 times the expected average.

We consider now an attack scenario in which all the events are generated by one sensor (or by several sensors connected to the same low-level manager).

---

[1] The used formula is $f(P_i) = \frac{c}{i}$, where $i$ is the rank, $P_i$ indicates the event which occupies the i-th rank (the i-th most frequent), $f(P_i)$ is the number of times (frequency) that the $P_i$ event occurs, $c$ is a constant equivalent to $P_i$.

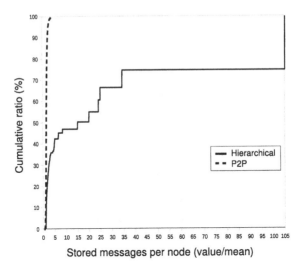

**Fig. 4.** Load balancing with Zipf distribution based inserts

This is the worst case for a hierarchical architecture, because all the alerts must be handled by the same manager node. The results are impressively poor: one manager has to sustain a load 420 higher than that related to the expected average. On the other hand, the proposed architecture preserves the same load distribution of the other two scenarios. This shows the robustness of the peer-to-peer architecture with respect to any attack scenario.

## 6   Conclusion

In this paper we propose an innovative architecture for collaborative and distributed intrusion detection, malware analysis and alert dissemination.

With respect to previous work in the same field, our architecture represents a more flexible cooperation infrastructure that is able to manage heterogeneous information, ranging from IP addresses to the complete binary code of malware specimens and IDS alerts. Moreover, our architecture is not focused on a specific analysis algorithm and can leverage heterogeneous analysis engines. Finally, a publish/subscribe scheme is used for efficient and timely dissemination of relevant analysis results to all the collaborative nodes.

Being based on a DHT overlay, the proposed architecture avoids single points of failures, guarantees high scalability, load balancing and self organization capabilities, that allows us to implement a system that may integrate several thousand collaborative nodes. The viability of the proposed architecture is demonstrated through a prototype based on open source software, while large scale results referring to up to 10,000 nodes have been obtained through simulations.

We are now working on a hardened version of the distributed architecture that uses digital certificates and secure communications among the peers. We are also improving the detection algorithm in order to detect polymorphic and metamorphic malware where the cryptographic hashes of the binary code are different for every sample thus preventing the identification of the same threat.

## Acknowledgments

This research has been funded by the IST-225407 EU FP7 project CoMiFin (Communication Middleware for Monitoring Financial Critical Infrastructures).

## References

1. Colajanni, M., Gozzi, D., Marchetti, M.: Collaborative architecture for malware detection and analysis. In: Proc. of the 23rd International Information Security Conference, Milano, Italy (September 2008)
2. Druschel, P., Rowstron, A.: Past: A large-scale, persistent peer-to-peer storage utility. In: 8th Workshop on Hot Topics in Operating Systems, Schoss Elmau,Germany (May 2001)
3. Malan, D.J., Smith, M.D.: Host-based detection of worms through peer-to-peer cooperation. In: Proc. of the 2005 ACM Workshop on Rapid Malcode, Fairfax, VA, USA (November 2005)
4. Dumitrescu, C.L.: Intctd: A peer-to-peer approach for intrusion detection. In: Proc. of the 6th IEEE International Symposium on Cluster Computing and the Grid, SMU Campus, Singapore (May 2006)
5. Zhou, C.V., Karunasekera, S., Leckie, C.: A peer-to-peer collaborative intrusion detection system. In: Proc. of the 13th IEEE International Conference on Networks, Kuala Lumpur, Malaysia (November 2005)
6. Yegneswaran, V., Barford, P., Jha, S.: Global intrusion detection in the domino overlay system. In: Proc. of the ISOC Symposium on Network and Distributed Systems Security (February 2004)
7. Janakiraman, R., Waldvogel, M., Zhang, Q.: Indra: A peer-to-peer approach to network intrusion detection and prevention. In: Proc. of the 12th IEEE International Workshops on Enabling Technologies, Linz, Austria (June 2003)
8. Stoica, I., Morris, R., Karger, D., Kaashoek, F., Balakrishnan, H.: Chord: A scalable peer-to-peer lookup service for internet application. In: Proc. of the ACM SIGCOMM 2001, San Diego, CA, USA (August 2001)
9. Locasto, M.E., Parekh, J.J., Keromytis, A.D., Stolfo, S.J.: Towards collaborative security and p2p intrusion detection. In: Proc. of the IEEE Information Assurance Workshop, Maryland, USA (June 2005)
10. Rowstron, A., Kermarrec, A., Castro, M., Druschel, P.: Scribe: The design of a large-scale event notification infrastructure. In: Proc. of the 3rd International Workshop on Networked Group Communication, UCL, London, UK (November 2001)
11. Snort Homepage: Snort - the de facto standard for intrusion detection/prevention, http://www.snort.org
12. Nepenthes Homepage: Nepenthes - finest collection, http://nepenthes.mwcollect.org
13. Prelude IDS Homepage: Prelude, http://www.prelude-ids.com/en/welcome/index.html

14. IETF Intrusion Detection Working Group: Idmef standard described in rfc4765, http://www.ietf.org/rfc/rfc4765.txt
15. MySQLAB: Mysql, http://www.mysql.com
16. Rowstron, A., Druschel, P.: Pastry: Scalable, decentralized object location, and routing for large-scale peer-to-peer systems. In: Guerraoui, R. (ed.) Middleware 2001. LNCS, vol. 2218, p. 329. Springer, Heidelberg (2001)
17. Castro, M., Druschel, P., Kermarrec, A.M., Rowstron, A.: One ring to rule them all: Service discovery and binding in structured peer-to-peer overlay networks. In: Proc. of the 10th SIGOPS European Workshop, Saint-milion, France (September 2002)
18. Castro, M., Druschel, P., Hu, Y.C., Rowstron, A.: Exploiting network proximity in distributed hash tables. In: Proc. of the International Workshop on Future Directions in Distributed Computing, Bertinoro, Italy (June 2002)
19. Castro, M., Druschel, P., Ganesh, A., Rowstron, A., Wallach, D.S.: Security for structured peer-to-peer overlay networks. In: Proc. of the 5th Symposium on Operating Systems Design and Implementaion, Boston, MA, USA (December 2002)
20. Mahajan, R., Castro, M., Rowstron, A.: Controlling the cost of reliability in peer-to-peer overlays. In: Proc. of the 2nd International Workshop on Peer-To-Peer Systems, Berkeley, CA, USA (February 2003)
21. Rowstron, A., Druschel, P.: Storage management and caching in past, a large-scale, persistent peer-to-peer storage utility. In: Proc. of the 18th ACM Symposium on Operating Systems Principles, Chateau Lake Louise, Banff, Canadav (May 2001)
22. Castro, M., Jones, M.B., Kermarrec, A., Rowstron, A., Theimer, M., Wang, H., Wolman, A.: An evaluation of scalable application-level multicast built using peer-to-peer overlays. In: Proc. of the Infocom 2003, San Francisco, CA, USA (April 2003)
23. W3C: Extensible markup language (xml), http://www.w3.org/XML/
24. Sun: The java database connectivity (jdbc), http://java.sun.com/javase/technologies/database/index.jsp
25. Norman SandBox Homepage: Norman sandbox information center, http://sandbox.norman.com
26. CWSandbox Homepage: Cwsandbox, behavior-based malware analysis remote sandbox service, http://www.cwsandbox.org

# F3ildCrypt: End-to-End Protection of Sensitive Information in Web Services

Matthew Burnside and Angelos D. Keromytis

Department of Computer Science
Columbia University in the City of New York
{mb,angelos}@cs.columbia.edu

**Abstract.** The frequency and severity of a number of recent intrusions involving data theft and leakages has shown that online users' trust, voluntary or not, in the ability of third parties to protect their sensitive data is often unfounded. Data may be exposed anywhere along a corporation's web pipeline, from the outward-facing web servers to the back-end databases. The problem is exacerbated in service-oriented architectures (SOAs) where data may also be exposed as they transit between SOAs. For example, credit card numbers may be leaked during transmission to or handling by transaction-clearing intermediaries.

We present F3ildCrypt, a system that provides end-to-end protection of data across a web pipeline and between SOAs. Sensitive data are protected from their origin (the user's browser) to their legitimate final destination. To that end, F3ildCrypt exploits browser scripting to enable application- and merchant-aware handling of sensitive data. Such techniques have traditionally been considered a security risk; to our knowledge, this is one of the first uses of web scripting that *enhances* overall security.Our approach scales well in the number of public key operations required for web clients and does not reveal proprietary details of the logical enterprise network. We evaluate F3ildCrypt and show an additional cost of 40 to 150 ms when making sensitive transactions from the web browser, and a processing rate of 100 to 140 protected fields/second on the server. We believe such costs to be a reasonable tradeoff for increased sensitive-data confidentiality.

## 1 Introduction

Recent intrusions resulting in data leakages [1,2] have shown that online users simply cannot trust merchants to protect sensitive data. Security incidents and theft of private data are frequent, often in spite of the best intentions of corporate policy, faithful compliance to standards and best practices, and the quality of security/IT personnel involved. Data may be exposed anywhere along a web-driven pipeline, from the outward-facing web servers to the back-end databases, so security personnel must protect a wide front. Furthermore, in service-oriented architectures (SOAs), data may also be exposed as they transit between SOAs, and, of course, the remote SOAs must also be configured and administered safely.

P. Samarati et al. (Eds.): ISC 2009, LNCS 5735, pp. 491–506, 2009.

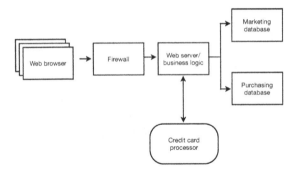

**Fig. 1.** A simple e-commerce server pipeline

Data leakage can be very expensive to the parties involved. It was recently reported that an attacker compromised the networks of clothing retailer TJ Maxx and stole credit card information for 45.6 million customers, dating back to December 2002 [1]. It is estimated that this breach will cost TJ Maxx approximately $197 million. Another attacker stole 4.2 million credit card numbers from grocery store chain Hannaford [2] with an unknown cost to the company, though a recent study [3] estimated an average cost of $197 per compromised customer record.

For a corporation to safeguard sensitive user information, it must be protected in an end-to-end fashion [4], in transit from the web browser to the back-end databases, and during storage at the database. Protecting the back-end databases may come in the form of legal [5] or technical [6,7] protection. F3ildCrypt focuses on the technical protection of data in transit. While a protocol like SSL provides adequate protection for data on the wire, it provides no protection for transitions. Even with SSL protection between a web browser and web server, and between the web server and back-end database, sensitive data may still be revealed during the transition across the web server.

Consider a simple e-commerce website for a widget store, as in Figure 1. The website uses an Asynchronous Javascript and XML (Ajax)-based shopping cart [8] and XML-formatted content, served from PHP-based business logic. All customer communication with the website takes place over SSL. Customer data, including name, address, and order history are processed by the business logic and stored in a back-end MySQL database. As new orders for widgets arrive, the business logic transmits order information to the website's credit card processor.

An order consists of an XML document[1] containing the customer's name, address, email address, a list of each type of widget to be purchased, and the customer's credit card details. Note that each field is useful to only a subset of applications in the website pipeline. That is, multiple machines have access to data for which they have no need – a violation of the principle of least privilege

---

[1] The choice of XML is not integral to our scheme; we can just as easily use JSON or any other data encoding/representation. XML was selected for convenience in prototyping, and because of its wide use in SOA environments.

[9]. For example, there is no reason to expose the credit card data to the web server – in fact, it should only be revealed to the credit card processor – and there is no reason to reveal the customer's email address to the purchasing database.

To use this website safely, a customer must trust that both the widget store and the credit card processor are taking appropriate steps to protect his credit card information. Additionally, as far as the user is concerned, any protection derived from the SSL session is lost in the pipeline downstream from the web server, since the SSL tunnel ends at the web host. There is no guarantee to the customer or to the corporation that the downstream machines are not currently compromised and that they are suitably protected against future compromise (since downstream machines may be located in SOAs operated by third-party corporations).

The goal of F3ildCrypt is to guarantee that data are encrypted end-to-end, as they traverse an SOA and its children SOAs. F3ildCrypt is based around three components: an XML gateway, an in-line proxy re-encryption engine [10], and a Javascript policy and Java applet cryptographic engine.

We use an Ajax-based approach where fields are encrypted at the customer's web browser. In the straightforward approach, this would require that the Ajax application be bundled with certificates containing public keys for the internal web-pipeline components, so it can encrypt the information appropriately. However, this approach may not be appealing to a corporate entity, since it requires, for example, revealing the name and public key of the corporation's credit card processor. In general, it exposes the internal logic of the enterprise (including external business relationships, processing intermediaries, and the internal pipelines, which may change at any time) to the customer. A key contribution of this work is to use proxy re-encryption at the gateway to map fields encrypted by the user to the individual internal components or partner SOAs, without exposing clear-text at the gateway, and without revealing those partner relationships to the end-user.

Ajax-like techniques (and, more generally, web browser scripting) have long been considered a security risk, for good reasons. To our knowledge, our approach is one of few that enhances overall security through use of such techniques. Although the use of public key cryptography inevitably increases the overall latency and processing cost of any given web transaction, we experimentally demonstrate that the costs in this case are reasonable. Furthermore, these costs need only be incurred when sensitive information is being transmitted; in our widget-store testbed, the costs are only incurred when the user makes a purchase. The preceding portion of the session, while the user is browsing in the store, does not incur *any* additional performance impact.

## 2   Related Work

Proxy re-encryption [11,10] allows a third-party to transform a ciphertext for Alice into a ciphertext for Bob, without revealing the plaintext to the third party.

Consider Alice, with key pair $(pk_A, sk_A)$, and Bob with key pair $(pk_B, sk_B)$. A re-encryption key from Alice to Bob $rk_{A \to B}$ has the following property for all plaintext $P$:

$$pk_B(P) = rk_{A \to B}(pk_A(P)) \tag{1}$$

If, for example, Alice wishes to temporarily re-direct her encrypted email to Bob, but she does not wish to reveal her secret key, she can generate a re-encryption key $rk_{A \to B}$ and deliver that key to a proxy. (See [10] for the details on generating this key; it is a function of Alice's private key and Bob's public key.) The proxy can then re-encrypt messages destined for Alice so that Bob may read them. The plaintext is never revealed to the proxy.

The authors in [10] demonstrate a unidirectional, *single-hop* scheme, while the scheme proposed in [11] is bidirectional and *multi-hop*. Meaning, essentially, that $rk_{A \to B} = rk_{B \to A}$, and a ciphertext can be re-encrypted from Alice to Bob to Carol. The algorithm from [10] is partially implemented in the JHU-MIT Proxy Re-cryptography Library (PRL) [12], which we use in our prototype.

XML is fast becoming a standard for document transfer on the web, and there is a body of work on securing those documents. Element-level encryption of XML fields was first proposed by Maruyama and Imamura [13]. There are now several XML-based firewalls on the market including the Cisco ACE XML Gateway [14] and the IBM XS40 Security Gateway [15]. These devices allow field-level transforms, including cryptographic primitives, of XML content as it traverses the firewall. Appliances like these provide high performance, but do not provide end-to-end protection of the individual fields.

There have been a number of proposals for XML-based access control systems [16,17,18]. One of the most popular is the eXtensible Access Control Markup Language (XACML) [19]. It provides XML-based standards for defining policies, requests, and corresponding responses. An XACML policy consists of a list of subjects, actions and resources, followed by a list of rules for which subjects may apply which actions to which resources.

W3bCrypt [20] first introduced the notion of end-to-end encryption of data in a web pipeline. The W3bCrypt system consists of a Mozilla Firefox extension that enables application-level cryptographic protection of web content. Web content is encrypted or signed in the browser before being delivered to the web application. This provides field-level end-to-end protection for user data, but does not protect the corporate network from information revealed by the key distribution. That is, in order to use W3bCrypt across an entire web pipeline, including multiple possible calls to external SOAs, the logical architecture of the server network must be revealed to the client in the form of pairwise key sharing. By providing this protection, F3ildCrypt may be viewed as a successor to W3bCrypt.

Li *et al.* use *automaton segmentation* [21] to explore privacy notions in distributed information brokering systems. The authors model global access control policies as a non-deterministic finite automata, and divide those automata into segments for evaluating network components. The automaton segmentation system considers privacy for users, data, and meta-data, but does not consider privacy notions with respect to the logical network layout and corporate inter-actions between service providers.

Sun Microsystems has implemented the Java WSDP 1.5 XWS-Security Frame-work [22] to assist programmers in securing web services. However, this scheme does not extend to the client (browser). Singaravelu and Pu [23] demonstrate a secure web services system based on the WS-Security framework. The key dis-tribution mechanism used by this system requires pairwise shared keys between endpoints, potentially revealing the internal logical architecture and SOA de-pendencies. Chafle *et al.* [24] use data flow constraints to protect web services, but this requires complete, centralized control of all SOAs involved.

## 3   Architecture

In this section, we describe our network and threat models, and our design requirements. We then examine several design alternatives, before explaining the overall F3ildCrypt architecture.

### 3.1   Network Model

We consider service-oriented architecture (SOA)-style networks where external requests to the network have a single entry point and request-handling takes the form of a tree. A single parent SOA may make requests on multiple child SOAs in the course of processing a request.The SOAs may each operate under different administrative domains, with varying legal and corporate policies toward the privacy and protection of data traversing their networks. There may also be political, corporate, or technical pressure to prevent disclosure of the logical architecture of each SOA, and the identities of their children SOAs.

### 3.2   Threat Model

A corporation whose business model requires handling sensitive user information (*e.g.*, credit cards, Social Security numbers, *etc.*) has both financial and political incentives to protect those data as they traverse its network. There are commonly used mechanisms, like SSL, for protecting the data point-to-point, but this does not protect against data leakage at compromised intermediate hosts.

Thus, our threat model encompasses large-scale networks of inter-operating SOAs where multiple internal hosts or networks may be compromised. These nodes may cooperate to extract and reveal data from transient information flows. We focus particularly on those information flows containing sensitive data related to, *e.g.*, identity theft. Our approach does not protect against the compromise of a node that *legitimately* has the need to view a specific piece of sensitive information.

Additionally, the logical architecture of the corporate network, along with any SOA peering agreements, is sensitive. Information of this nature should be protected from disclosure.

### 3.3 Requirements

Our goal is to provide XML-field granularity end-to-end protection of data transmitted from a web browser to each field's destination end-host within the web pipeline of an e-commerce site. The web pipeline may encompass multiple remote SOAs, and the end-to-end property must hold across SOA boundaries. Additionally, the confidentiality of the logical internal architecture of each SOA must be respected. That is, no architecture details should be disclosed to the web clients or across parent or children SOA boundaries.

### 3.4 Design Alternatives

An XML firewall, like those marketed by IBM [15] or Cisco [14], or a similar proxy, sited at an SOA's network edge, can provide some protection. The proxy or firewall encrypts individual fields of each document to the fields' destination host within the SOA. However, this is not an end-to-end solution and an end-user has no way of verifying that an XML firewall or proxy is in place, let alone operating correctly. The customer must simply trust the SOA beyond the narrow confines of the commercial transaction.

Another approach is to generate a public key pair at each host in the web pipeline, use a trusted third party (VeriSign, GeoTrust, *etc.*) to sign certificates for each, and deliver the certificate set to each web browser or SOA client. In the event that a document containing fields with sensitive data must be delivered to the website, the web browser (or a browser-embedded crypto engine) can then encrypt each field directly to its destination end host.

There are several serious flaws in this design. If the e-commerce site links to external SOAs, the keys for each host in each external SOA must also be delivered to the web client. Thus, this solution does not necessarily scale well in the number of certificates. As more SOAs become involved, a cache of hundreds or thousands of certificates would have to be provided to each new web client, and the certificate caches for existing web clients would have to be updated each time the internal architecture of the SOA or any of its dependent SOAs changed. This solution also has the unfortunate consequence of revealing details to the end user (and thus to competitors) about the logical architecture of the e-commerce site and its SOA partners. By collecting and correlating the certificate sets, an adversarial client may be able to identify individual hosts in an SOA. Furthermore, this technique reveals the identities of the SOA partners. These details may encompass trade secrets and other confidential information.

### 3.5 F3ildCrypt Architecture

Our proposed solution is based on the technique of proxy re-encryption. Each SOA publicizes a certificate containing a public key, called the external key, $pk_E$. This key is used by the SOA's clients, either web browsers or other SOAs. Before sending a document containing sensitive data fields to an SOA, a client cryptographically transforms each field containing sensitive data, using the external key. The client chooses which fields to transform based on an XACML client policy delivered from the SOA.

Meanwhile, each host or application in the SOA has an associated public key pair. This set of public keys is the internal key set $pk_{I_0}...pk_{I_n}$. These keys are used for communication internal to the SOA.

The external key $pk_E$ is generated at a host called the external-key holder. The public keys of the internal applications $pk_{I_0}...pk_{I_n}$ are delivered to this host and used, in concert with the external secret key $sk_E$ to generate the re-encryption keys $rk_{E \to I_0}...rk_{E \to I_n}$, as in [10]. The fundamental property of proxy re-encryption holds that, for any plaintext $P$ and internal application $j$:

$$pk_{I_j}(P) = rk_{E \to I_j}(pk_E(P)) \tag{2}$$

The re-encryption keys are installed at a host called the *proxy re-encryption engine*. Fields from documents arriving at the SOA have been encrypted under $pk_E$ and are handled by the proxy re-encryption engine. The latter re-encrypts each field under re-encryption key $rk_{E \to I_j}$, where $j$ is the individual host within the web pipeline designated to process that field, based on a XACML server policy. The plaintext *is not revealed* until it arrives at the intended destination host.

This solution requires an SOA to deliver to its clients a certificate containing only the single external key $pk_E$, avoiding the problem of sending what could be a set of hundreds or thousands of certificates. Furthermore, no logical infrastructure details are revealed to the client. With the exception of the external-key holder, any subset of intermediate hosts between the client and end-host – including the proxy re-encryption engine itself – can be compromised without leaking any sensitive user data.

Compromise of the external-key holder, however, could be dangerous, requiring that special care be taken to secure this machine. Luckily, the bandwidth requirements on the external-key holder are extremely low. It is only used to generate the re-encryption keys so, after initial setup, its use is only required when adding new internal hosts. Thus, in the extreme, it is possible to keep the external-key holder offline at all times, and distribute keys through it by hand.

## 4 Implementation

Our implementation of F3ildCrypt consists of a Javascript-based policy engine and a Java-based cryptography engine delivered to each web browser. The web server connects to the server using SSL. On the server side, we provide a Python-based XML gateway with in-line proxy re-encryption engine for each SOA, and a Python-based XML proxy at each internal application. These proxies store the key pairs for their respective applications, and perform decryption and XML unwrapping on behalf of the application.

The Java cryptography engine and in-line proxy re-encryption engine use the proxy re-encryption algorithm described in [10]. This algorithm is based on bilinear maps [25], and is partially implemented in the JHU-MIT Proxy Re-cryptography Library (PRL) [12]. For our implementation, we ported the PRL to both

Java and Python. We note that the JHU-MIT PRL supports only single-hop re-encryption, thus limiting the recursive depth of our implementation until such time as an implementation of the multi-hop algorithm from [11] is available.

F3ildCrypt setup in an SOA begins by designating an offline host as the external-key holder and generating the external key pair. The public key $pk_E$ is signed by a trusted third party and the certificate is made available to the public. This is the key with which all clients will encrypt sensitive data sent to the SOA.

At each application inside the SOA we install an XML proxy which serves as that application's entry point into the F3ildCrypt network. This proxy stores the internal key pair $(pk_{I_j}, sk_{I_j})$ associated with the application. Any documents with encrypted fields arriving at the application are intercepted and decrypted by the XML proxy before delivery to the application proper.

Each internal public key is delivered in offline fashion (hand-delivered via USB key, for example) to the external-key holder, where the re-encryption keys are generated. The re-encryption key for proxy $j$ is $rk_{E \to j}$ and it is a function of the external secret key $sk_E$ and $pk_{I_j}$. The re-encryption keys are then hand-delivered to the proxy re-encryption engine.

The proxy re-encryption engine operates as a client to the XML gateway. The XML gateway stores a set of XSLT stylesheets. Each stylesheet describes the transformation to be applied to a given field type in a document. The XSLT implementation is extended with the proxy re-encryption function, so applying the cryptographic transformations becomes an application of a stylesheet, as in W3bCrypt [20]. The specific stylesheets are chosen based on a system administrator-defined XACML policy.

The XML gateway uses the XSLT transforms to re-encrypt designated fields, targeting them to the appropriate internal hosts. It processes incoming documents containing fields encrypted under $pk_E$. These fields are re-encrypted under the various re-encryption keys $rk_{E \to I_0} ... rk_{E \to I_n}$, in accordance with the XACML policy, before forwarding the document on to the web pipeline.

When a client connects to the SOA over SSL, the SOA responds with the contents of an Ajax web application, implementing, for example, a shopping cart application. Packaged along with the application is the Javascript-based policy engine and an applet containing the Java cryptography engine. At the browser, the package then downloads from the SOA an XACML policy document to be applied to uploaded documents, and a certificate store containing the signed certificate for the SOA's external key. When, in the course of user interaction with the application, an XML document must be uploaded, the Javascript engine applies the XACML client policy. This policy describes which fields of the document should be encrypted. The cryptography engine encrypts the designated fields with the external key, and then the document is uploaded to the SOA.

Now consider the case of a parent SOA, with external key $pk_{E_p}$ making requests on a child SOA with external key $pk_{E_c}$. The child SOA implements the F3ildCrypt architecture, with internal key pairs for its own internal applications. As in the parent case, and given the appropriate proxy re-encryption algorithm,

XML documents arriving at the child SOA's XML gateway are re-encrypted by the proxy re-encryption engine.

To make use of the child SOA, the system administrator at the parent uses the publicly known $pk_{E_c}$ and its secret key $sk_{E_p}$ to generate a re-encryption key $rk_{p \to c}$. Fields within a document sent to the parent SOA, but destined for the child SOA, are re-encrypted under $rk_{p \to c}$ at the parent XML gateway. When the fields arrive at the child XML gateway, they may be re-encrypted again, to the end-hosts within the child SOA.

## 4.1 Example

In this section we will describe a sample application of the F3ildCrypt architecture. It is based on the network for a small e-commerce site selling widgets, called Widgets4Cheap. The site consists of a firewall, web server with business logic, and back-end databases for marketing and purchases, as was shown in Figure 1. Widgets4Cheap also makes use of an external credit card processor.

The website presents to the user a web page with a simple catalog and shopping cart application, where the user may browse widgets and select items to purchase. When the customer makes an order, the order is delivered to the web server in the form of an XML document. An order consists of the customer's name, physical address, email address, a list and count of each model of widget to be purchased, and the customer's credit card information.

Customer data, including name, billing address, and order history are stored in the purchasing database. The customer's email address is stored in the marketing database. As orders arrive, the business logic transmits order information and credit card details to the website's credit card processor.

Revealing the internal architecture of the Widgets network is undesirable, as it may reveal business or trade secrets (this is exacerbated in more sophisticated networks). Additionally, even with an SSL connection between the client and the web server, the compromise of any internal host in the Widgets4Cheap pipeline could be catastrophic to the company and its customers, since every internal host, particularly the firewall and web server, has access to all the customer information in transit.

To protect this network, we define a high-level security policy. The customer's billing address, and order details may only be revealed to the purchasing database, while the email address may only be revealed to the marketing database. The credit card information and total payment is revealed only to the credit card processor.

Before implementing this policy, we deploy the F3ildCrypt infrastructure, as shown in Figure 2. Co-located with each internal application is an XML proxy which stores the key pair for that application. This XML proxy serves to decrypt the incoming XML documents, and unwrap the XML as necessary. On a separate offline machine (the external-key holder) the system administrator generates the external key pair which will be presented to remote users. A certificate for this key is signed by a third-party certificate authority. In the case of this example implementation, this was an in-house certificate authority.

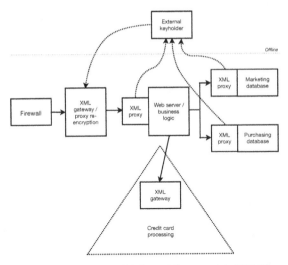

**Fig. 2.** Diagram of the network for Widgets4Cheap with F3ildCrypt installed

```
<rule ruleid="creditcard_transform" effect="permit">
    ...
        <attributevalue datatype"string">
        order/creditcard
        </attributevalue>
    ...
</rule>
<obligation obligationid="reencrypt_on_transit"  fulfillon="permit">
  <attributeassignment attributeid="reencrypt" datatype="string">
    ccn_reencrypt.xsl
  </attributeassignment>
</obligation>
```

**Fig. 3.** A rule from the XACML server policy file. When the gateway receives an XML document, the rule attempts to match the XPath `order/creditcard`. When this rule fires, the obligation indicates that the XSLT transform `ccn_reencrypt.xsl` should be applied.

The external-key holder is then used to generate re-encryption keys for each internal application and the credit-card processor, and these are delivered to the XML gateway, thereby allowing the gateway to re-encrypt traffic to the internal applications and credit-card processor SOA.

At the XML gateway we place a set of XACML policy files that describe the transformations to be applied to documents in transit, an example rule of which is shown in Figure 3. The XML gateway also contains a set of XSLT documents for implementing those transforms, an example of which is shown in Figure 4.

Meanwhile, the Javascript policy engine and Java crypto engine applet are incorporated into the Ajax application viewed by customers, along with a XACML client policy file and a certificate store containing the Widgets4Cheap external key. After browsing the catalog and selecting his items, the customer makes

```
<xsl:template match="creditcard">
  <xsl:copy-of
      select="encrypt:reencrypt(., reencrypt_key[7]')"/>
</xsl:template>
```

**Fig. 4.** An XSLT snippet for re-encrypting the credit card information. Demonstrates usage of the XSLT extension function **reencrypt()**. It applies proxy re-encryption to the matched XML field using the re-encryption key **reencrypt_key[7]**.

```
<order>                                    <items>
<date>1 January 2008</date>                  <item>
<name>H. Simpson</name>                        <quantity>1</quantity>
<address>                                      <detail>Big red widget</detail>
  <street>742 Evergreen Ter.</street>          <cost>69.96</cost>
  <city>Springfield</city>                   </item>
  <state>USA</state>                         <item>
  <zip>12345</zip>                             <quantity>1</quantity>
</address>                                      <detail>Blue suede widget</detail>
<email>homer@springfield.com</email>           <cost>109.95</cost>
<creditcard>                                  </item>
  <payment>179.90</payment>                 </items>
  <issuer>American Express</issuer>        </order>
  <number>1234-5678-1234-5678</number>
  <expiration month="10" year="2010"/>
</creditcard>
```

**Fig. 5.** A purchase order for two pairs of widgets from Widgets4Cheap

```
<rule ruleid="creditcard_rule" effect="permit">
    ...
        <attributevalue datatype"string">
          order/creditcard
        </attributevalue>
    ...
</rule>

<obligation obligationid="encrypt_on_send" fulfillon="permit">
  <attributeassignment attributeid="encrypt" datatype="string">
    encrypt(key[n])
  </attributeassignment>
</obligation>
```

**Fig. 6.** The XACML client rule, abridged for clarity and space. This rule and obligation describes the action to be taken on the credit card section of the XML document: encrypting it with a key obtained from the certificate store.

his purchase as in Figure 5. Before transmitting this document, the application applies the XACML client policy. The XACML client policy file describes which fields in the order document should be encrypted. A snippet from the Widgets4Cheap client policy is shown in Figure 6. When the policy is evaluated, the cryptography engine encrypts the necessary fields, resulting in a new, field-encrypted order document.

When the now-transformed document arrives at the Widgets4Cheap website, it is processed by the XML gateway/proxy re-encryption engine, which applies the server XACML policy to determine which XSLT transforms to apply. The

XSLT transforms apply the proxy re-encryption to the document, re-targeting the field encryptions that were originally applied by the client. The business logic then processes the order, delivering the various XML fields to their intended targets. The individual XML fields are intercepted by the XML proxies at each application and decrypted before being passed on to the application proper. The re-encrypted credit card information is passed to the credit-card processor, who may recursively apply this system, distributing the received information through its network.

## 5    Evaluation

We evaluated the performance of F3ildCrypt by measuring its impact on the web browser clients, on the XML gateway, and on the XML proxies at each host. We performed micro-benchmarks at the individual hosts, as well as throughput measurements on the servers.

Our experimental setup consisted of the network described in Figure 2. Each server application ran on a Dell PowerEdge 2650 Server, with a 2.0GHz Intel Xeon processor, 1GB RAM, and 36GB Ultra320 SCSI hard drive. All machines ran OpenBSD 4.2. and were linked via Gigabit Ethernet. The applications included OpenBSD PF on the firewall, Apache 1.3.29/PHP 4.4.1 on the business logic server, and MySQL 5.0.45 on the database servers.

The client ran on a MacBook Pro, with a 2.4 GHz Intel Core 2 Duo, 2 GB RAM, and 150GB 5400 RPM Fujitsu hard drive. The machine used OS X 10.5.2 with Darwin kernel version 9.2.2. The web browsing platform installed on this computer was Mozilla Firefox 2.0.0.13.

The extra work incurred on the web browsing client consists of applying the XACML policy followed by application of the appropriate cryptographic transformations. We used a Java port of the JHU-MIT Proxy Re-cryptography Library [12], running as an applet in the browser, which implements the proxy re-encryption scheme described in [10]. The Java cryptographic engine applet and Javascript policy engine together are approximately 25KB. We measured the performance of the client by encrypting multiple 128-byte fields, as shown in Figure 7a. After processing, most XML documents increase in size between 10% and 30%.

The most common sizes for identity-related sensitive data (*e.g.*, credit card numbers, birth dates, *etc.*) are less than 1K, so the cost incurred at the browser in these cases will range from 40 to 150 ms. Of course, this cost is only incurred when sensitive data requiring encryption is actually transmitted.

The additional work incurred at the XML gateway consists of parsing the incoming XML documents and applying proxy re-encryption; Figure 7b shows the combined cost. We isolated the re-encryption cost per field in Figure 8a. An XML proxy decrypts the encrypted fields from incoming documents; we isolated the decryption cost at the XML proxy in Figure 8b.

These results show that fields from XML documents can be processed at a rate of 100 to 140 fields/second, and the majority of the processing time is dedicated

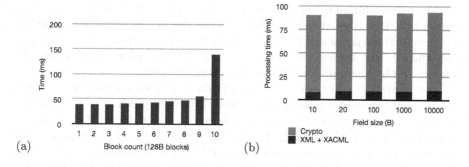

**Fig. 7.** (a) Time (ms) to encrypt multiple 128-byte fields on the client. (b) Processing time on a document containing a single field of 20 bytes. Shows the relative time devoted to cryptography versus the XML and XACML processing.

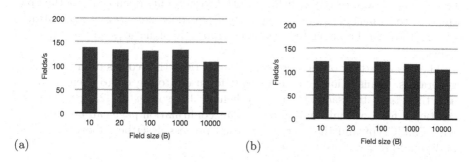

**Fig. 8.** (a) Re-encryption rate (fields/s) at the XML gateway vs. incoming field size. As field size grows, the processing rate decreases. (b) Decryption rate at an XML proxy vs. incoming field size.

to the re-encryption process. This time can be significantly improved through software optimization; the JHU-MIT PRL is not optimized for execution time. The re-encryption cost can be further substantially reduced through the addition of a hardware cryptographic accelerator [26].

## 6   Discussion

The F3ildCrypt system is designed to assist an online entity in protecting its users' sensitive information. The user must not longer collectively trust the web application, the back-end databases, and the system administrators with each sensitive item he provides. Now, for that same item, he only has to trust its intended destination.

F3ildCrypt is designed to assist the system administrators in making the end-user's trust well-founded. However, to provide further assurance to the user, this approach may be combined with a P3P-like policy [27] working in concert with a browser-based cryptography engine like W3bCrypt [20]. Additional protection may come from obtaining a signature on the Ajax application itself from a trusted third party. This trusted third party (*e.g.*, the Better Business Bureau) would certify that the Ajax is encrypting or protecting data to the correct recipients. Regardless of the means, the user, or a trusted third party, must verify the contents of the Ajax application and the associated policy.

For a motivated adversary attacking a F3ildCrypt-enabled system, note that the external-key holder possesses the secret key corresponding to the external public key. Whoever possesses of the secret key is capable of decrypting all messages to that SOA, making the external-key holder a desirable target for attackers. However, it is infrequently used and has low bandwidth requirements. This machine can operate entirely offline, with the occasional generation of a re-encryption key taking place via diskette or USB key.

We also note that, within the network of the F3ildCrypt-equipped SOA, like in a traditional network, an adversary who has compromised an intermediate machine may swap or replay fields, or otherwise modify documents as they pass through that machine's possession. F3ildCrypt does not prevent such attacks, though they can be alleviated via timestamps and signatures on the individual fields.

There is an attack on web browsing transactions that comes from transaction generators. Transaction generators wait for users to log on to their accounts, and then issue transactions on their behalf. Jackson *et al.* [28] propose as a solution a form of confirmation page. This confirmation page can be integrated with F3ildCrypt and the user-certification process described above to provide additional protection to the user.

# 7    Conclusion

The F3ildCrypt system provides end-to-end protection to users and SOAs by encrypting XML fields at the client web browser. The SOA protects its internal architecture by using proxy re-encryption to re-target the XML fields at the SOA edge. The processing cost at the web browser ranges from .5 to 1 second when making sensitive transactions, and a processing rate of 100 to 140 XML fields/second on the server, of which the latter could be easily improved through software optimization and hardware acceleration.

Future work on F3ildCrypt includes integration of the proxy re-encryption engine with the web browser itself, and extensions to the browser to assure the user that the correct transformations have been applied.

# Acknowledgements

This work was supported in part by the National Science Foundation through Grant CNS-07-14647 and by ONR through Grant No. N00014-07-1-0907. Any opinions, findings, conclusions, and recommendations expressed in this paper are those of the authors and do not necessarily reflect the views of the NSF, ONR or the US Goverment.

# References

1. Lemos, R.: TJX theft tops 45.6 million card numbers (March 2008),
   http://www.securityfocus.com/news/11455
2. Card data stolen from grocery chain (March 2008),
   http://www.securityfocus.com/brief/704
3. Institute, T.P.: 2007 Annual Study: Cost of a Data Breach(November 2007),
   http://www.ponemon.org/press/PR_Ponemon_2007-COB_071126_F.pdf
4. Saltzer, J.H., Reed, D.P., Clark, D.D.: End-to-end arguments in system design.
   ACM Transactions on Computer Systems (TOCS) 2(4), 277–288 (1984)
5. Regulation (EC) No 45/2001 of the European Parliament and of the Council of 18
   December 2000. Official Journal of the European Communities (December 2001)
6. Cai, L., Yang, X.: A reference model and system architecture for database firewall.
   In: Proceedings of IEEE SMC 2005, pp. 504–509 (2005)
7. Bai, K., Wang, H., Liu, P.: Towards Database Firewall: Mining the Damage Spreading Patterns. In: Proceedings of ACSAC 2006, pp. 178–192 (2006)
8. Garrett, J.J.: Ajax: A New Approach to Web Applications (February 2005),
   http://www.adaptivepath.com/ideas/essays/archives/000385.php
9. Saltzer, J.H., Schroeder, M.D.: The protection of information in computer systems.
   Proceedings of the IEEE 63(9), 1278–1308 (1975)
10. Ateniese, G., Fu, K., Green, M., Hohenberger, S.: Improved proxy re-encryption
    schemes with applications to secure distributed storage. In: Proceedings of the 12th
    Annual Network and Distributed Systems Security Symposium, NDSS 2005 (2005)
11. Blaze, M., Bleumer, G., Strauss, M.: Divertible protocols and atomic proxy cryp-
    tography. In: Nyberg, K. (ed.) EUROCRYPT 1998. LNCS, vol. 1403, pp. 127–144.
    Springer, Heidelberg (1998)
12. JHU-MIT Proxy Re-cryptography Library (March 2008),
    http://spar.isi.jhu.edu/~mgreen/prl/
13. Maruyama, H., Imamura, T.: Element-Wise XML Encryption(April 2000),
    http://lists.w3.org/Archives/Public/xml-encryption/2000Apr/att-0005/
    01-xmlenc
14. Cisco ACE XML Gateway (March 2008),
    http://www.cisco.com/en/US/products/ps7314/index.html
15. WebSphere DataPower XML Security Gateway XS40 (March 2008),
    http://www-306.ibm.com/software/integration/datapower/xs40/
16. Damiani, E., di Vimercati, S.D.C., Paraboschi, S., Samarati, P.: A fine-grained
    access control system for XML documents. ACM Transactions on Information and
    System Security (TISSEC) 5(2), 169–202 (2002)
17. Luo, B., Lee, D., Lee, W.C., Liu, P.: QFilter: fine-grained run-time XML access
    control via NFA-based query rewriting. In: The Thirteenth ACM International
    Conference on Information and Knowledge Management, pp. 543–552 (2004)
18. Fundulaki, I., Marx, M.: Specifying access control policies for XML documents with
    XPath. In: Proceedings of the ninth ACM symposium on Access control models
    and technologies, pp. 61–69 (2004)
19. OASIS eXtensible Access Control Markup Language (XACML) (2005),
    http://www.oasis-open.org/committees/security/
20. Stavrou, A., Locasto, M., Keromytis, A.: W3bcrypt: Encryption as a stylesheet.
    In: Zhou, J., Yung, M., Bao, F. (eds.) ACNS 2006. LNCS, vol. 3989, pp. 349–364.
    Springer, Heidelberg (2006)

21. Li, F., Luo, B., Liu, P., Lee, D., Chu, C.H.: Automaton segmentation: A new approach to preserve privacy in XML information brokering. In: Proceedings of the 14th ACM conference on Computer and Communications Security (CCS) (2007)

22. Mahmoud, Q.H.: Securing Web Services and the Java WSDP 1.5 XWS-Security Framework (March 2005),
http://java.sun.com/developer/technicalArticles/WebServices/security/

23. Singaravelu, L., Pu, C.: Fine-grain, end-to-end security for web service compositions. In: IEEE International Conference on Services Computing (SCC 2007), pp. 212–219 (2007)

24. Chafle, G., Chandra, S., Mann, V., Nanda, M.G.: Orchestrating composite web services under data flow constraints. In: Proceedings of the IEEE International Conference on Web Services, pp. 211–218 (2005)

25. Boneh, D., Franklin, M.: Identity-based encryption from the Weil Pairing. SIAM Journal of Computing 32(2), 586–615 (2003)

26. Keromytis, A.D., Wright, J.L., de Raadt, T.: The Design of the OpenBSD Cryptographic Framework. In: Proceedings of the USENIX Annual Technical Conference, June 2003, pp. 181–196 (2003)

27. Cranor, L., Langheinrich, M., Marchiori, M., Presler-Marshall, M., Reagle, J.: The Platform for Privacy Preferences 1.0 (P3P1.0) Specifcation (April 2002)

28. Jackson, C., Boneh, D., Mitchell, J.: Transaction generators: Root kits for the web. In: Proceedings of the 2nd USENIX Workshop on Hot Topics in Security (2007)

# Author Index